AF772517

WITHDRAWN
FROM THE VPI & SU
LIBRARY COLLECTION

Ocean Forecasting

Springer

Berlin
Heidelberg
New York
Barcelona
Hong Kong
London
Milan
Paris
Tokyo

Nadia Pinardi • John Woods
Editors

Ocean Forecasting

Conceptual Basis and Applications

With 154 Figures and 7 Tables

Springer

EDITORS:

Professor Dr. Nadia Pinardi
University of Bologna
Corso di Scienze Ambientali
Piazza J.F. Kennedy 12
Palazzo Rasponi
48100 Ravenna
Italy
E-mail: *n.pinardi@ambra.unibo.it*

Professor Dr. John Woods
Imperial College of Science, Technology & Medicine
Department of Earth Science and Engineering
Royal School of Mines
Prince Consort Road
London SW7 2BP
United Kingdom
E-mail: *j.woods@ic.ac.uk*

GC
28
O22
2002

ISBN 3-540-67964-2 Springer-Verlag Berlin Heidelberg New York

Library of Congress Cataloging-in-Publication Data
Ocean forecasting : conceptual basis and applications; with 7 tables / Nadia Pinardi; John Woods (eds.). - Berlin; Heidelberg; New York; Barcelona; Hong Kong; London; Milan; Paris; Tokyo: Springer 2002
ISBN 3-540-67964-2

This work is subject to copyright. All rights are reserved, whether the whole or part of the material is concerned, specifically the rights of translation, reprinting, reuse of illustrations, recitations, broadcasting, reproduction on microfilm or in any other way, and storage in data banks. Duplication of this publication or parts thereof is permitted only under the provisions of the German Copyright Law of September 9, 1965, in its current version, and permission for use must always be obtained from Springer-Verlag. Violations are liable for prosecution under the German Copyright Law.

Springer-Verlag Berlin Heidelberg New York
a member of BertelsmannSpringer Science+Business Media GmbH
http://www.springer.de
© Springer-Verlag Berlin Heidelberg 2002
Printed in Germany

The use of general descriptive names, registered names, trademarks, etc. in this publication does not imply, even in the absence of a specific statement, that such names are exempt from the relevant protective laws and regulations and therefore free general use.

Cover Design: Erich Kirchner
Typesetting: Camera-ready by the editors

SPIN: 10679411 32/3130/ym – 5 4 3 2 1 0 – Printed on acid free paper

Foreword

The Advanced Study Courses in the field of Marine Science and Technology were part of the training programme developed from 1989 until 1999 within MAST, the Marine Science and Technology Programme of the European Union. They were related to the core topics of MAST Programme, marine systems research, extreme marine environments, regional Sea research, coastal systems research and engineering, and marine technology. The main objectives of these study courses were to further advance education in topics at the forefront of scientific and technological development in Europe, and to improve the communication between students and experienced scientists on a European and international level. Over the years the Programme sponsored around 30 Advanced Study Courses. They took place in ten different member states of the European Union and their contribution in the formation of a European scientific community was significant. They also encouraged exchanges and contacts with several countries around the world such as United States, the third Mediterranean countries and others. The Course on Ocean Forecasting was one of the most successful with regard to its content, number of applications for participation and students satisfaction.

When considering the need for the Advanced Study Course on Ocean Forecasting, it is important to remember that the Oceans and Seas have always played a central role throughout the history of mankind. This is seen from the times of the ancient civilizations of Egypt and Greece with the Phoenician traders, to the Viking voyages of exploration and discovery in medieval times.

In the Middle Ages, just as in ancient and medieval times, the sea remained the major route for trade. The naval fleets of the middle ages were first used to protect national interests and then in the centuries that followed to voyage beyond European shores to explore and colonise lands far overseas. Trading ships followed, now crossing vast oceans, as well as local and regional seas. Today, the oceans remain one of the major trading routes for world goods in what is now a truly global economy.

The oceans have also been a major source of food for the human race, with fishing being one of the oldest established professions known to man. The seas and oceans now yield other vital supplies such as oil and natural gas, with new areas opening up for exploitation, particularly in the field of biotechnology. With this increased exploitation of the seas' resources come new pressures. These are, to name but a few; over fishing of fish stocks; pollution in many forms, for example from sewage outlets along coastlines, nitrate and phosphate run-off from the land and from oil spillage; increased vessel traffic and issues of safety at sea.

Then there is the growing awareness of the central role played by the oceans in controlling the global climate. If proof is needed, one only has to compare the warm temperate climate of Western Europe with that of the cool temperate climate of Canada, both lying at the same latitude but enjoying very different climates under the influence of the ocean currents in the North Atlantic. Understanding the role of the oceans in climate change is therefore a key element in being able to forecast the likely evolution of the world's climate over the coming century. There has never been, therefore, a greater requirement than at present to be able to forecast the state of the sea and ocean, both over short, medium and long time scales.

Operational oceanography is one of the key sciences that address this vital need for ocean forecasts. Operational oceanography can be seen as the systematic and routine measurement of the seas, oceans and atmosphere, with the immediate interpretation of the data collected and the rapid interpretation and dissemination of the forecast results. The observation and monitoring networks linked to operational oceanography operate over all scales, from the local and regional, to basin-wide such as for the Mediterranean and on to the global scale when climate change and ocean circulation issues are being considered.

Operational oceanography is therefore a subject with a true European dimension and the subject has been supported both in the previous Research Framework Programme (FP) through the Marine Science and Technology Programme (MAST), and in the present FP as a part of the Key Action studying Sustainable Marine Ecosystems.

The projects supported by the European Union range from: pilot pre-operational forecasting systems covering regional seas such as the Mediterranean, with the Mediterranean Forecasting System Pilot Project (MFSPP); to projects on vessel traffic safety using radar ocean sensing (EuroROSE); to those looking at preparing analysis tools for the operational forecasting of nutrients in estuaries of European rivers (PIONEER). This support has continued into the V FP with, amongst others, projects being funded to develop satellite-based ocean forecasting systems (SOFT) and to forecast and assess rough waves and their impact on marine structures (MAXWAVE).

The expected results from these projects will facilitate safe, sustainable marine activities in coastal, shelf and slope waters and also in the deep ocean. The benefits of this research will contribute to the sustainable development of the many marine activities carried out in European waters. The success of the Advanced Study Course on Ocean Forecasting helps to ensure that the young scientists and technologists of today are trained and prepared to meet the challenges of tomorrow in this exciting field of research.

Alan Edwards *Elisabeth Lipiatou*
Sustainable Marine Ecosystems
European Commission
Brussels

Preface

Atmospheric forecasting reached maturity in the late 20th century. Oceanic forecasting will do so early in the 21st century. Much has been learnt from the techniques of weather forecasting, but the ocean presents its own scientific and technical problems, which are being vigorously addressed. The World Ocean Circulation Experiment in the 1990s opened the way to the new strategy of collecting data to be exploited in numerical models. Operational Forecasting and Nowcasting spring from those advances. A new generation of oceanographers growing up in this post-WOCE era is studying the modern typos of monitoring, modeling and assimilation. This book provides the new generation with an up-to date account of these powerful techniques, showing what they contribute to ocean forecasting. The chapters are updated versions of lectures given at the Ocean Forecasting Advanced Study Course, held at the International Marine Centre, Oristano (Italy) in 1997 and funded by the European Union's Fourth Framework Programme.

Ocean Forecasting has an ancient history, but it has developed rapidly in the last two decades of the 20th century, spurred on by the Rio Conference Earth Summit in 1992, and the publication of Agenda 21, with its important chapters on the ocean. The Conference realized and recommended that ocean would be monitored and predicted in the way as the atmosphere in order to study the impact of anthropogenic change on the earth ecosystems. The Global Ocean Observing System (GOOS) was launched by the Inter-Governmental Commission of Unesco in 1990 to coordinate and focus efforts to monitor and forecast the ocean. The aim is to provide a worldwide observing system that supports the monitoring of the state of the oceans, distributes the information, and uses the observations to forecast the future evolution in support of actions planned to deal with problems of climate, fisheries, pollution, ship safety, offshore operations and coastal protection.

Since 1994, EuroGOOS (http://www.eurogoos.org) has provided a focus for its members (30 national research and operational agencies) to foster ocean forecasting in Europe. The main activity of EuroGOOS focused in the European regional Seas, from the Arctic region to the Mediterranean and Black Sea. Many of the EuroGOOS regional projects are discussed in this volume and have been a substantial component of the operational oceanography activities in Europe in the last five years.

Ocean forecasting is helping to answer urgent societal questions such as the efficient exploitation of marine resources, the sustainable development of coastal areas and the mitigation of adverse effects due to pollution on marine ecosystems and human health. Ocean forecasting will in addition give a revolutionary view of the ocean and its dynamics at unprecedented accuracy and resolution to advance our basic knowledge of the earth climate system. The design and implementation of ocean nowcasting and forecasting activities will in addition create public awareness of marine environmental problems and will create new public services and private industry sectors.

Ocean forecasting is already operational at leading meteorological agencies around the world (including the US National Center for Environmental Prediction, the European Centre for Medium range Weather Forecast and the UK Met Office). It has been contributed to the prediction of El Niño-La Niña events on time scales of several months. Among the challenges now being addressed by ocean forecasting are the extension of predictions to the coastal areas, the forecasting of chemical elements cycling in the marine environment and the prediction of sea level changes due to global change scenarios

The school would not have been possible without the support of the staff of IMC, in particular Dr. S. Vallerga, Dr. R. Covacci, Dr. A. Perilli, Dr. A. Ribotti, Dr. R. Sorgente and Mr. F. Angotzi. We would like to acknowledge the skillful editorial work of Dr. C.Cesarini of CLU srl. Last but not least, we must thank the endless work of Dr. L. Amato who coordinated the scientific reviewing process. One of the editors, N. Pinardi, acknowledges the support of the MFSPP EU contract (MAS3-CT98-0171).

Nadia Pinardi
John Woods
Editors

Table of Contents

List of Contributors

Dr. Francis Auclair
GRGS/LEGOS
Observatoire Midi-Pyrénées
14, Av. Edouard Belin
31401 TOULOUSE
France
Tel: (33) (0)5 61 33 30 07
FAX: (33) (0)5 61 25 32 05
E-mail: Francis.Auclair@cnes.fr

Dr. Mounir Benkiran
LEGOS
18 Avenue Edouard Belin
31401 Toulouse Cedex 4
France
Tel. 0033-5 61-332915
Fax: 0033-5 61-253205
E-mail: Mounir.Benkiran@cnes.fr

Dr. Claudia Cesarini
ARPA
Agenzia Regionale per la Protezione dell'Ambinete
Viale Silvani, 6
40100 Bologna, Italy
Tel: 39-051-284535
Fax: 39-051-284664
E-mail: c.cesarini@smr.arpa.emr.it

Dr. Hans Dahlin
Swedish Meteorlogical and Hydrological Institute
5-601 76 Norrkoping
Sweden
Tel: 46 11 15 83 05
Fax: 46 11 15 83 50
E-mail: hdahlin@smhi.se

Dr. Pierre De Mey
Groupe de Recherche de Geodesie Spatiale
CNRS/UMR39
18 Avenue Edouard Belin
F-31055 Toulouse Cedex
France
Tel. 0033-5 61-332901
Fax: 0033-5 61-253205
E-mail: demey@nanook.cst.cnes.fr

Dr. Encho Demirov
Istituto Nazionale di Geofisica e Vulcanologia
Via di Vigna Murata 605
Roma, Italy
Mailing Address:
ISAO-CNR
Via Gobetti 101
40129 Bologna Italy
Tel: 39-51-6398006
Fax: 39-51-6398132
E-mail: demirov@ingv.it

Dr. Christian Eckert
Max-Planck-Institut fur Meteorologie
Bundesstrasse, 55
D-2000 Hamburg 13
Federal Republic Of Germany
Fax: +49 40 41173 366

Prof. Geir Evensen
Nansen Environmental and
Remote Sensing Center
Edvard Griegsviei 3A
N-5037 Solheimsviken
Norway
Tel.: +47 55 297288
Fax: +47 55 200050

Dr. Tal Ezer
Atmospheric and Oceanic Sciences Program
Sayre Hall, Forrestal Campus
Princeton University
Princeton NJ 08544-0710
Phone: 609-258-1318
Fax: 609-258-2850
E-mail: ezer@splash.Princeton.EDU

Dr. Martin Fischer
IMGA-CNR
Via Gobetti, 101
40129 Bologna
Tel. +39 51 6398060
Fax +39 51 6398132
E-mail: fischer@imga.bo.cnr.it

Dr. Nicholas C. Flemming
EUROGOOS
Room 346/18
Southampton Oceanography Centre
Empress Dock
European Way
Southampton SO14 3ZH
Tel.: ++44 1703 596242
Fax: ++44 1703 596399
E-mail: n.flemming@soc.soton.ac.uk

Prof. Serena Fonda Umani
Laboratorio di Biologia Marina
Strada Costiera 336
34010 Santa Croce (TS)
Tel.: 040 224400-224464-224620-224621
Fax : 040 224437
E-mail: labbioma@univ.trieste.it

Prof. Lennart Funkquist
Swedish Meteorological and Hydrological Institute
Department of Research and Development
SE-601 76 Norrkoeping
Schweden
Tel: 46-11-495 8289
Fax: 46-11-495 8001
E-mail: lfunkquist@smhi.se

Dr. Michele Giani
ICRAM
Istituto Centrale per la ricerca scientifica e tecnologica applicata al mare
Struttura tecnico scientifica in Chioggia
Viale Stazione, 5
30015 Chioggia (VE)
Tel.: 041 5500636
Fax: 041 5500626
E-mail: ICRAMVE@ux1.unipol.it

Dr. Anselm Grötzner
Max-Planck-Institut für Meteorologie
Room no. 119
Bundesstr. 55
D-20146 Hamburg
Germany
Phone: +49-40-41173-241
Fax: +49-40-41173-298
E-mail: groetzner@dkrz.de

Prof. Keith Haines
University of Edinburgh
Department of Meteorology
James Clerk Maxwell Buildings
Mayfield Road
Edinburgh, EH9 3JZ
England
Tel. 0044-131-6505096
Fax: 0044-131-6624269
E-mail: kh@ed.ac.uk

Dr. Sirpa M. Häkkinen
NASA, Goddard Space Flight Center
Mailstop 971.0
Greenbelt, MD 20771, USA
Tel: +1 301 614-5712
Fax: +1 301 614-5644
E-mail: sirpa@fram.gsfc.nasa.gov

Dr. Gerbrand J. Komen
KNMI
Wilhelminalaan 10
P.O.Box 201
3730 AE De Bilt
The Netherlands
Tel.: 0031 30 220 69 11/2206676
Fax. 0031 30 221 04 07

Dr. Mojib Latif
Max-Planck-Institut fur Meteorologie
Bundesstrasse, 55
D-2000 Hamburg 13
Federal Republic Of Germany
Tel.: +49 40 41173-248
E-mail: latif@dkrz.a400.dc

Dr. Pierre-Yves Le Traon
Collecte Localisation Satellites
CLS Space oceanography Division
18, Avenue Edouard Belin
31055 Toulouse Cedex
France
Tel: +33 5 61 28 16 89
Fax: +33 5 61 28 14 32
E-mail: Letraon@cls.cnes.fr

Dr. Andrew C. Lorenc
Meteorological Office
London Road
Bracknell, Berkshire RG12 2SZ
United Kingdom
Tel.: ++44 1344 8562227
Fax: ++44 1344 854026
E-mail: aclorenc@meto.gov.uk

Dr. Giuseppe M. R. Manzella
ENEA - CREA
S. Teresa
P.O. Box 316
I. 19100 La Spezia, Italy
Tel. 187-536246
Fax. 187-536213

Prof. George L. Mellor
GFDL
Atmospheric and Oceanic Program
Sayre Hall, Forrestal Campus
Princeton University
Princeton, NJ 08544-0710, USA
Tel. 001-609-2586570
Fax: 001-609-2582850
E-mail: glm@splash.princeton.edu

Dr. Giuseppe Montanari
Arpa-Daphne
Viale Vespucci 2
47042 Cesenatico, Italy
Tel.: 0547 83941
Fax: 0547 82136
E-mail: sod2rn.arpa.emr.it

Prof. Kikuro Miyakoda
George Mason University, COLA
Maryland, USA
4041 Powder Mill Rd, Suite 302
Calverton, MD 20705-3106 USA
Email: miyakoda@cola.iges.org
Tel. 301-595-7000
Fax 301-595-9793

Dr. Antonio Navarra
Istituto Nazionale di Geofisica e Vulcanologia
Via di Vigna Murata 605
Roma, Italy
Mailing Address:
ISAO-CNR
Via Gobetti 101
40129 Bologna Italy
Tel: 39-51-6398014
Fax: 39-51-6398132
E-mail: navarra@ingv.it

Dr. Paolo Oddo
Laboratorio di Simulazioni Numeriche del Clima e degli Ambienti Marini
Corso di Laurea in Scienze Ambientali
University of Bologna
Piazza J.F. Kennedy,12
48100, Ravenna, Italy
Tel. +39 54 484269
Fax +39 54 484268
E-mail: p.oddo@ambra.unibo.it

Dr. Richard C. Patchen
Dynalysis of Princeton,
219 Wall Street,
Princeton, New Jersey 08540
Phone: (609) 924-3911
Fax: (609) 924-8793
E-mail: rcp@dynalysis.com

Prof. Nadia Pinardi
Laboratorio di Simulazioni Numeriche del Clima e degli Ambienti Marini
Corso di Laurea in Scienze Ambientali
University of Bologna
Piazza J.F. Kennedy,12
48100, Ravenna, Italy
Tel. +39 54 484269
Fax +39 54 484268
E-mail: n.pinardi@ambra.unibo.it

Prof. David Prandle
Proudman Oceanographic Laboratory
Bidston Observatory, Birkenhead
Merseyside UK L43 6RA
Tel. 0044-51-653-8633
Fax: 0044-51-6536269
E-mail: d.prandle@pol.ac.uk

Prof. Allan R. Robinson
Harvard University
(Pierce Hall 100C)
Division of Applied Sciences
29 Oxford Street
Cambridge, MA 02138 USA
Tel. (617) 495-2819
Fax: 617-495-5192
E-mail: robinson@pacific.harvard.edu

Dr. Jurgen Sellschopp
NATO SACLANT Undersea Research Centre
Viale San Bartolomeo, 400
19138 La Spezia Italy
Tel: +39 0187 527 256
Fax: +39 0187 527 331
E-mail: sellscho@saclantc.nato.int

Dr. Axel Timmermann
Department Predictability of Weather and Climate
Royal Netherlands Meteorological Institute (KNMI)
P.O. Box 201, 3730 AE De Bilt, The Netherlands
Tel: +31-30-2206377
Fax: +31-30-2202570
E-mail: timmera@knmi.nl

Dr. Marina Tonani
Istituto Nazionale di Geofisica e Vulcanologia
Via di Vigna Murata 605
Roma, Italy
Mailing Address:
ISAO-CNR
Via Gobetti 101
40129 Bologna Italy
Tel: 39-51-6398006
Fax: 39-51-6398132
E-mail: m.tonani@isao.bo.cnr.it

Dr. Reinhard Voss
Max-Planck-Institut für Meteorologie
Bundesstr. 55, D-20146 Hamburg
Germany
Tel: + 49 40 41173 - 0 or - 106
Fax: + 49 40 41173 - 298 or - 106
E-mail: reinhard.voss@dkrz.de

Prof. John Woods
T.H.Huxley School of Environment, Earth Science and Engineering
Imperial College
London SW7 2BP
Tel: +44 (0) 20 7594 7414
Fax:+44 (0) 20 7594 7444
email: j.woods@ic.ac.uk

Dr. Marco Zavatarelli
Dipartimento di Fisica
Università di Bologna
Viale Berti-Pichat 6/2
40127 Bologna, Italy
Tel: +39-051-5195213
Fax: +39-051-249644
E-mail:M.Zavatarelli@ambra.unibo.it

1 Strategic Planning for Operational Oceanography

NICHOLAS C. FLEMMING
Southampton Oceanography Centre, Empress Dock, Southampton SO14 3ZH, UK

1.1 Introduction

The question that I will try to answer in this paper is " how can national, regional, and international agencies rationally develop a strategy or linked strategies which will tell them whether it is worth investing in an operational ocean observing and forecasting system, and if the answer is positive, how should this investment be planned, implemented, and managed?" We will consider first the motives for establishing the Global Ocean Observing System (GOOS) and its regional components.

Operational oceanography is the provision of scientifically based information and forecasts about the state of the sea (including its chemical and biological components) on a routine basis, and with sufficient speed, such that users can act on the information and make decisions before the relevant conditions have changed significantly, or become unpredictable. National meteorological offices have been providing such a service for the atmospheric weather for many years, and military and fisheries agencies have been providing some information on marine sub-surface conditions since the 1960s. Since the late 1980s it has been possible to envisage a global system, similar to the World Weather Watch, and linked to national and regional observation and modelling services, which could provide an analogous service for many parameters of the marine water column and coastal zone.

Operational oceanography has been defined at some length in the publication devoted to setting out the long term plans for GOOS (IOC, GOOS 1998 Prospectus, p. 7-10) and in the EuroGOOS Strategy (Woods et al., 1996, p. 10). Although the emphasis is on obtaining and processing data in real time, or near real time, the products required by customers may be nowcasts, hindcasts, or forecasts, all based on the most recently obtained data. The EuroGOOS Data Requirements Survey (Fischer and Flemming, 1999, p. 22) shows that these products are required in approximately equal proportions.

The intellectual realisation in the 1980s that civilian marine forecasting might be possible at a useful variety of time and space scales, and that it depended critically on a global closure to eliminate unknown open boundaries, did not lead automatically to an agreement on development or implementation. It was obvious to any working group, committee, or UN Agency, that the practical problems of funding, management, political agreements, sensitivity about national waters, instrumentation, data processing, computing power, and so on, were almost insuperable, even if the scientific feasibility were certain, which it was not. The global oceanographic science experiments like the Tropical Ocean Global Atmosphere experiment (TOGA) and the World Ocean Circulation Experiment (WOCE) promised to strengthen the scientific justification for a Global Ocean Observing System (GOOS), but the necessary political and social justification was still missing. How could any country or UN Agency justify such expenditure without having any certainty of the benefits? Could the benefits be measured in financial terms at all, or were they principally social and environmental, which might be desperately important, but almost impossible to quantify? By the mid-1980s there was much popular and political concern about global warming and climate change, but even these threats could not be quantified, and so investment in GOOS was still postponed.

Above all the question of costs and benefits plagued the subject and made it almost impossible to proceed. If there was no single over-riding reason, no easily identifiable single customer for the information, why was an ocean observing system needed? If there were thousands - or millions - of small customers, how could one identify the products they needed, and how would they pay? One can see that national navies can justify a limited regional service based on local conditions needed to plan and control submarine warfare, and major national factory fishing fleets can use satellite images and operational oceanography to predict fronts and fish shoals. But beyond this the case was difficult to define. Pure scientific research, driven by curiosity to understand the ocean, did not justify permanent investment on the scale which would probably be needed.

In this section I will examine the way in which different organisations at national, regional, and global level, have evolved strategies for solving these problems, and for committing themselves at an appropriate level for the development of operational oceanography. I will make some comparisons to the development of meteorological services, and conclude with some pointers to the future.

1.2 Early developments, the comparison with meteorology

National meteorological offices existed in many countries a hundred years ago. WMO was established in 1950 and the World Weather Watch in 1962. The value of good meteorological forecasts was manifest from the start, since anybody could see the value of forecasts to farmers wishing to harvest their crops, sailors trying to avoid a storm, or road-builders planning to pour concrete. Commercial aviation at a trans-oceanic scale demanded accurate forecasts which could make the difference between success and failure, life or death, on a daily basis. The national meteorological offices therefore had an immediate, dominant, and politically visible customer from the 1950s onwards. Today the world weather forecasting system and services costs about $2bn per year, and produces proven benefits which are many times that.

Attempts to justify expenditure on meteorology in the 1960s and 70s were successful, showing that the benefits to aviation, shipping safety, agriculture and construction far outweighed the cost to the taxpayer. Furthermore, the aviation industry, and behind them the military, were prepared to pay substantially for the safety and efficiency which were dependent upon good forecasts. Modern analysis of the economic value of weather forecasting is very sophisticated, and there are numerous works on the subject, for example Katz and Murphy (1997).

By comparison, measurement, modelling and forecasting the oceans and coastal seas are more complex because of the scales of mesoscale eddies and fronts, indicating a computer requirement 1000 times more powerful than that needed for global meteorology (Pollard, 1994; OECD, 1994), because the instrumentation is more technologically challenging, the scientific processes are less well understood, and the benefits are, at first sight, less obvious. The revolution in computing power and numerical modelling during the last ten years has made ocean modelling practical, and the availability of satellite remote sensing with steadily reducing costs provides a valuable global data set for assimilation into routine models. But obtaining an adequate description of the internal state of the ocean depends upon in situ instrumentation, which, in many cases, is still in a stage of prototype development.

Since 1989 there has been a steadily increasing effort to evaluate the economic and social value of the marine and coastal industries and services, and the implicit value of the marine environment (OECD, 1994; Glantz, 1992; National Research Council, 1989; Woods Hole Oceanographic Institution, 1993; Sassone and Weiher, 1996; Adams et al., 1995; Woods et al., 1996; Costanza, 1997; Pugh and Skinner, 1996; IFREMER, 1999; Weiher, 1999; RASCL, 1999). Recent work by EuroGOOS (Fischer and Flemming, 1999) shows that, from a list of 136 categories of activity, representatives of almost every industrial, commercial, and environmental and conservation service in the marine environment can specify marine operational data which would improve the efficiency of their activity. The only activities which did not respond to the survey were deep sea mining and extraction of minerals from sea water. If we add to this range of environmental and commercial activities the objectives of managing, predicting, and possibly controlling climate variability and climate change, it is clear that the base now exists for justifying operational oceanography in the same way that meteorology was justified 20-30 years ago. Flemming (1999) has summarised the value of trying to assess the aggregate economic and social benefits from an ocean observing system which will provide forecasts on timescales from hours to decades.

Different regions identify different timescales of the whole spectrum of forecastable marine processes as potentially important. In the USA and central and southern Americas the impact of the El Niño/Southern Oscillation (ENSO) dominates economic and social value assessment. In China, South East Asia, and southern Asia it is the Monsoon cycle. In Europe it is the impact of short term ocean and coastal processes on marine industries and services, followed by potential climate forecasts of the impact of fluctuations in the North Atlantic Oscillation and the Gulf Stream. Every human being stands to benefit from understanding and forecasting the variability of global climate on a timescale of decades, but that benefit is so far into the future, and so uncertain, that it is still difficult to quantify.

Oceanography is following rapidly along the course taken decades ago by meteorology, but we still require a great deal of strategic planning on timescales of 5 to 10 years, with a steadfast consistency of purpose, to reap the full potential benefits.

1.3 Origins of GOOS

The realisation that a civilian Global Ocean Observing System (GOOS) was a feasible proposition grew steadily during the preparatory work for the Second World Climate Conference in October 1990, (Houghton et al., 1990; Jaeger and Ferguson, 1991), and the UN Conference on Environment and Development (UN, 1992). The threat of global climate change and possible sea level rise provided a unifying theme and objective which was recognised by all nations. Previous discussions about an operational observing scheme for the ocean had concentrated on the benefits for fisheries, engineering, offshore oil exploration, shipping, and coastal environmental management, and different countries have justifiably different views and priorities on these matters. There had seemed to be no way to tie all the different interests together in order to obtain coherent action for the common good. Additionally, prior to the 1990s, any observing scheme faced the apparently insoluble problems of high cost, inadequate instrumentation, insufficient scientific understanding of ocean processes, and insufficient computing power to handle the data or make forecasts.

Numerous workshops and study groups met during 1989-1992 to investigate the potential reality of GOOS. It became clear that the problems were solvable, at least in principle. Computing power has increased so rapidly in the 1980s and 1990s that this is not now a serious limitation, and the situation continues to improve. For the most advanced high resolution ocean models scientists are now considering the benefits of using petaflops computers. Experience of ocean modelling, climate modelling, and modelling of coastal seas and processes has resulted in a wide range of suitable programmes which can be applied, with further development, to GOOS. Thanks to the innovation of new principles in sensors, and the use of microprocessors embedded in instrument packages, *in situ* instruments are now enormously more powerful, can measure a far wider range of variables, and are much more durable and reliable than previously. It is possible to consider the development of operational instrumentation which can be used by relatively untrained personnel, based on existing proven instruments designed for scientific research. Many *in situ* instruments will have to be left to drift, or operate on fixed moorings untended for months or years at a stretch, with data being returned by satellite link. This principle is exhibited in the ARGO project (ARGO, 1998), www.argo.ucsd.edu. The experience gained from ocean observing satellites (SEASAT, GEOSAT, TOPEX-Poseidon, ERS-1 and 2, MOS, SEAWIFS) has established procedures and products so that it is now possible to consider a planned succession of customised operational satellites to monitor the ocean on a permanent basis.

During the 1980s scientific understanding of the ocean progressed rapidly, especially through the large scale integrative experiments such as TOGA, WOCE, and JGOFS, which provide a picture of the way in which all the parts of the ocean interact on each other as a global machine. There are still many mysteries and uncertainties about processes in the ocean, but there was by 1990 the confidence that it was at least possible to start constructing a scientifically designed observing scheme which will provide data in a logical way, and with sufficient accuracy and sample spacing to define the processes at work. If such a system starts by observing those factors which are best understood, (sea surface temperature field, wind field, upper ocean currents, sea surface topography, upper ocean thermal structure, floating sea ice, and sea surface wind-gravity wave spectrum), it will lead to a progressive improvement both of the observing scheme itself, and of the science which will permit further improvements of the system. The key is to start off with a firm scientific grounding, and this was provided by the Report on an Ocean Observing System, prepared by the Ocean Observing System Development Panel (OOSDP, 1995). The OOSDP was set up by the Committee for Climate Change in the Ocean (CCCO) and the Joint Scientific Committee of the World Climate Research Programme (JSC). The OOSDP Report emphasises the physical variables required to detect or predict climate change, but many of the same observations would provide benefits in the shorter term also.

During the mid-1990s in the period after the 1992 Rio Convention, plans for GOOS continued to evolve, while the Conventions and Treaties initiated at Rio strengthened the demand for ocean observations at the global level. GOOS itself was established by agreements between UN Agencies in late 1993. The Framework Climate Convention, the Convention on Biodiversity, and the Convention on Sustainable Development all require greatly increased knowledge of ocean processes, and specific understanding of the present state of the ocean and the trends of change through time. These issues were analysed in a meeting organised by OECD in spring 1998 (OECD, 1998).

Although the most technically feasible components of GOOS are the physical observations and predictions, the biological requirements are both scientifically important and politically essential. Biological processes, particularly phytoplankton growth, strongly influence the carbon cycle and the penetration of light and heat into the upper layers of the ocean. Biological productivity, water quality, and public health issues are of vital public interest in the coastal zone. International global science programmes such as JGOFS and GLOBEC (www.pml.ac.uk/globec) provide some of the basic understanding needed in ecosystem modelling.

The potential cost of GOOS, or a regional component of GOOS, is a major obstacle to the development and implementation of GOOS, and the political and institutional commitment to investing in GOOS only occurs when it can be demonstrated in a thoroughly sound analysis that the economic and social benefits from the use of the data will exceed the costs by an acceptable multiple, using strict principles of estimation, accounting, and discounting to net present value. Such calculations must not be based on an idealised estimate of the way in which the data from

GOOS would be used, but must include a justified estimate of the actual take-up of the data by industry and regulatory authorities, and their ability to use the data and information to make beneficial decisions. These considerations have been analysed by Weiher (1999) and Flemming (1999), and are further discussed by Brown (2000) and Castelucci (2000).

The GOOS plans identify five modules of application, that is, sectors of the marine user community which could benefit from improved forecasts and monitoring. The 5 modules are: Climate; Living Marine Resources; Ocean Services; Health of the Ocean; and Coastal and Shelf Seas. During 1992, while the UNCED Rio meeting was in session, there was a great deal of controversy about the economics of mitigating actions and preventative actions which might be carried out to reduce the causes and impact of climate change. The principle risk or threat arising from climate change will probably become seriously damaging in about 30-50 years due to the steady build up of anthropogenic "greenhouse" gases in the atmosphere. Assuming a discount rate of the order of 7%, $1000 in 30 years time is worth about $130 now, discounted over 50 years it is worth only $34 now. Applying these figures to the concept of GOOS produced an unfortunate effect, since GOOS requires considerable initial investment in hardware, communications, and data product distribution in its early stages. The potential benefits or disasters avoided in 30-50 years time had to be of enormous financial scale to justify spending $millions now. Although the worst scenarios for climate change might justify such expenditure if the disasters were certain, it is more difficult to make such a justification when the disaster is an unquantifiable probability. Many publications in the USA during 1992, the time of the Rio UNCED Conference, sought to show that public expenditure on science or other projects to avoid or mitigate the impact of climate change were a waste of money, because of the discount factor. It would be better to wait, accept any damage that occurred, if it occurred, or fix it later. The justification for GOOS was thus forced full circle back to the consideration of all possible economic and social benefits on every timescale from days to decades. Many of these potential benefits are outlined in Chapter 17 of Agenda 21 (UN, 1992). The acute concern with climate change and global environmental change had created the political cohesion and sense of common interest which gave GOOS its popularity, but by late 1992 it was clear that all possible benefits must be aggregated to show how GOOS might produce benefits year by year. One of the key medium-term benefits is in the prediction of seasonal and year-to-year climate variability, such as the El Niño-Southern Oscillation, or ENSO. This led to the successful conversion of the TOGA TAO array and the associated modelling system into an operational forecasting system after the completion of the TOGA experiment.

If all the short- to medium-term benefits from monitoring and predicting the ocean and coastal seas provide a sufficient economic justification for GOOS, then the continuous accumulation of data year after year will eventually produce the long time-series and monitoring statistics which will permit the detection of climate change, and even prediction of climate change. The over-arching human need to understand and predict climate and global environmental change still provides an intellectual and logical link which holds GOOS together in the long term. In the design of the observing system and data archiving, care must be taken to ensure that the accuracy required for climate monitoring and prediction is guaranteed, even if it is a greater level of accuracy than that needed for the shorter term services. Through the development of GOOS every coastal state should gain practical benefits for the management of its resources and marine environment, and in the long term the threat of climate change will be detected at the earliest possible moment. This aggregation process is analysed by Flemming (1999).

The renewed emphasis on the short-term economic benefits on a prediction timescale of weeks to months and applications to many different industries does not mean a return to the disparate problems of the early 1980s. The advance in numerical modelling techniques for major ocean basins and the coastal shelf seas has created a synthesising force which enables scientists, planners and data managers to integrate the sum of input data gathered from many different sources, and to generate informative products which can be used by hundreds of different customers for different reasons. It is not easy to achieve this goal, but it is possible. Thus observations of the physical parameters of the Atlantic will be used in numerical models which, after the addition of some other data streams, may result in products as diverse as flood warnings in the USA, estimates of fisheries stock migration on the continental slopes off the west coast of Ireland, and prediction of dangerous currents and internal waves threatening oil platforms off the Niger delta. It is possible to use one on-going programme of, for example, sea surface temperature measurements, to support the generation of data products with myriad applications. GOOS therefore offers an attractively efficient way of reducing costs and maximising benefits for many customers.

When it become clear that the shorter term economic benefits provided the main justification for GOOS, particularly in Europe where the ENSO effect is minimal, it was then necessary to find a way of evaluating and summing the benefits which would accrue to a wide range of maritime and coastal industries. In September 1993 the Organisation for Economic Co-operation and Development (OECD) convened an Experts Meeting of the Megascience Forum in Tokyo to discuss the economics of GOOS, and from this a book was published in August 1994 (OECD, 1994). The Megascience Forum established for the first time the approximate scale of the aggregate of all maritime industries and services, and their contribution to national and global GNP. Since 1993-4 several countries have set up national committees to develop policies and projects in support of GOOS, and analysis of the benefits and economics of GOOS has been a high priority. In 1999 27 countries signed the GOOS commitments agreement in Paris, confirming their intention to make nationally funded observations available to GOOS.

The OECD Megascience Forum concluded that during the next ten years, during the development of the early phases of GOOS, considerable investment is needed in basic oceanographic science and pre-operational research into the design and implementation strategy of GOOS. Thus some sub-systems can be implemented quite quickly, while research is still needed on others. OECD Megascience Forum concluded that operational satellites are needed for ocean observations, and it was thus significant that at the User Consultation Meeting on Earth Observing Satellites convened by ESA in October 1994 the importance of planning ocean observation and coastal monitoring was given the highest priority for the Earth Watch operational series of missions.

EuroGOOS was formed in December 1994. and MedGOOS held its first formal meeting in March 1999, although preliminary meetings had been held earlier (MedGOOS, 1999). These two organisations exemplify the development of interest in regional co-ordination for GOOS. Another successful region of GOOS was also set up in 1994 North East Asia, and known as NEAR-GOOS (NEAR-GOOS, 1999). GOOS itself is now managed through the GOOS Steering Committee, which meets annually; the GOOS International Project Office based at IOC Paris, and the GOOS Intergovernmental Committee, which meets periodically, usually in conjunction with the IOC Assembly.

In the next section I will review the way in which strategic planning has influenced the development of EuroGOOS, and more recently MedGOOS.

1.4 Regional Operational Ocean Systems

Notwithstanding the consensus on the importance of global ocean measurements that developed in the early 1990s, and the establishment of GCOS, the actual priorities that would motivate one country or agency to invest in GOOS varied considerably. During the late 1980s and early 1990s different regions, governments, and agencies identified different priority objectives or justifications for GOOS within the general complex of climate forecasting and shorter term economic and social benefits. In the USA, the benefits of forecasting the 3-7 year ENSO cycle and its impacts throughout the tropical zone quickly emerged as a priority. India and China both have interest in improving the forecasting of the monsoon system and large scale rainfall patterns, linked to the dynamics of the Indian Ocean, and possibly the ENSO cycle and the Pacific Decadal Oscillation. The heavy flooding in China during 1998 illustrates yet again the importance of this factor.

Australia, through its Heads of Marine Agencies structure has co-ordinated its commitments to GCOS and GOOS. Several national studies have shown the overall importance of marine resources and trade to Australia, including a large offshore oil industry, and so their policy tends to favour a general purpose observing and information product approach designed to provide a wide range of services. Netherlands conducted a survey of their options for commitment to any or all of the G30Ss, and identified a lead agency for each global system. RIKZ was designated as the lead agency for GOOS and EuroGOOS. In the UK the Inter Agency Committee for Marine Science and Technology (IACMST) made preliminary assessment as early as 1993 of the national value of participation of GOOS, and set up a sub-committee on GOOS and EuroGOOS which is still active.

EuroGOOS was established in 1994 and for the first two years devoted its activities to strategic analysis of the needs of Europe for operational ocean services, and defining objectives. This period culminated in the publication of the Strategy for EuroGOOS (Woods et al., 1996) and the First EuroGOOS Conference held at the Hague in 1996, with a Conference Proceedings volume published in 1997 (Stel et al., 1997).

Two key elements were created during that phase, which have stood the test of time. These were the sub-regional sea Task Teams, each responsible for identifying needs and developing strategies for implementation in the sea areas of the Baltic, North West Shelf Seas, Arctic, Atlantic, and Mediterranean. These areas are so radically different in terms of oceanographic conditions and economic exploitation of the sea and coastal zone that they have to be treated as separate and distinct units. The scale of such subdivisions would naturally vary in different regions of the globe, but it is equally clear that the Mediterranean itself would benefit from detailed analysis at the sub-regional level, and then implementation of observations at an integrated level.

The second key principle established early on was the so-called Virtuous Circle, whereby sectors of activity are continuously reviewed and progressed through a changing suite of projects. The sectors of activity are Customer and User analysis; Economic assessment; Scientific research to remove obstacles; Technology development; Trials and case studies; Operational Observing Systems; Numerical assimilation of data, models and forecasting; Product design and services; and thence back to the Customers and Users.

From 1997 to 1999 EuroGOOS developed a range of projects on different scales related to various problems, technological requirements, regional studies, designed to test pre-operational systems. At the same time, existing operational services provided by national agencies were analysed to see how they could fit together or reinforce each other on a European scale. In March 1999 EuroGOOS held its Second Conference in Rome, and the development of operational and pre-operational observing systems and products was manifest. The speed of development has been greatly assisted by the increased speed and range of services on the Internet. At the end of 1999 EuroGOOS published a short document entitled "The EuroGOOS Forward Look" (EuroGOOS, 1999). This sets out in a brief way the targets which EuroGOOS hopes to achieve by the year 2008. The next few years will be dedicated to implementing operational and pre-operational systems to achieve those targets.

Analysis of the reasons for investing in EuroGOOS, and justifying the expenditure has proceeded continuously but in a somewhat erratic fashion. Very broad brush figures were set out in the Strategy for EuroGOOS (Woods et al., 1996), but far more detailed work is needed, and the EuroGOOS Economics Working Group has held two meetings in order to develop further methodology (EuroGOOS, 1998). This is a subject which requires more work.

EuroGOOS now includes 28 Government Agencies from 15 European countries. Between 3% and 5% of input to the European GNP (EU) is generated directly by marine-based industries and services (Woods et al., 1996, p.21). The value added directly by these activities is of the order of $140-230bn/yr. The industries and services are subject to uncertainty, loss of efficiency, and direct costs and damage caused by the unpredictable forces of the marine environment such as storms, sea level surges, waves, erosion, transport and resuspension of pollutants, shifts in fish stock migration, and toxic algal blooms. The temperature and salinity of the north Atlantic determines the weather and climate of Europe, Russia, and the Mediterranean. Europe has a great need to understand, monitor and predict the state of its coastal seas, the Mediterranean, and the adjacent oceans, Atlantic and Arctic. The monitoring and forecasting of the North Atlantic Oscillation, the fluctuations of the Gulf Stream, convection and deep water formation, and Arctic sea ice formation, are particularly important at the oceanic scale (Broecker, 1991; Parrilla et al., 1994; Schott et al., 1994; DYNAMO, 1997; Sutton and Allen, 1997; Le Provost and Flemming, 1998; EuroCLIVAR, 1998; Wood et al., 1999; Shindell et al., 1999).

Improvement of the short- to medium-term prediction services for maritime conditions would improve the value of maritime industries and services by a few percent. If we accept 1% as a most conservative estimate, the value added to the GNP of the EU by a prediction system is of the order of $1.4-2.3bn/yr (Woods et al., 1996). This is a minimum which should be exceeded by a factor of 2-3. In addition there are the longer term benefits of climate prediction, and its impact on agriculture, energy generation, water supply management, land use and other social activities, which would be of the same order. More recent economic studies under way now are suggesting the methods for quantifying in comparable ways the benefits from short term commercial activities, through specific short and medium term public good benefits, to long term public good, environmental, and climate scale risks and benefits.

Europe possesses the wealth, institutions, and expertise to benefit strongly from operational oceanography on a European, Mediterranean, and North Atlantic-Arctic scale. The same institutions permit Europe to contribute to and participate in the Global Ocean Observing System (GOOS), (IOC, 1998, p.114). European agencies recognise their obligation to contribute resources and skills pro rata to the global observing system, and it is also important to Europe that the global infrastructure is designed so as to guarantee the required data products and benefits needed by Europe. Most of the countries with EuroGOOS Members attended the 4th meeting of the Intergovernmental GOOS Committee (I-GOOS-IV) in June 1999 (IOC, 1999), and the GOOS Commitments meeting in July 1999.

European national marine research institutes, operational establishments (fisheries agencies, meteorological offices, environmental protection agencies), and transnational agencies and bodies (ESA, ICES, EUMETSAT, ESF, CEC, ECMWF, Eureka EUROMAR, and the pollution and dumping Conventions) in combination possess most of the scientific research basis, expertise, engineering ability, and data processing ability needed to install and manage operational oceanography on a European scale, and to benefit substantially from global operational oceanography. However, prior to 1994 there was no declared objective, and no mechanism, to coordinate the collaboration between these bodies. It is the objective of EuroGOOS to promote this co-ordination.

Europe is dependent upon and influenced by marine conditions more than any other developed continental region. Changes in mean sea level, changes in storm conditions and coastal erosion, have a greater impact on shelf-seas and oceanic fisheries, tourism, land use, shipping and ports, and offshore oil than in other continents. Concern about possible global sea level rise and its combination with regional earth movements is of high priority, as evidenced by the activity of the European component of the Global Sea Level Observing System (GLOSS). The Mediterranean is an almost closed basin with unique circulation which requires a relatively high resolution observation and modelling scheme (Golnaraghi et al., 1996; Jeftic et al., 1993; Jeftic et al., 1996). The Mediterranean Forecasting System proposed by EuroGOOS in collaboration with the Mediterranean GOOS organisation (MedGOOS) emphasises regional nested modelling and forecasting (Pinardi and Flemming, 1998). This modelling is important both for the Mediterranean coastal states, and to provide accurate assessment of the Mediterranean outflow of dense water into the Atlantic. The Baltic region has a highly developed set of collaborational structures, with a sophisticated programme of linked hydrodynamic models, interfacing with biological productivity models (Buch and Dahlin 2000).

The ability to monitor and predict North Atlantic changes on the multi-year timescale would provide the boundary conditions for models of the shelf seas, and hence permit prediction of conditions in the coastal areas. In addition to the Atlantic variability it is important to understand the variability of Arctic sea ice, and the variability of the ocean waters under the ice. European countries will benefit greatly from a programme of remote sensing of the Atlantic and Arctic Oceans, combined with in situ instrumentation. The ocean scale models should be run so as to provide seasonal, inter-annual and multi-year predictions. Most EuroGOOS activities to date, and the forecasts and models run by Member agencies, are intended to provide short to medium term forecasts of coastal and marine processes. The full value of incorporating seasonal and inter-annual ocean models into climate variability forecasting has not yet been tested. This is likely to occur soon, as basin scale models such as FOAM (Forecast Ocean Atmospheric Model) and MERCATOR are implemented at higher resolution, and integrated with climate models. Forecasts on this scale will influence the planning of fresh water utilisation, management of dams and reservoirs, river management, agriculture, and energy planning. In 1997 EuroGOOS held a joint workshop with participants from the USA to analyse requirements for operational modelling of the Atlantic (Le Provost and Flemming, 1998). EuroGOOS Member Agencies are prominent in the planning of the Atlantic ARGO Pilot Experiment, jointly with the USA.

European countries have policies on aid and assistance to developing countries, and agreements for joint programmes and collaboration or partnership on projects. GOOS is a natural vehicle for collaboration in capacity building. As a general structure for EuroGOOS is set up, the common or collective view on the involvement of developing countries is an important element. EuroGOOS actively collaborates with the other GOOS Regions, especially with Mediterranean GOOS (MedGOOS) (Drago, 1998), and Africa GOOS, and Black Sea GOOS.

There are therefore several fundamental assumptions in the design of EuroGOOS activities, and its justification as a worthwhile investment.

- The maximum benefit of operational oceanography can be developed only if we include regional and sub-regional components.
- Effectiveness depends strongly on the ability to obtain data fast and run real time models.
- Models need to be nested and interfaced on different scales, and representing different variables.
- Europe must play a proportionate role in the global observing system.
- The missing factors which EuroGOOS can provide are to promote the transition from research mode to operational mode, and to promote focused collaboration on agreed objectives.
- Present scientific knowledge enables us to start designing a fully operational system, but continuing research is needed to exploit the full possibilities in terms of resolution, forecast horizon, and variables.
- New technology is needed to enable more data to be obtained routinely without increased personnel and ship requirements.

Civilian operational oceanography has existed in one form or another for a decade or more, and increasingly complex proposals are being developed now. It therefore seems embarrassing to have to admit that a thorough costing of the proposals, at either national or regional level, has not been carried out. There is a genuine problem here of which we should not be ashamed. Various rough estimates have been made (e.g., IOC, 1993; Flemming, 1994; IOC, 1998, p.30-33). The following discussion is based on consideration of the costs of a global operational observing system such as GOOS, but the same caveats apply to attempting an over-simple analysis of the costs of EuroGOOS. Since the European region tends to gain benefits from investing in an efficient global system, as well as from applications directly within the European sea areas, we can consider comments on the global system and European components as having equivalent force.

An attempt to sum the cost of global or regional ocean observations, or a complete operational observing system including modelling and forecasting, naturally involves summing a wide range of sub-costs, purchases of equipment, deployment costs, ship operations, maintenance and replacement of equipment, satellite launch costs, equipment planning and design costs, communications and data processing, modelling centres, computers, product delivery, and so on. Each component has a different proportion of capital investment, duration, and running costs. Aggregation of these costs requires that each component should be analysed to see whether there are multiple beneficiaries who should share the cost, hidden overheads, sunk costs, or whether it might be developed anyway for other purposes, and thus have a zero cost to GOOS.

Depending upon assumptions made in solving the ambiguities, the apparent cost of operational oceanography can be made to appear larger or smaller through an order of magnitude. Accounting procedures need to be agreed and transparent, at least at the national level, and preferably at the regional level.

This is not just a question of trying to limit the misleading impression which can be given by presenting accounts in a deliberately distorted way. The assumptions that are needed before costs of different systems can be aggregated are fundamental. For example, to what extent should the cost of remote sensing by satellite be included, when all satellites so far have been funded for scientific research purposes, and the same sensors sometimes fulfil different functions over land and sea, thus satisfying different user communities? (How are we going to fund truly operational ocean observing satellites?). How do we cost the deployment of instruments which are dropped from ships of opportunity? How do we account for data, such as bathymetric charts, or other services which may be provided by the military? How do we reconcile costs between agencies or countries which use different principles for estimating total or marginal costs, or different rates of amortising capital? If there were no planning for GOOS or EuroGOOS at all there would be a hap-hazard development of numerous observing systems and models producing a very poor overall performance. The apparent "cost" of GOOS/EuroGOOS would be zero, since all the costs would be incurred for other reasons. Once GOOS or EuroGOOS exists the costs become apparent, but so does the gain to be achieved by analysis, planning and design. In the extreme case the introduction of a planned and designed system might actually reduce the true costs of an observing system while increasing the benefits. This may be overly optimistic, but it illustrates the difficulty in identifying the cost of GOOS and EuroGOOS or MedGOOS.

This section does not provide answers to the problem of trying to predict the costs of operational oceanography, but it does illustrate that a sophisticated cost model will be required, fully taking into account standard procedures for coping with sunk costs, shared overheads, shared benefits, self-funded new technology, rented and leased equipment, etc. While such an approach may be justified, in practice agencies seem to be adopting a pragmatic step-by step process, committing to successive stages of ocean observation as each phase appears to provide benefits.

Surveys in France and the UK (IFREMER, 1997, 1999; Pugh and Skinner, 1996) have estimated the scale of different government marine activities, commercial operations, commercial services, living and mineral resources, and social activities such as tourism and enjoyment of parks and wildlife. These studies have not attempted to evaluate the more qualitative benefits such as impacts on public health, or the longer term effects of marine observations on climate, agriculture, and energy or water utilisation. One side effect of these studies is to provide a set of categories of activity in which the organisations are more or less intensive users of marine information and forecasts. Starting from this base, and using lists of members of trade associations, government departmental mailing lists, university departments, and exhibitors in various marine engineering and trade exhibitions, it is possible to construct mailing lists of many hundreds or thousands of potential beneficiaries from a marine observing and forecasting service at the national level. A study of this kind has been carried out for the users of the data products from the MFSPP.

EuroGOOS has designed a survey of customers for operational ocean data and their data requirements, which has been run in 6 countries (Fischer and Flemming, 1999). The survey was conducted in Denmark, The Netherlands, UK, Italy, Spain, and Greece. This provides a reasonable balance between northern, temperate, Atlantic operators and southern, warm climate, Mediterranean operators. The survey identifies classes of hundreds of serious users of operational marine data and forecasts, and prioritises by frequency of requirement the variables which are most in demand, their geographical scale, accuracy, temporal and spatial resolution, product type (raw data, processed, statistics, forecast, hindcast, nowcast) and latency and medium of delivery. A side effect of such surveys is to build up a familiarity and working relation between the agencies planning new observing systems and their potential customers. Presentation of papers and reports on these themes at trade and industry conferences provides a very positive feedback.

There is the further range of additional benefits which are more difficult to quantify, but which all add value to the benefits from EuroGOOS. Firstly there are the long term benefits from improved climate prediction which are quantifiable, though reduced in value by discounting. These are being examined by GCOS, and there is no doubt that this will show increased benefits attributable to GOOS (GCOS, 1994).

Then there are a host of rather intangible benefits which may be measurable in economic terms, but which at first sight are intractable to measurement in money. Econometric techniques have been developed in this field, but they have not yet been applied to marine and coastal activities. These benefits include the aspects of conserving biodiversity, protecting wildlife, preserving the aesthetic appearance of the coastal zone and wetlands, preserving ecological balance even when it is not shown to jeopardise fisheries or other living resources, and minimising the public sense of disturbance or insecurity which may be caused by climate change or rise in sea level. In its extreme form, this type of analysis computes the overall economic benefit from the existence of different natural environments (Costanza et al., 1997).

The costs of EuroGOOS and the benefits achieved may be balanced in various ways. If EuroGOOS Member agency activities were regarded as a pure public good, then a computation of the notional value of the benefits would be sufficient, and there would need to be no actual charging of customers or direct recovery of costs. If EuroGOOS Member activities were regarded as a commercial exercise, all the costs would have to be recovered from the sale of services and products. It is premature to make judgements on these factors, but already it is possible to see that parts of the infrastructure, parts of the remote sensing, deep ocean sampling, and long term climate predictions tend to have the characteristics of public good economics, while the short term coastal engineering, navigational, and fisheries predictions can probably be marketed as cash services. A full study of the costs and benefits of EuroGOOS Member activities would have to break down the services into a range of categories between these two extremes.

1.5 Conclusions

Regional strategy is an essential component of GOOS, and operational oceanography in general. With few exceptions the nation state is too small, and has too small an area of EEZ within its control, to install a range of observing systems which will be meaningful on their own. Regional sea areas such as the Baltic, the North Sea, or the Mediterranean are natural strategic planning and development areas for operational oceanography. The adjacent coastal states then have an incentive to collaborate, as provided by the examples of the MFSPP (Pinardi and Flemming, 1998) or the BOOS Plan (Buch and Dahlin, 2000).

The rate of development of instruments, computers, satellite sensors, and communications through the internet is such that strategic thinking and planning must be pragmatic. Some things are going to happen anyway, both good and bad. Successful planning for operational oceanography depends upon identifying those developments which we can influence, and concentrating effort on these factors, while exploiting other development as well as we can.

By definition, operational oceanography requires investment to achieve a purpose, and the benefits of the outcome must have greater value than the cost. Different countries and regions have different needs, both sort term and long term, depending upon their local industries, services, social needs, and environmental conditions. A great deal can be learned by example, comparing previous studies with new areas, but it is almost always necessary to conduct new analysis to identify local priorities.

Europe is an archipelago continent, and this includes the north shore of the Mediterranean. The region depends to a high degree on marine transport and marine resources while being prone to a wide range of maritime hazards and the effects of climate change. It is therefore practical to develop marine observing systems on all scales from coastal to shelf to oceanic, in order to obtain short, medium, and long term benefits. The shorter timescale products and benefits provide an economic and social return which counter-balances the negative cash flow in the long term system which has to be built up over decades.

Annex: Acronyms

ARGO Array for Real-time Geostrophic Oceanography
BOOS Baltic Operational Oceanographic System
BSH Bundesamt für Seeschiffahrt und Hydrographie (Germany)
CCCO Committee for Climate Change in the Ocean
CEC Commission of the European Communities
ECMWF European Centre for Medium Term Weather Forecasting
EEZ Exclusive Economic Zone
ENSO El Niño Southern Oscillation
ERS Earth Resources Satellite
ESA European Space Agency
ESF European Science Foundation
EU European Union
EUMETSAT European Meteorological Satellite Organization
EUREKA European Research and Co-ordination Agency
EuroCLIVAR European Climate Variability and Predictability
EuroGOOS European (component) Global Ocean Observing System
EUROMAR European Marine Research Programme within EUREKA
FOAM Forecasting Ocean Atmosphere Model
G30Ss The three Global Observing Systems (GCOS, GOOS and GTOS)
GCOS Global Climate Observing System
GEOSAT Ocean observing satellite
GLOBEC Global ocean Ecosystems dynamics
GLOSS Global Sea Level Observing System
GNP Gross National Product
GOOS Global Ocean Observing System
GTOS Global Terrestrial Observing System
IACMST Inter-Agency Committee on Marine Science and Technology (UK)
ICES International Council for the Exploration of the Seas
IFREMER French institute for research and exploitation of the sea
IOC Intergovernmental Oceanographic Commission (UNESCO)
JGOFS Joint Global Ocean Flux Study
JSC Joint Scientific Committee of the World Climate Research Programme
MedGOOS Mediterranean Global Ocean Observing System
MERCATOR French operational high-resolution global ocean prediction project
MFSPP Mediterranean Forecasting System Pilot Project
MOS Ocean observing satellite
NOAA National Oceanic and Atmospheric Administration (US)
OECD Organisation for European Co-operation and Development
OOSDP Ocean Observing System Development Panel
RASCL Roger A Stacey Consultants Ltd.
RIKZ Netherlands Institute for Coastal and Marine Management
SEASAT Ocean observing satellite
SEAWIFS Ocean observing satellite

TOGA Tropical Oceans and the Global Atmosphere
TOPEX/POSEIDON Joint US/French Ocean Topography Experiment
UNCED United Nations Conference on Environment and Development
UNEP United Nations Environment Programme
WMO World Meteorological Organisation
WOCE World Ocean Circulation Experiment

2 Satellite Oceanography for Ocean Forecasting

PIERRE-YVES LE TRAON
CLS Space Oceanography Division

2.1 Overview of space oceanography

This lecture aims at providing a general introduction to satellite oceanography in the context of ocean forecasting. Satellite oceanography is an essential component in the development of operational oceanography. Major advances in sensor development and scientific analysis have been achieved in the last 20 years. As a result, several techniques are now mature (e.g. altimetry, infra-red imagery) and provide quantitative and unique measurements of the ocean system.

We begin with a general overview of space oceanography, summarizing why it is so useful for ocean forecasting and briefly describing satellite oceanography techniques, before looking at the status of present and future missions. We will then turn to satellite altimetry, probably the most important and mature technique currently in use for ocean forecasting. We will also detail measurement principles and content, explain the basic data processing, including the methodology for merging data sets, and provide an overview of results recently obtained with TOPEX/POSEIDON and ERS-1/2 altimeter data. Lastly, we will focus on real-time aspects crucial for ocean forecasting. Perspectives will be given in the conclusion.

2.1.1 Why do we need satellites for ocean forecasting?

An ocean hindcasting/forecasting system must be based on the assimilation of observation data into a numerical model. It also must have precise forcing data. The ocean is, indeed, a turbulent system. "Realistic" models of the ocean are impossible to construct owing both to uncertainty of the governing physics and of an initial state (not to mention predictability issues). Continuous observations are required to drive the model towards a realistic state. Ocean forecasting therefore calls for an ocean observing system which should also include forcing data. Both in-situ and satellite data are needed:

- The usefulness of in-situ data is limited by poor space/time coverage, and access to remote regions (e.g. southern oceans) is often difficult. Conventional techniques (ship measurements, ship deployment) also may not be suitable for an operational system. Reliable, autonomous techniques have to be used. Such techniques are already being used or starting to emerge (e.g. XBT, TOGA/TAO, PIRATA, profiling floats such as P-ALACE or PROVOR) and promising new techniques (e.g. gliders, acoustic tomography) are under development. Global operational monitoring of the ocean with in-situ data is still, however, to be developed. The Argo project (Roemmich et al., 1999) will be a major step towards this objective.

- Satellite data, on the other hand, can provide long-term, repeated (synoptic) and global measurements of key parameters (e.g. sea level, SST, winds). These are surface parameters but they may be representative of the deep ocean (e.g. sea level from altimeters). An additional advantage is that measurements can be acquired and processed in near-real time.

There is an obvious complementarity between satellite and in-situ data. Satellite and in-situ measurements have different contents (e.g. vertical structure from in-situ data, surface information from satellites) and very different space/time sampling. In-situ data are also required for calibration and long-term validation of satellite data.

2.1.2 Space oceanography techniques

Overview

Many space oceanography techniques have been developed in the last two decades. I will base my discussion on a broad definition of space oceanography encompassing direct observation of the ocean, ice and ocean forcing terms (wind, heat flux, precipitation, evaporation). Technologies range from passive radiometers measuring the natural electromagnetic signal which radiates from the sea to active radiometers (radars) which transmit a signal and measure the modified echo signal. Passive radiometers allow us to look at different oceanic signals depending on the wavelength. Wavelength bands available are actually limited by absorption of the electromagnetic signal by molecular aerosols and water vapor.

A brief overview of space techniques and of the retrieved oceanographic parameters is given here. For a more detailed description the reader is referred, for example, to Robinson (1985), Weller and Taylor (1993) and Ikeda and Dobson (1995).

In the visible band, between 0.4 μm and 0.8 μm, we find measurements of ocean color (see below) and surface solar radiation flux [one component of the net heat flux at the ocean surface which can be well retrieved from space data (Planton, 1994)]. In the infrared band (0.8 μm to 100 μm), we find, of course, the Sea Surface Temperature (SST) and the infrared flux, although, due to cloud cover, the latter is difficult to determine precisely from satellite measurements. In the visible and infrared bands, the main limitation is cloud cover. SST and ocean color measurements from space can thus only be obtained for clear skies. Sea ice surface characteristics can also be distinguished in the visible and infrared bands (AVHRR).

In the microwave band (0.1 cm to 100 cm), cloud cover is no longer a problem. Main parameters which can be measured in the microwave band are the integrated atmospheric water vapour (not directly relevant here), sea ice and surface wind speed. The absence of cloud cover is particularly important for sea ice studies (measurement of ice concentration and ice extent). Surface precipitation is a non-radiative parameter which can be inferred from the satellite radiative measurements (microwave radiation is related to atmospheric liquid water content, not directly to rain rate, but a good correlation between the two parameters exists-besides, precipitation is poorly known and satellites provide very useful information). Note that visible and infrared techniques can also provide useful information on precipitation. Latent heat flux or evaporation rate can also be inferred from SST, wind speed and integrated water vapour (although data on surface level humidity would be needed). Finally, one should note that SST can also be retrieved from microwave radiation but with a lower resolution and accuracy compared to infrared radiation. There is also an interesting capability for measuring Sea Surface Salinity from microwave measurements.

Main active microwave radiometers or radars for satellite oceanography are: altimeters, scatterometers and Synthetic Aperture Radars. Satellite altimetry provides measurements of the sea surface topography, significant wave height and wind speed modulus. It also provides measurements of ice sheet topography (e.g. mass balance of Antarctica and Greenland ice sheet). We shall describe altimeter measurement principles in the next section. A scatterometer measures the electromagnetic signal backscatter coefficient in three directions. The backscatter coefficient depends on wind speed and direction. By analyzing backscatter in different directions, we can estimate the wind speed direction. There is a 180° ambiguity because differences in backscatter 180° apart are very small. This ambiguity is generally removed through consistency check from previous measurements and/or using atmospheric ocean models. Scatterometers thus provide wind speed and wind direction measurements. They are also useful for measuring sea ice, although not as useful as microwave radiometers. Synthetic Aperture Radar (SAR) is used mainly to measure the wave spectrum. These measurements may also be used to map the winds and ocean currents at small scales but the signal is complex and difficult to invert (Johannessen, 1995). SAR images also provide useful measurements of sea ice parameters (Johannessen et al., 1995).

Finally, one has to consider satellite data collection and/or location systems for in-situ measurements. They are vital for operational oceanography since they allow the near real time data collection and, if needed, the location of drifting instruments for velocity estimations (e.g. surface drifters, ALACE, profiling floats). In particular, it includes the well known ARGOS system (www.cls.fr/html/argos).

Altimetry, SST and ocean color

A ranking of satellite oceanography techniques for ocean forecasting objectives would probably give altimetry, SST, scatterometers and probably ocean colour. The reason why altimetry is so useful will be detailed in the next section. We will not discuss further scatterometry as we focus here on ocean parameters. The availability of winds with high space and time resolution is, however, a critical requirement for ocean forecasting (see Milliff et al, 1999 for a review).

It is clear that SST measurements are crucial for understanding and predicting the ocean/atmosphere system. SST is an important factor that influences (and is influenced by) the ocean and atmosphere circulation. The heat fluxes at the ocean/atmosphere interface are also strongly linked to SST. This parameter is operationally measured by the NOAA satellites (AVHRR), and useful data sets are already available. Geostationary satellites (e.g. GOES, METEOSAT) provide a much better temporal sampling but with a degraded spatial resolution and accuracy. SST anomalies are key indicators of changes in the environment (e.g. El Niño events). At smaller scales, SST measurements can provide useful information on the mesoscale ocean flow field and position of the main current system (e.g. Gulf Stream and ocean color measurements) and regions of upwelling. Surface currents may also be derived from the analysis of successive SST images (e.g. Kelly and Strub, 1992).

The usefulness of ocean color measurements was demonstrated by the CZCS instrument on board the Nimbus-7 satellite (1978-1986). There are now several ocean color missions flying, and more are scheduled for the years ahead. Ocean color provides an indirect way of measuring phytoplankton concentration (chlorophyll-a pigment, for the so-called Case 1 waters) but is also sensitive to the influence of sediments, particulate and dissolved organic matter (so-called Case 2 waters found mainly in coastal areas) (see Bricaud, 1995). Phytoplankton is the primary food and energy source for the ocean ecosystem, so it is very important for ecosystem studies. It also plays a significant role in the carbon cycle since phytoplankton dissolves CO_2 through photosynthesis into organic components (biological primary production). Global mean chlorophyll pigment concentration derived from CZCS data shows, for example, regions of high phytoplankton concentration in the upwellings and at high latitudes. Near the coast, the signal cannot be interpreted in terms of phytoplankton only. From the chlorophyll pigment concentration maps, we can derive maps of primary production (production of organic carbon through photosynthesis). Quantitative estimation of phytoplankton and primary production via ocean color measurements is difficult, however, as the retrieved signal is not always dominated by chlorophyll (dissolved organic matter, other pigments). The absorption coefficient of phytoplankton is also variable (according to how it is aggregated). To estimate primary production, we must assume a vertical distribution of chlorophyll and primary production also depends on species, environmental and physiological conditions, etc. There is clearly a need for better sensors and sampling, more wavelengths (SEAWIFS, OCTS, POLDER) but also comprehensive calibration, validation, and new bio-optical algorithms before ocean color measurements can be used quantitatively. These data sets are nevertheless very useful for verifying coupled ecosystem models. Chlorophyll concentration observation is also a useful parameter to be assimilated in these models.

2.1.3 Space oceanography missions

Tables below provide an overview of existing and approved satellite missions for altimetry, SST, scatterometry and ocean color (courtesy of A. Ratier) for the next decade [before and during the Global Data Assimilation Experiment (GODAE), Smith and Lefebvre (1997)] (Ratier, 1999). As can be seen, there are a substantial number of missions although only SST could be described as operational. There is

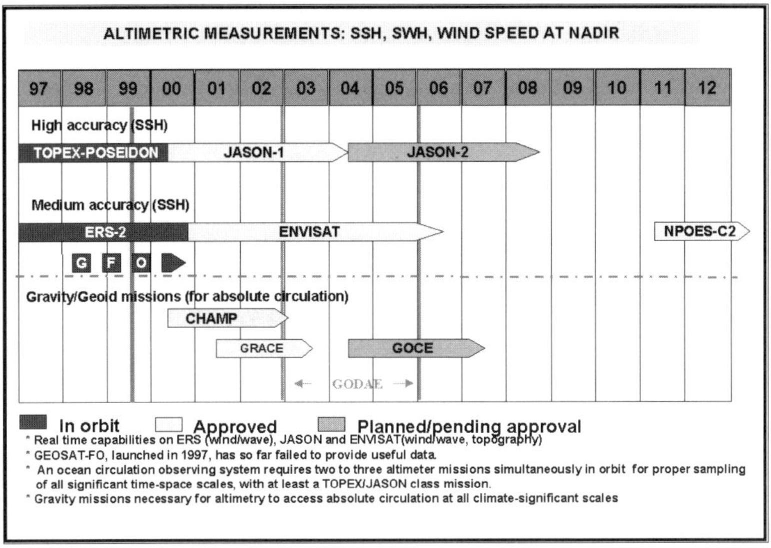

Table 2.1 Schedule of altimeter and gravity missions (courtesy of A. Ratier)

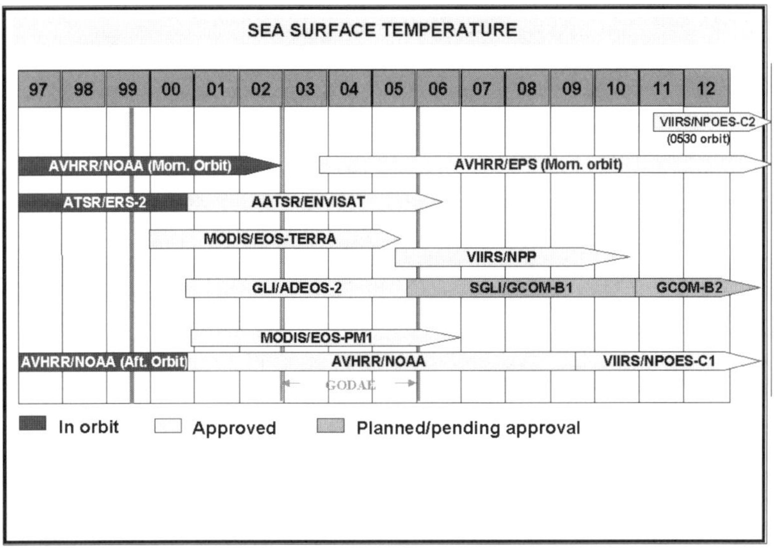

Table 2.2 Schedule of SST missions (courtesy of A. Ratier)

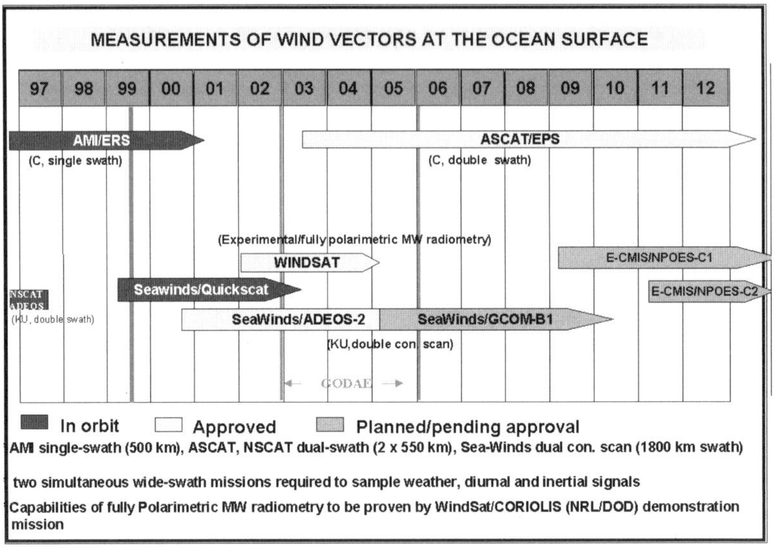

Table 2.3 Schedule of scatterometry missions (courtesy of A. Ratier)

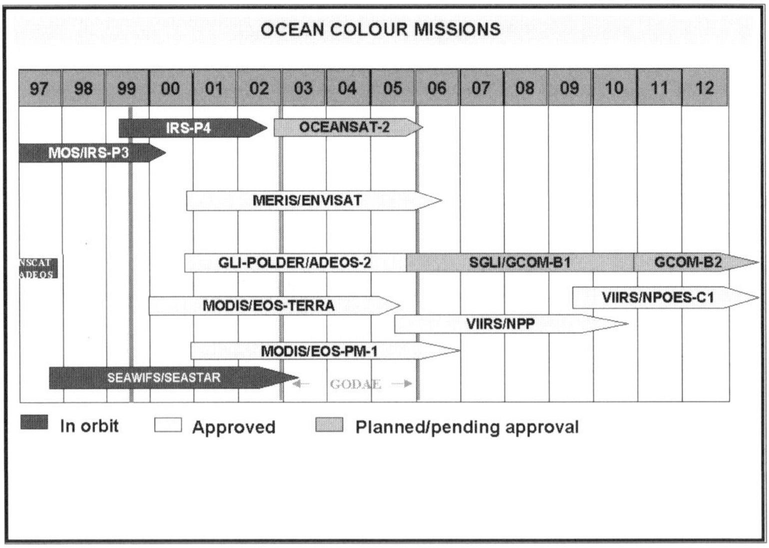

Table 2.4 Schedule of ocean color missions (courtesy of A. Ratier)

not yet a commitment for long term, operational measurements. Note that, as far as altimetry is concerned, gravimetric missions (CHAMP, GRACE and GOCE) were added (see Johanessen et al., 1999). They should allow the estimation of a precise geoid. As we will see later, without precise knowledge of the geoid , altimetry can only provide data on the variable part of the ocean dynamic topography.

Finally, among promising future techniques, one should mention the possibility of measuring sea surface salinity from space. The concept has been demonstrated from plane measurements and there are projects in the Europe (SMOS) and in the USA for flying such a mission in the coming years. Although the precision will be somewhat limited, this will be a very important contribution to the ocean observing system (see Johanessen et al., 1999).

2.2 Satellite Altimetry

2.2.1 Overview

There are two main reasons for focusing on satellite altimetry. First, it is probably the most important satellite technique for oceanography, particularly for ocean forecasting. Satellite altimetry has unique capabilities for providing global synoptic view of the ocean circulation. It provides measurements of sea surface topography which is an integral of the ocean interior. These measurements are a strong constraint for the 3D ocean circulation estimation. Second, satellite altimetry is a very mature technique. It started more than 20 years ago with GEOS-3 (1975) and SEASAT (1978) missions. After a seven year gap, these were followed by the US Navy's GEOSAT mission (1985-1989) and the European Agency ERS-1 mission (1991-1996). Two satellites are now operating : ERS-2, the successor of ERS-1 and the US/French TOPEX/POSEIDON (T/P) mission. For the next decade, future missions have already been decided : GEOSAT Follow On was launched in early 1998 and ENVISAT (successor of ERS-1/2) and Jason-1 (successor of T/P) missions are scheduled for launch in early 2001.

Satellite altimetry is one of the most complex and challenging techniques in terms of accuracy. It requires the range between the satellite and the sea surface to be measured to within a few cm, i.e. assuming a typical satellite orbit altitude of 1000 km a relative accuracy of 10^{-8}. There have been major advances in sensor and processing algorithm performance over the last 20 years. It is important to realize that these advances were made possible through continuous cooperation between engineers and scientists. As a result, accuracy has progressed from several meters to a few cm only. T/P marked a major improvement in accuracy. Its orbit is known with an accuracy of about 2 cm rms and the satellite - ocean surface distance is determined to within 2 cm.

We shall now describe measurement principles, content and errors, and explain data processing techniques. More details can be found in Stewart (1985), Chelton (1988), Rummel and Sanso (1993) or Le Traon (1995). We will then look briefly at results recently obtained with T/P and ERS-1/2 altimeter data before discussing real-time aspects.

2.2.2 Measurement principle

The principle of altimetry measurement is simple (although the system is complex). The altimeter measures the range from the satellite to the ocean surface by determining the time taken by the radar pulse to travel from the satellite to the ocean surface. Using a precise orbitography system, we can determine the position of the satellite relative to a reference ellipsoid. Combining these two measurements yields an estimation of the sea level relative to a reference ellipsoid. This estimation comprises the geoid (an equipotential of the earth gravity field to which a motionless ocean would exactly conform) and the ocean dynamic topography. The geoid can vary as much as 100 m for only 1 m for the dynamic topography (the parameter of interest here).

Altimetry also provides measurements of significant waveheight (the radar pulse is first reflected by crests and then by troughs) and wind speed (the power of the returned signal depends on wind speed). These parameters are also used to estimate electromagnetic bias correction (interaction of the electromagnetic wave with the ocean surface).

The satellite repeats exactly the same ground track every cycle. For T/P and Jason-1, this cycle is 10 days, for ERS-1/2 multi-disciplinary phase and for ENVISAT it is 35 days, and for GEOSAT and GFO 17 days. Every cycle, it thus observes the same geoid signal (which does not vary over time) and the dynamic topography (which varies over time).

2.2.3 Content of altimetric measurement

Geostrophy

Assuming the geoid is known (we will discuss this particular problem later on), satellite altimetry provides measurements of ocean surface dynamic topography, which is directly related to ocean currents through the geostrophic approximation. Simple scaling of the Navier Stokes equations shows that for space scales larger than a few tens of km and time scales longer than a few days (except for a thin layer at the surface which is directly driven by winds-the so-called Ekman layer), the circulation is in geostrophic equilibrium (e.g. Pedlosky, 1979). This means that the pressure force is balanced by the Coriolis force. This yields equations 1 and 2:

$$\begin{cases} fv = \dfrac{1}{\rho_0}\dfrac{\partial P}{\partial x} \\ -fu = \dfrac{1}{\rho_0}\dfrac{\partial P}{\partial y} \end{cases} \qquad f = 2\Omega\sin\theta \qquad (1)$$

$$(2)$$

where Ω is the angular rotation of the earth, θ is latitude and P the pressure.

At the surface, P is equal to $\rho\, g\, \eta$, where η is the dynamic topography (sea level above the geoid, local vertical), ρ is density and g is gravity. The dynamic topography and oceanic currents are thus simply related:

$$\begin{cases} fv = g\dfrac{\partial \eta}{\partial x} & (1') \\ -fu = g\dfrac{\partial \eta}{\partial y} & (2') \end{cases}$$

$$\left\{ \frac{\partial p}{\partial z} = -\rho g \right. \qquad (3)$$

Equation 3 represents hydrostatic balance.

This is, of course, only a diagnostic relationship. It does not tell us what causes the ocean motion (which would require the other terms in the Navier Stokes equations to be taken into account). What it does tell us is that where there is motion there must be a deviation relative to the geoid as expressed in equations 1' and 2'. The physical explanation for this is simple. Pressure forces will tend to cause water to flow to a sea level low. However, because the motion is relatively large-scale, the Coriolis force is large and will deviate the water to the right (left) in the Northern (Southern) Hemisphere. The water will be deviated up to 90° and an equilibrium between pressure and Coriolis forces will be reached. As a result the flow will be cyclonic (clockwise in the N.H.) for the low and anticyclonic for the high. The same holds for the atmosphere: ocean dynamic topography is related to ocean surface current in oceanography in the same way as atmospheric pressure is related to surface wind in meteorology.

A map of the mean dynamic topography as derived from T/P data and a geoid model (e.g. Stammer et al., 1996) reflects the main features of the ocean circulation. Ocean subtropical gyres which are characterized by high dynamic topography and anticyclonic motions, subpolar gyres and the Antarctic Circumpolar Current which is characterized by a sea level drop of more than one meter from north to south. Using satellite altimetry, we can thus obtain pictures of the ocean circulation on a regular basis (e.g. every 10 days for T/P).

Baroclinic and barotropic motions, steric height

The content of the ocean dynamic topography signal is now detailed to show how it is related to the ocean interior. More details can be found in the Gill and Niiler (1973) paper.

Let us first assume a homogeneous ocean [ρ = constant or $\rho = \rho\,(z)$] :

$$fv = \frac{1}{\rho_0}\frac{\partial P}{\partial x} \quad \Rightarrow \quad f\frac{\partial v}{\partial z} = -\frac{g}{\rho_0}\frac{\partial \rho}{\partial x} \quad \Rightarrow \frac{\partial v}{\partial z} = 0$$

There will be the same velocity over the total water column. These are the so-called barotropic motions which do not depend on the density field.

The density actually depends on x, y, z and t [$\rho = \rho(x,y,z,t)$]. Let us define ρ' as the density anomaly given by $\rho = \langle\rho(z)\rangle + \rho'$, where $\langle\rho\rangle$ is the water density at a given temperature and salinity (e.g. T = 0°C , S = 35‰). From equation 3, we derive :

$$f\frac{\partial v}{\partial z} = -\frac{g}{\rho_0}\frac{\partial \rho}{\partial x} \qquad (4)$$

This is the thermal wind equation. Horizontal density variations are associated with vertical shear (so-called baroclinic motions). Integrating (4) from z_0, the reference level, to z_1, the surface, yields :

$$v(z_1) = v(z_0) - \frac{g}{f}\int_{z_0}^{z_1}\frac{1}{\rho_0}\frac{\partial \rho'}{\partial x}dz \qquad (5)$$

If density is known, we only need to know the velocity at the reference level, z_0, to calculate the velocity over the total water column. From equation (5), we can easily obtain equation (6):

$$v(z_1) = v(z_0) - \frac{g}{f}\frac{1}{\rho_0}\frac{\partial}{\partial x}\left(\int_{z_0}^{z_1}\frac{\rho'}{\rho_0}dz\right) \qquad (6)$$

$$\Rightarrow v(z_1) = v(z_0) + \frac{g}{f}\frac{\partial \eta_s}{\partial x}$$

with $\eta_s(z_0, z_1) = -\int_{z_0}^{z_1}\frac{\rho'}{\rho_0}dz$ and η_s is the steric height. Thus, at the surface, $v(z_1) = \frac{g}{f}\frac{\partial \eta}{\partial x}$ where η is the sea level and from (6):

$$v(z_0) + \frac{g}{f}\frac{\partial \eta_s}{\partial x} = \frac{g}{f}\frac{\partial \eta}{\partial x}$$

Let's now write $v(z_0) = \frac{1}{f\rho_0}\frac{\partial P_{z_0}}{\partial x}$ where P_{z_0} is the pressure at the reference level, usually defined as the bottom.

The same expression holds for u (with y derivative) which means that:

$$\Rightarrow \eta = \eta_s + \frac{P_{z_0}}{\rho_0 g}$$

The dynamic topography η (measured by altimetry) is the sum of the steric height (baroclinic component) and of a bottom pressure term, sometimes incorrectly referred to as the barotropic component. The exact definition of a barotropic motion is a motion which does not depend on z; a barotropic motion will induce bottom pressure changes, but density changes, if not compensated over the water column, can also induce bottom pressure changes. This shows that the dynamic topography is an integral of density (steric height - baroclinic part), with a contribution from bottom pressure. It is really a measurement representative of the total water column (although it is directly related to surface geostrophic current). This is why satellite altimetry is so useful for constraining the 3D ocean circulation.

Numerical applications

Influence of latitude on sea level slope and ocean current

The Coriolis parameter f is equal to $2\Omega \sin \phi$ (Ω is the angular rotation of the earth, ϕ is latitude). This gives $f \approx 10^{-4}\ s^{-1}$ (at 45°N/S) and $f \approx 10^{-5}\ s^{-1}$ (at 5°N/S). From equation 1' (and 2'), a slope of 1 meter over 100 km will give a current of 1 m/s at 45°N/S. A slope of 10 cm over 100 km only will give the same current speed at 5°N/S.

Influence of temperature gradient on dynamic topography

If we assume a reference level at 1000 m and no salinity influence, the dynamic topography change measured by altimetry for a track crossing the Gulf Stream (or a ring) with a temperature gradient of 10°C over 100 - 200 km and over a depth of 1000 m will be:

$\eta = \eta_s = - \int_{1000} \rho' / \rho_0\, dz \Rightarrow \Delta \eta_s = \Delta \eta = 1000\, \Delta\rho / \rho_0 = 1000\, \alpha\, \Delta T \approx 1.5$ m
with $\alpha = -1/\rho\, \partial\rho/\partial T$ (thermal expansion coefficient) equal to $1 - 2\ 10^{-4}\ s^{-1}$

Equatorial regions

At the equator, the Coriolis parameter f is equal to zero. This means we can no longer assume geostrophy or use dynamic topography to infer the oceanic circulation. It still reflects, of course, the change in density field (e.g. thermocline depth or heat content variations). In particular, the sea level variations (η) in the equatorial regions are closely related to the variations of the depth of the thermocline (H) ($\eta \approx -\Delta\rho / \rho_0 H$, where $\Delta\rho$ is the density difference between the deep and surface layers). In practice, geostrophy works up to about ± 2 degrees (although the estimation is more sensitive to noise). A relationship between zonal current and dynamic topography has been proposed, however, by Picaut et al. (1990) ("equatorial geostrophy"). This relationship is derived as follows:

$$f u = - \partial\eta/\partial y \quad \text{(geostrophy)}$$

$$\partial f/\partial y\, u + f\, \partial u/dy = -g\, \partial^2\eta/\partial y^2$$

At the equator $\beta u = -g\, \partial^2\eta/\partial y^2$.

This relationship is valid at low frequencies but the estimation is very sensitive to noise (because of the second derivative). It has been shown to provide good results with altimetry (compared to currentmeter data) but altimeter data need to be filtered both in space and time.

Steric effect related to heat fluxes

The dominant signal in seasonal large-scale variations in sea level is due to heat fluxes. This has been well illustrated with T/P (e.g. Stammer, 1997). It is important to accurately estimate this effect before analyzing other dynamically more significant effects, like the influence of wind forcing on circulation, for example.

As previously explained, the steric height is given by: $\eta_s = -\int_{bottom}^{surface} \rho'/\rho_0 dz$

Density is a function of temperature, salinity and pressure. For small changes of T and S, the variations of density are given by:

$$\frac{\Delta\rho'}{\rho_0} = \frac{\partial\rho}{\partial T}\frac{\Delta T'}{\rho_0} + \frac{\partial\rho}{\partial S}\frac{\Delta S'}{\rho_0} = -\alpha\Delta T' - \beta\Delta S'$$

where α is the thermal expansion coefficient and β the equivalent for salinity. The contribution of temperature to density in the mixed layer is generally much more important than salinity (except at high latitudes).

The equation of heat (or potential temperature) conservation at the ocean surface reads:

$$\frac{\partial T}{\partial t} + u\frac{\partial T}{\partial x} + v\frac{\partial T}{\partial y} + w\frac{\partial T}{\partial z} = K_v\frac{\partial^2 T}{\partial z^2}$$

with at $z = 0$, $Q = \rho_0 C_p K_v \frac{\partial T}{\partial z}$ with Q=air/sea heat flux.

It is easy to show (e.g. Gill and Niiler, 1973) that at large scales (> 1000 km), advection is negligible. Integrating over H, the mixed layer depth, this yields:

$$\int_{-H}^{0} \frac{\partial T'}{\partial t} dz = \frac{\partial}{\partial t}\int_{-H}^{0} T' dz = \frac{Q}{\rho_0 C_p}$$

Ignoring the salinity contribution on density, we obtain the following approximation:

$$\Rightarrow \frac{\partial}{\partial t}\eta_s' = \frac{\alpha}{\rho_0 C_p} Q$$

The net heat flux induces (steric) sea level changes according to the simple equation given above. This also means that sea level can provide useful information on heat fluxes. Thus, if properly used, satellite altimetry should allow us to correct for ocean heat flux errors (e.g. by assimilating altimetry and SST data). For ocean forecasting, this also shows that ocean dynamic topography is a superposition of different signals with very different vertical and horizontal scales. The complexity of the measurement content must be taken into account in the assimilation procedure.

2.2.4 Altimeter data processing

So far, we have considered that altimetry provides measurements of the surface dynamic topography. We now describe the basic altimeter data processing techniques used to extract the dynamic topography from measurements of sea surface topography. Our analysis includes a discussion of measurement errors.

Oceanic signal extraction from altimetry and measurement errors

The altimetric observation of the sea surface topography S can be described by:

$$S = N + \eta + \varepsilon$$

where N is the geoid, η the dynamic topography and ε the measurement errors (orbit error, propagation effects in the troposphere and ionosphere, tides, electromagnetic bias, inverse barometer effect, altimeter measurement noise).

Present geoids generally are not accurate enough to estimate the absolute dynamic topography η globally, except at very long wavelengths (> 2000 km) (see also discussion in section 2.2.3). The variable part of the dynamic topography η' ($\eta' = \eta - \langle\eta\rangle$) (or Sea Level Anomaly, hereafter SLA) is, however, easily extracted since no prior knowledge of the geoid height is needed. The most commonly used method is the so-called repeat track method (collinear analysis), suitable for satellites whose orbits repeat their ground tracks (to within ± 1 km) at regular intervals. For a given track, the variable part of the signal is thus obtained by removing the mean profile, which contains the geoid and the quasi-permanent dynamic topography, from each profile.

Altimeter measurements of sea surface topography are affected by a large number of errors. These include errors on the range measurement due to propagation effects in the troposphere and ionosphere; electromagnetic bias; errors due to inaccurate ocean and terrestrial tide models; cross-track geoid errors, and inverse barometer effect. Most are large-scale, and do not limit the use of altimetry for ocean mesoscale studies. For studies of large-scale oceanic variability, most of these corrections must be taken into account, however, since they can significantly contaminate the oceanic signal. Some of these errors can be corrected with dedicated instrumentation (dual-frequency altimeter for ionospheric corrections, microwave radiometer for wet tropospheric corrections). Electromagnetic bias can be deduced by analyzing altimeter data.

The inverse barometer effect is actually a large, real oceanic signal which reflects the response of the sea level to changes in atmospheric pressure. The static response assumes a non-dynamic adjustment of the ocean due to atmospheric pressure (about 1 cm for 1 mb change in atmospheric pressure). Recent numerical and empirical studies suggest that the ocean generally responds as an inverted barometer except over very short time scales and in semi-enclosed seas (e.g. Gaspar and Ponte, 1997). This correction should thus generally be applied when studying the dynamic response of the ocean.

The tidal signal is the most important variable signal in altimetric data. This signal is only partially corrected using global ocean tide models. The residual errors are then aliased at certain periods depending on the repeat-period of the satellite. The satellite repeat-period should be chosen to avoid aliasing near dominant oceanic periods (e.g. annual or semi-annual periods). The M2 tide is thus aliased near 60 days for T/P. This is much less of a problem than the GEOSAT 317-day aliasing period which can cause major problems for the interpretation of altimetric signals. More accurate global tide models based on hydrodynamic models and/or T/P data are now available and should be used. Still, aliasing may occur and this has to be taken into account when interpreting the data.

Last, but not least, comes the orbit error. This error is caused by imperfect knowledge of the spacecraft position in the radial direction. It is actually the largest error on altimetric measurements of sea surface topography. It depends on the quality of the satellite tracking system. For T/P, precise orbit determination is achieved via three distinct tracking systems (DORIS, Laser, GPS) providing almost global coverage of satellite orbits. The radial orbit error obtained is thus accurate to about 2 cm, compared to 10 cm accuracy for the most recent GEOSAT and ERS-1/2 orbits available. Orbit errors are long-wavelength errors (about 40 000 km) that can be reduced by analyzing the altimeter data. This can be done via global or regional minimization of crossover or repeat-track differences (relative to a mean or to a given cycle). These differences do not contain any geoid signal and are dominated by the orbit error. Crossover or repeat-track differences also contain the oceanic signal. The main problem is thus to separate the orbit error from this oceanic signal. We must first assume an a priori analytical form or an a priori spectrum of the orbit error. A more empirical approach, commonly used, is to approximate the orbit error by fitting a first or second degree polynomial over a given arc length. However, this method also removes the large-scale oceanic signal and is only suitable for mesoscale studies. To minimize ocean signal removal, more complex methods using cross-track information are required. Global crossover minimization can thus be used to estimate the orbit error without removing too much of the large-scale oceanic signal. More generally using inverse techniques, the orbit error signal could be obtained through a global adjustment taking into account not only the spatial but also the temporal characteristics of the orbit error and oceanic signal.

Mapping

Most applications need maps (and associated formal error) of the altimetric signal (SLA) on regular space/time grids. This can be done using optimal interpolation methods which use a priori knowledge of the space and time scales of the ocean signal. It is preferable, particularly in low eddy energy regions and when several altimeter data sets are merged, to take into account an along-track long wavelength error (correlated noise, e.g. due to orbit, tidal or inverse barometer residual errors) in the method (Le Traon et al., 1998).

Merging of multiple-satellite altimeter data

The merging of multi-satellite altimetric data sets is necessary for a better mapping of sea level and oceanic circulation variations. To merge muti-satellite altimetric missions, it is first necessary to have homogeneous and inter-calibrated data sets. Homogeneous means that same geopotential model and reference systems for the orbit and same (as far as possible) instrumental and geophysical corrections (e.g. same tidal models, same meteorological models, etc) should be used. Inter-calibrated means that relative biases and drifts be corrected for at the mm level and also that the orbit be reduced (to have consistent data sets). A methodology we proposed and successfully tested is to use the most precise mission (T/P, Jason-1) as a reference for the other satellites (Le Traon et al., 1995; Le Traon and Ogor, 1998).

Once altimetric data have been homogenized and inter-calibrated, the next step is to extract the Sea Level Anomaly for the different missions and to merge the data using a mapping or assimilation technique. To extract the SLA, it is preferable to use a common reference surface (e.g. a very precise mean sea surface or mean profiles consistent between the different missions) to obtain the SLA relative to the same ocean mean. Then data can be merged using a mapping technique as previously described.

2.2.5 Results from T/P and ERS-1/2

An overview of results obtained with T/P and ERS-1/2 is now given to illustrate the contribution of altimetry to ocean forecasting. It contains T/P results on mean sea level variations, large scale sea level and oceanic circulation variations, Kelvin/Rossby waves, El Nino and the comparison with global eddy resolving models. T/P and ERS-1/2 results deal with the combination of T/P and ERS-1/2 for regional and mesoscale analyses. The reader is referred to the two T/P (1994, 1995) and the ERS-1/2 (1998) Journal of Geophysical Research special issues for a more complete overview.

To forecast possible global sea level rise, you need to observe the mean sea level trend over a very long period with accuracy better than 1 mm/year. To achieve this accuracy from 1000 km altitude is a real challenge. T/P is the first altimetric mission which enables us to observe the global mean sea level variations with such an accuracy. The signal over the first five years is between 1 and 2 mm/year (10 to 20 cm per century), consistent with results derived from tide gauges which were (and still are) extremely useful for validating altimeter measurements. The main advantage of altimetry is global coverage, which makes measurements much more representative of the global mean sea level. Satellite altimetry also offers a unique ability to observe mean sea level variations at smaller regional scales. We know that global warming will not induce a uniform rise in the mean sea level. T/P estimations show that the sea level rise can be much higher in some places (a few cm/year) although there are some regions where the signal is clearly related to an interannual ocean signal (not a secular drift). This also shows why it is so difficult to monitor the global mean sea level from a limited number of tide gauges. T/P data have also provided very useful estimates of mean sea level variations in regional seas (e.g. Mediterranean, Black Sea, Caspian Sea). The ability to forecast the mean sea level change in these high density regions is of outmost importance.

One of the main achievements of T/P has been its ability to measure large-scale sea level variability. Seasonal variations are mainly due to steric effects related to heat fluxes (see section 2.3.3). They are 180° phase-shifted between the two hemispheres and have a typical amplitude of 3 to 5 cm. These variations are correlated with sea surface temperature (expansion and contraction of surface waters) but with a phase lag of 1 to 2 months. The amplitude of the signal is smaller in the southern hemisphere, because of its greater expanse of ocean, and because of the role of continental air masses in the northern hemisphere. In the tropical oceans, seasonal variations mainly result from the seasonal variation of the circulation in response to wind forcing. Interannual variations can also be as large as seasonal variations. El Niño events and associated Kelvin waves are also well observed by T/P (Fall 1992 and Winter 1993, Fall 1994 and Winter 1995 and mainly Fall 1997 and Winter 1998). Several studies of tropical Kelvin and Rossby waves using T/P data have shown that they are a potentially important factor in understanding the coupled ocean/atmosphere system related to the El Niño Southern Oscillation (ENSO). Rossby waves and their propagation characteristics have also been studied in the mid-latitudes. One of the main results was to show that the propagation was faster than that of free linear waves (Chelton and Schlax, 1996). This may have major effects on the way the ocean adjusts to atmospheric forcing.

T/P also provides a unique data set for validating global ocean models and understanding their main deficiencies (Stammer et al., 1996; Fu and Smith, 1997). A detailed model/data comparison is a fundamental step before assimilation. Comparison of T/P results with global high-resolution (1/4 or 1/6 degree) simulations shows that the models still underestimate variability in mid- and high-latitude regions. The seasonal signal is qualitatively well reproduced. At high frequencies (higher than 100 days^{-1}), the barotropic response of the ocean to wind forcing has been characterized globally, for the first time, by comparing T/P and model results.

We have also learned a lot from altimetry about the mesoscale ocean circulation variations (e.g. Gulf Stream, Kuroshio, Antarctic Circumpolar Current, Canary Basin, Mediterranean sea). T/P alone cannot provide a good mapping of mesoscale signals: at least two satellites are needed. The contribution of merged T/P and ERS-1/2 data in providing a detailed description of mesoscale dynamics has been illustrated in, for example, the Mediterranean sea (Ayoub et al., 1998). The ability to map mesoscale signals was also very well demonstrated by the comparison of T/P and ERS-1 data with in-situ measurements in the Canary basin during the Semaphore experiment (Hernandez et al., 1995). The agreement was better than 3 cm rms, which is about the accuracy we expect from in-situ measurements. Satellite altimetry is actually the only way of precisely monitoring and possibly forecasting mesoscale signals.

2.2.6 Real-time aspects

Real-time aspects are crucial for operational oceanography. Altimeter data (if required during the mission design) can be acquired and processed in near-real time (2 days) (e.g. ERS-1/2, Jason-1, ENVISAT). However, the data are less accurate because orbit computation requires environmental parameters which cannot be obtained in real time. Dedicated processing is thus needed to correct the orbit error, along with continuous comparison of real-time products with precision delayed-time products to assess accuracy. Results recently obtained with T/P and ERS-2 near-real time data are almost as accurate as with delayed-time data thanks to dedicated processing techniques (Boone and Le Traon, 1997; Cheney et al., 1997). T/P data are now used in near-real time for El Niño prediction for which they were shown to be very useful (Cheney et al., 1997).

2.3 Prospects

Space oceanography provides unique tools for ocean forecasting. Some techniques are very mature (altimetry, SST) and can be readily used in models. However, new space oceanography missions (e.g. for measuring salinity), and better sampling of sea level and winds are required. Operationality for space oceanography missions is also not yet guaranteed (as it is for meteorological satellites), although it is hoped that this may be achieved in the foreseeable future.

In the future, ocean forecasting, through data assimilation, will be an appropriate and powerful means to assess the impact of the satellite (and in-situ) observing systems, to identify gaps and to improve the efficiency/effectiveness of the observing system. A better use of space (and in-situ) data using effective data assimilation techniques and a better assessment of the ocean observing system components are among the main objectives of GODAE, the Global Ocean Data Assimilation Experiment (Smith and Lefebvre, 1997; Le Traon et al., 1999).

Acronyms

ALACE Autonomous LAgrangian Circulation Explorer
ARGO Array for Real-time Geostrophic Oceanography
AVHRR Advanced Very High Resolution Radiometer
CLS Collecte Localisation Satellites
CTD Conductivity Temperature Depth
CZCS Coastal Zone Color Scanner
DUACS Developing Use of Altimetry for Climate Studies
GFO GEOSAT Follow On
GODAE Global Ocean Data Assimilation Experiment
OCTS Ocean Color and Temperature Scanner
P-ALACE Profiling Autonomous LAgrangian Circulation Explorer
PIRATA PIlot Research moored Array in the Tropical Atlantic
POLDER Polarization and Directionality of the Earth's Reflectance
PROVOR (www.brest.ird.fr/goos/provor.html)
SAR Synthetic Aperture Radar
SLA Sea Level Anomaly
SST Sea Surface Temperature
SMOS Soil Moisture Ocean Salinity
TAO Tropical Atmosphere-Ocean
TOGA Tropical Ocean Global Atmosphere (Program)
WOCE World Ocean Circulation Experimen
XBT EXpendable BathyThermograph

3 A Marine Information System for Ocean Predictions

GIUSEPPE M.R. MANZELLA*, AND MFS-VOS GROUP[1]
*ENEA Marine Environment Research Centre

3.1 Introduction

The quality and management of the coastal and marine environments are particularly delicate issues in current European social, economical and political agendas. However, genuine environmental concerns relating to the sea are often obfuscated and obscurely formulated because of, among other causes, the absence of a steady supply of reliable information.

Nowday, there is an increasing demand of oceanographic and environmental data by many different users, whose primary objectives concentrate on the requirements of data and information, as well as a simplification of their exchange. As a consequence, different institutions assure a 'demand driven' programming of their activities and show their commitment to be 'result oriented'. Toward this aim the institutions involved in such activities, establish in their work programmes a number of detailed projects, including specific products and services to be delivered. Particular examples are provided by those programmes aiming to environmental trends assessment and trends due to human activities or natural changes.

The environmental problems, such as climate change, loss of biodiversity, coastal zone threats and management, chemical risks, are at a different stage of recognition. To tackle these problems there is a need of objective, reliable and comparable information to be used for the identification of priorities and management options. This implies that there is a need of qualified long-term monitoring systems and forecast models which can allow the management of the sea and the prediction of environmental changes.

The intrinsic interdisciplinary nature of the marine science (UNESCO, 1984) demands that the management of the information and data cannot be made only by software engineers experts, but need a systematic approach involving experts in the relevant marine fields. System analysis, optimisation, numerical models, graphical presentation, multi-disciplinary data bases, remote sensing, on-line monitoring and computer technology are facilitating marine and coastal resources management. At the same time, the necessity to provide data and information to a wide variety of users (e.g. scientific community, environmental managers) is shifting the activities

1. MFS-VOS Group is formed by: A. Bruschi, A. Cruzado, G. Fusco, M. Gacic, G. P. Gasparini, T. Gervais, V. Kovacevic, C. Millot, M. Tonani, C. Tziavos, Z. Velasquez, A. Walne, V. Zervakis, G. Zodiatis

of data managers and data analysts toward the development of more and more complex and comprehensive Marine Information Systems.

An important effort of the data managers is nowadays dedicated to the management of real time data, as part of a wider operational ocean/coastal forecasting system of the ocean circulation and ecosystem variability. The successes of various

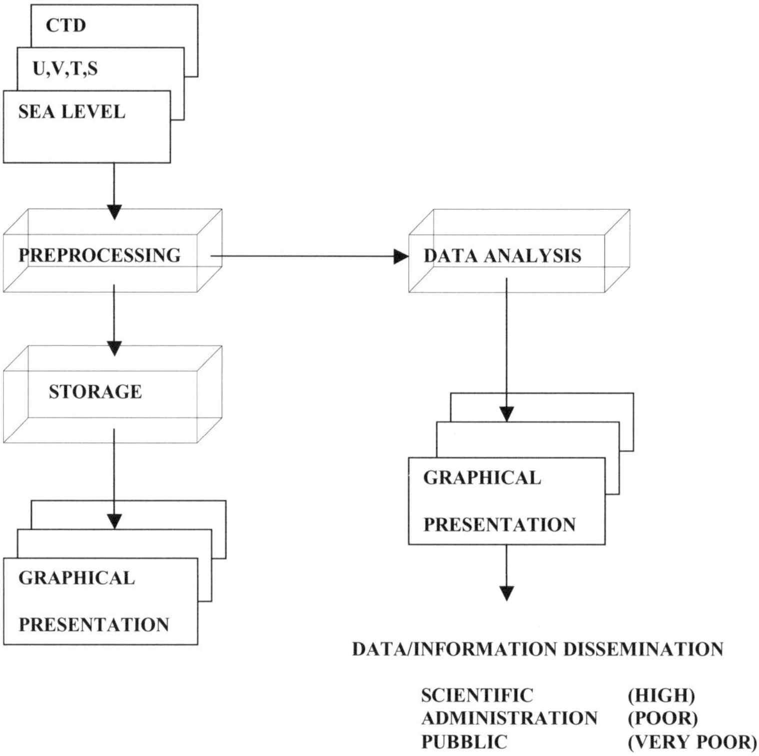

Fig. 3.1 Former Data Banking, analysis and dissemination system. The new data were compared with the historical data, then archived for internal use. The level of dissemination was high to the scientific community and very poor to the general pubblic.

research programmes, such as the Tropical Oceans Global Atmosphere (e.g. Hayes et al., 1991) and World Ocean Circulation Experiment (e.g. Schott et al., 2000) have demonstrated the feasibility of global operational observing and forecasting systems. Operational coastal activities are being set up in limited marine areas (Malone et al., 1999). In the Mediterranean a basin wide operational system has been recently developed and has assured data and forecast of the physical properties (circulation, temperature and salinity fields) of the sea.

This paper will present a general overview of marine information system concepts and then will give details on the operational observing system set up in the Mediterranean as part of the Mediterranean Forecasting System project (Pinardi and Flemming, 1998) developed in the framework of the EuroGOOS activities. In section 3.2 the main concepts of information systems are presented, followed by some general ideas on sampling requirements (section 3.3). A general overview of the operational observing systems are presented in section 3.4, followed by a presentation of concepts developed for the Mediterranean Sea (section 3.5). The discussion (section 3.6) include also the lesson learned from the Mediterranean experience.

3.2 Marine Information System

The first significant examples of Marine Information Systems were developed during the '60s. Until the '80s the data management was composed of many independent processes (Fig 3.1) the data collection, the pre-processing by technicians, the storage in files whose access was different from one system to another (e.g. sequential, direct access, etc). Data analysis, graphical representation and dissemination were part of the scientific study of the ocean dynamics.

During the last 50 years, monitoring systems and oceanographic cruises have generated a large suite of processed samples, analyses and products. An efficient information strategy was then necessary for the use and dissemination of these data among users and the public. Information Systems were built with the overall goal to provide the data management and communication facilities that were necessary for (e.g.) oceanographic data distribution from collecting centres to users, routine monitoring, analysis and assessment of the state of the marine and coastal environment (UNESCO, 1999a, 1999b, 1999c).

An "Information System" is then defined as a system that manages all information and not just that directly associated with measured variables (Fig. 3.2). More generally, "information" includes raw and processed data, information on measurements methods, information on analysis and data analysis (metadata); "management" includes communications from instruments to responsible centres, quality control, data storage and exchange, data sharing policies and dissemination.

All these aspects are becoming more and more important for the 'near-real-time' data, which need to be analysed in a short delay after data collection.

The understanding of environmental changes requires the establishment of effective mechanisms for the request and distribution of data as well as for the quality assurance determination. Since the early 1960's data centres and designated national agencies were established in the framework of IOC programmes, to support both national and international organizations with oceanographic data and information services (UNESCO, 1997).

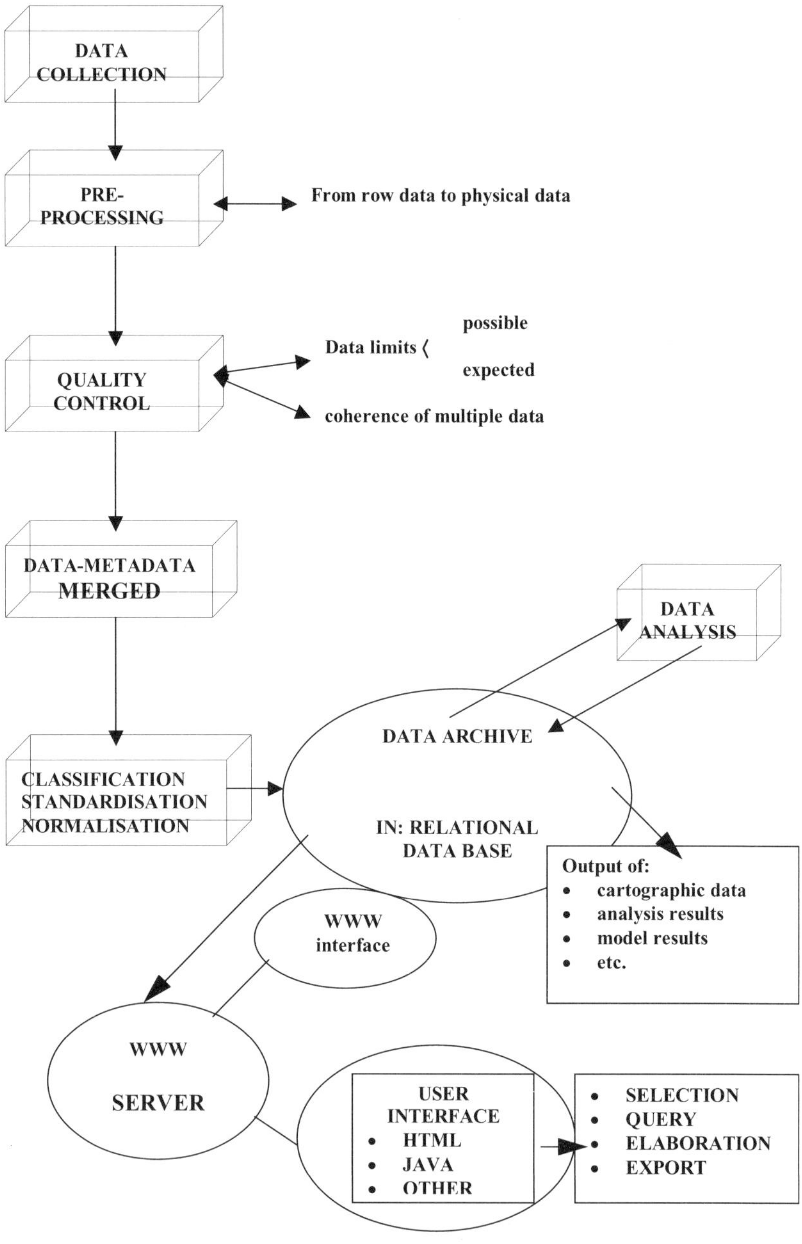

Fig. 3.2 Scheme of a Marine Information System. The new data is compared with climatology. Flags are used to indicate the quality of the profiles. More information on data collection programmes, instruments used, methodologies applied, etc. are added (metadata). Internet connections allow a wide dissemination.

The importance of data management has been stressed in many reports and international meetings at national, European and global level. In particular, the 1992 Earth Summit in Rio requested States to co-operate with international organisations in order to: '*consider mechanisms to develop comparable and compatible techniques, validate methodologies and measurements, organise regular scientific reviews, develop options for corrective measures, agree on formats and storage, and communicate the information gathered to potential users*'. The interpretation of this Agenda 21 statement has important consequences on the organisation of the data management organization in terms of (Manzella and Nair, 1998):

- data collection
- data and information flow
- storage
- quality assurance protocols
- analysis and dissemination

In practice the quality of data is based on concepts which comprises the quality assurance of fieldwork, the quality control of the data and their proper use (IOC, 1995).

A particular attention must be devoted to data quality which is depending on:

- precision of instruments
- calibration of instruments
- in situ sampling methodology
- intercalibration/intercomparison
- data quality check/regional validation
- trained personnel.

There is a wide range of potential usage of a data management system: archiving, dissemination, development of management plans, support to major experiments, development monitoring systems, developments of standard for data processing, etc. It is then vital to clearly determine the objective of the data management system and the level of activity to be achieved. There are several groups of potential clients that need support from an information system, but all their needs cannot be generally satisfied by a unique, centralised system. The potential users of the information system are:

- The research community. This community need to have data processed and documented properly in order to improve the effectiveness of scientific programmes.

- The environmental authorities. They need to regulate activities such as marine resources exploitation, ferry and ship operations, ocean dumping, etc. Data and information are then required for an assessment of the state of the system, the trends and the effect of the human activities on the marine environment.

- Marine engineering communities. They need data available for (e.g.) building and operating marine facilities.

-The forecasting agencies. They need observations to initialise forecasts of the marine weather, water level, storm surges and severe ocean events.

Such a variety of potential clients with different needs poses the question on how to organise an efficient data management system.

Before the advent of high speed data communication and modern information systems, the main option was the establishment of centralised data centres. These were acquiring, were possible, copies of data, processing and quality controlling them, building data bases. For those data not entering in the system, inventories were prepared and maintained, giving the opportunity to users to 'refer' to the organisation or person holding the data.

The communication technology now in place permits the creation and maintenance of distributed data bases, which can develop system more 'client oriented' and develop a closer working relationship with both data suppliers and potential clients.

In this chapter we consider monitoring as the basic data collection methodology which can provide observations to the different users of a data management system.

3.3 Requirement for monitoring

One of the main questions to be answered in future years is whether the observed environmental changes are natural or man-induced. A precise answer requires long term reliable observations and statistical tools, as well as modelling efforts able to produce realistic scenarios of the marine environment. Monitoring is the basic approach to collect relevant data in order to achieve these objectives.

After setting the environmental objective, the monitoring will be used for scientific and/or management purposes. Fundamental differences exist between the two monitoring purposes, however, there is a complementarity that can be of mutual benefit.

In the case of monitoring for management, the system consist of repeated measurements of (JAMP, 1995):

- the quality of marine environment and each of its compartments, i.e. water, sediments and biota;
- the natural and anthropogenic inputs which may affect the quality of the marine environment;
- the effects of anthropogenic activities and inputs.

In this case, the monitoring contributes effectively to the assessment of the marine environment and management providing the information necessary to the description of the spatial and temporal distribution of physical, chemical and biological parameters, as well as for the establishment of relationships between anthropogenic activities and observed changes. This means that assessment criteria and procedures are mainly based on the use of statistics.

The 'scientific monitoring' is different from the previous example, since it involves more complex post-processing work for the purpose of identifying causes of environmental changes and relationships among the physical, chemical and biological processes. This means that new hypothesis are tested, and complex models are used for simulation and understanding of processes.

In general the monitoring for management addresses specific, limited objectives and selected areas, parameters and/or indicators are sampled. The scientific assessment (research project) is usually regional or local in scope (UNESCO, 1997). In the past twenty years monitoring has become also the method to obtain observations for operational oceanography. The monitoring activity for operational oceanography serves the acquisition of real time data to be used for assimilation into forecast models and provides information with a short delay to interested users.Thus operational oceanography introduces the concept of timelinees in the data acquisition part of the data management system. In the following sections data management is described for its usage in operational oceanographic systems.

3.4 Near-Real time data collection and management for Operational Oceanography

Data and products requirements for an operational system vary from area to area and the range of time and space scales. These include seasonal climate analysis and prediction, near-surface condition for the open sea (for transports, safety, fishery, etc.), boundary conditions for regional and coastal forecast systems. The operational data are also the base for the assessment of long-term variability.

Operational Oceanography requires quality-controlled data in near real time to be used for (e.g.) forecasting and alert. The provision of data includes the following steps:

- Data collection by automatic instruments
- Data control and sub-sampling
- Data transmission to collecting centre
- Quality check based on statistics
- Data dissemination for analysis, assimilation and forecast.

We concentrate here on in situ data sets leaving the problems of data management of satellite data to another section of this book. In situ data are necessary in order to provide information on the internal structure of the ocean, such as temperature profiles, upper thermocline depth, etc. The main problems related to in situ data collection are related to available technology and instrument suitability. Data collection is a service that must be performed in an 'operational way', which means that observing systems must assure data collection with regular timing and spacing in a cost effective way. The instrumentation installed on board these operational ships must assure the minimum use of personnel.

Although physical, chemical and biological data could be included in the operational acquisition system, the technology for the different parameters acquisition is at a different stage of development. In principle, it is possible to acquire physical, chemical and biological data in the upper layer (few hundred meters), however, instrument suitability varies greatly, and cost effectiveness suggest to limit the amount of data to few parameters. While there are few problems in the temperature and salinity profiles by using expendable probes, for what concern the chemical and biological data (e.g. nutrients and chlorophyll) only surface data can be collected at a low cost, although the technology is increasing our capabilities to collect this data at depth (Molinari, 1999).

The acquisition system onboard the ship could also include a rough quality control procedure and may also contain a software for data decimation, which is due to data transmission limitations. In fact there is, actually, a limitation on satellite capability to store and transmit data, which has to be taken into account.

3.4.1 Definitions

A number of working definitions are given in oceanography (UNESCO, 1999):

-for the oceans operational data are in 'real time' when delivered before 30 days from the collection time,

-'timely data' are those which continue to be representative of the observed environmental conditions,

-'operational products' are those prepared for users in a timely way on a regular basis.

3.4.2 Sampling strategies

Sampling strategies for in situ near real time monitoring have progressed quite rapidly in the past decade (Molinari, 1999). One example of global ocean monitoring network is the TAO/TRITON array (Mc Phaden, 1995) which consists of several surface buoy deep mooring stations in the tropical Pacific. Data are available in near real time on internet and Web as well as products for different users and forecasts.

Together with the fixed buoy mooring strategy there is also a network of meteorological drifting buoys which transmit in real time and the Ship of Opportunity Network which is the major concern of this paper (Rossby et al., 1995). This system uses commercial ship routes to monitor surface parameters and temperature profiles in the first kilometer of the water column. From the ships, the data, through the satellites, reaches the collecting centres, where a quality control must assure the dissemination of quality data to users (Fig. 3.3).

The sampling strategies that have been developed for the ships of opportunity network have some well defined scientific rationales. Among these there is the acknowledgement that the environment is changing and that there is an evaluation the network. Three modes of sampling have attracted the main attention (Smith et al., 1999):

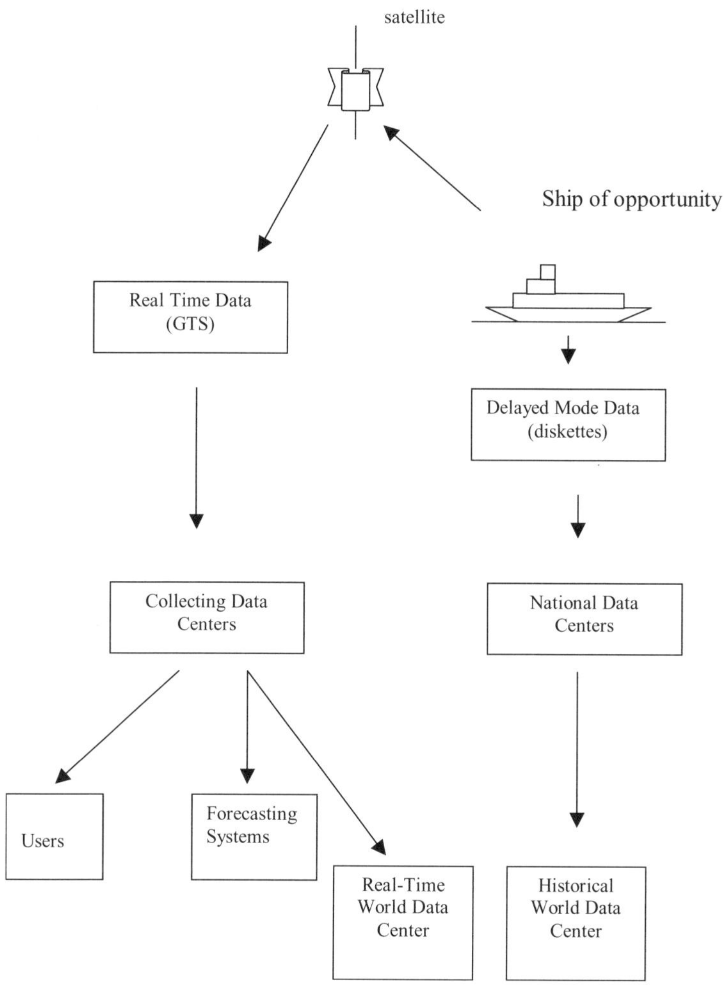

Fig. 3.3 Flow of real time data from sensors to users for the Ship of Opportunity Network. The data are collected, and a gross range check is performed before the decimation of the original profile and transmission. The data colleting centre provide a more accurate quality control before the data dissemination, which must be assured in a timely fashion.

-Low sampling resolution: this mode is comprising networks with monthly repetition period and 4 XBT drops per day; it is used to measure the seasonal and inter-annual variability of the upper ocean ;

-Frequently repeated sampling: tracks are covered 18 times per year with a spatial resolution of 100 - 150 km; at an ocean level, this mode is used in presence of infra-seasonal oscillations and other small scale variability;

-High density sampling: is done in tracks from boundary to boundary with closely spaced XBTs, in order to resolve mesoscale, fronts and boundary currents; the repetition time is 3 months, the spatial resolution 10-50 km.

It is interesting to note the difference between high sampling and the sampling requirements for the Mediterranean Sea, where for resolving the mesoscale and the synopticity the temporal sampling must be 15 days and the spatial one 15 - 20 km. In the Mediterranean case the 'real time' data must be delivered in 2 - 7 days from collection time. Table 3.1 shows the differences in time and spatial resolution required in the open ocean and in the Mediterranean, where the relatively small Rossby radius requires high density sampling both in time and space.

	Temporal resolution (days)	Spatial resolution
Low resolution XBT	30	4 drops/day
Frequently Repeated XBT	20	100-150 km
High Density XBT	90	10-50 km
MFS-VOS	15	12-20 km

Table 3.1 Comparison between the temporal and spatial resolution of the World Ocean (the first three lines) and the Mediterranean Ship of Opportunity Programs.

3.5 The Ship of Opportunity Programme in the Mediterranean Sea

In the framework of the Mediterranean Forecasting System (MFS, Pinardi and Flemming, 1998) it was set up, for the first time, a basin wide, ship of opportunity programme in the Mediterranean, in order to provide:

- An operational monitoring having a high temporal and spatial resolution
- A near real time data delivery
- An open, user friendly access to decimated data through a web page
- A Mediterranean module to the IOC-WMO Ships of Opportunity Programme - Implementation Panel (SOOP-IP)

The main objectives of this programme, which was called MFS - VOS (Volunteer Observing Ships) were:

- deploy elements of a monitoring system for the ocean upper layer
- develop methods/techniques to analyse real time data
- improve real time data exchange capabilities
- provide data for marine forecast

Institution	Ship name	Track number	Initial port – Final port	Distance (nm)
SAHFOS	City of Dublin	1.	Haifa Gibraltar	780
CEAB	Isabella Annabella	2.1	Barcelona Arzew	300
CEAB	Isabella Annabella	2.2	Barcelona Skikda	340
LOB	Cap Canaille	3.	Sete Tunis	450
IOF-CRAM	Majestic Splendid	4.	Genova Palermo	410
OGS	Lipa	5.	Ploce Malta	520
NCMR	Mariska Sariska	6.	Pireus Alexandria	400
DFMR	Princesa Victoria Princesa Mariska	7.	Limassol Port Said	180

Table 3.2 The MFS-VOS transects along which data were collected from September 1999 to May 2000.

There are many elements which are novel in the MFS-VOS programme with respect to the similar programmes in the World Ocean. MFS-VOS is focusing on the mesoscale variability and on 'line sampling'.

The transects along which XBT data were collected, are from boundary to boundary, and were chosen in order to detect some of the major dynamical structures in the Mediterranean Sea. The Table 3.2 provide some details on the ship tracks selected in the MFS-VOS system (see also Fig. 3.4 which provide an overview of the launch positions from September 1999 to May 2000).

3.5.1 The collection/transmission system

The monitoring programme had very strong requirements on data delivery: the VOS data were to be controlled for their quality and released two days after the collection. Other requirements on temporal and spatial sampling (see Table 3.1) forced frequent XBT launches. In fact, with ships having speeds of 12 - 22 knots, the time interval between the launches was of about 0.5 - 1 hour.

Temperature profiles were collected by means of a standard XBT system composed by hand launchers, data processing/recording devices and expendable probes.

Data were transmitted by using the ARGOS system, which assured the deliver of data in real time to the MFS-VOS data centre (in ENEA-CRAM, La Spezia, Italy) and to the meteorological services through the Global Telecommunication System (UNESCO, 1999). The system does not allow the transmission of full resolution data, since only a message of 512 bits can be sent though the ARGOS satellite system; for this reason an ARGOS data collection center in Toulose developed new

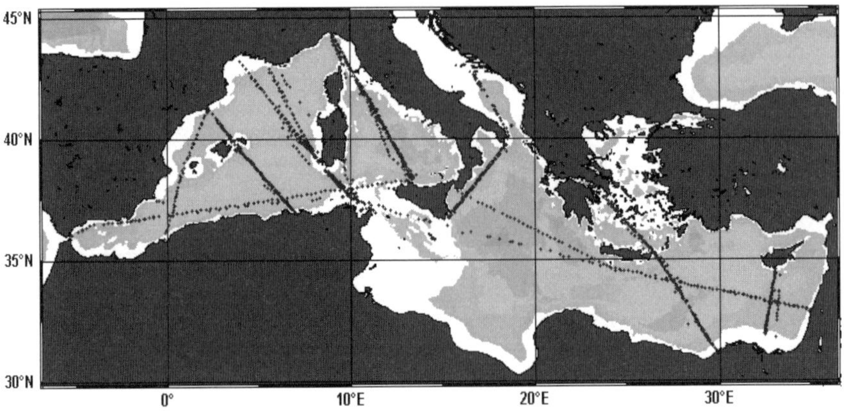

Fig. 3.4 MFS-VOS tracks during the period September 1999 - May 2000.

strategies and software for MFS-VOS. Instead of transmitting a unique message containing position and data, it was decided to transmit two messages of 512 bits, the first containing a series of positions from which the one corresponding to the XBT launch can be calculated by interpolation, the second containing the launch time and data. The old ARGOS software was able to select up to 15 significant points of a XBT profile, and each temperature value point was constituted by 8 bits. The new software was calculating always 15 significant points from surface up to 460 m. The surface temperature value was coded in 9 bits. The other points were calculated as difference with respect to the previous value. In this way it was possible to code this difference with 7 bits. A CRC, a software check, was added in order to check the accuracy of the entire message. A certain care was taken in order to avoid the rounding errors. Table 3.3 provide a comparison between the accuracy of the new code with respect to the previous one.

ARGOS transmission	Old software	MFS-VOS software
Depth accuracy (0 – 200m)	2 m	1 m
Depth accuracy (below 200m)	2 m	2 m
Time	2min 48sec	2min
Sea surface temperature	0.1333 °C	0.1 °C
Temperature	0.1333 °C	0.1 °C

Table 3.3 Comparison between the previous ARGOS software and the one developed for MFS-VOS.

The ARGOS software was originally developed for the World Ocean areas and their temperature characteristics. In particular, the Mediterranean water masses were considered as a 'noise' and the quality control included in a such software was discarding the XBT profiles collected in Atlantic areas with the Mediterranean outflow. It was then necessary to modify all the quality control criteria of the software, in order to retain all those kind of profiles typical of the Mediterranean area (e.g. homogeneous vertical profiles, temperature at surface lower than at depth, high gradients even below the surface layer, etc.). Although the software was highly improved it is still not adequate for a controlled decimation of the temperature profiles in the Mediterranean.

3.5.2 The data collection procedures and quality control on board

The MFS-VOS programme has implemented a series of quality assurance concepts which assured the success of the programme. Protocols were developed for the assurance of quality of:

- the measurement programme strategy
- the field work
- the collected data
- the use of data.

The goal of the MFS-VOS was the delivery of data for data assimilation in forecast models. In order to reach this goal, the measurement programme was first designed to capture the main dynamical features of the Mediterranean, then those commercial ship tracks which assured a regular sampling of these features were selected.

All the technicians on board the different ships assured identical field operations. For this aim, the protocol on quality assurance of field work provided many information on:

- the acquisition/transmission system and its theoretical framework,
- expected results,
- responsibility,
- working procedures,
- safety instructions,
- data quality control,
- deviations with respect to the sampling design and remedies.

A personnel training was done during an 'ad hoc' meeting. Another training period was done during the first three months of the MFS-VOS programme (September - November 1999). In this period only one cruise per month was carried out by each ship and a continuous contact between the programme responsible and the MFS-VOS participants allowed the correction of other possible error.

3.5.3 Data quality control

The first control was visually done by the personnel on board the ships of opportunity. In case of a wrong profile, another XBT was launched only in the case that the time interval with the preceding launch was less than 15 minutes. The software developed in collaboration with ARGOS system engineers performed another check before the data transmission. The parameters controlled by the software are presented in the Table 3.4.

Quality Control on board (data between 0 – 460 m)	
Outside range temperature	2 – 34 °C
Temperature inversion above 200 m	4.5 °C
Temperature inversion below 200 m	1.5 °C
Upper 5 metres	No control

Table 3.4 Parameters controlled for the quality control on board. The check was done automatically by the ARGOS software

During the three month training period, it was noted that the sea surface temperature was presenting some errors, due to causes such as (e.g.) the different initial temperature of the XBT sensor, the non vertical drop of the probe, or others. For this reason it was decided to assume as sea surface temperature the value at 5 m.

The data transmitted to the MFS-VOS data centre were again controlled with a software which included the gross range check developed during the MEDATLAS project (www.ifremer.fr/sismer/program/medar). A final check was done before the assimilation in the forecast model.

3.5.4 Near real time data flow

The data (decimated) were transmitted via satellite to the ARGOS centre in Toulouse, where they were decoded and transmitted by e-mail to the MFS-VOS data centre in ENEA CRAM (La Spezia), which provided the dissemination through a ftp site accessible through a project WWW. The data dissemination is completely open to all kind of users and to the general public.

The MFS-VOS WWW offer also the access to Ocean Data View software graphical representation of the data (Schlitzer, 2000).

3.6 Discussion and conclusions

During the last decade, an important change in the concept of data management has occurred. In the past the data manager was a technician charged to control the quality of the data and prepare the data for the analysis, which was a duty of a scientist. In an efficient dissemination system, the data analysis precedure is now part of the information released to the external users. This means that the scientific control of the Information Systems is more and more demanding.

Considering the scope of this book, some special issues must be discussed. In the case of forecasting systems, data must move rapidly from data collection platforms (ships, buoys, satellite) to the collecting centre. The data must be processed to combine them from different sources, to assess their quality, to add information about the data collection process, to generate products from the data and to move the products and data to end users.

The entire process from data collection to dissemination, must be monitored in order to ensure that no data are lost, and that all data reach the users and the international data archives. Equally important is to assure that data transmission takes place quickly and efficiently.

However, there are many decisions that must be taken and require scientific judgement of the problems to be afforded. These decisions include:

- the level of information in the scoping stage,
- the selection of processes to be sampled,
- the management objective,
- the management schemes,
- the projects and plan.

There are many documents that have analysed the European Environmental Action Programme and have proposed measures to strengthen the institutional infrastructure for developing effective environmental policies. These includes:

- improvement of environmental monitoring and information systems;
- widening access to environmental information;
- establishment of environmental impact assessment systems;
- more research on eco-systems;
- development of systems to interpret environmental data.

There is the understanding that management is a dynamic process and that management measures need to be regularly assessed, revised and adapted where necessary. Then management of marine sites must be flexible, adapting to the changing condition of each site such as:

- changes on human pressures on the environment
- natural changes in marine resources
- increase in knowledge and understanding of the marine ecosystems through monitoring and research.

This paper provide information on concepts adopted for the development of a monitoring system in the Mediterranean basin. Although only the temperature parameter was considered, such concepts can be easily transferred to multiparametric monitoring. In particular it is important establish protocols which must assure efficient data flow, quality control and graphics options.

Although the main goal of the MFS-VOS programme is to deliver data for forecasting, there are also some other important results, which will be more deeply exploited in forecoming papers, demonstrating how the real time data deliver give a rapid assessment of the environmental status. For example, MFS-VOS monitoring of the temperature allow to detect changes in the Mediterranean Sea, in comparison with the climatological values. The sampling intervals (both in time and space) allow also a study of the stratification in some key areas. In particular, it was seen that during the winter 1999-2000 the convection in some dense water formation areas (e.g. the Liguro-Provencal basin and the Adriatic Sea) was not deep as in previous years. During the period September 1999 - May 2000, it was also noted that the eastern Mediterranean Sea had higher temperatures than climatology (~2 °C higher).

An improved monitoring system capable of detecting physical, chemical and biological parameters could allow an effective management of the marine environment, assess in real time the status of it, provide the necessary data for its forecast.

The lesson learned from the first experience of an operational observing system in the Mediterranean is of paramount importance for the future implementation of the future data collecting activities:

- due to the relatively high costs it is necessary to maintain a consortium of institutions belonging to different countries,
- the institutions must be coordinated by a panel, in the same manner as the global VOS XBT community is coordinated by the Ship of Opportunity Programme Implementation Panel (SOOPIP),
- it is critical to ensure that the data collected gets to the users in a timely fashion,
- as a consequence, an efficient flow of data and information must be ensured,
- the dissemination of data in a timely way limits the time available for quality control, which cannot be completely automated. MFS-VOS has realised a 3-step control, the first being on board by a 'customised' software, the second in the Near-Real-Time data center, the third before the assimilation into forecast model,
- still supplementary end-to-end information management system must be established in order to include the wide public among the 'privileged' users.

It must be underlined that the success of MFS-VOS is due to the fact that the observational network is maintained by the research community, which defined the important signals to observe and the spatial and temporal resolution needed. This led to the final conclusion of this paper: the Marine Information System is an valuable tool for oceanographic studies and decision making processes, however, any decision must be based on a deep scientific knowledge of the marine environment.

Acknowledgement

This work was supported by the DG XII MAST III *Mediterranean Forecasting System Pilot Project* and MURST *Ambiente Mediterraneo*.

4 A Generalization of a Sigma Coordinate Ocean Model and an Intercomparison of Model Vertical Grids

GEORGE L. MELLOR[1], SIRPA M. HÄKKINEN[2], TAL EZER[1] AND RICHARD C. PATCHEN[3]

[1]*Atmospheric and Oceanic Sciences Program Princeton University, Princeton, NJ*

[2]*NASA Goddard Space Flight Center, Greenbelt, MD*

[3]*Dynalysis of Princeton, Princeton N.J.*

4.1 Introduction

Numerical ocean models increasingly make use of σ - coordinate systems. A paper by Gerdes (1993) shows that these coordinate systems can be more general; he termed the generalized form an "s - coordinate" system. The main advantage of the σ or s - system is that, when cast in a finite difference form, a smooth representation of the bottom topography is obtained; one can also easily incorporate a bottom boundary layer as well as a surface boundary layer in those coordinate systems. This is intuitively appealing and Gerdes has shown that superior numerical results are obtained relative to a z - level system. However, in regions of steep topography and crude resolution - a limiting case would be a seamount represented by a single grid point surrounded by a flat bottom - the so-called sigma coordinate pressure gradient error exists (Haney 1991, Mellor et al. 1994, 1998) and at least locally a z - level coordinate system might be preferred. On the other hand, in a recent study, Bell (1997) has shown that the step structure of z - level models lead to vorticity errors and consequent errors in the barotropic component of the flow which, he reports, cause rather large temperature errors (3 to 4° C) on a 1° x 1° grid of an Atlantic Ocean model after 3 months of integration. And it is difficult to model bottom boundary layers in a z - level model (Winton et al. 1998).

The pioneer Bryan-Cox model (Bryan 1969, Cox 1984) is a z - level model. A modification by Spall and Robinson (1990) is termed a "hybrid" coordinate system; they describe it as a z - level system in the region, $0 > z > - z_c$ = constant, and a σ system when $z_c > z > - H(x, y)$ where the transformed sigma equations apply. Presumably, this system is adopted so that surface mixed layers, which do not scale on depth, may be best represented. However, the hybrid system would appear to require separate numerical implementations for the two regions; their objectives can be realized more simply with the s - coordinate system described here.

For the sigma coordinate system of the Princeton Ocean Model (henceforth POM), the top numerical level, $k = 1$, follows the free sea surface and the lowest numerical level, $k = kb$, follows the bottom depth; for $1 < k < kb$, the distance between levels are in fixed proportion to each other independent of elevation or depth. In this paper we revise POM so that it is basically an s - coordinate system (although the variable, s, is, for very good reason, different from that used by Gerdes) and the proportionality constraint is removed. However, unlike the Gerdes formulation, the derivation and implementation permit a free surface and the distance between levels can change in time. Also the structure of the numerical algorithm is designed to support a wide variety of vertical coordinate systems including a z - level system. The present model generalization permits a z - level representation everywhere in the model domain or locally in selected regions of the domain.

Song and Haidvogel (1994) have adopted many of the characteristics of POM but have also generalized to an s - coordinate system. Our formulation and implementation strategy, as described below, differs from theirs and we include the z - level option. Intercomparisons between different models have been the focus of recent projects (DYNAMO 1997, Willems et al. 1994, Chassignet et al. 2000).

The intercomparisons of this paper isolate the effect of different vertical grids; otherwise model physics and numerics are identical.

4.2 The Governing Equations

We first restrict attention to the analytic description of the basic equations although it is our intention that they be cast in finite difference form and we condition the description with that in mind. The basic equations described in an (x,y,z) Cartesian coordinate system for the velocity components, U, V and W, and for potential temperature and salinity, T and S, are

$$\Im(1) = 0 \tag{1}$$

$$\Im(U) - fV + g\frac{\partial \eta}{\partial x} + \frac{g}{\rho_o}\int_z^\eta \left[\frac{\partial \rho'}{\partial x}\right] dz' = \frac{\partial}{\partial z}\left[K_M \frac{\partial U}{\partial z}\right] + F_x \tag{2}$$

$$\Im(V) + fU + g\frac{\partial \eta}{\partial y} + \frac{g}{\rho_o}\int_z^\eta \left[\frac{\partial \rho'}{\partial y}\right] dz' = \frac{\partial}{\partial z}\left[K_M \frac{\partial V}{\partial z}\right] + F_y \tag{3}$$

$$\Im(T) = \frac{\partial}{\partial z}\left[K_H \frac{\partial T}{\partial z}\right] + F_T - \frac{\partial R}{\partial z} \tag{4}$$

$$\Im(S) = \frac{\partial}{\partial z}\left[K_H \frac{\partial S}{\partial z}\right] + F_S \tag{5}$$

and for twice the turbulence kinetic energy, q^2 and length scale, l , we have

$$\Im(q^2) = \frac{\partial}{\partial z}\left[K_q\frac{\partial q^2}{\partial z}\right] + 2\mathrm{K}_M\left[\left(\frac{\partial U}{\partial z}\right)^2 + \left(\frac{\partial V}{\partial z}\right)^2\right] + \frac{2g}{\rho_o}K_H\frac{\partial\tilde{\rho}}{\partial z} - \frac{2q^3}{B_1 l} + F_q \tag{6}$$

$$\Im(q^2 l) = \frac{\partial}{\partial z}\left[K_q\frac{\partial q^2 l}{\partial z}\right] + E_1 l\left(\mathrm{K}_M\left[\left(\frac{\partial U}{\partial z}\right)^2 + \left(\frac{\partial V}{\partial z}\right)^2\right] + E_3\frac{g}{\rho_o}K_H\frac{\partial\rho^2}{\partial z}\right) - \frac{q^3}{B_1}\tilde{W} + F_l \tag{7}$$

$$\Im(\varphi) \equiv \frac{\partial\varphi}{\partial t} + \frac{\partial U\varphi}{\partial x} + \frac{\partial V\varphi}{\partial y} + \frac{\partial W\varphi}{\partial z} \tag{8}$$

In the above, x and y are horizontal coordinates (to be cast in a curvilinear, orthogonal coordinate system in section 4.4.) and z is the vertical coordinate; g is the gravity constant and f is the Coriolis parameter. The density is ρ whereas $\partial\tilde{\rho}/\partial z$ is the vertical density gradient corrected for adiabatic lapse rate. $\partial R/\partial z$ on the right hand side of (4) is the divergence of radiation flux, R. The vertical mixing coefficients are K_M, K_H, and K_q and are functions of a Richardson number dependent on density stratification (Mellor and Yamada, 1982, Galperin et al, 1988 and Mellor, 1996). $\tilde{W}$ in (7) is a "wall proximity function". Horizontal diffusion fluxes are represented by F_α ($\alpha = x, y, T, S, q, l$) .

4.3 Transformation to the *s*-coordinate system

The Cartesian coordinate system, (x, y, z, t), is transformed to the s - coordinate system, (x^*, y^*, k, t^*), according to

$$x = x^* \tag{9a}$$

$$y = y^* \tag{9b}$$

$$t = t^* \tag{9c}$$

$$z = \eta(x^*, y^*, t^*) + s(x^*, y^*, k, t^*) \tag{9d}$$

where k is a continuous variable in the range, $1 \le k \le kb$, but when the differential equations are discretized, k will also be discrete and will be the label of the numerical level. This usage is quite convenient and obviates the need for an intermediate variable [which Gerdes called s, and is not the usage in (9d)]. We define η to be the surface elevation so that we will require $s = 0$ at $k = 1$. The key attribute of the transformation is (9a,b); i.e., the horizontal coordinates in the Cartesian coordinate system are equated to the horizontal coordinates in the sigma system. Otherwise, the transformation is fairly general and includes, for example, a z - level system, $s = \sigma(k)(H_{max} + \eta(x,y,t))$, and the sigma system, $s = \sigma(k)(H(x,y) + \eta(x,y,t))$; $H(x,y)$ is the bathymetry and H_{max} is its maximum value. In both cases, $\sigma(1) = 0$ and $\sigma(\mathrm{kb}) = -1$. In the case of the z - level system (really a *quasi* z - level system because s is a function of time, but it is a weak function of time since, generally, $\eta/H \ll 1$), levels near the bottom are masked to approximate the bathymetry and, thus, kb is a function of the horizontal coordinates. The distinguishing feature of both of these limiting cases is that k is functionally separated from x, y and t. In general they need not be separable as we shall demonstrate.

There are two ways of deriving the s - coordinate equations. The first way is to apply integral (control volume) equations to a volume element bounded by x^* ,y^* , k and $x^* + \Delta x^*$, $y^* + \Delta y^*$, $k + \Delta k$. The second way is to formally transform equations (1)-(5). Thus, if ϕ is any dependent variable such that $\phi(x, y, z, t) = \phi^* (x^*, y^*, k,t^*)$ then

$$\frac{\partial\phi}{\partial x} = \frac{\partial\phi^*}{\partial x^*} + \frac{\partial\phi^*}{\partial k}\frac{\partial k}{\partial x}, \qquad \frac{\partial\phi}{\partial y} = \frac{\partial\phi^*}{\partial y^*} + \frac{\partial\phi^*}{\partial k}\frac{\partial k}{\partial y} \tag{10a,b}$$

$$\frac{\partial\phi}{\partial z} = \frac{\partial\phi^*}{\partial k}\frac{\partial k}{\partial z}, \qquad \frac{\partial\phi}{\partial t} = \frac{\partial\phi^*}{\partial t^*} + \frac{\partial\phi^*}{\partial k}\frac{\partial k}{\partial t} \tag{10c,d}$$

From the fact that $\partial z/\partial x = 0$, we obtain $\partial k/\partial x = -(\eta_x + s_x)/s_k$ where $s_x \equiv \partial s/\partial x^*$ and $\eta_x \equiv \partial\eta/\partial x^*$. Similarly, $\partial k/\partial y = -(\eta_y + s_y)/s_k$, $\partial k/\partial z = -1/s_k$ and $\partial k/\partial t = -(\eta_t + s_t)/s_k$. (In discretized form, $s_k \equiv \delta s/\delta k$ is simply the distance, δs, between levels since $\delta k = 1$.) Therefore,

$$\frac{\partial\phi}{\partial x} = \frac{\partial\phi^*}{\partial x^*} - \frac{\partial\phi^*}{\partial k}\frac{s_x + \eta_x}{s_k}, \qquad \frac{\partial\phi}{\partial y} = \frac{\partial\phi^*}{\partial y^*} - \frac{\partial\phi^*}{\partial k}\frac{s_y + \eta_y}{s_k} \tag{11a,b}$$

$$\frac{\partial\phi}{\partial z} = \frac{\partial\phi^*}{\partial k}\frac{1}{s_k}, \qquad \frac{\partial\phi}{\partial t} = \frac{\partial\phi^*}{\partial t^*} - \frac{\partial\phi^*}{\partial k}\frac{s_t + \eta_t}{s_k} \tag{11c,d}$$

We define a new variable, ω , such that

$$\omega \equiv W - (s_x + \eta_x)U - (s_y + \eta_y)V - (s_t + \eta_t) \tag{12}$$

Note that the surface kinematic condition, $W = \eta_x U + \eta_y V + \eta_t$, is satisfied if we require that $\omega = 0$ where $s = 0$. Similarly, at the bottom, $W = -H_x U - H_y V$, and we set $\omega = 0$ where $s = -(H + \eta)$.

If we use (11a, b, c, d) and (12) in equations (1) through (9) and drop all of the asterisks from independent and dependent variables, we obtain,

$$\Im(1) = 0 \tag{13}$$

$$\Im(U) - fVs_k + gs_k\frac{\partial\eta}{\partial x} + g\frac{s_k}{\rho_o}\int_k^0\left[s_k\frac{\partial\rho'}{\partial x} - (s_x+\eta_x)\frac{\partial\rho'}{\partial k'}\right]dk'$$
$$= \frac{\partial}{\partial k}\left[\frac{K_M}{s_k}\frac{\partial U}{\partial k}\right] + \left(\frac{\partial s_k\tau_{xx}}{\partial x} + \frac{\partial s_k\tau_{xy}}{\partial y}\right) \tag{14}$$

$$\Im(V) + fUs_k + gs_k\frac{\partial\eta}{\partial y} + g\frac{s_k}{\rho_o}\int_k^0\left[s_k\frac{\partial\rho'}{\partial y} - (s_y+\eta_y)\frac{\partial\rho'}{\partial k'}\right]dk'$$
$$= \frac{\partial}{\partial k}\left[\frac{K_M}{s_k}\frac{\partial V}{\partial k}\right] + \left(\frac{\partial s_k\tau_{xy}}{\partial x} + \frac{\partial s_k\tau_{yy}}{\partial y}\right) \tag{15}$$

$$\Im(T) = \frac{\partial}{\partial k}\left[\frac{K_H}{s_k}\frac{\partial T}{\partial k}\right] + \left(\frac{\partial s_k q_T}{\partial x} + \frac{\partial s_k q_T}{\partial y}\right) - \frac{\partial R}{\partial k} \tag{16}$$

$$\Im(S) = \frac{\partial}{\partial k}\left[\frac{K_H}{s_k}\frac{\partial S}{\partial k}\right] + \left(\frac{\partial s_k q_S}{\partial x} + \frac{\partial s_k q_S}{\partial y}\right) \tag{17}$$

$$\Im(q^2) = \frac{\partial}{\partial k}\left[\frac{K_q}{s_k}\frac{\partial q^2}{\partial k}\right]$$
$$= \frac{2\mathrm{K}_M}{s_k}\left[\left(\frac{\partial U}{\partial k}\right)^2 + \left(\frac{\partial V}{\partial k}\right)^2\right] + \frac{2g}{\rho_o}K_H\frac{\partial\tilde{\rho}}{\partial k} - \frac{2s_kq^3}{B_1 l} + F_q \tag{18}$$

$$\Im(q^2 l) = \frac{\partial}{\partial k}\left[\frac{K_q}{s_k}\frac{\partial q^2 l}{\partial k}\right]$$
$$+ E_1 l\left(\frac{\mathrm{K}_M}{s_k}\left[\left(\frac{\partial U}{\partial k}\right)^2 + \left(\frac{\partial V}{\partial k}\right)^2\right] + E_3\frac{g}{\rho_o}K_H\frac{\partial\rho^2}{\partial k}\right) - \frac{q^3}{B_1}\tilde{W} + F_l \tag{19}$$

where now we define

$$\Im(\varphi) \equiv \frac{\partial s_k\varphi}{\partial t} + \frac{\partial U s_k\varphi}{\partial x} + \frac{\partial V s_k\varphi}{\partial y} + \frac{\partial W\varphi}{\partial k} \tag{20}$$

In the above, $\tau_{xx} = A_M(2\partial U/\partial x)$, $\tau_{yy} = A_M(2\partial V/\partial y)$, $\tau_{xy} = A_M(\partial U/\partial y + \partial V/\partial x)$, $q_{Tx} = A_H(\partial T)/\partial x$, $q_{Ty} = A_H\partial T/\partial y$ and $q_{Sx} = A_H\partial S/\partial x$, $q_{Sy} = A_H\partial S/\partial y$. Horizontal diffusion for q^2 and $q^2 l$ is the same form as that for T and S. We now note that the gradients on the right sides of the momentum and scalar fluxes are reckoned along k surfaces. However, any (simple) alternative, say, to transform so that gradients are taken along horizontal or isopycnal surfaces, leads to problems in properly modeling the bottom boundary layer (Mellor and Blumberg, 1985) and possibly the surface layer. In the case of isopycnal surfaces large vertical flux terms normal to top or bottom surfaces can be created and can compete with the vertical mixing parameterization denoted by K_M and K_H; these recognize that turbulent motion normal to surfaces are small whereas A_M and A_H are not. On the other hand, mixing along k surfaces creates problems in, say, the interior ocean in the presence of vertical temperature or salinity gradients; this will be discussed below.

Values for the horizontal viscosity, A_M, and diffusivity, A_H, will play an important role in the solutions to be discussed below.

4.4 Finite Difference Equations

The differential equations are first written in finite volume form, that is, the equations are integrated over a finite cell volume. In particular for the advective operator,

$$\iiint \Im(\varphi)\delta V = \frac{\partial}{\partial t}\iiint \varphi \delta V + \iint \varphi U \cdot \mathbf{n}\delta A$$

where δV is an elemental volume, δA an elemental area and $\mathbf{n}$ a unit vector normal to the area. Then

$$\Im(1) = 0 \tag{21}$$

$$\begin{aligned}\Im(U) - fAV\delta s + g\delta y\delta s\delta_x\eta + g\delta y\delta s\sum_k[\delta s\delta_x\rho - (\delta_x s + \delta_x\eta)\delta_k\rho] \\ = A\delta_k\left[K_M\frac{\delta_k U}{\delta s}\right] + \delta_x(\tau_{xx}\delta s\delta y) + \delta_y(\tau_{xy}\delta s\delta x)\end{aligned} \tag{22}$$

$$\begin{aligned}\Im(V) + fAU\delta s + g\delta x\delta s\delta_y\eta + g\delta x\delta s\sum_k[\delta s\delta_y\rho - (\delta_y s + \delta_y\eta)\delta_k\rho] \\ = A\delta_k\left[K_M\frac{\delta_k V}{\delta s}\right] + \delta_x(\tau_{xy}\delta s\delta y) + \delta_y(\tau_{yy}\delta s\delta x)\end{aligned} \tag{23}$$

$$\Im(T) = A\delta_k\left[K_H\frac{\delta_k T}{\delta s}\right] + \delta_x(q_{Tx}\delta s\delta y) + \delta_y(q_{Ty}\delta s\delta x) \tag{24}$$

$$\Im(S) = A\delta_k\left[K_H\frac{\delta_k S}{\delta s}\right] + \delta_x(q_{Sx}\delta s\delta y) + \delta_y(q_{Sy}\delta s\delta x) \tag{25}$$

where now we define

$$\Im(\varphi) \equiv \frac{A\delta_t(\delta s\varphi)}{\delta t} + \delta_x(\delta s\delta y U\varphi) + \delta_y(\delta s\delta x V\varphi) + A\delta_k(\omega\varphi) \tag{26}$$

This formulation accommodates an orthogonal, curvilinear, horizontal coordinate system; the elements of each cell have the horizontal dimensions, δx, δy and the area, $A \equiv \delta x\delta y$. The vertical dimension is δs. Formally, there are additional horizontal viscous terms in (22) and (23) due to curvature, but these have been neglected since the terms themselves are highly empirical and we prefer the values, A_M and A_H, be as small as possible. In the calculations discussed below, the horizontal coordinates will be spherical, a special case of more general orthogonal curvilinear coordinates accommodated by the model.

4.5 Model Strategy

The algorithmic structure of the generalized model does not differ greatly from the previous POM (Blumberg and Mellor 1987, Mellor 1996) in that it has a free surface, split internal, external time steps and second order horizontal differencing on a staggered C grid (although work on other differencing schemes is in progress) and an imbedded turbulence closure scheme which primarily governs surface and bottom boundary layers. The new features are that the more general vertical increment, $\delta s(i,j,k,n)$, where i,j,k are the spatial indices and n is the time index, has replaced the sigma increment, $\delta s(k)(H(i,j) + \eta(i,j,n))$ and the previous two-dimensional land mask is now three-dimensional to accommodate the z - level grid when that system is invoked. All of the code except for one subroutine is universal. The exceptional subroutine is crafted to produce whichever coordinate system is desired. In the examples discussed below, there are three such subroutines (SUBROUTINE MAKSIG, for example), one for each of the three grids we have chosen to test in this paper. The active subroutine is called at each internal mode time step.

4.6 Numerical Simulations Comparing Three Vertical Coordinate Systems

We will examine calculations using three vertical grids, a z - level grid, a σ grid and a generalized σ grid. Fig. 4.1 illustrates the domain used in the calculations which includes a bowl shaped deep basin in the north, east and west and a vertical wall at the equator; the basin connects to a shallow continental shelf in the north. Fig. 4.1a represents both the sigma and generalized sigma grids. The horizontal resolution of all the experiments is about 1°x1° on a 51 x 76 horizontal grid. There are 21 vertical grid levels with spacing decreasing near the surface; of course, for the z - level grid, levels are truncated for all but the deepest depths (Fig. 4.1b). All vertical walls including those of the z - level steps are adiabatic and frictionless.

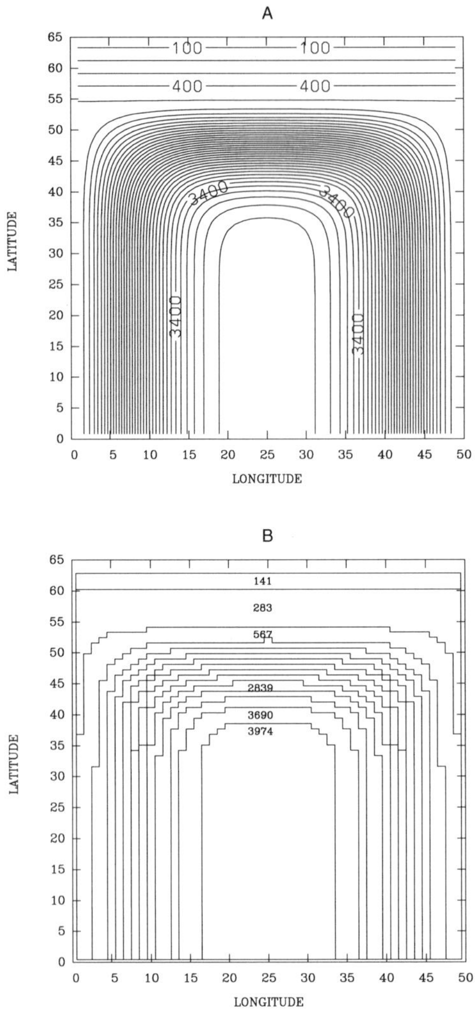

Fig. 4.1 Bottom topography of the model domain, a bowl shaped deep basin with a shallow shelf in the northern part, as represented by different vertical grids: (a) The conventional and the generalized sigma coordinate grids use a continuous bottom depth; the contour interval used in the plot is 100 m. (b) The z-level grid has a step-like topography so that the lines show the location of vertical walls and the numbers show the depth of the bottom of the steps.

The initial temperature is $T(x,y,z) = T(z) = 5 + 15\exp(z / 1000m)$ and the applied zonal wind stress field is $(\tau_{xo}, \tau_{yo}) = (-.0002\ m^2s^{-2}\ \sin\{2\pi(\phi/60°)\}, 0)$ where ϕ is latitude. The surface heat and salinity flux and penetrative radiation are nil.

Fig. 4.2 shows the three vertical grids in a north-south cross section at the center of the domain shown in Fig. 4.1. For the z-level grid (Fig. 4.2a), we have a conventional grid with the characteristic step structure at the bottom. For the conventional sigma grid (Fig. 4.2b), the spacing between the different layers have the same proportion independent of depth, and the lowest layer smoothly follows the bottom topography. For the generalized sigma grid (Fig. 4.2c), we have devised a grid which is mostly a z - level grid, but departs from the z - level constraint near the bottom boundary such that the lowest layer follows the bottom topography. Layers therefore converge near the bottom and allow for good resolution of the bottom boundary layer in most of the domain. The minimum layer thickness is constrained to be 5 cm at the bottom.

In the first set of experiments the three models were integrated for one year using viscosity and diffusivity values of $A_M = 10^5$ m^2 s^{-1} and $A_H = 0.2\ A_M$, respectively. This level of horizontal viscosity is that used in the GFDL, Bryan-Cox applications for this resolution (Cox 1975, Bryan and Lewis 1979, Rosati and Myakoda 1988, Derber and Rosati 1989, Bell 1997) and a reduced diffusivity is also typical; we have found that calculations with $A_H = A_M$ do not differ significantly from those discussed below. However, the reduced diffusivity is generally favorable to the sigma systems to reduce a vertical component of the along - sigma diffusion. Another way to counter along-sigma diffusion is to subtract the (initial in the present case) climatological temperature fields before reckoning along sigma gradients of temperature and salinity and fluxes, and this has been invoked in all three model runs (but is inconsequential for the z - level run). The ultimate solution is to set the along-sigma diffusivity equal to zero which we find possible in many applications of the sigma grid (Mellor et al. 1998, Ezer and Mellor 2000).

Surface elevations, representing the surface velocities, are shown in the left panels and transport streamfunctions, representing the depth integrated velocities, are shown in the right panels in Fig. 4.3 for all three grids. The two sigma models are almost identical. The z - level calculations are also quite similar to the two sigma models. However, the z - level surface velocities in the shallow northern region are much stronger. In Fig. 4.4, the temperature cross sections of the three experiments appear to be virtually identical in most of the domain, producing similar upper ocean mixed layers, except near the continental shelf, where only the two sigma models appear to produce also bottom boundary layers. In Fig. 4.5, velocity cross sections of the three experiments are fairly similar in the deep ocean, but significantly differ from each other near the continental slope and shelf where the z - level grid produces noisy velocity fields.

If the horizontal viscosity and diffusivity are reduced by an order of magnitude, $A_M = 10^3$ m^{2}s^{-1} and $A_H = 0.2\ A_M$, the situation changes. The sigma models tolerate the reduced values; the z - level model does not as seen in Figs. 4.6, 4.7 and 4.8. At this reduced viscosity and diffusivity, the sigma models exhibit Rossby waves near the equator (Figs. 4.6d and 4.6f) that are absent from the high viscosity calculations (Fig. 4.3); additional plots (not shown) reveal that the waves propagate westward at a speed of about 20 km d^{-1} , about 30% higher than the speed of the simplest baro-

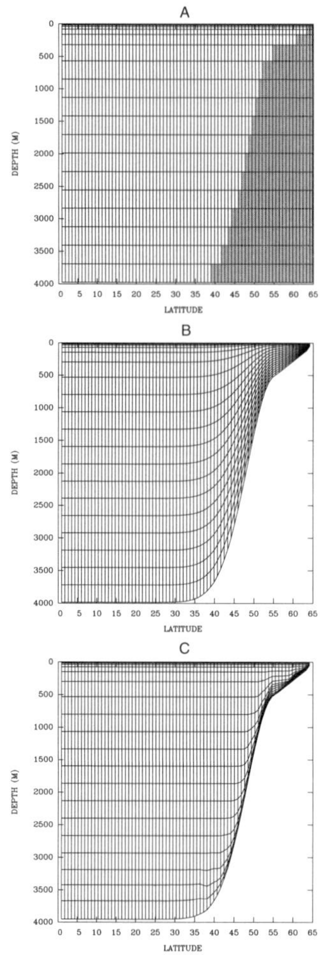

Fig. 4.2 Vertical grids in a nort-south cross section at the center of the domain: (a) the z - level grid, (b) the conventional sigma coordinate grid and (c) the generalized sigma grid. The grid in (c) is dominated by z coordinates in the deep ocean, but follows the topography like sigma coordinates near the sloping bottom with layers converging to create a well resolved bottom boundary layer.

tropic Rossby wave theory. Experiments (not shown) with even smaller viscosity of $A_M = 10^2\ m^2s^{-1}$, still show similar results for the sigma grid calculations, but with somewhat more intense Rossby waves.

Other calculations have been performed. They include a lower resolution, 2° x 2°, on a 26 x 38 x 11 grid and the same topography, initial conditions and surface forcing as before. The viscosity was raised to $A_M = 5 \times 10^5\ m^2s^{-1}$ and $A_H = 0.2\ A_M$.

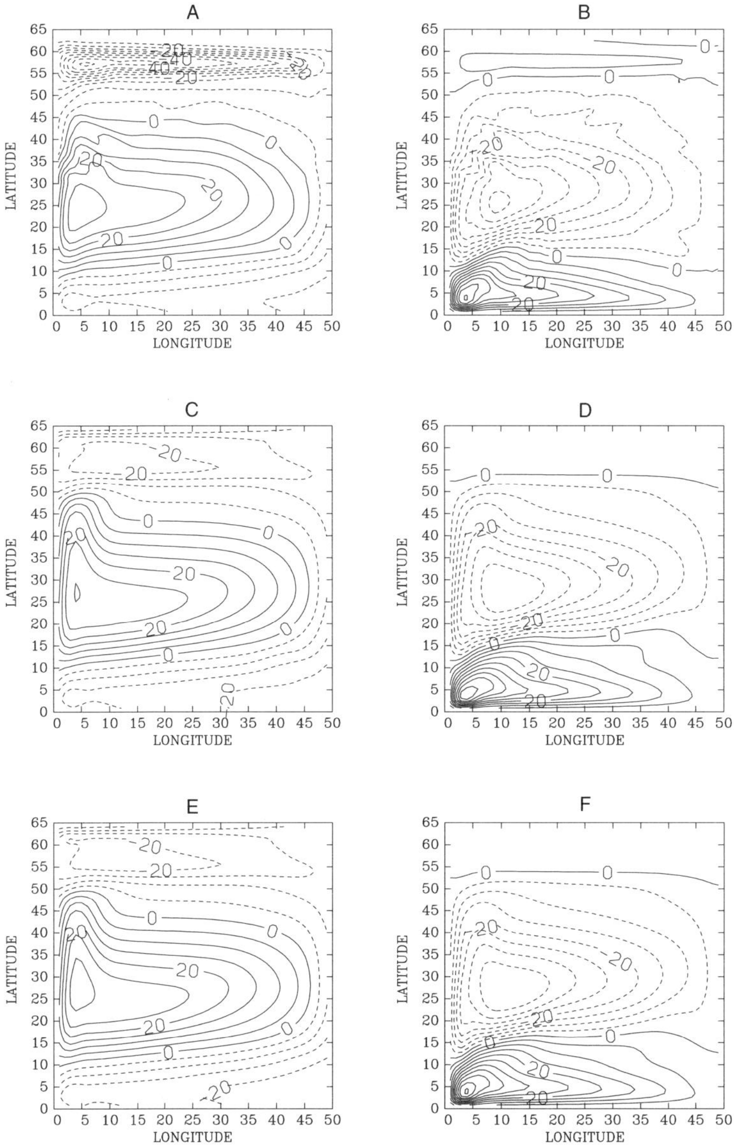

Fig. 4.3 The surface elevation (left panels) and the total stream function (right panels) after 360 days of calculation using a constant viscosity of $A_M = 10^4\ m^2\ s^{-1}$; results are shown for the z- level grid (top panels a and b), the s grid (middle panels c and d) and the generalized s grid (bottom panels e and f). Contour intervals are 5 cm in (a), (c) and (e) and 5 Sv (1 $Sv = 10^6\ m^3\ s^{-1}$) in (b), (d) and (f).

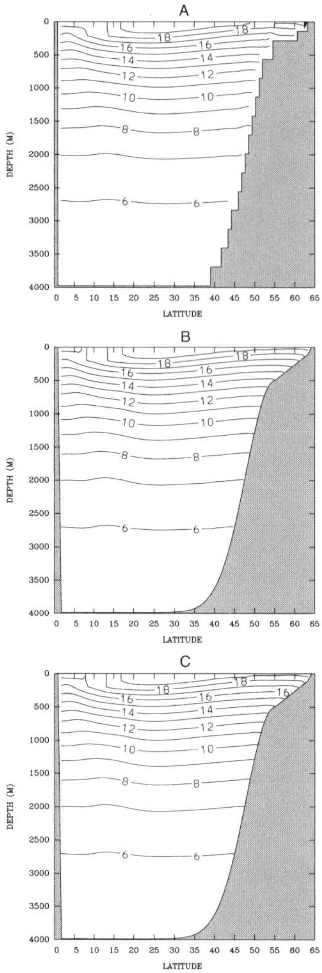

Fig. 4.4 North-south temperature cross sections at the center of the domain after 360 days for (a) the z - grid , (b) the s grid and (c) the generalized s grid; $A_M = 10^4\ m^2\ s^{-1}$.

The results were quite similar to the higher resolution, high viscosity results. However, the z - level grid produced numerically divergent results which was repaired by increasing the number of vertical grid points from 11 to 21.

This is the place to also mention that POM usually operates with a non-linear, Smagorinsky viscosity and diffusivity which has the advantage that its empirical constant is non-dimensional and nearly resolution independent and yields low values in regions where velocities are low. As noted by Derber and Rosati (1989), the

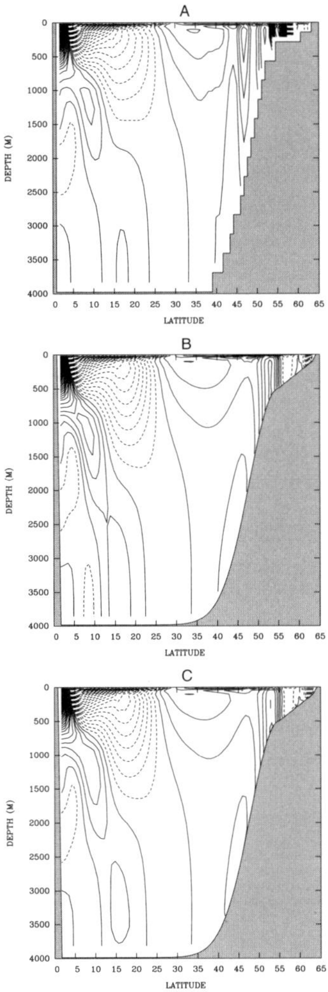

Fig. 4.5 North-south zonal velocity cross sections at the center of the domain after 360 days for (a) the z - grid , (b) the s grid and (c) the generalized s grid; $A_M = 10^4$ m ^{2} s $^{-1}$. Contour interval is 0.5 cm s $^{-1}$; dashed contours represent negative values and westward flows.

GFDL model cannot tolerate the Smagorinsky formulation and we have also found this to be the case in the present *z* - level model.

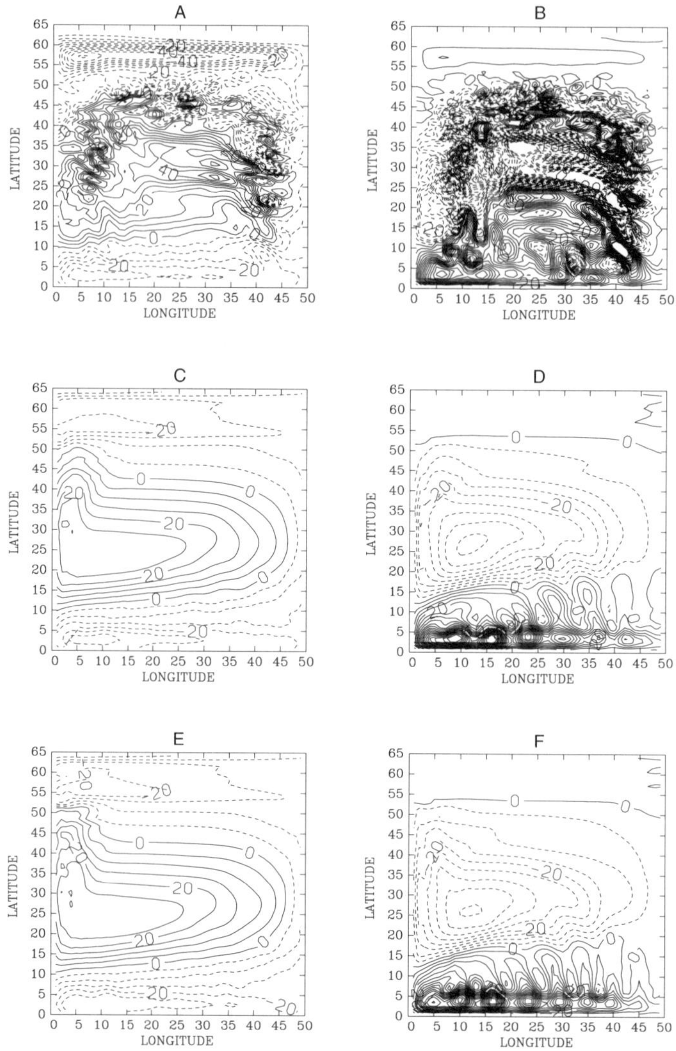

Fig. 4.6 Same as Fig. 4.3, but for the low viscosity case with $A_M = 10^3\ m^2\ s^{-1}$.

4.7 Errors

The source of sigma coordinate pressure gradient error is easily identified (Haney, 1991) although the error itself has not been fully understood until recently (Mellor et al. 1994, 1998). It is known that the error can be greatly reduced by sub-

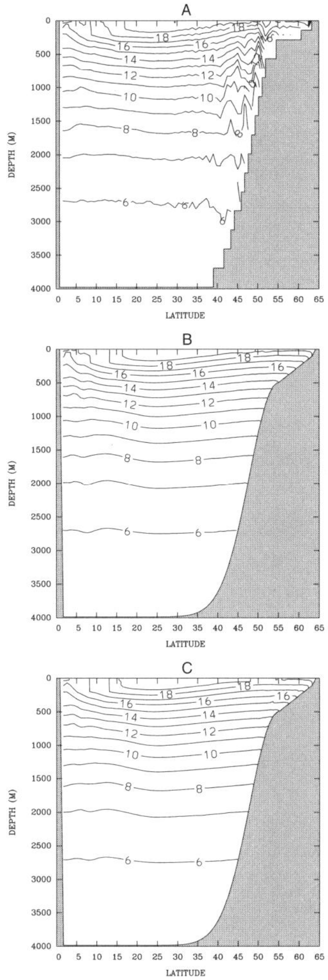

Fig. 4.7 Same as Fig. 4.4, but for the low viscosity case with $A_M = 10^3\ m^2\ s^{-1}$.

tracting the area averaged density before evaluating density gradients and performing the baroclinic integrals in (14) and (15) or (22) and (23). By running the straight σ model with no wind forcing and with *no* subtraction of the area averaged density, the maximum velocity and rms errors were approximately 2.5 cm s^{-1} and 0.15 cm s^{-1} in the high viscosity case and 5.0 cm s^{-1} and 0.30 cm s^{-1} in the low viscosity case; the maximum errors were located in the shallowest regions along the eastern and western boundaries. The maximum velocity errors in the generalized sigma model were about the same although the prior expectation was that they would be smaller. In both cases, the maximum vertically averaged velocities were

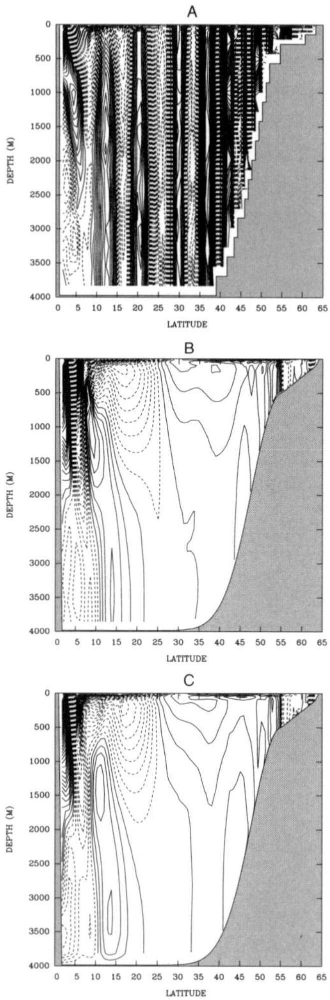

Fig. 4.8 Same as Fig. 4.5, but for the low viscosity case with $A_M = 10^3\ m^2\ s^{-1}$.

similar to the rms velocity errors. The corresponding errors in the z - level case were, of course, nil.

After subtraction of the area averaged density, the errors in the sigma models were also nil. Of course, after the model has run with wind forcing and baroclinic density structure has formed, there must exist some quite small but hard-to-evaluate sigma errors.

However, it is apparent from our results that, with wind forcing, errors in the z - level model, may be much larger than in the σ grid, in particular, in shallow regions and if viscosity is not high enough. Applying a computational stability analysis to one - dimensional, advection-diffusion equation,

$$\phi_{i+1} - \phi_{i-1} = 2R^{-1}(\phi_{i+1} - 2\phi_i + \phi_{i-1})$$

(Bryan et al. 1975) derived the stability condition $R < 2$, where $R = u\Delta x/A$ is the grid cell Reynolds number. Apparently, the step structure of the z - level grid excites this mode (R varies spatially but is of the order 2 for the high viscosity runs, but is much larger for the low viscosity runs) whereas the σ grids do not. Bell (1997) has produced a more complex analysis which identifies vorticity errors due to step structure.

4.8 Other Grids

We have looked at and compared only two conventional grids and a third somewhat unconventional, generalized sigma, or s - coordinate grid. (It should be mentioned however, that in the particular application tested here, the superiority of the s - coordinate system over the standard σ - coordinate system has not been demonstrated). However, the possible grid structures are infinite. For example, there is no reason not to have a sigma-like grid in one portion of a model domain and a z - level in another portion where, for example, the topography in a low resolution model might be especially steep. Given the model structure described in section 4.5, it would be possible to create an isopycnal model by creating a single enabling subroutine. The attendant problems of surface and bottom layer representation seem daunting to us, however. Other possibilities are an adaptive vertical coordinate system to resolve the surface mixed layer or an underlying sharp spring and summertime thermocline such as observed in the Great Lakes (Bedford and Schwab 1998).

4.9 Summary

There are two main thrusts to this paper. The first thrust is the creation of a model algorithm whose vertical coordinate system is quite general; this algorithm invites future inventiveness with other grids beyond the preliminary tests presented in this paper. The second thrust is a model intercomparison study of conventional z - level and σ - coordinate vertical grids and one unconventional generalized sigma grid. The intercomparisons are unique in that all aspects of the models are identical except for the vertical grid specification. One finding is that the sigma grids can tolerate much smaller levels of horizontal viscosity and diffusivity. In shallow water, errors arise in the z - level grid which, depending on viscosity (which itself may be considered a source of error), are of a different nature and obviously much larger than errors due to the sigma pressure gradient error.

In the final analysis, the straight σ coordinate grid may be a good choice for many ocean applications, including basin-scale climate calculations (Ezer 1999, Ezer and Mellor 1997). Song and Haidvogel (1994) claim that adequate resolution of the surface layers are a significant advantage of the s - coordinate grid. However, if the vertical grid spacing is, say, logarithmically reduced near the surface to adequately resolve the surface layer in the deepest portions of the model domain, then the only penalty of the σ grid is that, in the shallower portions, the surface layers are over resolved. In any case, the new algorithm, developed here for a widely used sigma coordinate ocean model (POM), provides a tool to further investigate the differences between different grids, and provides more flexibility for users to optimize their grids for particular applications.

Acknowledgments

This research was supported by the Office of Naval Research grant N0014-93-0037 and by the computational support and facilities of NOAA's Geophysical Fluid Dynamics Laboratory.

5 Atmospheric Data Assimilation and Quality Control

ANDREW C. LORENC
The Met. Office, Bracknell, England

5.1 Introduction

In this paper we discuss the basic physics of the atmospheric data assimilation problem, in order to understand the important factors to be considered in its mathematical solution. The key mathematical technique, the optimal combination of information, is also approached from its Bayesian basics. Much of this is based on earlier papers (e.g. Lorenc, 1986). The novelty of this paper is its bringing together of these in a simple didactic form, following the agreed notation of Ide et al. (1997), with very simple examples to aid in the physical interpretation of the analysis equations.

In the second half of the paper the analysis is extended to handle gross observational errors, described by a non-Gaussian error distribution. Again, much of this is based on earlier papers (Lorenc and Hammon, 1988, Dharssi et al., 1992, Ingleby and Lorenc, 1993). Various practical methods can be derived using Bayesian ideas; they are compared using a very simple example.

5.2 What is Data Assimilation?

There are insufficient observations at any one time to determine the state of the atmosphere. So if we want a detailed complete picture, we need additional information. This is available as knowledge of the behaviour and probable structure of the atmosphere. For instance the knowledge of the typical structure of a frontal depression enables a human to draw an "analysis" of the atmospheric state, based on scattered observations. To advance beyond this subjective approach, the behaviour of the atmosphere is embodied in a computer model. In particular, knowledge of the evolution with time is embodied in a forecast model. This enables us to use observations distributed in time. The model also provides a consistent means of representing the atmosphere. *Assimilation is the process of finding the model representation which is most consistent with the observations.*

Usually, data assimilation proceeds sequentially in time. The model organises and propagates forward the information from previous observations. The information from new observations is used to modify the model state, to be as consistent as possible with them and the previous information. It is the experience with operational assimilation for Numerical Weather Prediction (NWP) that there is usually more information in the model state, from previous observations, than there is in a new batch at a single synoptic time. Thus it is important to preserve this in the assimilation process; it is not just a question of fitting the new data. Since all information has to be represented within the model, it is important that the model should be of sufficiently high resolution, with physically realistic detail, to represent the information observed.

Some research is investigating non-sequential data assimilation methods, especially four-dimensional variational assimilation. This paper is designed to provide the groundwork for this, without going into detail.

5.3 Useful prior knowledge about the atmosphere

We need first to decide what aspects of the atmosphere we are modelling. Certain scales, including of course all those not resolved by the model, and types of motion, e.g. sound waves, are not represented. We define our "true" state, the target of an ideal assimilation, as the atmospheric state with these filtered out. They then contribute to the error of representativeness of the model, but not to the analysis error (i.e. if the analysis fits our "true" state, its analysis error is zero, even if it does not represent all the detail of the real atmosphere).

For the scales considered in NWP, the atmosphere is usually smooth, slowly varying, and close to horizontal non-divergence. We can observe these properties and quantify them. The smoothness can be described by a correlation function, or power spectrum. The slowly varying property leads to useful balance relationships. For instance by assuming that the rate of change of wind, and other residual terms, are negligible compared to the pressure gradient and Coriolis forces, we get the geostrophic relationship. Unfortunately none of these simple relationships is exact, and imposing them would probably damage our prior estimate of the atmospheric state, x^b. To avoid this, we apply the same balance arguments to x^b, and subtract. The residual terms which lead to inaccuracy in the simple relationships then approximately cancel. It is thus appropriate to apply the smoothness, geostrophic balance, and non-divergence relationships to the residuals, $x\text{-}x^b$. They are used in modelling the structure of the background covariance **B**.

The observation that the atmosphere (for those aspects we are trying to represent) is slowly varying, can be used to derive balance relationships involving diabatic processes, such as latent heating, and the circulations of the associated weather systems. However these effects are not easily described mathematically. An alternative approach is to use directly the "slowly varying" property by *initialising* a numerical model so that its initial evolution is slowly varying.

By far the most useful and accurate prior knowledge that we have, is of the equations governing the atmosphere's evolution in time, enabling us to build NWP models. The simplest use of models is to carry information forward in time from a past analysis, to provide the background for a new analysis. This process is the core of all operational NWP systems; without it analyses and forecasts would be substantially less accurate. Computationally feasible methods for integrating this knowledge more effectively into the assimilation, treating it as a single four-dimensional problem, are the subject of current research.

5.4 Products and Uses of Assimilation

Assimilation produces a convenient, comprehensive, high resolution, representation of the atmosphere. It has been clearly demonstrated that the use of a computer model is usually better (i.e. leads to better forecasts) than the subjective human approach. The main practical use of these assimilated "analyses" is for initialising numerical weather prediction forecasts. They are also useful for climate and general circulation studies, for instance in the calculations of fluxes, which make use of their high resolution and comprehensive coverage. However it must be remembered that the blend of observed and modelled information will vary according to the accuracy and coverage of the observations. So they must be used with great care for model validation, and climate change detection.

Very useful secondary products of a data assimilation system are the statistics on the (mis–)fit of observations to model. These can be more directly used for model (in–)validation, and for the monitoring of observing systems.

5.5 The Optimal Combination of Information

The history of this goes back to Gauss (1809, 1823), who in his study of the motion of the planets developed methods for the weighted combination of observations with errors. We will approach it from a Bayesian standpoint, following Lorenc (1986) and Lorenc and Hammon (1988). The resulting equations are equivalent to Gauss's minimum variance approach, for observations with a Gaussian error distribution.

5.5.1 Bayes' Theorem for Discrete Events

The Bayesian approach is to use probabilities to describe the accuracy of our knowledge about past events. We then have a formalism for modifying the probabilities in the light of new knowledge; exactly what we need to do in sequential data assimilation. We introduce this with a discrete example for events A and B. $P(A)$ is the probability of A occurring (this is the usual use of probabilities), or a measure of our certainty that A occurred in the past (this is the Bayesian use of probabilities). Then $P(A \cap B)$ is the probability that A and B both occurred, and $P(A \mid B)$ is the conditional probability of A given B has occurred. We have two ways of expressing $P(A \cap B)$:

$$\begin{aligned} P(A \cap B) &= P(B)P(A|B) \\ &= P(A)P(B|A) \end{aligned} \tag{1}$$

This leads directly to Bayes' Theorem:

$$P(A|B) = \frac{P(B|A)(P(A))}{P(B)} \tag{2}$$

So we start with a prior probability of *A*, and add the information that B has occurred, to give us the posterior probability of *A* given *B*. To evaluate this we can calculate *P*(*B*) from:

$$P(B) = P(B|A)P(A) + P(B|\bar{A})P(\bar{A}) \tag{3}$$

where $\bar{A}$ means *not* A.

5.5.2 Single-variable Bayesian Analysis with Gaussian probability distribution functions

For continuous variables we use probability distribution functions (pdfs):

$$p(x)dx = P(x \le x^t < x + dx) \tag{4}$$

where x^t is the true value. Bayes' Theorem becomes:

$$p(x|y^o) = \frac{p(y^o|x)p(x)}{p(y^o)} \tag{5}$$

Let us start thinking of *x* as the model state, which accumulates our knowledge. $p(x)$ is the prior distribution; our knowledge from previous observations. $p(x \mid y^o)$ is the posterior distribution, after adding the information from the observation y^o. $p(y^o \mid x)$ is the probability density of getting the observation y^o, given our previous knowledge. Note that this is a density in *y*-space. Regarded as a function of *x*, $p(y^o \mid x)$ is not a probability density (its integral is not necessarily one); it is called the likelihood function for *x*.

Let us assume that our prior knowledge is that *x* is near x^b, and that the variance of its deviation from x^b is V^b. The usual way of modelling such a distribution is as a Gaussian:

$$p(x) = N(x|x^b, V^b) = (2\pi V^b)^{-1/2} \exp\left(\frac{1}{2}\frac{(x-x^b)^2}{V^b}\right) \tag{6}$$

If our observation directly measures the variable x, with observational error variance V^o, then the probability of observed value y^o, given the true value is x, can also be modelled by a Gaussian pdf:

$$p(y^o|x) = N(y^o|x, V^o) = (2\pi V^o)^{-1/2}\exp\left(\frac{1}{2}\frac{(y^o - x)^2}{V^o}\right) \tag{7}$$

We can get $p(y^o)$ by integrating over all x:

$$p(y^o) = \int p(y^o|x)p(x)dx \tag{8}$$

For Gaussians (6) and (7) this gives:

$$p(y^o) = N(y^o|x^b, V^o + V^b) \tag{9}$$

Substituting in (5) gives:

$$p(x|y^o) = N(x|x^a, V^a) \tag{10}$$

where

$$\begin{aligned} \frac{x^a}{V^a} &= \frac{y^o}{V^o} + \frac{x^b}{V^b} \\ \frac{1}{V^a} &= \frac{1}{V^o} + \frac{1}{V^b} \end{aligned} \tag{11}$$

This is the standard formula for the combination of observations with error, known since the work of Gauss. $p(x)$ —the prior distribution, $p(x\,|\,y^o)$ —the posterior distribution, and $p(y^o\,|\,x)$ —the likelihood function, are plotted in Fig. 5.1 for four different values for y^o. A unique property of the Gaussian pdfs can be seen in them, and in the above equations: the shape of the posterior distribution, and its relative position to x^b and y^o, are both independent of the observed value. They only depend on the error variances.

The reason for this becomes apparent if we take logarithms of both sides of (5). The Bayesian analysis equation becomes:

$$-\ln[p(x|y^o)] = -\ln[p(y^o|x)] - \ln[p(x)] + \text{constant}. \tag{12}$$

Gaussians become quadratics, which are summed to give another quadratic:

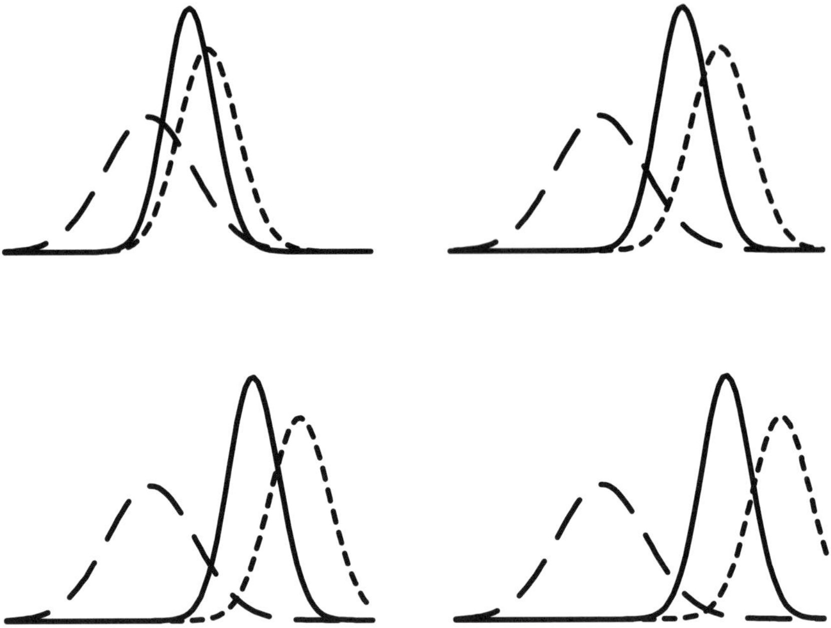

Fig. 5.1 Prior pdf $p(x)$ (*dashed line*), posterior pdf $p(x \mid y^o)$ (*solid line*), and likelihood of observation $p(y^o \mid x)$ (*dotted line*), plotted against x for various values of y^o. (Adapted from Lorenc and Hammon 1988).

$$\begin{aligned} -\ln[p(x|y^o)] &= \frac{1}{2}\frac{(x^b - x)^2}{V^b} + \frac{1}{2}\frac{(y^o - x)^2}{V^o} + \text{constant} \\ &= \frac{1}{2}\frac{(x^a - x)^2}{V^a} + \text{constant}. \end{aligned} \tag{13}$$

The Gaussian curves of Fig. 5.1 become the quadratics of Fig. 5.2, with the most probable values for x being at the minimum of the total curve. So the Bayesian combination of Gaussian pdfs gives the same "best" analysis as a weighted least-squares best fit to the data.

5.5.3 One-dimensional Bayesian Analysis

To extend the above to data assimilation of observations into a model, we need to introduce several new ideas. We do this in the context of the simplest relevant example: .

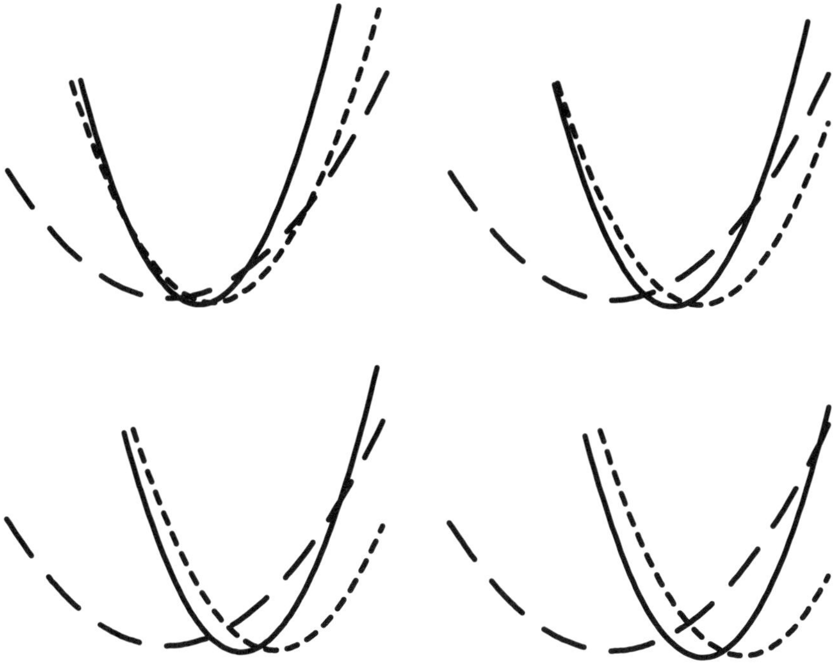

Fig. 5.2 As Fig. 5.1 for -log(probabilities).

-A model which represents the "world" with two grid points: $\boldsymbol{x} = \begin{pmatrix} x_1 \\ x_2 \end{pmatrix}$

-An instrument which gives one observed value y^o midway between the points. We can interpolate a model estimate y of the observed value:

$$\begin{aligned} y = H(\boldsymbol{x}) &= \frac{1}{2}x_1 + \frac{1}{2}x_2 \\ &= H\boldsymbol{x} = \begin{pmatrix} \frac{1}{2} & \frac{1}{2} \end{pmatrix} \begin{pmatrix} x_1 \\ x_2 \end{pmatrix} \end{aligned} \tag{14}$$

We need a probability distribution (the background pdf) to describe what we know about $\boldsymbol{x}$, before getting the observation. Let us suppose we have a prior estimate $x_1{}^b$ with error variance V^b:

$$p(x_1) = N(x|x_1^b, V^b) = (2\pi V^b)^{-1/2} \exp\left(\frac{1}{2}\frac{(x_1 - x_1^b)}{V^b}\right) \tag{15}$$

with a similar equation for $p(x_2)$. Because $\boldsymbol{x}^b$ is produced by a single process, usually a forecast, errors in $x_1{}^b$ and $x_2{}^b$ are correlated:

$$\langle(x_1^b - x_1^t)(x_2^b - x_2^t)\rangle = \mu V^b \tag{16}$$

where μ is the correlation coefficient.

The corresponding joint pdf can be modelled by a multi-dimensional Gaussian:

$$\begin{aligned} p(x_1 \cap x_2) &= p(x) = N(x|x^b, \boldsymbol{B}) \\ &= (2\pi^2|\boldsymbol{B}|)^{-1/2}\exp\left(\frac{1}{2}(x - x^b)^T \boldsymbol{B}^{-1}(x - x^b)\right) \end{aligned} \tag{17}$$

where $\boldsymbol{B}$ is the covariance matrix:

$$\boldsymbol{B} = V_b\begin{pmatrix} 1 & \mu \\ \mu & 1 \end{pmatrix}. \tag{18}$$

$p(\boldsymbol{x})$ is assumed to be the combination of all our prior knowledge about the model state. A schematic contour plot, for our simple two-parameter model, is shown by the long dashed ellipses in Fig. 5.3.

We now consider the new information provided by the observation. A perfect instrument would measure a "true" value y^t. Real instruments are not perfect; they have *instrumental error*, which we model by a Gaussian. (In our example this is one-dimensional, but we use a more general notation which can be applied for more than one observation.)

$$\begin{aligned} p(y^o \cap y^t) &= N(y^o|y^t, \boldsymbol{E}) \\ &= (2\pi^2|\boldsymbol{E}|)^{-1/2}\exp\left(\frac{1}{2}(y^o - y^t)^T \boldsymbol{E}^{-1}(y^o - y^t)\right) \end{aligned} \tag{19}$$

We cannot use this directly in Bayes' theorem because it is a function of y^t, rather than $\boldsymbol{x}$. We need to specify also the probability of a perfect instrument observing y^t, given that the true model state is $\boldsymbol{x}$. Because we have to interpolate from $\boldsymbol{x}$ to y, we cannot know y^t exactly; the *representativeness error* can be thought of as the error in the interpolation operator $\boldsymbol{H}$ in (14).

$$\begin{aligned} p(y^t|x) &= N(y^t|\boldsymbol{H}(x), \boldsymbol{F}) \\ &= (2\pi|\boldsymbol{F}|)^{-1/2}\exp\left(\frac{1}{2}y^t - \boldsymbol{H}(x^t)^T \boldsymbol{F}^{-1} y - \boldsymbol{H}(x^t)\right) \end{aligned} \tag{20}$$

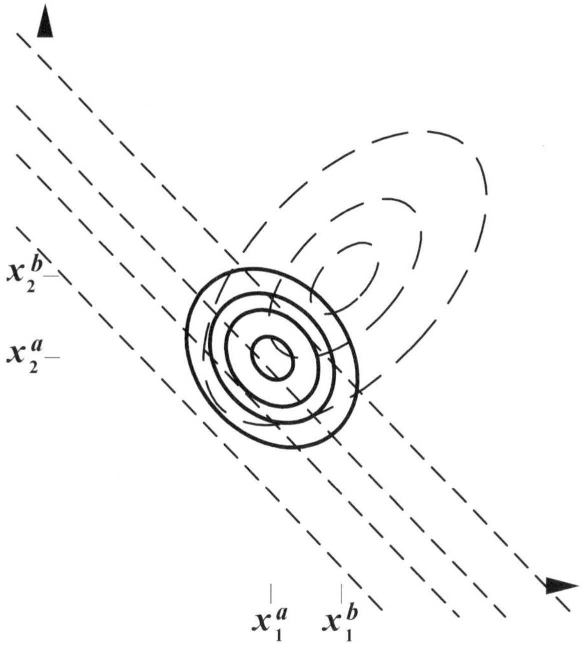

Fig. 5.3 Contour plots for the simple example of section 5.5.2: long dashes: the prior pdf $p(x_1,x_2)$, a two-dimensional Gaussian with centre (x_1^b,x_2^b), short dashes: the likelihood function $p(y^o \mid x_1,x_2)$, a Gaussian ridge about the line $1/2(x_1+x_2)=y^o$, thick lines: the posterior pdf $p(x_1,x_2 \mid y^o)$, a two-dimensional Gaussian with centre (x_1^a,x_2^a).

If we assume that instrumental and representativeness errors are independent, we can convolve them to get a combined *observational error*:

$$\begin{aligned} p(y^o|x) &= \int p(y^o|y^t)p(y^t|x)dy^t \\ &= N(y^o|\boldsymbol{H}(x), \boldsymbol{E}+\boldsymbol{F}) \qquad (21) \\ &= (2\pi|\boldsymbol{E}+\boldsymbol{F}|)^{-1/2}\exp\left(\frac{1}{2}(y^o-\boldsymbol{H}(x))^T(\boldsymbol{E}+\boldsymbol{F})^{-1}(y^o-\boldsymbol{H}(x))\right) \end{aligned}$$

The sum of the instrumental and representativeness error covariances, $\boldsymbol{E}+\boldsymbol{F}$, is often written as a single observational error covariance $\boldsymbol{R}$. The above derivation shows that it is composed of two parts, $\boldsymbol{E}$ which is a function of the instrument characteristics, and $\boldsymbol{F}$ which is a function of the model resolution. For instance, for a wind observation from a radiosonde, the errors in tracking the balloon might lead to an instrumental error of about 1ms^{-1}. The error of representativeness due to trying to predict such a wind from a model with horizontal grid-length 200 km would be about 3ms^{-1}, and a grid-length on 20 km would reduce the error of representativeness to about 1ms^{-1}.

$p(y^o | x_1,x_2)$ is plotted as a function of (x_1,x_2) in the short dashed lines in Fig. 5.3. It is called the likelihood function[1]. Note that the ridge extends to infinity; all values on the ridge line are, as far as the observed information is concerned, equally likely. To get a unique "best estimate", we need to combine this with the prior information.

We can substitute (17) and (21) into the Bayesian analysis equation:

$$\begin{aligned} p(x|y^o) &= \frac{p(y^o|x)p(x)}{p(y^o)} \\ &\propto N(y^o|\boldsymbol{H}(x), \boldsymbol{E}+\boldsymbol{F})N(x|x^b, \boldsymbol{B}) \end{aligned} \tag{22}$$

It is a property of Gaussians that, if H is linearisable:

$$\begin{aligned} & N(y^o|\boldsymbol{H}(x), \boldsymbol{E}+\boldsymbol{F})N(x|x^b, \boldsymbol{B}) \\ = \; & N(y^o|\boldsymbol{H}(x), \boldsymbol{E}+\boldsymbol{F}+\boldsymbol{HBH}^T)N(x|x^a, \boldsymbol{A}) \end{aligned} \tag{23}$$

where $\boldsymbol{x}^a$ and $\mathbf{A}$ are defined by:

$$\begin{aligned} \boldsymbol{A} &= \boldsymbol{BH}^T(\boldsymbol{HBH}^T+\boldsymbol{E}+\boldsymbol{F})^{-1}\boldsymbol{HB} \\ x^a &= x^b + \boldsymbol{BH}^T(\boldsymbol{HBH}^T+\boldsymbol{E}+\boldsymbol{F})^{-1}(y^o + -\boldsymbol{H}(x^b)) \end{aligned} \tag{24}$$

Cancelling the constant of proportionality gives:

$$p(x|y^o) = N(x|x^a, \boldsymbol{A}) \tag{25}$$

So the posterior probability, after adding the observational information, is a Gaussian with mean $\boldsymbol{x}^a$ and variance $\mathbf{A}$. As in Fig. 5.1, the posterior pdf in Fig. 5.3 (solid lines) is narrower and taller than the prior. Adding information from the observation increases our confidence, and reduces the error variance from **B** to **A**.

For our simple example the algebra is easily done by hand, giving:

$$x^a = \begin{pmatrix} x_1^a \\ x_2^a \end{pmatrix} = \begin{pmatrix} x_1^b \\ x_2^b \end{pmatrix} + \frac{\left(V^b\left(\frac{1+\mu}{2}\right)\right)^2}{\boldsymbol{E}+\boldsymbol{F}+V^b\left(\frac{1+\mu}{2}\right)}\left[y^o - \frac{x_1^b+x_2^b}{2}\right]\begin{pmatrix} 1 \\ 1 \end{pmatrix} \tag{26}$$

1. It does not integrate to one over x, so it is not a probability.

If this problem is solved using optimal interpolation (OI), the approximation is made that $\boldsymbol{HBH}^{\mathrm{T}}=V^b$, so the $\left(\frac{1+\mu}{2}\right)$ do not appear.

5.5.4 Practical Analysis Methods

We saw in (13), and Fig. 5.2, the effect of taking logarithms of the posterior probability. For multi-dimensional Gaussians, this becomes:

$$-\ln(p(x|y^o)) = \frac{1}{2}(x^b - x)^T\boldsymbol{B}^{-1}(x^b - x)$$
$$+\frac{1}{2}(y^o - \boldsymbol{H}(x))^T(\boldsymbol{E}+\boldsymbol{F})^{-1}(y^o - \boldsymbol{H}(x)) + \text{constant.} \tag{27}$$

So the most probable $\boldsymbol{x}$ minimises

$$J(x) = \frac{1}{2}(x^b - x)^T\boldsymbol{B}^{-1}(x^b - x)$$
$$+\frac{1}{2}(y^o - \boldsymbol{H}(x))^T(\boldsymbol{E}+\boldsymbol{F})^{-1}(y^o - \boldsymbol{H}(x)) \tag{28}$$

For linearisable $\boldsymbol{H}$ we can solve this explicitly, giving (24). This is known as the "OI" (Optimal Interpolation) analysis equation; but since it is only optimal if the assumptions about the pdfs are correct, a preferred name is Statistical Interpolation. The same equations appears in the analysis step of the Extended Kalman Filter (Ghil and Malanotte-Rizzoli, 1991).

The OI equation (24) requires the solution of a linear system of order the number of observational data. Meteorological observations are so numerous that practical simplifications are required; for instance by partitioning into analysis volumes, and modelling $\boldsymbol{HBH}^{\mathrm{T}}$ by a correlation function (Lorenc 1981). Alternatively, iterative methods can be used. One such, the "Successive Correction Method", was developed pragmatically, before OI. It can be derived from (24) if we define:

$$W = \boldsymbol{BH}^T(\boldsymbol{E}+\boldsymbol{F})^{-1}$$

$$\boldsymbol{Q} \approx (\boldsymbol{I}+\boldsymbol{WH})^{-1} \tag{29}$$

$$x^{[u+1]} = x^{[u]} + \boldsymbol{Q}(\boldsymbol{W}(y^o - \boldsymbol{H}(x^{[u]})) + x^b - x^{[u]})$$

Lorenc (1992) showed that, particularly for cases where the correlation function used to model $\boldsymbol{B}$ has a simple form, an easily calculated approximation to $\boldsymbol{Q}$ is sufficient to get iteration (29) to converge to the "OI" result.

Variational Analysis (Var) methods search directly for the minimum of (28), using a descent algorithm which needs the gradient of J:

$$\left(\frac{\partial J}{\partial x}\right)^T = -\boldsymbol{B}^{-1}(x^b - x) - \boldsymbol{H}^T(\boldsymbol{E}+\boldsymbol{F})^{-1}(y^o - \boldsymbol{H}(x)) \quad (30)$$

If (30) is calculated accurately, they can use more sophisticated and efficient descent algorithms than the pre-conditioned steepest descent which is used in (29). The prediction (H) of observed values y from a model state x, can be generalised from a simple interpolation, as in (14), to a more general prediction using a forecast model and observation operators, allowing Var to be applied to observations distributed in time, and indirect observations (e.g. satellite radiances).

5.6 Quality Control

5.6.1 Why Quality Control?

The quality control we do in data-assimilation has two reasons:

1) We have physical reasons for believing certain events may occur which affect the observed value. We wish to detect these events.
2) The distribution of errors associated with a datum is such that there is a non-negligible probability of errors that would be unacceptably large *for the use we are making of the datum.*

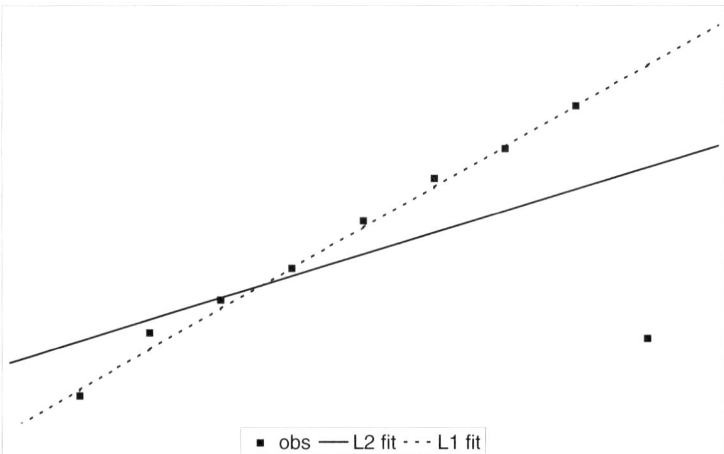

Fig. 5.4 Best fit straight lines to data including a gross error: solid line using a quadratic (L2) norm, dotted line using a mean absolute (L1) norm. (Based on Tarantola 1987).

Note that 2 depends on our use of the observation. If we are using an analysis method based on a quadratic penalty function, it is linear in the observed values. A single large error can then be disastrous (Fig. 5.4 solid line). However if instead we minimise the mean absolute deviation (We shall see that this is the correct norm if the observational error probability distribution function is proportional to an exponential of the absolute deviation), the bad datum is ignored (Fig. 5.4 dotted line). Analysis methods designed to ignore such outliers are also considered to be quality control methods.

5.6.2 Discrete Bayes Theorem Applied to Gross Observational Errors

I have two dice. One is weighted towards throwing sixes. I have performed some experiments with them, and have the prior statistics that:

for the weighted (W) die,	$P(6 \mid W)$	$= 8/60$
for the good (G) die,	$P(6 \mid G)$	$= 10/60$
I choose one at random:	$P(W) = P(G)$	$=1/2$

I throw this die, and it shows a six. Now:

$$\begin{aligned} P(6) &= P(6 \mid W)\, P(W) + P(6 \mid G)\, P(G) \\ &= 58/60 \;\; 1/2 \; + 10/60 \;\; 1/2 \\ &= 34/60 \end{aligned}$$

We can now apply Bayes' Theorem:

$$\begin{aligned} P(G \mid 6) &= P(6 \mid G)\, P(G) \,/\, P(6) \\ &= 10/60 \;\; 1/2 \;\; / \, 34/60 \;=\; 5/34 \\ P(W \mid 6) &= P(6 \mid W)\, P(W) \,/\, P(6) \\ &= 58/60 \;\; 1/2 \;\; / \, 34/60 \;= 29/34 \end{aligned}$$

The information that I have thrown a six has added to my knowledge, so that the posterior probability that the chosen die is weighted has increased. If I were to throw again, and get another six, the probability would increase again.

5.6.3 Non-Gaussian Model for Observational Errors

The simplest model that allows for the observed fact that observational errors are not in practice Gaussian, is to assume that a small fraction of the observations are corrupted, and hence worthless. The others have Gaussian errors. For each observation we have:

$$p(y^o|x) \;=\; p(y^o|G \cap x)P(G)p(y^o|\overline{G} \cap x)P(\overline{G}) \tag{31}$$

G is the event "there is a gross error" [2], and $\overline{G}$ means not G.

$$\begin{aligned} p(y^o|\overline{G} \cap x) &= N(y^o|\boldsymbol{H}(x), \boldsymbol{E} + \boldsymbol{F}) \\ p(y^o|G \cap x) &= \begin{cases} k & \text{over the range of plausible values} \\ 0 & \text{elsewhere} \end{cases} \end{aligned} \tag{32}$$

2. We assume that G is independent of $\boldsymbol{x}$, so that $P(G|\boldsymbol{x}) = P(G)$.

5.6.4 Analysis, allowing for gross errors

The normal way of dealing with non-Gaussian, long-tailed distributions such as these is to detect and reject probable gross errors in a quality control step prior to the analysis, and then to assume that the errors of the remaining observations come from a Gaussian distribution. The quality control decision can either use pragmatic criteria, or apply the discrete Bayes' theorem to the event G "there is a gross error" (Lorenc and Hammon, 1988):

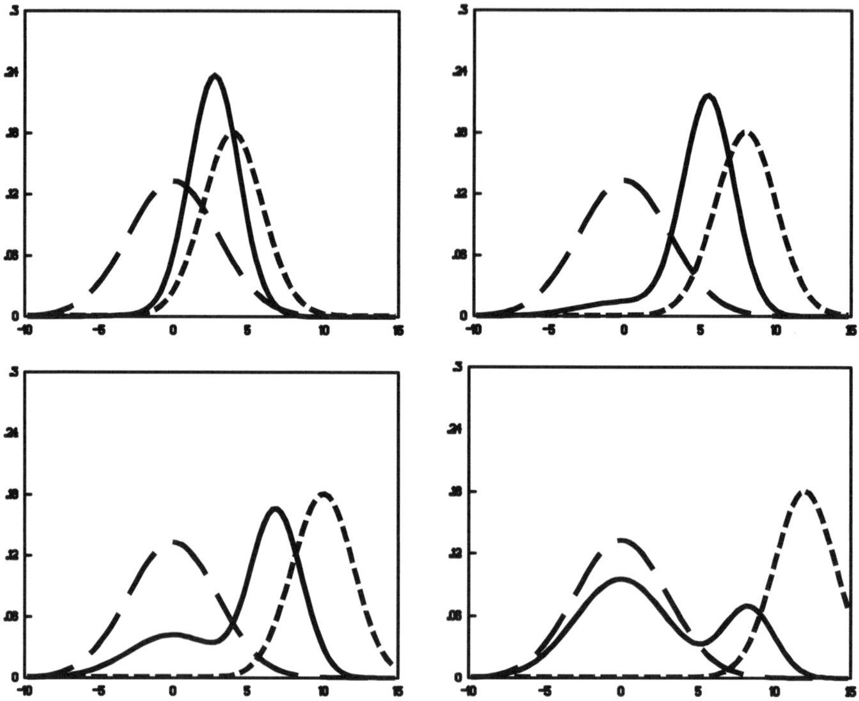

Fig. 5.5 As Fig. 5.1 for an observation with a 5% chance of gross error.

$$
\begin{aligned}
P(G|y^o) &= \frac{P(y^o|G)P(G)}{P(y^o)} \\
&= \frac{kP(G)}{kP(G) + N(y^o|\boldsymbol{H}(x^b), \boldsymbol{R} + \boldsymbol{HBH}^T)P(\overline{G})}
\end{aligned} \tag{33}
$$

We can derive the formula for the posterior pdf $p(x|y^o)$ either directly from the continuous Bayes theorem [3], or else from:

$$\begin{aligned} p(x|y^o) &= p(x|y^o \cap G)P(G|y^o) + p(x|y^o \cap \overline{G})P(\overline{G}|y^o) \\ &= N(x|x^b, B)P(G|y^o) + N(x|x^a, A)P(\overline{G}|y^o) \end{aligned} \tag{34}$$

The posterior pdf is the weighted sum of two Gaussians, corresponding to accepting or rejecting the observation. The weights given to each are the posterior probabilities of G and $\overline{G}$. When the peaks are distinct, these correspond to the areas under each. The results of allowing for gross errors in this way can be quite dramatic, even if $P(G)$ is small. Fig. 5.5 shows the equivalent of Fig. 5.1, with errors appropriate for pressure observations from ships, which have about 5% gross errors. When the observation and the background agree, there is little difference from Fig. 5.1. But when they disagree, the posterior distribution becomes bi-modal. It is worth noting that, in contrast to the statement at the end of section 5.5.3, the posterior variance can be greater than the prior variance - a strange result for those used to dealing with Gaussian distributions.

Fig. 5.6 shows the same examples in the log(probability) form of Fig. 5.2. The observational penalty is not quadratic; it has plateaux away from the observed value. Adding this to a quadratic background penalty can give multiple minima.

5.6.5 What is the Best Analysis?

In section 5.6.4 we saw how the Bayesian approach can, in theory, give us the posterior pdf, describing how likely it is that each atmospheric state is correct. However to start a forecast we need a single best analysis[4]. In our simple example of Fig. 5.5, is the best analysis in the tallest peak, or in the peak with the largest area? The theoretical approach to this is outlined here. But in practice this is impracticable; more pragmatic judgements of what is best are implicit in the practical schemes described in section 5.7.

To approach the best analysis objectively, we have to define how much it costs us to be wrong, or alternatively how much we benefit from being nearly right. If we have a quadratic cost function, then the best analysis is the mean of the posterior pdf. If we have a spike (delta function) benefit function, the best analysis is at the maximum of the posterior pdf. Probably the best simple model is between the two: a Gaussian shaped benefit function such that analyses close to the truth are useful, while those a long way from correct are equally worthless. A Gaussian shaped benefit function, with width specified by pseudo-covariance $\boldsymbol{C}$, has the advantage of facilitating algebraic calculation of the benefit; convolving it with the posterior pdf from (34) we get:

$$\boldsymbol{benefit}(x) = N(x|x^b, \boldsymbol{B}+\boldsymbol{C})P(G|y^o) + N(x|x^a, \boldsymbol{A}+\boldsymbol{C})P(\overline{G}|y^o) \tag{35}$$

---

3. By substituting (31) and (32) into (5), and assuming that $p(\boldsymbol{x})$ is negligible outside of the range of plausible values.
4. The choice of the best ensemble of analyses, to span the range of possible forecasts, is a major research area beyond the scope of this paper.

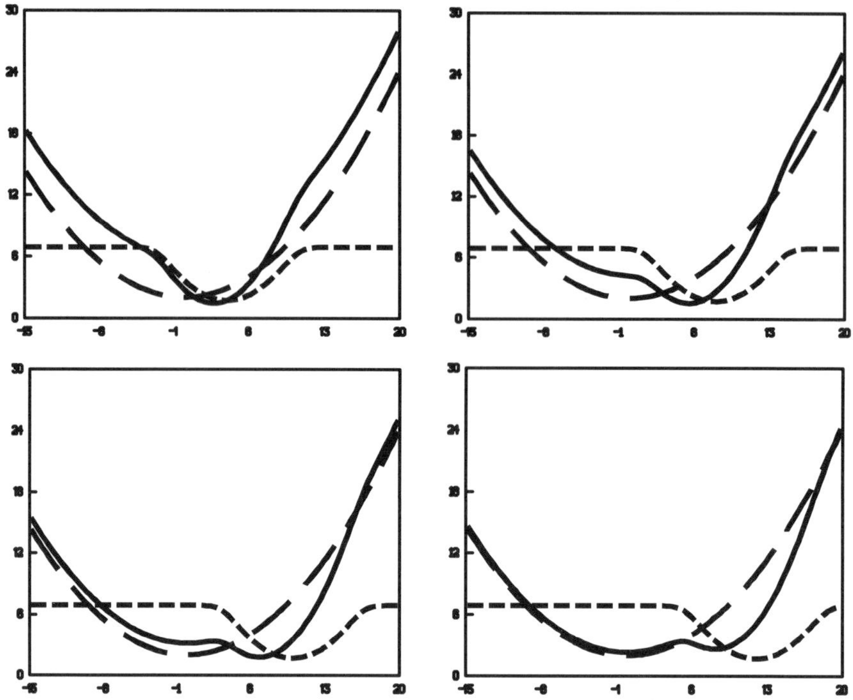

Fig. 5.6 As Fig. 5.4 for -log(probabilities).

Fig. 5.7 shows the expected benefit calculated using (35), for different values of C. C=0 corresponds to a delta function benefit; the curve is identical to the posterior pdf (similar to those shown in Fig. 5.5; this example is for a more accurate but equally unreliable observation). The analysis value which gives the greatest expected benefit is shown by ×, and corresponds to the maximum of the posterior pdf. In practice, an analysis which is as accurate as the background $\boldsymbol{x}^b$ is still of some use. So it is more plausible that the width of the benefit function should be similar to that of the background pdf (which has variance 9 in our example). The expected benefit curve for this case has its maximum at +. It has chosen the peak from the posterior pdf which has the largest area (0.61 compared to 0.39). Finally, if we expand the width of the benefit function so that it is large compared to the separation between the peaks, then we get the curve with maximum at □. For very large C the Gaussian benefit function becomes a quadratic, and the maximum is always at the mean value of the posterior pdf.

5.7 Practical Methods of Quality Control

5.7.1 Individual Quality Control

Lorenc and Hammon (1988) extended (33) to two observations with independent gross errors, and background values. This can be treated exactly by first calculating $P(G_i \mid y^o)$ for each, using (33), and then modifying them by a buddy check factor:

$$
\begin{aligned}
P(G_1|y_1^o \cap y_2^o) &= P(G_1|y_1^o)\frac{P(y_1^o)P(y_2^o)}{P(y_1^o \cap y_2^o)} \\
P(G_2|y_1^o \cap y_2^o) &= P(G_2|y_2^o)\frac{P(y_2^o)P(y_2^o)}{P(y_1^o \cap y_2^o)}
\end{aligned}
\qquad (36)
$$

where

$$
\frac{P(y_1^o)P(y_2^o)}{P(y_1^o \cap y_2^o)} = \left[1 - P(\bar{G}_1|y_1^o)P(\bar{G}_2|y_2^o)\left\{1 - \frac{P(y_1^o \cap y_2^o|\bar{G}_1 \cup \bar{G}_2)}{P(y_1^o|\bar{G}_1)P(y_2^o|\bar{G}_2)}\right\}\right]^{-1} \qquad (37)
$$

The algebra, and computation, to extend this exact calculation to n observation goes as 2^n. However it has been found in practice that sequentially applying the two-observation buddy check is a reasonable approximation. (See Lorenc and Hammon (1988) appendix C for more details of the pairwise buddy check). Finally, each observation i is used if $P(\bar{G}_i|y^o) > P(G_i|y^o)$. This is a generalisation of the decision made by the **C**=9 curve in Fig. 5.7.

The main weakness of this approach as implemented is its sequential nature. Nearby observations are not combined before checking an observation, but rather they are used one-by-one. So the way they support, or contradict, each other is only approximately allowed for. Note that for each observation the posterior pdf is split into two different peaks[5], leading to independent decisions for each observation which may not be consistent, as we shall see in 5.7.4 .

5.7.2 Simultaneous Quality Control

Lorenc (1981) introduced the Optimal Interpolation (OI) analysis method used operationally at ECMWF (until replaced by 3DVAR). This performs an explicit solution of a quadratic variational problem. The solution is calculated in boxes for as many observations as we can afford to handle at once.

A key feature of the ECMWF system is the use of the same methodology for quality control. Lorenc (1981) shows how, once the inverse of the OI matrix **M** (=$\mathbf{HBH}^{\mathrm{T}}$+**R** in our current notation) has been calculated, then it is possible with relatively few operations to solve the system of equations involving a smaller

5. Actually, they may not be distinct peaks.

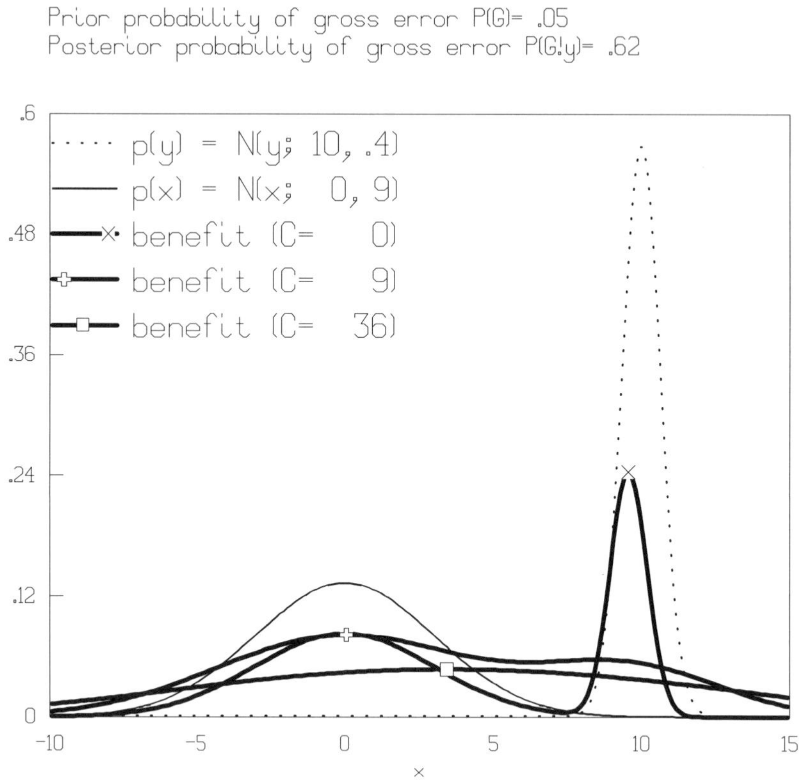

Fig. 5.7 Expected benefit as a function of analysed value, from a case similar to those in Fig. 5.5. Curves are plotted for three different benefit functions, with widths C=0 (maximum at ×), C=9 (maximum at +), and C=36 (maximum at □). Shown for reference are the background pdf (with x^b=0, B=9), and the observational pdf (with y^o=10, R=0.5). The prior probability of gross error was assumed to be P(G)=0.05, and the posterior probability was calculated to be P(G | y^o)=0.61.

matrix omitting one (or a few) observations. He used this to check each observation in turn against a value analysed using all the other observations. An observation fails if:

$$(y^o - y^a)^2 > T^2(V_o + V_a) \tag{38}$$

where y^a, with error variance V_a, is the analysis obtained using the OI equations, at the position of the observation being checked, omitting the observation being checked and other rejected observations.

In the Lorenc (1981) paper the tolerance (T) was set in a somewhat empirical manner to 4.0. When, as the scheme developed, we tried to account for the better quality of weather ship observations by reducing their observational error V_O, we found that this resulted in more being rejected:- not what we wanted. (It was this behaviour, and the subjective tolerance in what was otherwise an objective analysis, that induced me to study the Bayesian approach.) It is shown in Lorenc and Hammon that, to match the criterion , $P(\bar{G}_i|y^o) > P(G_i|y^o)$ the tolerance T should be given by:

$$T^2 = 2\ln\left[\frac{P(\bar{G})}{P(G)}\right] + \ln\left[\frac{k^2}{2\pi(V_o + V_a)}\right] \tag{39}$$

where k is the probability density of observations with gross error, as defined in (32). T is shown in Fig. 5.8.

In applying this method, observations have to be either included in, or excluded from, the analysis. While an observation is checked, the decisions on other observations are frozen. The ECMWF scheme follows a pragmatic approach of rejecting the worst, then rechecking the others, iteratively until no more fail.

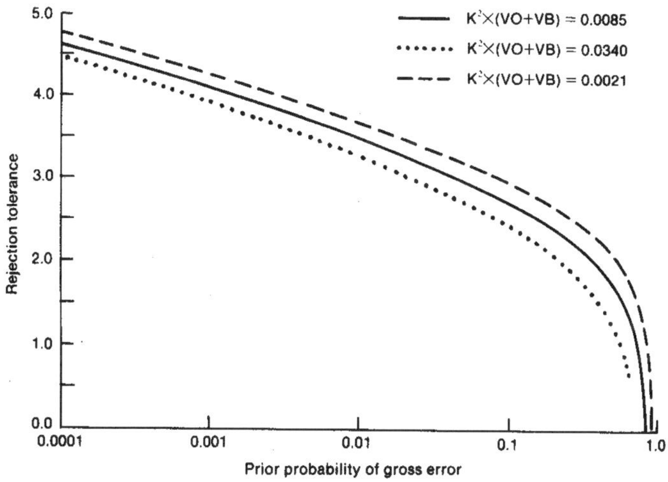

Fig. 5.8 Rejection tolerance T, plotted against prior probability of gross error (from Lorenc and Hammon 1988).

Ingleby and Lorenc (1993) present equations for extending the Bayesian approach of Lorenc and Hammon (1988). From the n gross error events G_i, they define 2^n new combined events C_α each corresponding to a particular set of rejections:

$$\begin{aligned} C_0 &= G_n \cap G_{n-1} \ldots \cap G_2 \cap G_1 \\ C_1 &= G_n \cap G_{n-1} \ldots \cap G_2 \cap \bar{G}_1 \\ C_2 &= G_n \cap G_{n-1} \ldots \cap \bar{G}_2 \cap G_1 \\ &\vdots \\ C_{2^n-1} &= \bar{G}_n \cap \bar{G}_{n1} \ldots \cap \bar{G}_2 \cap \bar{G}_1 \end{aligned} \tag{40}$$

Bayes theorem can be applied to each of these combined events:

$$P(C_\alpha | y^o) = \frac{P(y^o | C_\alpha) P(C_\alpha)}{\sum_{\alpha'=0}^{2^n-1} P(y^o | C_{\alpha'}) P(C_{\alpha'})} \tag{41}$$

Note that the denominator is the same in all the expressions; if we only want to find the most likely C_α it need not be evaluated. Evaluation of one $P(y^o | C_\alpha)$ involves evaluating only a single multi-variate Gaussian. In fact the ECMWF OI method (with Bayesian tolerance T) is deciding which is more likely out of two C_α which differ just by the observation being tested. Because it is judging between sets of quality control decisions, we call this approach *Simultaneous Quality Control.*

The C_α can be regarded as being the vertices of an n-dimensional hyper-cube. The method starts from a first-guess set of rejections, and tests each observation in turn. This is equivalent to searching for the most probable of the adjacent vertices. It then iterates, re-testing some vertices adjacent to the new C_α. This strategy is like the SIMPLEX algorithm of linear programming, but applied to a non-linear problem.

The main weakness of this method as implemented is that its cost prohibits evaluation of more than a few of the possible combinations. The SIMPLEX-like algorithm is not guaranteed to find the absolute maximum, but depends on a good initial estimate as to which observations are correct. This is illustrated in section 5.7.4.

5.7.3 Variational Analysis with non-Gaussian Errors

It is possible to use our model of observational errors directly in a variational algorithm. Dharssi et al. (1992) did this for simulated windlidar observations. Instead of the quadratic $\frac{1}{2}(y^o - H(x))^T (\mathbf{E}+\mathbf{F})^{-1} (y^o - H(x))$, the observational penalty becomes (for diagonal $\mathbf{E}+\mathbf{F}$):

$$J_o = \sum_i \ln[N(y_i^o | \boldsymbol{H}(x), V_{ri}) P(\bar{G}_i) + k_i P(G_i)] \tag{42}$$

where V_{ri} ($=\mathbf{E}_{ii}+\mathbf{F}_{ii}$) is the observational error of i if it has not a gross error.

Differentiating this gives:

$$\frac{\partial J_o}{\partial y_i} = \frac{(y_i^o y_i)}{V_{ri}}\left\{\frac{N(y_i^o|y_i, V_{ri})P(\bar{G}_i)}{N(y_i^o|y_i, V_{ri})P(\bar{G}_i) + k_i P(G_i)}\right\} \quad (43)$$

where y_i is the element of $H(x)$ corresponding to $\boldsymbol{x}$ interpolated to the ith observation position. The first term is just what we get if the observation error is Gaussian, as in variational methods with a quadratic penalty function. The term in braces is equal to the probability that observation i has not a gross error, given that $\boldsymbol{x}$ is exactly correct. So for each iteration u of the descent algorithm, all one has to do to allow for gross errors is to replace the observational error for each observation in the formulae for the standard gradient of a quadratic penalty by:

$$E_{o[u]} = \frac{V_r}{P(\bar{G}|y^o \cap \boldsymbol{H}(x_{[u]}))} \quad (44)$$

i.e. the observational error variance should be inflated by one over the probability (given the current best estimate $x_{[u]}$) that the observation has not a gross error.

Note that this only gives the correct gradient of J_o; it does not give the correct penalty (for which we need (42)) nor the correct second derivative. In general, analysis error estimates based on the second derivative, valid for Gaussian distributions (e.g. in (23)) will be over-optimistic for long-tailed distributions.

We saw in Fig. 5.5 that if errors are non-Gaussian, the penalty function is non-quadratic and can have multiple minima. So the end point of a descent algorithm iteration will depend on the first-guess. In a set of variational assimilation experiments with a one-dimensional shallow water model and its adjoint, I found that (for the example studied) the first-guess had to be very good to get convergence to the best solution (Lorenc 1988, Fig. 13).

Dharssi et al. (1992) studied various approaches for overcoming this problem for a simple example with two observations, so that the penalty function can be plotted as contours. Ordinary descent algorithms did not always find the lowest minimum (Fig. 5.9). Better results could be obtained by artificially increasing the assumed observational error for early iterations, slowly reducing it to its true value. Alternatively, one can artificially decrease the prior $P(G)$ to zero for early iterations. Neither approach always worked. In fact we saw in 5.6.5 that the extremum of the posterior pdf is not necessarily the best analysis, particularly for observations with low observational error but some gross errors. Increasing the observational error makes the modified posterior pdf more like the expected benefit curve for C=9 shown in Fig. 5.7. In simulations using rather dense observations with large probabilities of gross error (up to 50%), the method with increased observational error worked satisfactorily.

The main weakness of this method is its use of a variational descent algorithm in situations where a discrete decision between distinct possibilities is needed. It is dependent on a good starting point initial guess. Another theoretical weakness is its

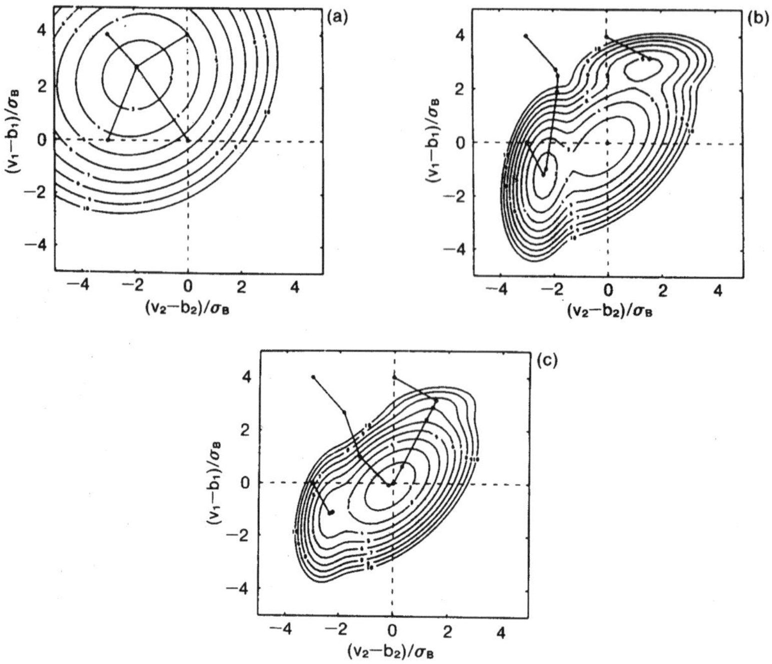

Fig. 5.9 Fig.5.7 from Dharssi et al. (1992).

search for the maximum of the posterior pdf, rather than the estimated benefit. We saw in 5.6.5 how an accurate but unreliable observation can give a tall but narrow peak, which is not actually the best analysis. This effect can be partially alleviated by artificially increasing the observational error.

5.7.4 Comparison

Ingleby and Lorenc (1993) compared Individual Quality Control (IQC), Simultaneous Quality Control (SQC), and non-Gaussian Variational Analysis (VAN) for some simple examples. One is shown in Fig. 5.10. There are two observations and thus four combinations C_α, each corresponding to a dotted Gaussian curve in Fig. 5.10. The table below shows their posterior probabilities.

	G_2	$\overline{G}_2$	$G_2 \cup \overline{G}_2$
G_1	.393	.191	.584
$\overline{G}_1$	.003	.413	.416
$G_1 \cup \overline{G}_1$	.396	.604	1.0

The most likely combination is for them both to be correct: the table shows $P(\overline{G}_1\cap\overline{G}_2)$=.413. This is the SQC result. Note that the simplex-like search algorithm would not work in this case: if we start at $G_1\cap G_2$, both $\overline{G}_1\cap G_2$ and $G_1\cap\overline{G}_2$ are less likely, so we do not get to $\overline{G}_1\cap\overline{G}_2$.

To get the probabilities for IQC we have to sum rows and columns: the table show $P(\overline{G}_1)$=.416, $P(\overline{G}_2)$=.604. So observation 1 probably has a gross error, and observation 2 is probably correct. Note however that if we make these decisions individually, and then draw the analysis assuming $G_1\cap\overline{G}_2$, then we are choosing a rather unlikely combination.

The variational analysis method chooses the highest point on the solid curve in Fig. 5.10. Note that a descent algorithm starting from the background would not have found this. If we convolve in a benefit function, as discussed in section 5.6.5, the best analysis will be near the mean of the larger peak in the posterior pdf, i.e. between the IQC and VAN/VQC/SQC results.

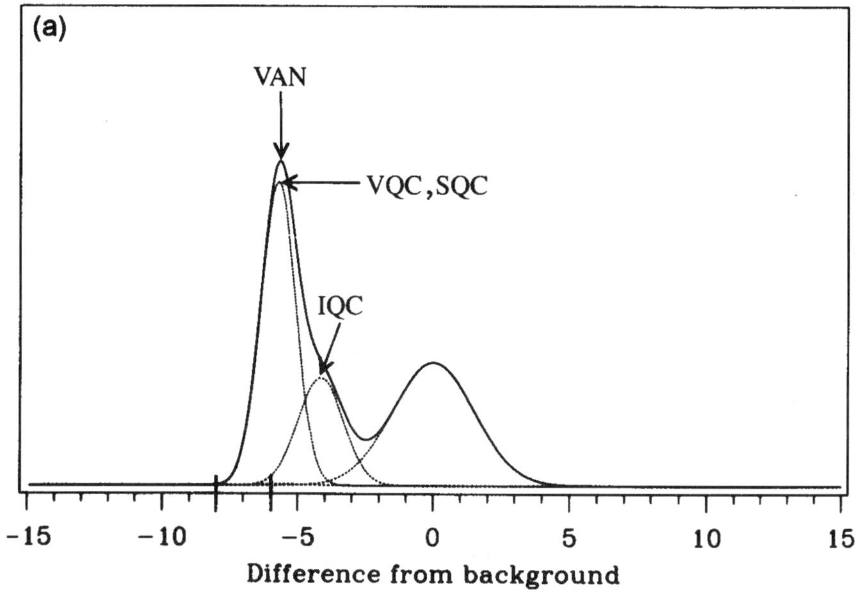

Fig. 5.10 Posterior pdf for collocated observations differing by -8 mb and -6 mb from the background. The prior $P(G)$=.05 for each, the error variance of good observations is V_e=1.0 mb^2, the background error variance is V_b=(1.5 mb)^2, and the probability density of observations with gross errors is k=.043 mb^{-1}. (from Ingleby and Lorenc 1993).

5.8 Monitoring

To the manager of a manufacturing company, Quality Control has a different meaning to that we have implied so far; he wants to know about and prevent errors. In NWP we call the equivalent process "monitoring". The purpose is to collect statistics on the performance of observing and processing systems, to detect systems that are not performing as expected, and to feed this information back so the deficiency is corrected at source. To do this we need:

- a comprehensive database of basic and processed observed values, independent estimates of the same quantities, and parameters affecting the processing
- software for categorising, sorting, and analysing the database
- effort to try categorisations and look for "unexpected" behaviour
- communications, willpower and persistence, to get errors from stages out of your direct control rectified.

Design of the monitoring system is as important as design of the data-assimilation scheme; it should not be added on as an afterthought. Monitoring by NWP centres of the operational World Weather Watch observations is probably the major cause of the significant increase in observational quality which has been seen in radiosondes (S. Uppala, personal communication), ships (C. Heasman, personal communication), and cloud track winds.

Another important product of monitoring is a good description of the observational error characteristics. If we are using the gross-error model, we need to know the prior probability of error, and the error distributions of gross-error and of "good" observations. Without these, the quality control is not objective[6]. Lorenc and Hammon (1988) showed how the statistics of observations processed by their quality-control scheme could be used (with some added "judgement") to "bootstrap" the assumed prior distributions.

For some observation types, more complicated error models are called for. For instance there are many different ways that a radiosonde temperature and geopotential report can be corrupted. Gandin et al. (1993) have devised a "Comprehensive Quality Control" scheme which looks for sixteen. Because of the redundancy of information in a radiosonde message, it is often possible to correct errors.

Many observations have bias errors, which monitoring statistics are useful in detecting and correcting. For instance many ships and buoys have mean surface pressure errors which persist until the instrument is recalibrated; the Met Office routinely updates a list of corrections for them. The "observational" errors in satellite radiance soundings are biased by errors in the radiative transfer calculations used in H; all successful methods for using the radiances use empirical bias corrections obtained from a monitoring process.

---

6. By "objective" I mean more than the automatic application of *ad hoc* rules, rather that the rules themselves have some statistical foundation.

6 Sequential Data Assimilation for Nonlinear Dynamics: The Ensemble Kalman Filter

GEIR EVENSEN
Nansen Environmental and Remote Sensing Center, Bergen, Norway

6.1 Introduction

Sequential data assimilation methods have proven useful for many applications in meteorology and oceanography. For example are most operational weather prediction systems applying a sequential data assimilation technique where observations are "assimilated" into the model whenever they are available.

For linear dynamics the optimal sequential technique is the Kalman filter. In the Kalman filter an additional equation for the second-order statistical moment is integrated forward in time to predict error statistics for the model forecast. The error statistics are then used to calculate a variance-minimizing estimate whenever measurements are available.

For nonlinear dynamics the extended Kalman filter may be applied, in which an approximate linearized equation is used for the prediction of error statistics. The implementation of the extended Kalman filter for data assimilation in a multilayer quasi-geostrophic (QG) model has previously been discussed by Evensen (1992). The main result from this work is the finding of an apparent closure problem in the error covariance evolution equation. The extended Kalman filter applies a closure scheme where third- and higher-order moments in the error covariance evolution equation are discarded. This simple closure technique results in a unbounded error variance growth caused by the linearization performed when higher-order moments are neglected. Thus, it has been shown that the error covariance equation is based on a too simplified closure approximation and this may lead to a nonphysical error variance evolution (see e.g., Evensen (1992), Miller et al. (1994), Gauthier et al. (1993), and Bouttier (1994)).

The Ensemble Kalman Filter (EnKF) was introduced by Evensen (1994b) as an alternative to the traditional Extended Kalman Filter (EKF). It was shown that if the dynamical model is written as a stochastic differential equation, one can derive the Fokker–Planck equation for the time evolution of the probability density function which contains all the information about the prediction error statistics. The EnKF is a sequential data assimilation method where the error statistics are predicted by solving the Fokker–Planck equation using Monte–Carlo or ensemble integrations. By integrating an ensemble of model states forward in time it is possible to calculate statistical moments like mean and error covariances whenever such information is required. Thus, all the statistical information about the predicted model state which are required at analysis times are contained in the ensemble.

In Evensen (1994b) an analysis scheme was proposed where the traditional update equation used in the Kalman Filter (KF) is applied, except that the gain is calculated from the error covariances provided by the ensemble of model states. It was also illustrated that a new ensemble representing the analyzed state could be generated by updating each ensemble member individually using the same analysis equation.

The EnKF is attractive since it avoids many of the problems associated with the traditional extended Kalman filter, e.g., there is no closure problem as is introduced in the extended Kalman filter by neglecting contributions from higher order statistical moments in the error covariance evolution equation. It can also be computed at a much lower numerical cost since only a few hundred model states may be sufficient for reasonable statistical convergence. For practical ensemble sizes, say O (100), the errors will be dominated by statistical noise, not by closure problems or unbounded error variance growth.

The EnKF has been further discussed and applied with success in a twin experiment in Evensen (1994a), in a realistic application for the Agulhas Current using Geosat altimeter data in Evensen and van Leeuwen (1996) and with the strongly nonlinear Lorenz equations in Evensen (1997).

This paper will briefly outline the ensemble Kalman filter (EnKF) and illustrate its properties with a few simple examples. Further, some preliminary results from an implementation with an OGCM will be presented.

6.2 Extended Kalman filter

It is instructive first to give a brief review of the extended Kalman filter algorithm. The derivation of the extended Kalman filter on matrix form can be found in a number of books on control theory (e.g., Gelb (1974), Jazwinski (1970)).

The evolution of the true state vector is described by a dynamical model,

$$\psi_k^t = \boldsymbol{f}(\psi_{k-1}^t) + \boldsymbol{q}_{k-1} \tag{1}$$

where $\boldsymbol{f}$ is a nonlinear model operator and $\boldsymbol{q}$ is a stochastic term representing model errors. A forecast is calculated from

$$\psi_k^f = \boldsymbol{f}(\psi_{k-1}^a) \ . \tag{2}$$

The error statistics is described by the error covariance matrix $\boldsymbol{P}_k = \overline{(\psi - \psi^t)(\psi - \psi^t)^T}$, which evolves according to the equation

$$\boldsymbol{P}_k^f = \boldsymbol{F}_{k-1}\boldsymbol{P}_{k-1}^a\boldsymbol{F}_{k-1}^T + \boldsymbol{Q}_{k-1} \ , \tag{3}$$

where $\boldsymbol{Q}_k = \overline{\boldsymbol{q}_k \boldsymbol{q}_k^T}$ is the model error covariance matrix, and $\boldsymbol{F}_k$ is the Jacobi matrix or tangent linear operator

$$\boldsymbol{F}_k = \left. \frac{\partial f(\psi)}{\partial \psi} \right|_{\psi = \psi_k} . \tag{4}$$

The equations (2) and (3) are used to generate a prediction or model forecast and a prediction of the corresponding error covariance estimate.

Whenever observations are available the analyzed estimate ψ^a is calculated as a linear combination of the vector of measurements $\boldsymbol{d}$ and the predicted model state vector ψ^f. The linear combination is chosen to minimize the variance in the analyzed estimate ψ^a and is given by the equation

$$\psi^a = \psi^f + \boldsymbol{K}(\boldsymbol{d} - \boldsymbol{H}\psi^f) , \tag{5}$$

where the Kalman gain matrix $\boldsymbol{K}$ is defined as

$$\boldsymbol{K} = \boldsymbol{P}^f \boldsymbol{H}^T (\boldsymbol{H}\boldsymbol{P}^f \boldsymbol{H}^T + \boldsymbol{R})^{-1} . \tag{6}$$

It is a function of the model state forecast error covariance matrix $\boldsymbol{P}^f$, the measurement error covariance matrix $\boldsymbol{R}$ and the measurement matrix $\boldsymbol{H}$ that relates the model state to the data.

The measurements are related to the true state by

$$\boldsymbol{d} = \boldsymbol{H}\psi^t + \boldsymbol{\varepsilon} , \tag{7}$$

with $\boldsymbol{\varepsilon}$ the measurement errors. In particular, the true model state is related to the true observations as

$$\boldsymbol{d}^t = \boldsymbol{H}\psi^t . \tag{8}$$

The measurement error covariance matrix is defined as

$$\begin{aligned} \boldsymbol{R} &= \overline{\boldsymbol{\varepsilon}\boldsymbol{\varepsilon}^T} \\ &= \overline{(\boldsymbol{d} - \boldsymbol{H}\psi^t)(\boldsymbol{d} - \boldsymbol{H}\psi^t)^T} \\ &= \overline{(\boldsymbol{d} - \boldsymbol{d}^t)(\boldsymbol{d} - \boldsymbol{d}^t)^T} \end{aligned} . \tag{9}$$

Note that

$$\psi^a - \psi^t = (\boldsymbol{I} - \boldsymbol{K}\boldsymbol{H})(\psi^f - \psi^t) + \boldsymbol{K}(\boldsymbol{d} - \boldsymbol{d}^t) , \tag{10}$$

which is obtained using (5) for replacing ψ^a and then adding $\boldsymbol{K}(\boldsymbol{d}^t - \boldsymbol{H}\psi^t) = 0$ from (8). The error covariance of the analyzed model state vector then becomes

$$
\begin{aligned}
P^a &= \overline{(\psi^a - \psi^t)(\psi^a - \psi^t)^T} \\
&= I - KHP^f I - H^T K^T + KRK^T \\
&= (I - KH)P^f
\end{aligned}
\qquad (11)
$$

The analyzed model state is the best linear unbiased estimate. This means that ψ^a is the linear combination of ψ^f and d that minimizes $TrP^a = \overline{(\psi - \psi^t)^T(\psi - \psi^t)}$, if model errors and observation errors are unbiased and not correlated.

The dynamical model (2) and the error covariance equation (3) together with equations for the analysis scheme, constitutes the so-called extended Kalman filter. Equations (2) and (3) are integrated forward in time and at analysis times the observations are used to update the model solution and its error covariance estimate.

It should be noted that the error covariance equation (3) is only approximate. It results from a linearization of an equation which references infinitely many higher order statistical moments. This statistical linearization lead to serious problems for strongly nonlinear dynamics. In Miller et al. (1994) it was shown that with the Lorenz equations, the error covariance equations under-estimated the predicted error covariance and this resulted in too low gain to keep the model close to the observed state. In Evensen (1992) the extended Kalman filter was used with a nonlinear QG model. Here the problem was that the error covariance equation provided unbounded error variance growth due to the lack of error variance saturation on a climatological level, an effect which requires a higher order closure scheme.

Another major problem of the Kalman filter is related to the storage and computation of the error covariance matrix. If the size of the state vector is n, the size of the error covariance matrix is n^2 and $2n$ model integrations are required to step it forward in time.

These problems with nonlinearities and computational load have lead to the search for alternative methodologies for predicting the error statistics. Currently, there are several approaches where one attempts to evolve the error covariance equation in a reduced state space to save computer time. However, this introduces additional approximations to an already approximate error covariance equation. In the next section the ensemble Kalman filter is introduced as an alternative to the traditional extended Kalman filter.

6.3 Ensemble Kalman Filter

The ensemble Kalman filter as proposed by Evensen (1994b) is now introduced. We will adapt a three stage presentation starting with the representation of error statistics using an ensemble of model states, then an alternative to the traditional error covariance equation is proposed for the prediction of error statistics and finally a consistent analysis scheme is presented.

6.3.1 Representation of error statistics

The error covariance matrices for the predicted and the analyzed estimate, $\boldsymbol{P}^f$ and , $\boldsymbol{P}^a$ are in the Kalman filter defined in terms of the true state as

$$\boldsymbol{P}^f = \overline{(\psi^f - \psi^t)(\psi^f - \psi^t)^T}, \tag{12}$$

$$\boldsymbol{P}^a = \overline{(\psi^a - \psi^t)(\psi^a - \psi^t)^T}, \tag{13}$$

where the overline denotes an expectation value, ψ is the model state vector at a particular time and the superscripts *f*, *a*, and *t* represent forecast, analyzed and true state, respectively. However, since the true state is not known, it is more convenient to consider ensemble covariance matrices around the ensemble mean, $\overline{\psi}$,

$$\boldsymbol{P}^f \cong \boldsymbol{P}^f_e = \overline{(\psi^f - \overline{\psi}^f)(\psi^f - \overline{\psi}^f)^T}, \tag{14}$$

$$\boldsymbol{P}^a \cong \boldsymbol{P}^a_e = \overline{(\psi^a - \overline{\psi}^a)(\psi^a - \overline{\psi}^a)^T}, \tag{15}$$

where now the overline denote an average over the ensemble. Thus, we can use an interpretation where the ensemble mean is the best estimate and the spreading of the ensemble around the mean is a natural definition of the error in the ensemble mean.

Now, since the error covariances as defined in (14) and (15) are defined as ensemble averages, there will clearly exist infinitively many ensembles with an error covariance equal to $\boldsymbol{P}^f$ and $\boldsymbol{P}^a$. Thus, instead of storing a full covariance matrix, we can represent the same error statistics using an appropriate ensemble of model states. Given an error covariance matrix, an ensemble of limited size will always provide an approximation to the error covariance matrix. However, when the size of the ensemble, *N*, increases the errors in the representation will decrease proportional to $1/\sqrt{N}$. Experience shows that we can represent an error covariance matrix with reasonable accuracy using only about 100–500 members in the ensemble.

Suppose now that we have a number *N* model states in the ensemble, each of dimension *n*. Each of these model states can be represented as a single point in an *n*-dimensional state space. All the ensemble members together will constitute a cloud of points in the state space. Such a cloud of points in the state space can be approximately described using a probability density function

$$\phi(\psi) = \frac{dN}{N}, \tag{16}$$

where dN is the number of points in a small unit volume and N is the total number of points. With knowledge about either ϕ or the ensemble representing ϕ we can calculate the statistical moments (mean, covariances etc.) we need whenever they are needed.

The conclusion so far is that the information contained by a full probability density function can equally well be represented by an ensemble of model states.

6.3.2 Prediction of error statistics

We start by writing the model dynamics as a stochastic differential equation

$$d\psi_k = \boldsymbol{f}(\psi_{k-1})dt + d\boldsymbol{q}_{k-1}, \tag{17}$$

where $d\boldsymbol{q} \in \Re^n$ is a vector of random white noise with mean zero. This equation is an Itô stochastic differential equation describing a Markov process. The evolution of the probability density for this equation is given by the Fokker–Planck equation

$$\frac{\partial\phi}{\partial t} + \sum_{i=1}^{n} \frac{\partial f_i \phi}{\partial \psi_i} = \sum_{i,j=1}^{n} \frac{Q_{ij}}{2} = \frac{\partial^2 \phi}{\partial \psi_i \partial \psi_j}, \tag{18}$$

where $\boldsymbol{Q} = \overline{\boldsymbol{q}\boldsymbol{q}^T}$ is the covariance matrix for the model errors. A derivation of this equation is given by Jazwinski (1970, p. 129).

The stochastic forcing, dq_{k-1}, introduces a diffusion term that tends to flatten the probability density function (spreading the ensemble) during the integration; that is, the probability decreases and the errors increase.

If this equation could be solved for the probability density function, it would be possible to calculate statistical moments of ϕ like the mean state and the error covariances at different time levels. However, a direct numerical integration of this equation becomes impossible for ocean circulation models.

By taking moments of the Fokker–Planck equation it is however possible to derive equations for the evolution of statistical moments like the mean and the error covariances. This is exactly the procedure used in the Kalman filter. Note also that for linear dynamics and with a Gaussian initial probability density, the probability density will be completely characterized by its mean and covariance matrix for all times. Thus, one can then use exact equations for the evolution of the mean and the covariance matrix as a simpler alternative than solving the Fokker–Planck equation.

Such moments of the Fokker–Planck equation, including the error covariance equation (3), are easy to derive, and several methods are illustrated by Jazwinski (1970, examples 4.19–4.21).

For a nonlinear model, the mean and covariance matrix will not in general characterize $\phi(\psi,t)$. They do, however, determine the mean path and the dispersion about that path, and it is possible to solve approximate equations for the moments, which is the procedure characterizing the extended Kalman filter.

Another alternative approach for solving the Fokker–Planck equation and predicting the error statistics is to use Monte–Carlo methods. If a probability density function is represented by a large ensemble of model states, it is possible to integrate each member forward in time using the stochastic model (17). Thus, integrating an ensemble of model states becomes equivalent to solving the Fokker–Planck equation using a Monte Carlo method.

The standard approach is to first calculate a best guess initial condition based on information available from data and statistics. An ensemble of initial states is then generated in which the mean equals the best guess initial condition and the variance is specified on the basis of knowledge of the uncertainty in the first-guess initial state. The covariance or smoothness of the ensemble should reflect the true scales of the system.

The effect of external error growth must be included to give reliable estimates for the evolution of errors. In the Kalman filter this can be done rather simply by adding the system error covariance matrix every time step. However, in the Monte–Carlo method, each ensemble member is integrated as a stochastic differential equation and is forced by smooth pseudo-random fields with a specified variance and covariance to simulate the model errors. This will provide a realistic increase in the ensemble variance provided that the estimate of the model error variance is reasonably good.

6.3.3 An analysis scheme

The Kalman Filter (KF) analysis scheme was based on the definitions of $\boldsymbol{P}^f$ and $\boldsymbol{P}^a$ as given by equations (12) and (13). We will now give a new derivation of the analysis scheme where the ensemble covariances are used as defined by (14) and (15). This is convenient since in practical implementations one is doing exactly this, and it will also lead to a consistent formulation of the EnKF.

As will be shown later it is essential that the observations are treated as random variables having a distribution with mean equal to the first-guess observations and covariance equal to $\boldsymbol{R}$. Thus, we start by defining an ensemble of observations

$$\boldsymbol{d}_j = \boldsymbol{d} + \boldsymbol{\varepsilon}_j, \tag{19}$$

where j counts from 1 to the number of model state ensemble members. Next we define the ensemble covariance matrix of the measurements as

$$\boldsymbol{R}_e = \overline{\boldsymbol{\varepsilon}\boldsymbol{\varepsilon}^T}, \tag{20}$$

and of course in the limit of an infinite ensemble, this matrix will converge towards the prescribed error covariance matrix $\boldsymbol{R}$ used in the standard Kalman filter.

The analysis step for the EnKF consists of the following updates performed on each of the model state ensemble members

$$\psi_j^n = \psi_j^f + \boldsymbol{K}_e(\boldsymbol{d}_j - \boldsymbol{H}\psi_j^f) \quad . \tag{21}$$

The gain matrix $\boldsymbol{K}_e$ is similar to the Kalman gain matrix used in the standard Kalman filter (6) and is defined as

$$\boldsymbol{K}_e = \boldsymbol{P}_e^f \boldsymbol{H}^T (\boldsymbol{H}\boldsymbol{P}_e^f \boldsymbol{H}^T + \boldsymbol{R}_e)^{-1} . \tag{22}$$

Note that equation (21) implies that

$$\overline{\psi}^a = \overline{\psi}^f + \boldsymbol{K}_e(\bar{\boldsymbol{d}} - \boldsymbol{H}\overline{\psi}^f) . \tag{23}$$

Thus, the relation between the analyzed and predicted ensemble mean is identical to the relation between the analyzed and predicted state in the standard Kalman filter in equation (5), apart from the use of $\boldsymbol{P}_e$ and $\boldsymbol{R}_e$ instead of $\boldsymbol{P}$ and $\boldsymbol{R}$. Note that the introduction of an ensemble of observations does not make any difference for the update of the ensemble mean since this does not affect equation (23).

In Evensen and van Leeuwen (1996) a more detailed discussion was given on the practical implementation of the analysis scheme. It is possible to avoid the computation and storage of the full error covariance matrix $\boldsymbol{P}_e$, by operating directly on the ensemble members or alternatively the influence functions defined as $\boldsymbol{H}\,\boldsymbol{P}_e$. In addition, for realistic systems with a large number of observations, the inversion in (22) becomes too expensive and poorly conditioned. One can then resort to an approximate algorithm where the analysis is computed grid point by grid point, using only observations within a specified radius from the grid point. This reduces the size of the matrix inversion in addition to allowing for an algorithm where the major storage requirement reduces to keeping the ensemble in memory.

If the mean is considered to be the best estimate, then the linearity of the analysis scheme makes it an arbitrary choice whether one update the mean using the first-guess observations, or if one update each of the ensemble members using the perturbed observations. However, it will now be shown that by updating each of the ensemble members using the perturbed observations one also creates a new ensemble having the correct error statistics for the analysis. The updated ensemble can then be integrated forward in time till the next observation time.

Moreover, the covariance of the analyzed ensemble is reduced in the same way as in the standard Kalman Filter. First, note that equations (21) and (23) are used to get

$$\psi_j^a - \overline{\psi}^a = (\boldsymbol{I} - \boldsymbol{K}_e\boldsymbol{H})(\psi_j^f - \overline{\psi}^f) + \boldsymbol{K}_e(\boldsymbol{d} - \bar{\boldsymbol{d}}) . \tag{24}$$

We then get

$$\begin{aligned} \boldsymbol{P} &= \overline{(\psi^a - \overline{\psi}^a)(\psi^a - \overline{\psi}^a)^T} \\ &= (\boldsymbol{I} - \boldsymbol{K}_e\boldsymbol{H})\boldsymbol{P}_e^f(\boldsymbol{I} - \boldsymbol{H}^T\boldsymbol{K}_e^T) + \boldsymbol{K}_e\boldsymbol{R}_e\boldsymbol{K}_e^T \\ &= (\boldsymbol{I} - \boldsymbol{K}_e\boldsymbol{H})\boldsymbol{P}_e^f \end{aligned} \tag{25}$$

Clearly the observations **d** must be treated as random variables to get the measurement error covariance matrix into the expression.

6.3.4 Summary

We now have a complete system of equations which constitutes the ensemble Kalman filter (EnKF), and the resemblance with the standard Kalman filter is maintained. This is also true for the forecast step. Each ensemble member evolves in time according to the model dynamics and the ensemble covariance matrix of the errors in the model equations, $\boldsymbol{Q}_e$, converges to $\boldsymbol{Q}$ in the limit of an infinite ensemble size. The ensemble mean then evolves according to an equation of the form

$$\begin{aligned}\overline{\psi_{k+1}} &= \overline{f(\psi_k)} \\ &= f(\overline{\psi_k}) + \text{n.l.}\end{aligned}, \tag{26}$$

where n.l. represents the terms which may arise if f is non-linear. One of the advantages of the EnKF is that the effect of these terms is retained since each ensemble member is integrated independently by the model.

The covariance of the ensemble evolves according to an equation

$$\boldsymbol{P}_e^{k+1} = \boldsymbol{F}\boldsymbol{P}_e^k\,\boldsymbol{F}^T + \boldsymbol{Q}_e + \text{n.l.}, \tag{27}$$

where $\boldsymbol{F}$ is the tangent linear model evaluated at the current time step. This is again an equation of the same form as is used in the standard Kalman filter, except of the extra terms n.l. that may appear if f is non-linear. Implicitly, the EnKF retains these terms also for the error covariance evolution. Thus, if the ensemble mean is used as the best estimate, with the ensemble covariance $\boldsymbol{P}_e^{f,a}$ interpreted as the error covariance $\boldsymbol{P}^{f,a}$, and by defining the observation error covariance matrix $\boldsymbol{R}_e = \boldsymbol{R}$ and the model error covariance $\boldsymbol{Q}_e = \boldsymbol{Q}$, the EnKF and the standard Kalman filter become identical. This discussion shows that there is a unique correspondence between the EnKF and the standard Kalman filter (for linear dynamics) and that one can certainly interpret the ensemble covariances as error covariances while the ensemble mean is used as the best guess trajectory.

The extended Kalman filter applies the evolution equations (26) and (27) with the n.l. terms neglected. However, the ensemble Kalman filter includes the full effect of these terms and there are no linearizations or closure assumptions applied. In addition, there is no need for a tangent linear operator or its adjoint, and this makes the EnKF very easy to implement for practical applications.

6.4 An example of the analysis scheme

An example is now presented which illustrates the analysis step in the EnKF. Further, as a validation of the derivation performed in the previous section the results are also compared with the standard Kalman filter analysis.

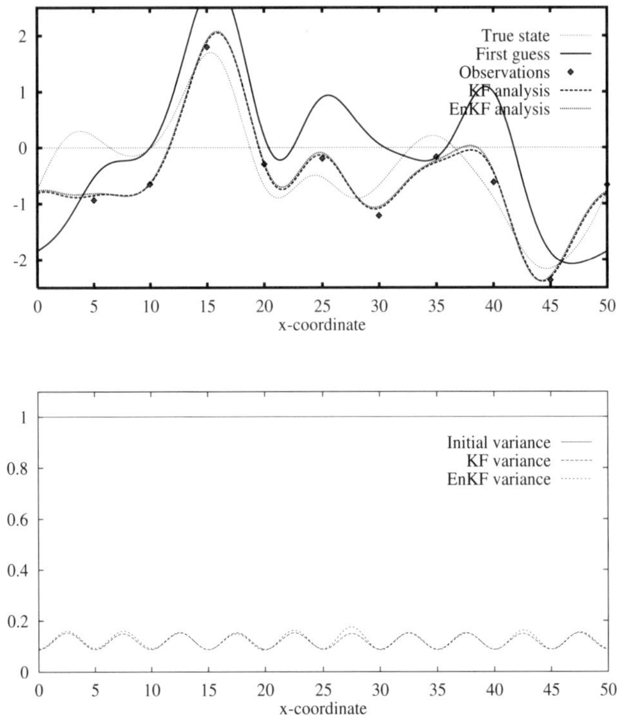

Fig. 6.1 Comparing results from the KF and the EnKF analysis schemes. On top the true reference state, the first-guess, and the analyzed estimate. The lower plot shows the corresponding error variance estimates.

For the experiment a 1–dimensional periodic domain in x, with $x \in [0,50]$, is used. We assume a characteristic length scale for the function $\psi(x)$ as $l = 5$. The interval is discretized into 1008 grid points which means there are a total of about 50 grid points for each characteristic length.

Using the methodology outlined in the Appendix of Evensen (1994b) we can draw smooth pseudo random functions from a distribution with zero mean, unit variance and a specified covariance given by

$$\boldsymbol{P}(x_1 - x_2) = \exp\left[-\frac{(x_1 - x_2)^2}{l^2}\right]. \tag{28}$$

This distribution will be called $\Phi(\psi)$ where the functions ψ have been discretized on the numerical grid.

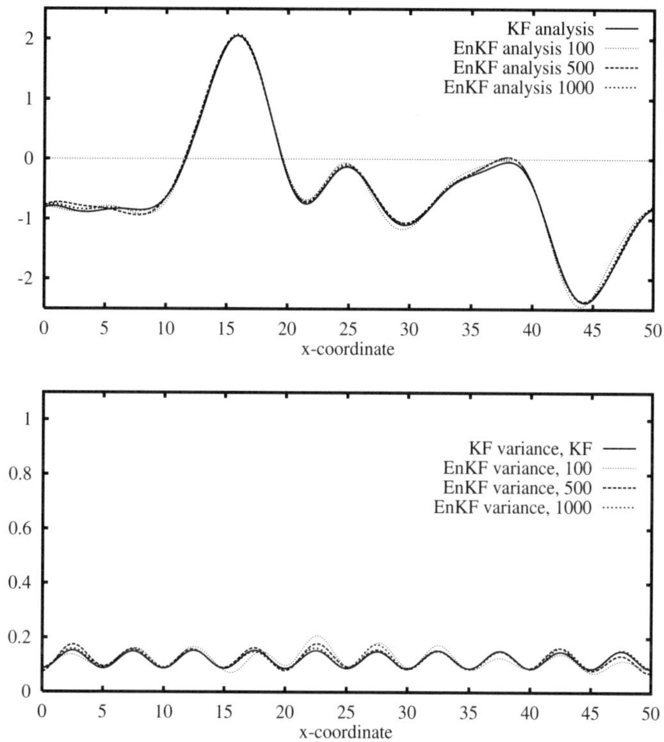

Fig. 6.2 Comparing results from the KF and the EnKF analysis schemes, using different ensemble sizes with 1000, 500, and 100 members. The upper plot shows the analyzed estimates. The lower plot shows the corresponding error variance estimates

A smooth function representing the true state ψ^t is picked from the distribution Φ and this ensures that the true state has the correct characteristic length scale l. Then a first-guess solution ψ^f is generated by adding another function drawn from the same distribution to ψ^t, i.e. we have assumed that the first-guess has an error variance equal to one and covariance functions as specified by (28).

The error covariance matrix for the first-guess is constructed by discretizing the covariance function (28) on the numerical grid to form $\boldsymbol{P}^f$.

There are 10 measurements distributed at regular intervals in x. Each measurement is generated by measuring the true state ψ^t and then adding Gaussian distributed noise with mean zero and variance 0.2.

An ensemble representing the error variance equal to one is now generated by adding functions drawn from Φ to the first-guess. Here 1000 members were used in the ensemble. Thus, we now have a first-guess estimate of the true state with the error covariance represented by the ensemble.

The results from this example are given in Fig. 6.1. The ensemble estimates are of course the means of the analyzed ensembles. By comparing the KF and the EnKF estimates it is clear that EnKF gives a consistent analysis ψ^a. The lower plot shows the corresponding error variances. The upper line is the initial error variance for the first-guess equal to one. Then there are two error variance estimates corresponding to the EnKF and the standard Kalman filter. Clearly, the EnKF analysis scheme provides an error variance estimate which is very close to the one which follows from the standard Kalman filter.

In Fig. 6.1 we examine the sensitivity of the analysis scheme with respect to the size of the ensemble. Clearly there is not a big difference in the estimates using 100 or 1000 ensemble members.

6.5 A highly nonlinear case: the Lorenz equations

An example is now given using the highly nonlinear and chaotic Lorenz equations. The celebrated Lorenz model has been subject of extensive studies motivated by its chaotic and strongly nonlinear nature. In the field of data assimilation, the model has served as a testbed for examining the properties of various data assimilation methods when used with strongly nonlinear dynamics. The results have been used to suggest properties and possibilities of the methods for applications with oceanic and atmospheric models which may also be strongly nonlinear and chaotic.

6.5.1 Model Equations

The Lorenz model consists of a system of three coupled and nonlinear ordinary differential equations, Lorenz (1963),

$$
\begin{aligned}
\frac{dx}{dt} &= \sigma(y - x) + q^x, \\
\frac{dy}{dt} &= \rho x - y - xz + q^y, \\
\frac{dz}{dt} &= xy - \beta z + q^z.
\end{aligned}
\tag{29}
$$

Here $x(t)$, $y(t)$, and $z(t)$ are the dependent variables and we have chosen the following commonly used values for the parameters in the equation; σ=10, ρ=28 and

β=8/3. The terms $q^x(t)$, $q^y(t)$ and $q^z(t)$ are assumed to represent the unknown model errors. Initial conditions for the model are given as

$$\begin{aligned} x(0) &= x_0 + a^x, \\ y(0) &= y_0 + a^y, \\ z(0) &= z_0 + a^z, \end{aligned} \tag{30}$$

where x_0, y_0, and z_0 are the first-guess values of the initial conditions and the terms a^x, a^y and a^z represent the errors in the first-guess initial conditions. If all the error terms were known or equal to zero, these equations would formulate a well-posed problem having a unique solution in a mathematical sense.

Now a set of measurements, $\boldsymbol{d} \in \Re^M$, of the true solution are assumed given and linearly related to the model variables by the measurement equation

$$\boldsymbol{d} = \boldsymbol{L}[x, y, z] + \boldsymbol{\varepsilon}, \tag{31}$$

where $\boldsymbol{L} \in \Re^M$ is a linear measurement functional, $\boldsymbol{\varepsilon} \in \Re^M$ is a vector of measurement errors and M is the number of measurements.

6.5.2 Discussion of cases

The initial condition for the reference case is given by

$$(x_0, y_0, z_0) = (1.508870, -1.531271, 25.46091)$$

and the time interval is $t \in [0,40]$. The observations and initial conditions are simulated by adding normal distributed noise with zero mean and variance equal to 2.0, to the reference solution. The initial conditions used are also assumed to have the same variance as the observations. These are the same values that were used in Miller et al. (1994) and Evensen and Fario (1997).

The following examples are discussed.

Experiment A: In this first experiment the distance between the measurements is $\Delta t_{obs} = 0.25$, which is the same as was used in Miller et al. (1994). Thus, it is possible to compare the results presented here with those presented in Miller et al. (1994) using the extended Kalman filter and a strong constraint variational method.

Experiment B: In order to examine the sensitivity with respect to measurement density, an additional experiment is now performed where the distance between the measurements is $\Delta t_{obs} = 0.5$.

The data assimilation estimate from the ensemble Kalman filter is given in Figs. 6.3 and 6.4 for the two cases. The ensemble size is 1000 members. The ensemble Kalman filter seems to do a reasonably good job in tracking the state transitions and also in reproducing the correct amplitudes for the peaks of the solution.

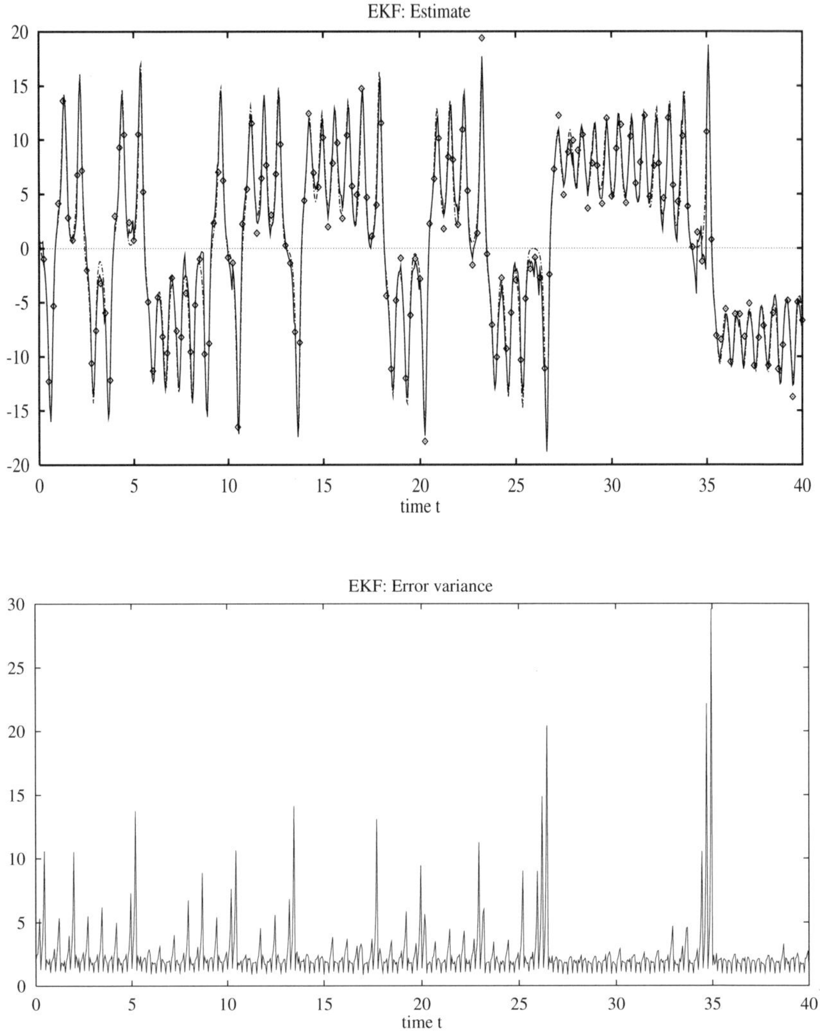

Fig. 6.3 Experiment A (Ensemble Kalman Filter): The upper plot shows the inverse estimate for x as the full line and the "true" reference state as the is the dash-dot line which is mostly hidden behind the estimate. The lower plot shows the corresponding error variance estimate.

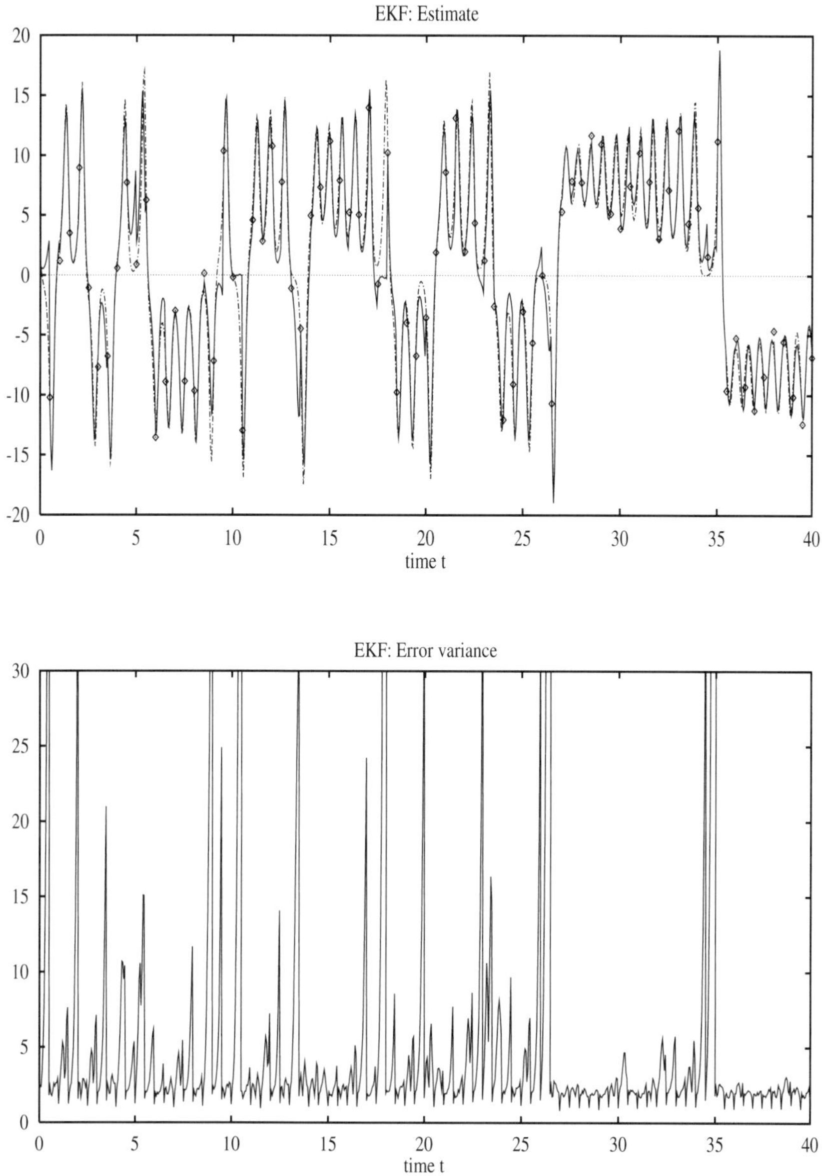

Fig. 6.4 Experiment B (Ensemble Kalman Filter): Same as Fig. 6.3 but for Experiment B.

There are a few locations where the filter estimate starts diverging from the reference solution, e.g. for $t = 26$ and $t = 34$. Note however that the ensemble Kalman filter recovers quickly and begins tracing the reference solution again. The error estimate given in the lower plot of Fig. 6.3 shows strong error growth at these particular locations and thus indicates that the ensemble is passing through a region in the state space which is associated with strong instability.The error estimates show the same behavior as was found by Miller et al. (1994) with very strong error growth when the model solution is passing through the unstable regions of the state space, and otherwise rather weak error variance growth in the more stable regions. Note for example the low error variance when $t \in [28,34]$ corresponding to the oscillation of the solution around one of the attractors.

Finally, it should be pointed out that in the ensemble Kalman filter a variance minimizing analysis is calculated at measurement times. Thus, even if the ensemble certainly is non-Gaussian due to the forward integration of nonlinear model equations, only the Gaussian part of the distribution is used. This is in contrast to the work by Miller et al. (1999) where the maximum-likelihood analysis is calculated by actually constructing the density function for the model evolution and then calculating the conditional density in terms of analytical functions. They found that this made a significant improvement on the analysis. However, it is still not clear how this approach can be used in a practical way for high dimensional state spaces.

6.6 An ensemble Kalman filter for an OGCM: Preliminary results

The ensemble Kalman filter has been implemented with the Miami Isopycnic Coordinate Ocean Model (MICOM) originally developed by R. Bleck at the University of Miami, [see e.g., Bleck and Smith (1990), and Bleck et al. (1992)]. The currently available data assimilation applications for Ocean General Circulation Models (OGCMs) have been based on rather simplistic assimilation schemes. None of these take proper error statistics into account and ad hoc approaches are used for the assimilation. Some examples are: Derber and Rosati (1989) who used an objective analysis technique to update the model temperature in a version of the Cox model (Rosati and Miyakoda, 1988). Mellor and Ezer, (1991) and Ezer and Mellor, (1994), used a univariate optimal interpolation algorithm with vertical projection of surface information in the model by Blumberg and Mellor (1987). Cooper and Haines (1996) used a vertical projection method based on water property conservation in the Cox model (Cox, 1987).

In the works by Mellor and Ezer, (1991), and Ezer and Mellor, (1994), sea surface height (SSH) observations were used to update the vertical density stratification to drive the geostrophic currents associated with gradients in the SSH. The density was implicitly updated by actually updating the vertical temperature profiles using estimated correlations between temperature and SSH. Clearly, this is an inconsistent approach, since the vertical density stratification depends on both the temperature and salinity. Updating only temperature will ultimately generate water masses with unrealistic T–S properties.

A method for overcoming this problem has recently been presented by Oschlies and Willebrand (1996), where both temperature and salinity were updated under a constraint of maintaining the T–S properties of the water masses. This approach gave satisfactory results and should now be extended to take proper error statistics into account in the analysis scheme.

In another recent work by Forbes and Brown (1996), a nudging method was used with the Miami Isopycnic Coordinate Ocean Model (MICOM) for assimilation of SSH observations from altimetry. They showed that the nudging approach was capable of pulling the model state towards the observations and that the model, over time, also propagated information from the surface into the lower layers.

Essential when developing an advanced data assimilation system with an OGCM is to estimate proper error statistics for the model prediction. That is, having a model forecast and an observation one needs to know the influence this particular observation will have on all the prognostic variables in the model. As an example, assume we have one observation of the sea surface or mixed layer temperature at a particular location. If the measured value is greater than the model predicted value this should imply that the model mixed layer temperature is increased by some amount in the analysis scheme. The update should of course be smooth in space. If the model mixed layer temperature is the only variable updated this will lead to an "unbalanced" model state. Clearly, the model density should be decreased accordingly to be consistent with the equation of state. Further if the mixed layer temperature is increasing we would expect a decrease of the mixed layer depth, which requires the full vertical density structure of the model to change.

Important for this study is that MICOM uses density as the vertical coordinate. Thus, in the vertical the model can be considered as a stack of layers each having a constant density. This leads to an interesting interpretation of how the vertical stratification should be updated by the analysis scheme. Instead of updating the thermodynamic variables themselves one rather change the locations of the model layer interfaces, which gives the same effect. Clearly, it becomes complicated to construct Otimal Interpolation schemes for performing such an analysis since at various locations and times different layer interfaces should be altered. As an example, the layers with low density water masses will outcrop below the mixed layer when going northward from the equator. However the ensemble Kalman filter holds all the required information about the space and time dependent error covariances between different model variables which are used to calculate the actual influence functions for the measurements.

Here, a preliminary example is presented from an ensemble Kalman filter implementation with a coarse resolution version of MICOM for the north Atlantic. A 256 member initial ensemble is created by perturbing layer interfaces in a model state resulting from a 10 years spin up run. Each of the ensemble members are generated by adding smooth pseudo random fields with a specified covariance and mean

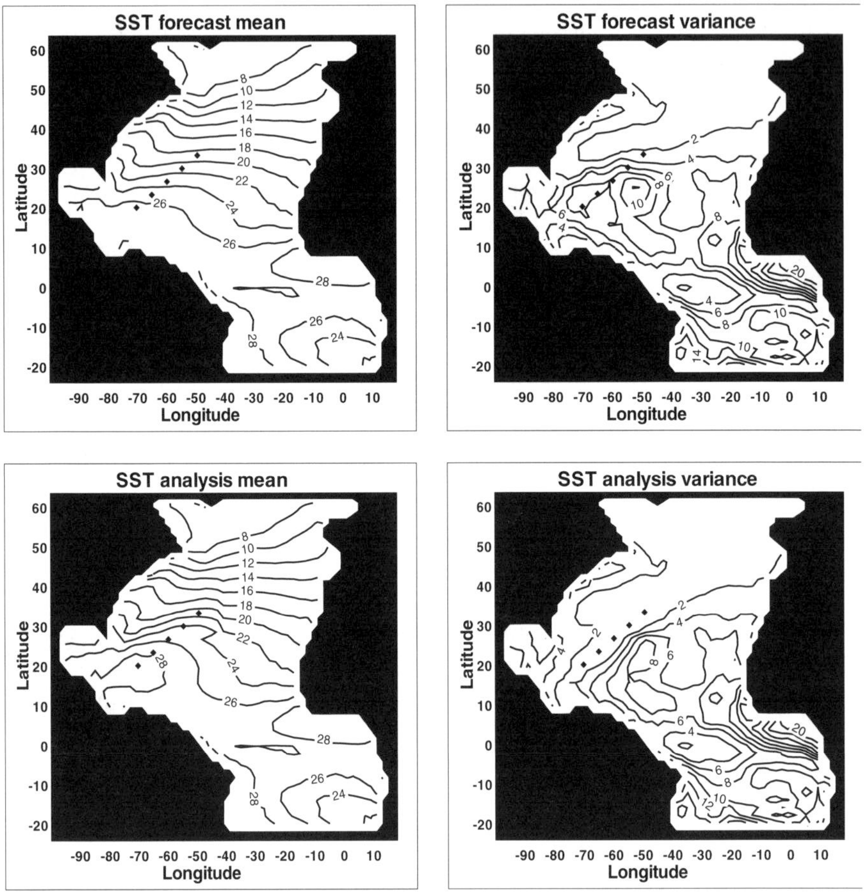

Fig. 6.5 Sea surface temperature prediction and analysis mean and variance.

equal to zero to the layer interfaces. Thus each of the ensemble members has a different vertical stratification at the initial time.

The ensemble is then integrated 10 years in a pure prediction mode with no assimilation of observations. Monthly means form the basis for the atmospheric forcing, however, to simulate realistic variability in the forcing fields and to account for errors in the monthly averages, the forcing was perturbed using pseudo-random fields. The perturbations were correlated in time to simulate the variability on a few days time scale of the real atmosphere, and of course different perturbations were used for each member of the ensemble. By perturbing the forcing fields we also include a realistic representation of the forcing errors which form an important part of the model errors. Note that time correlated model errors can be used in the ensemble Kalman filter contrary to the standard Kalman filter.

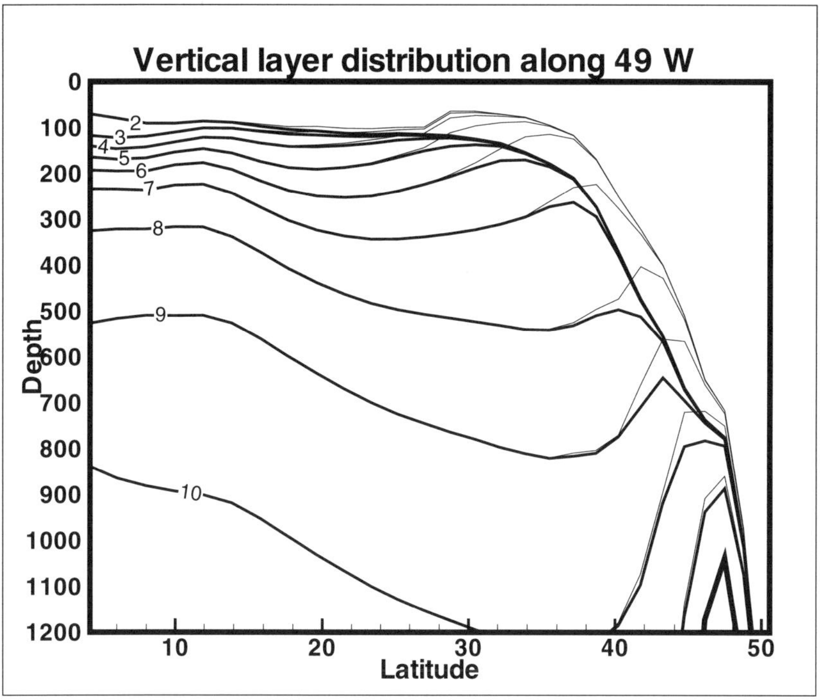

Fig. 6.6 Vertical section of layer thicknesses from South to North along longitude 49 W. The thick lines are the prediction before analysis and the thin lines denote the analysis.

This 10 year ensemble integration can be considered as a spin up of the ensemble before starting a real data assimilation experiment. Also the predicted ensemble allows us to examine in detail the ensemble statistics including the covariances developing between the different model variables.

In Fig. 6.5 the effect of an analysis step using the ensemble Kalman filter is shown. It is assumed that five observations of the SST are available (shown as diamonds in the plots). Using these observations which all have values equal to the ensemble mean plus 1.5 degrees an analysis is calculated. The upper left plot shows the smooth 10 years ensemble mean SST prediction in January and the lower left plot shows the resulting SST analysis. Clearly, the temperature is increased in the area close to the observations. The upper right plot shows the predicted error variance for the SST, and the lower right plot shows the variance for the analysis having a distinct reduction near the observations. These are results as would be expected.

Of greater interest is the influence the SST data will have in the vertical. In Fig. 6.6 the vertical layer distribution is shown before and after the analysis. The vertical section is taken from South to North along 49°W. The layer interfaces are shown with the thick lines before the analysis and the thin lines after the analysis. The contour labels is the number of the layer below that particular interface. The layer distribution shows the deepening of the mixed layer in the North which is MICOMs way of accounting for convection. Clearly, the increase in the SST, or model mixed layer temperature, has lead to a shallowing of the mixed layer. Further, the vertical influence is obvious for all of the 11 layers in the model, which has been moved upward in the water column. That is, the pycnocline has been lifted and the vertical density structure has been changed in a dynamically balanced manner.

A more extensive discussion of the EnKF implementation with MICOM and an application will be reported in a manuscript in preparation.

6.7 Summary

A general overview discussion has been given on sequential data assimilation for nonlinear ocean models. The emphasis is laid on the ensemble Kalman filter (EnKF) which has shown to be a promising approach suitable for the nonlinear models used in oceanography. The theoretical basis for the EnKF was discussed in detail and the capability of the filter to handle strongly nonlinear dynamics was demonstrated in an example with the Lorenz equations. Preliminary results were also presented from an implementation of the EnKF with the MICOM ocean general circulation model. This is the first reported implementation of an advanced sequential data assimilation technique, where proper evolution of error statistics are taken into account, with a realistic OGCM. It was illustrated how the ensemble statistics contain the required information for generating a dynamically balanced analysis and thereby also project surface information in the vertical in a consistent manner.

A general conclusion so far is that the EnKF should be further implemented and explored with other dynamical models. With a relatively acceptable numerical load which corresponds to about 100–250 model integrations, the method can be used for operational oceanography on extant computer resources. The parallel computers are also perfectly suited for ensemble integrations where each member can be integrated independently on a separate processor.

Acknowledgments

The work was supported by European Commission through the Environment and Climate program under contract ENV4-CT95-0113 (AGORA) and by the Nordic Council of Ministers contract FS/HFj/X-96001. Further, it received support from the Norwegian Super Computing Committee (TRU) through a grant of computing time.

7 Assimilation of Satellite Altimetry in Ocean Models

KEITH HAINES
University of Edinburgh, Department of Meteorology

7.1 Introduction

In this chapter we will look at some of the problems, both conceptual and practical, of assimilating one of the most widely available ocean data sets, that from satellite altimeters. The first sections will deal with conceptual issues. We consider how the ocean and atmospheric assimilation problems differ in some respects. We will look at the 4DVar (Four Dimensional Variational Assimilation) and 3D Sequential assimilation techniques to consider their merits from a physical perspective. Results are then shown from twin experiments of a 3D sequential assimilation approach which tries to account for the ocean physics when assimilating altimeter data. In the final sections we will show preliminary results from a study designed to assimilate TOPEX altimeter data into a global ocean model. Many new problems arise when real satellite data are used, some of which have been solved, and some not. We will summarise by describing the present state of the art and speculate on the likely developments over the next few years.[1]

7.2 Physical relationships between sea level and hydrography

One of the most important concepts we have in learning about ocean circulation is the concept of a water mass. The literature of physical oceanography contains many descriptions of water mass properties (temperature, salinity, and more recently, CFCs, nutrients, etc.) and their spatial distributions. The reason why water masses are so useful is because many of the properties which define them change only on very long timescales. They are the physical embodiment of the ocean's memory of past climate conditions at the surface where most water properties are determined at the time of formation. The air mass concept is also sometimes used in synoptic meteorological descriptions but air masses can mix with each other very rapidly, especially vertically during tropospheric convection events. In the ocean, such strong, deep convective mixing only occurs in a few places such as the Labrador, Greenland and Weddell seas. The result is that the

1. For up-to-date results on implementing the methods discussed here see, Fox et al. (2000a,b).

ocean has a long memory when viewed from a Lagrangian perspective, and this can, and should, be taken advantage of in constructing data assimilation systems.

Consider 2 states of the ocean separated in time by 1 year. Most of the differences in these ocean states are due to advection of water masses which have occured in the intervening period. The total volume and properties of each water mass has probably hardly changed. Because of this the second state can largely be obtained by an adiabatic and tracer preserving re-arrangement of the waters in the first state. Even during decadal timescale fluctuations, such as those seen in the strength and depth of the N Atlantic subtropical gyre, Levitus (1989), the water properties often remain fairly constant, as reflected by unchanging T-S relationships, although on this timescale the relative volumes of the water masses are likely to have changed. When low frequency variations are examined in ocean models, e.g. Cox (1987), the variations are often in the form of baroclinic Rossby waves propagating slowly westward in the subtropical gyre. The water mass variability associated with such wave propagation would also be well represented through displacements of in situ water masses. Such low frequency signals may also produce SST anomalies and therefore identifying such events in ocean data sets and initialising ocean models to contain them is probably one of the most important problems facing the ocean assimilation community if the ocean models are then to contribute to improving climate prediction.

Returning to the data assimilation problem, when we have observations of sea level this can tell us about the location of ocean eddies and Rossby wave positions, in other words the position of water mass fronts and distributions. It tells us very little about the water masses themselves. For this reason it is conceptually realistic to consider the altimeter assimilation problem as one of deriving a `Lagrangian' redistribution of water masses already present rather than an `Eulerian' alteration of existing water masses to make them consistent with the local sea level data. In practice of course the same analysed hydrography can be obtained by either method but it is not easy to ensure that integrated water masses are preserved using the Eulerian method, especially when local hydrography changes are based on statistical correlations.

It is interesting to view the physics of the 4Dvar and 3D sequential data assimilation methods in the light of the above arguements. The 4Dvar assimilation scheme is often set up to seek the best initial conditions which will allow a model trajectory launched from these conditions to pass close to some set of observations distributed over some time interval. If we accept that the most important processes taking place during this time interval are essentially advective then the 4Dvar scheme is seeking the correct initial spatial water mass distribution which will thereafter undergo dynamically defined displacements as the model integrates forward, in order to match the observations at subsequent times. In a 3D sequential scheme we initially have no accurate idea of the water spatial distribution although we might assume we already have a good knowledge of the integrated water properties, e.g. we might start from Levitus, or some better climatology if available. As each new data set comes in we can seek a redistribution of the existing water properties to be consistent with these observations. We have a better chance that the 3D assimilation scheme will converge if we are confident that at each assimilation time the post assimilation state could have been achieved by advection from the state prior to assimilation, which would guarantee that the analysed states are consistent in the important conserved properties of the ocean. Fig. 7.1 illustrates the nature of the convergence of the 2 assimilation schemes. Although the sequential scheme does not provide a good analysis of the ocean state before the end-time of the assimilation, it is possible to work backwards to obtain the solution at earlier times using the `Kalman Smoother' operation.

In Cooper and Haines (1996) a practical water redistribution assimilation scheme was developed for use with altimeter data. It is well known that in mid-latitudes a high sea level is usually correlated with a deeper thermocline, often with a warmer water layer at the surface, as exemplified in a warm core ocean ring. Similarly a low sea level usually indicates a raised thermocline, with cold water at the surface, as in a cold core ring. The Cooper and Haines (1996) assimilation scheme seeks a vertical displacement of water masses which would correspond to the observed change in sea level. Such a vertical displacement obviously satisfies the Lagrangian redistribution requirement although it is not as general as a full 3D rearrangement of water properties. The advantage is that a unique vertical displacement can easily be derived from an observed sea level anomaly with the imposition of a simple additional constraint. This constraint is: no change is permitted to the pressure at the ocean floor. Consider the hydrostatic equation for a given water column;

$$\eta \rho_0 g + \int_{-H}^{0} \rho g dz = p(-H) \tag{1}$$

where η is the sea level above a reference level 0, H is the depth of the ocean, $\rho(z)$ is the in situ water column density and $p(-H)$ is the pressure at the ocean floor. If an observation requires a change in the model sea level, $\Delta\eta$, and we place the constraint that $\Delta p(-H) = 0$ then we get;

$$\Delta\eta \rho_0 g + \int_{-H}^{0} \Delta\rho g dz = 0 \tag{2}$$

If $\Delta\rho(z)$ is required to represent a vertical displacement of the existing water column then the amount of vertical displacement, δh, is determined uniquely. When the water column is lifted, the bottom of the water column is extended with water of the same properties already present at the bottom, when the water column is lowered the surface water properties are again those already present at the surface. Fig. 7.2 illustrates this displacement, δh. This process conserves water properties and volumes except at the top and bottom of the column where the volume of water masses is clearly not conserved locally. The surface and bottom water volumes may still be conserved if the volume changes in all the surrounding water columns

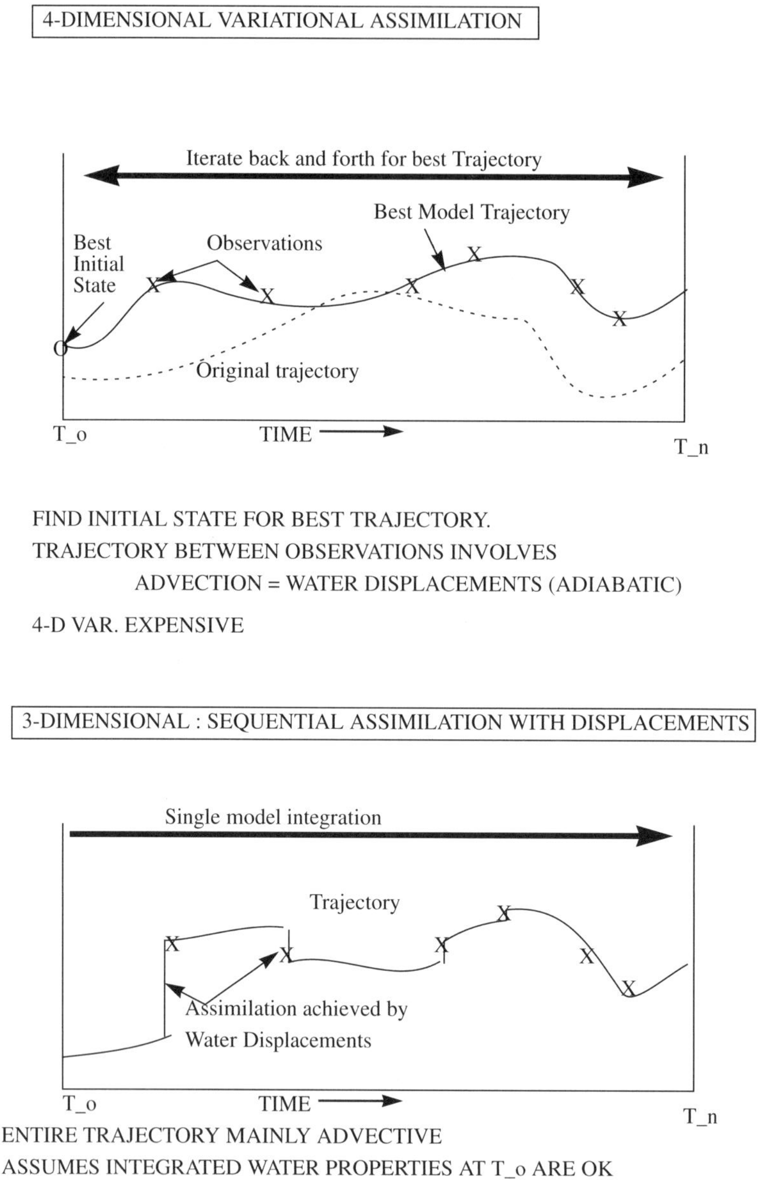

Fig. 7.1 Schematic of the 4DVar and 3D Sequential data assimilation methods. In the sequential method the adiabatic displacements at assimilation times are emphasised.

are also considered. Once the density field has been changed, velocity changes can be introduced which are in geostrophic balance with the density changes, at least away from the equator. The bottom constraint that $p(-H)$ does not change now ensures that the bottom torque, given by

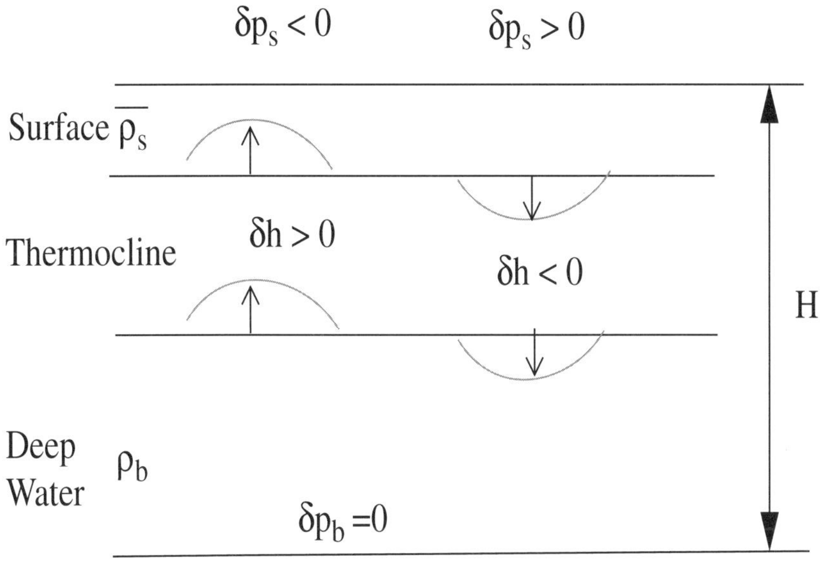

Fig. 7.2 Schematic of the vertical displacement altimeter assimilation method. Surface and bottom pressure are shown where $\delta p_s = \Delta\eta\rho_0 g$ from equation (2).

$$J(p(-H), H) = \frac{\partial p(-H)}{\partial x}\frac{\partial H}{\partial y} - \frac{\partial p(-H)}{\partial y}\frac{\partial H}{\partial x}$$

where J is the Jacobian, is not altered by the assimilation.

The advantages of this scheme include the following;

(i) It is local and derived purely dynamically, i.e. model independent, and it is easy and computationally cheap to apply.

(ii) It preserves the $T(S)$ relationship of the water column, i.e. the water properties.

(iii) It preserves the volume of each water mass and hence the stratification on potential density surfaces except at the top and bottom. It cannot therefore initiate convection when the model is restarted.

(iv) By not changing the bottom pressure it avoids changing bottom torques and hence reduces shocks in the model around steep topography.

(v) It should allow an independent assimilation scheme for temperature profile data (presumably available much more infrequently) to have a lasting impact on the ocean model.

Other altimeter assimilation methods which share some of these properties include, Oschlies and Willebrand (1996), Drakopoulos (1997) and Gavart and De Mey (1997). In the section which follows we show some results from applying this scheme in a twin experiment formulation.

7.3 Convergence in twin-experiment assimilation experiments

A number of twin experiment papers have appeared in the literature showing results from using the scheme described above, Cooper and Haines (1996), Drakopoulos (1997), Fox and Haines (1996), Fox et al. (2000). We will show some results from using a Global high resolution ocean model, OCCAM (Ocean Circulation and Climate Advanced Model), which is a free surface model with 1/4° horizontal resolution and 36 vertical levels. Performing assimilation with this model takes the assimilation scheme into new areas such as the tropics and the weakly stratified southern oceans where it has not been tested before. The OCCAM model was spun up for eight years with monthly varying SST and SSS relaxation and ECMWF wind stresses, and the ninth year was used as the `true' circulation, and the fields from the start of year 10 were used to initialise the assimilation run. Assimilation of sea surface height data was performed at 15 day intervals, with the first data being taken from day 1, year 9 of the truth run. This corresponds to model January 1st. Fields in the model were altered according to equation (2) and velocities were updated geostrophically, except within 3° of the equator where no velocity update was performed. It should be noted that the full height change field is assimilated, corresponding to the hypothetical case of perfect data, with known Geoid and zero observational error.

To track the convergence of the assimilation run towards the truth run, RMS errors (differences between values in the assimilation run and the model truth) of the density, temperature, salinity and velocity fields were calculated for various regions and depth ranges. Fig. 7.3 shows globally averaged RMS errors in all these quantities for both the upper water column (top 1000m) and below. Fig. 7.4 shows the same quantities for the Tropical Pacific only and Fig.7.5 is for the Antarctic Circumpolar current region. Errors in sea surface height before each assimilation step are reduced by more than 50 % after one assimilation and the following 15 days of model run, then continue to drop with maximum reductions of 80 % (and still increasing) achieved after 10 assimilation steps. Velocity errors undergo the largest reduction, particularly in the near-surface region. This is due to near-surface currents being strongly tied to the sea surface height gradients via geostrophy. The deeper currents are improved less in percentage terms by assimilation but interestingly convergence often continues during model runs. The improvement of deep sub-thermocline currents is an almost unique feature of this assimilation method,

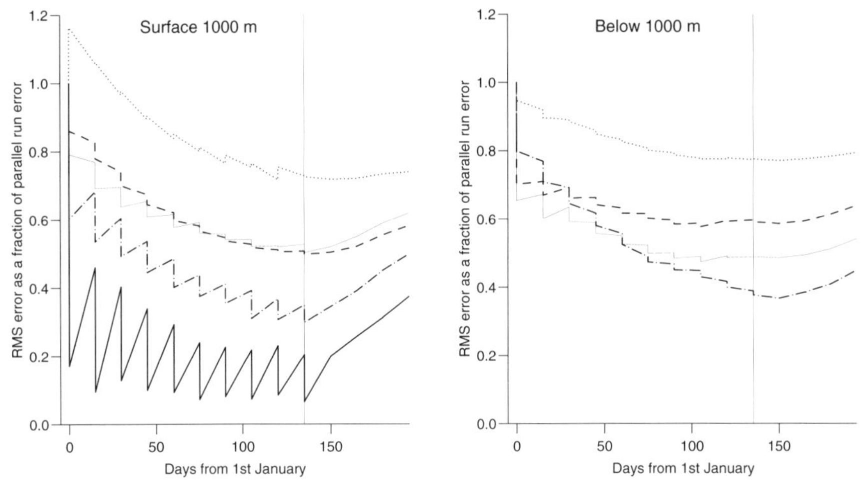

Fig. 7.3 Errors in the twin experiment: dark solid line---free surface height; pale solid line---density; short dashes---temperature; dotted ---salinity; dot-dashed line---velocity (both components). Left panel shows rms errors in the top 17 model levels (approx. 1000 m), right panel shows rms errors in levels 18--36.

Cooper and Haines (1996), Drakopoulos et al. (1997) (although verifying this with real data, outside the twin experiment framework, will be very difficult). Improvements in the density structure at all depths are also observed over the 10 assimilation steps. Global RMS density errors are reduced mainly at assimilation times, indicating the assimilation scheme for the projection of surface height changes onto the density structure is working well. Occasionally further reductions in density errors also take place during model runs between assimilations, but on average density errors do not change much as the model is run.

At the end of the assimilation experiment a 60 day forecast run was performed to get some insight into the rate of divergence of the assimilation run from the truth, and hence into the time-scale over which improved forecasts are likely to be possible (although meteorological data would not be forecastable over such times). The evolution of the RMS errors during this period are shown in Figs. 7.3 - 7.5. Divergence is most rapid near the surface, with the velocities generally worsening more rapidly than the tracers. In the tropical pacific, divergence is slowest with the deep layers actually continuing to improve over the 60 days of the forecast run. This supports the model of tropical dynamics as primarily wind-determined, a forecast made with a perfectly known wind field capable of producing very good results. The general picture is of the forecast run diverging from the model truth, but after running for 60 days a significant proportion of the gain from assimilation is retained, particularly for water properties and at deeper model levels.

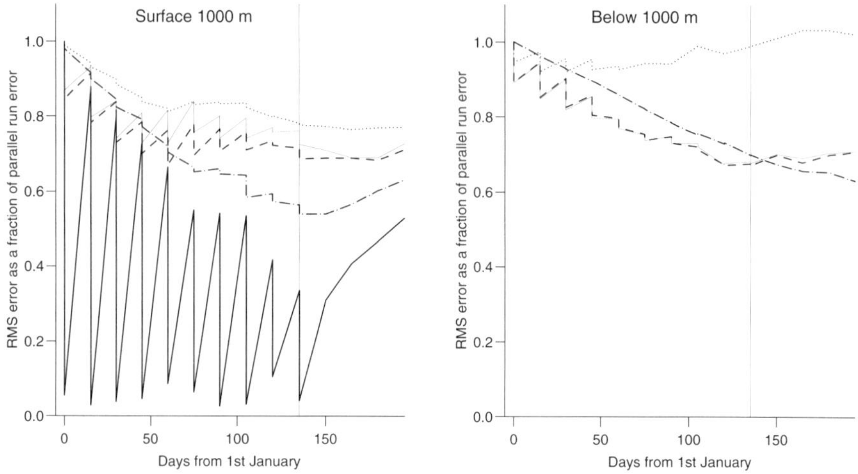

Fig. 7.4 As for Fig. 7.3 but calculated for the Tropical Pacific - here taken to be 140°E - 70°W, 15°S - 15°N.

These results show that the basic assimilation scheme works well both in the upper water column and at depth, and that once assimilation has been performed for some period it may be possible to make a considerably improved forecast of ocean currents and temperatures for some further period ahead. At this point the scheme needs to be tested with real data.

7.4 Assimilation of TOPEX data into the OCCAM model

Here we report on progress towards assimilating TOPEX maps for 1993 into the OCCAM model. Many new problems arise when assimilating real data. The results here are preliminary but they point to two encouraging possibilities. The first is that some errors in the model mean circulation, specifically the Gulf stream and Kuroshio separation paths, can be 'corrected' by assimilating appropriate mean sea level data. The second result points to the ability of the model to interpolate data effectively in time. This is highlighted by an ability to predict sea level anomalies more accurately that assuming persistence.

(i) Correcting mean sea levels

After 8 years of spin-up from Levitus, forcing with monthly mean winds, the OCCAM model was forced with 1992-1993, 6-hourly winds from ECMWF (European Center for Medium Range Weather Forecast). Surface temperature and salinity are still relaxed to climatological monthly mean values on a 30 day timescale.

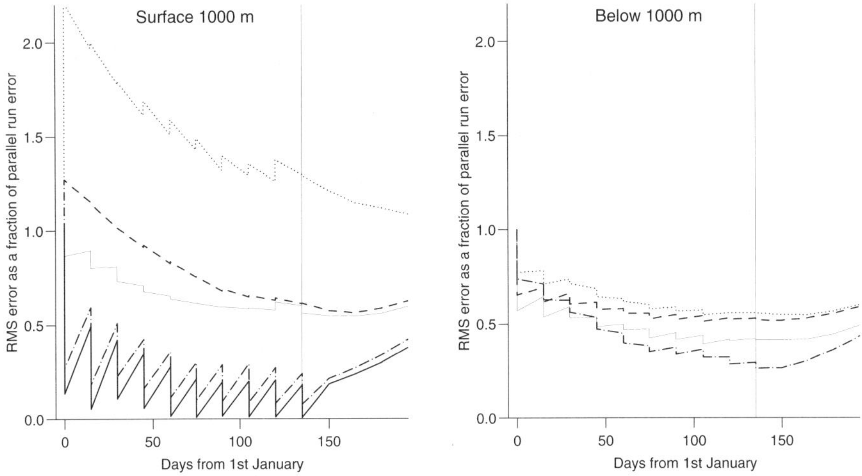

Fig. 7.5 As for Fig. 7.3 but calculated for the Southern Ocean/Antarctic Circumpolar Current region. 65-45°S.

Figures 7.6a,b show the mean sea level around the Gulf Stream and Kuroshio averaged over 1993 from the model. The common anti-cyclonic overshoot and standing Rossby wave can be seen in both regions. We attempted to reconstruct a more realistic mean sea level for both these regions using hydrography data, and then assimilated that mean along with the TOPEX anomaly maps. The new mean sea levels are shown in Figs. 7.6c,d. The change from Figs. 7.6a,b to 7.6c,d comes from calculating the difference in dynamic height between the model and Levitus (1994) hydrography, in the case of the Kuroshio, and between the model and an isopycnally mapped hydrography based on the Lozier et al. (1995) technique, for the Gulf stream. The modified mean sea levels match smoothly to the model derived mean for 1993 in all other regions.

The 1993 year model run was repeated with data assimilation. Every 10 days during this assimilation run the model sea level was corrected towards a sea level consisting of the annual mean (with corrections according to Figs. 7.6c,d) combined with a TOPEX anomaly map provided by the french CLS AVISO processing centre, "http://www-aviso.cls.cnes.fr". Fields other than sea level were updated by the Cooper and Haines (1996) vertical displacement method with a geostrophic update to velocities away from the equator. To demonstrate the effectiveness of the mean sea level correction, Fig. 7.6e,f show the sea surface temperature from this run on March 24th 1993. In both regions the western boundary current frontal position has separated from the coast at a lower latitude than in the run without assimilation and the sea surface temperatures clearly reflect this.

Correcting the path of these major current system may have important consequences if ocean models are to be used for climatic studies. There will, for exam-

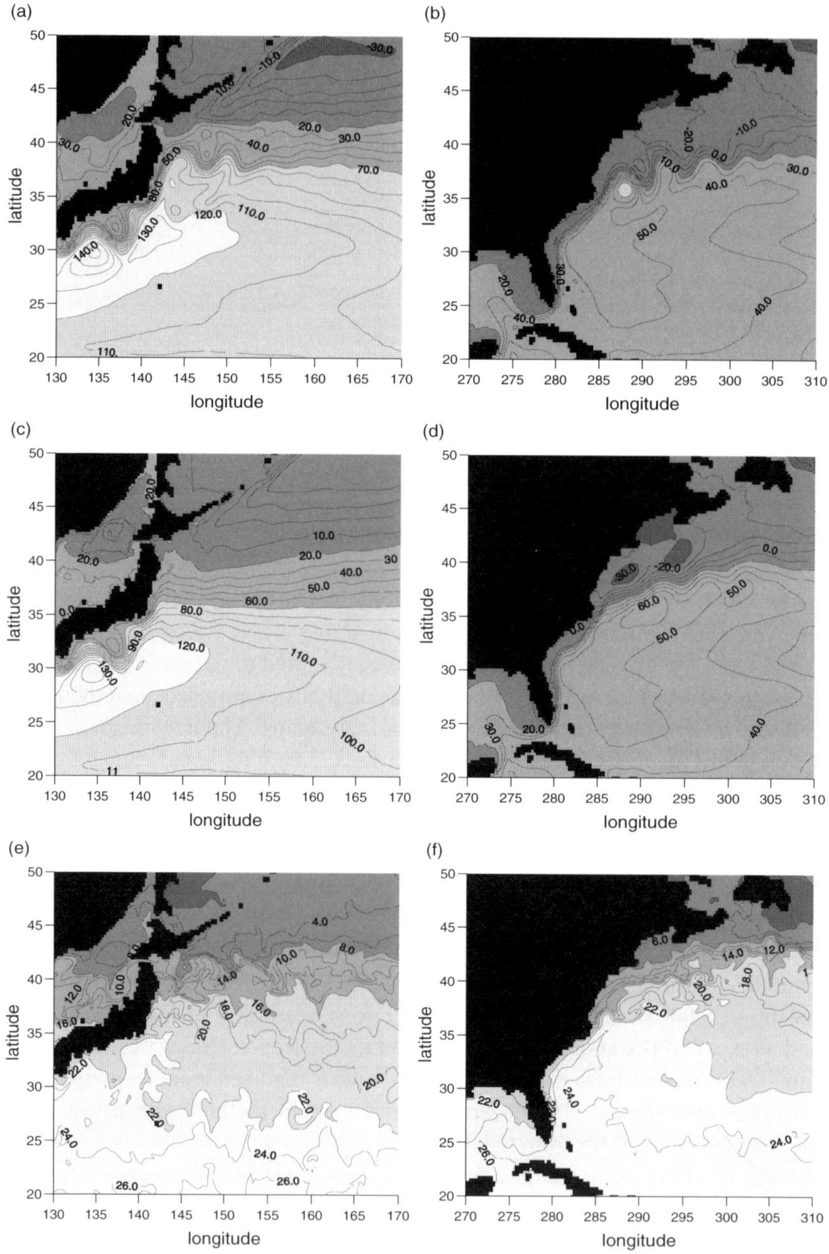

Fig. 7.6 (a,b) Model mean sea surface height, 1993. (c,d) Mean SSH after modifications according to climatology. (e,f) Sea surface temperature (°C) from the model with assimilation, 24th March 1993

ple, be large changes in surface heat flux associated with the altered frontal paths shown in Fig. 7.6e,f, although we have not yet diagnosed these from the run.

(ii) Sea level Tendency and error analysis

There is little point in using a model for data assimilation if that model is unable to interpolate data correctly in time over some period. To test this ability Fig. 7.7a shows the observed sea level change over the Atlantic during the 20 days from 30th January 1993 to 19th February 1993. This is simply the difference between the TOPEX maps from AVISO (without accounting for errors in the maps). Twenty days is the shortest period for which the maps contain entirely independent satellite data. Fig. 7.7b shows a model prediction of the same quantity (sea level change) based on assimilating data up to 30th January and thereafter allowing the model to run free for 20 days. If we compare Figs. 7a,b we find many features in common. Similar figures could be shown from any time during the assimilation run and also from most regions of the globe, even away from the Tropics. It is possible to define more quantitative comparison methods. The rms difference between the model predicted height anomaly and the observed TOPEX map on 19th February 1993 was 8 cm while the difference between the 30th January and 19th February TOPEX maps was 10 cm. The anomaly correlations for the above comparisons give 0.8 for the model prediction and 0.6 for the persistence field. Thus the model forecast can beat sea level persistence over 20 day periods.

The updating of model sea level was carried out as follows;

$$\eta_a = \eta_f + \frac{\sigma_f^2}{\sigma_f^2 + \sigma_T^2}(\eta_T - \eta_f), \tag{3}$$

where σ_T is the current TOPEX error map, σ_f is the model forecast error, η_T is the current TOPEX sea level anomaly and η_f is the model forecast sea level anomaly and η_a is the analysis sea level. Other fields are calculated from this update using the Cooper and Haines (1996) method. The model-data misfit, M, at analysis time is calculated as;

$$M = \sum \frac{1}{\sigma_T^2}(\eta_T - \eta_{f,a})^2 \tag{4}$$

where $\eta_{f,a}$ may be the forecast or the analysed model sea level, and the sum represents an area weighted average. Fig. 7.8a shows M through the 1993 run.

In the first few timesteps M is greatly reduced but thereafter the growth in M every 10 days is approximately equal to the reduction in M when new data is assimilated. The current σ_f is being set equal to the global mean of σ_T which is very crude and does not allow for a reduction in σ_f which would be necessary for convergence. An important future step will be to improve this error treatment.

Finally Fig. 7.8b shows the kinetic energy during the assimilation run. The assimilation is seen to increase the kinetic energy over the first 2 months but this levels off at about 20 % higher than the wind forced simiulation. Much of this difference seems to involve sea level variability in the quiet regions of the ocean such as the south Pacific where the OCCAM model has too little variability. In the active regions, such as the Gulf stream, the Aghulas and the ACC, the OCCAM

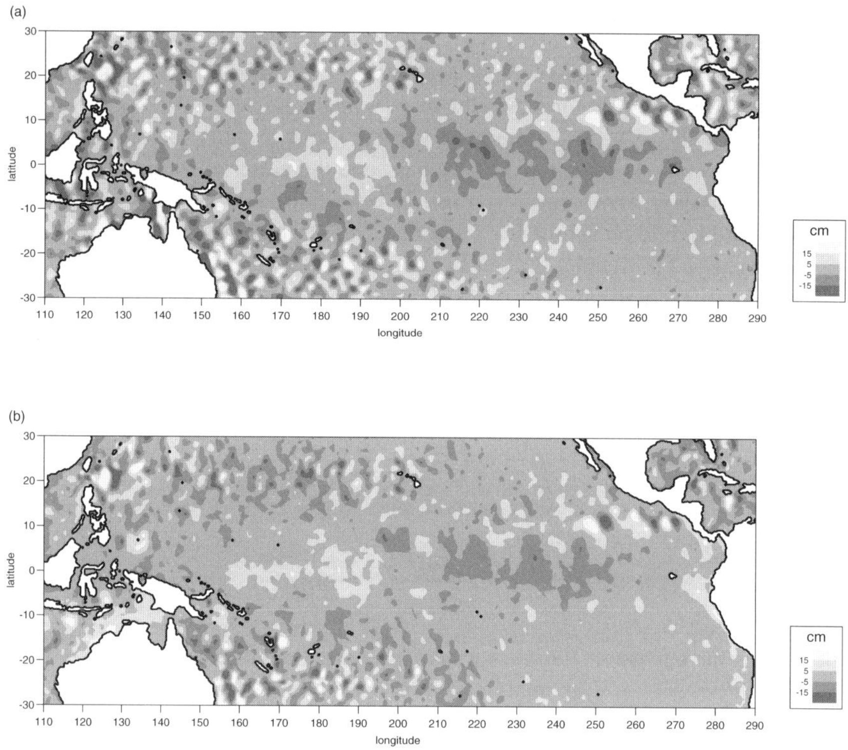

Fig. 7.7 Sea level change (cm) during 20 days from 30th January to 19th February, 1993. (a) as measured by TOPEX. (b) as predicted by OCCAM model, with assimilation up to 30th January.

model already simulates a realistic level of sea level variability. The reason for the discrepency in the other regions is not known at the present time.

7.5 Future development priorities

One of the biggest problems with the above approach is that it does not currently have an adequate error analysis. This is not only important to provide an estimate of forecast accuracy but it is also necessary in order to combine model with observational data at each sequential assimilation step. The required quantity is σ_f, the model forecast error, in equation (4). Of course it is simple enough to obtain the analysis error for η_a which (ignoring observation, forecast correlations) is;

$$\sigma_a^2 = \frac{\sigma_f^2 \sigma_T^2}{\sigma_f^2 + \sigma_T^2}, \tag{5}$$

but carrying this error forward to the next analysis time using the model physics is an extremely computationally intensive task. Ways must be found which simplify this process, probably for practical purposes to a point where it is considerably cheaper that performing a single model run. Only in this way can we continue to run the best physical models as part of the assimilation procedure. Recently we have moved to calculating σ_f at a low horizontal resolution using adaptive methods, without resorting to a Kalman filter type prediction of error evolution. This is appropriate because, in my opinion, it is likely to be the failings of the physical models or of the data quality going into them which will reduce the success of any ocean forecasting effort, rather than the failure of using a poor error analysis in the assimilation scheme.

A second major requirement is to initiate a multi-variable assimilation scheme. In particular there is a wealth of satellite Sea surface temperature and upper ocean temperature profile data from Bathythermographs, XBTs, which can be used in the assimilation process. In the first instance these data can be used for verification in order to assess whether the altimeter assimilation is improving the model fields. Later these data also need to be assimilated in order to provide the best possible ocean analyses, which should also lead to the best forecasts for use in operational oceanography.

7.6 Conclusions

This chapter has illustrated the application of one approach to the assimilation of satellite altimeter data to derive ocean currents and water properties. The vertical displacement method has been demonstrated both in twin experiments and has also been shown to be efficient and easy to apply for real altimeter data from TOPEX. The future prospects for ocean forecasting are bright, with the continuing global collection of accurate satellite data at high frequencies. There is still a long way to go before ocean forecasting becomes as sophisticated as atmospheric forecasting but there is every reason to believe that our understanding of the ocean circulation and thermodynamics will be greatly improved by these techniques over the next decade.

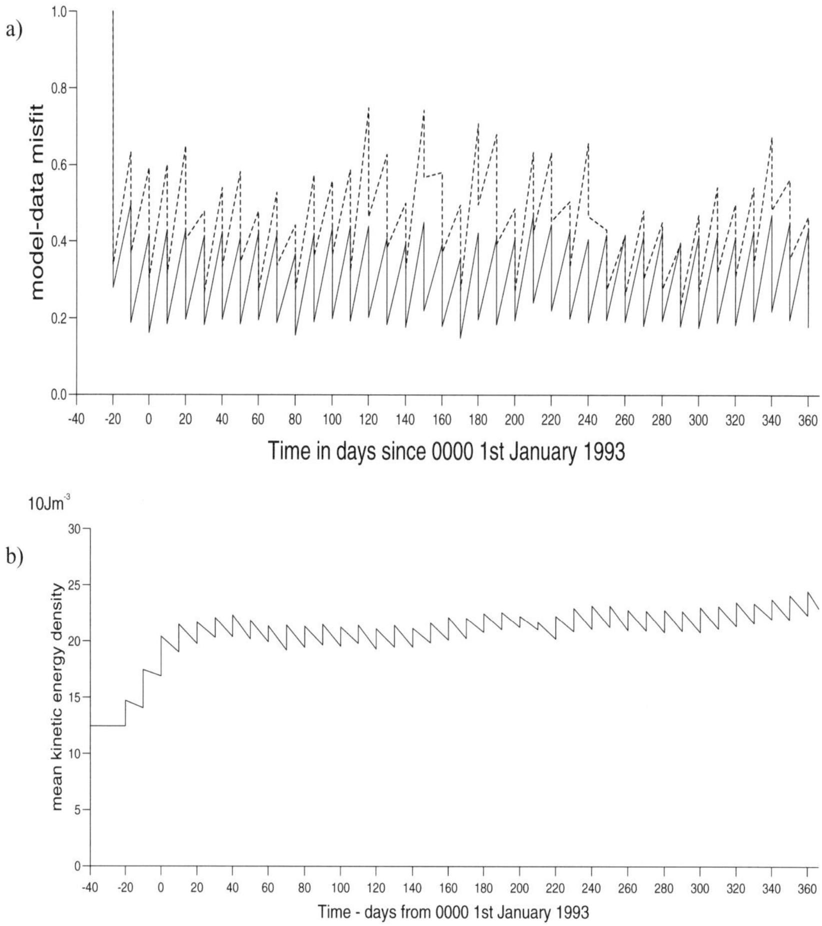

Fig. 7.8 (a) Model-data misfit averaged globally, solid line, and for the Gulf Stream, dashed line. Values are given as a fraction of the initial, pre-assimilation values: Global 73.1 cm^2, Gulf Stream 306.1 cm^2. (b) Evolution of volume average kinetic energy in the assimilating model. Note that assimilation begins at day -20 (11th Dec. 1992).

8 Ensembles, Forecasts and Predictability

ANTONIO NAVARRA
Istituto Nazionale di Geofisica, Rome, Italy

8.1 Introduction

The case of weather forecasts represents a mature example of prediction of a complete environmental system. The achievement of such an objective has required the overcoming of several rather formidable obstacles, some of them technical and organizational, but most of them conceptual and scientific, whose successful treatment has constructed the success story we know now.

The deployment of an adequate observational network, its maintenance and upgrading, the development of detailed numerical models and the availability of powerful super computers, are examples of some of the technical issues that have allowed progress. On the other hand, the understanding of the baroclinic life-cycles; the discovery of the interlocking areas of the planet, between middle latitudes and the tropics, between entrance and exit regions of the jet streams; the role of time-scale interactions, that has shown how the synoptic scale is strongly interacting with the monthly and seasonal time scales; the understanding of the role of diabatic processes, including most thermodynamical processes and land interactions, have allowed continuos progress and expanded the scope of the weather forecasts from the original limited application of aviation forecasts, i.e. upper air predictions, to predictions of an expanded weather system that today include the entire atmosphere, the soil, the upper ocean and can even include chemical constituents of the atmosphere as in the case of the experimental ozone predictions system.

The tremendous expansion of weather forecasts, supported by the increasing computing capabilities, terrific data assimilation techniques and the development of more accurate and detailed models, is opening the way that to what we can refer today as "environmental predictions", namely attempts to predict chemical and biological parameters not traditionally included in weather forecasts. The expansion of the predictive capabilities has been limited only by a fundamental properties of the meteorological equation, an extreme sensitivity to initial condition, sometimes known as the “butterfly problem”, or more scientifically as the fact that the equations describing the atmospheric motion are characterized by chaos.

The word chaos has been commonly used to describe this situation, but it is somewhat of a misnomer. According to the dictionary, chaos is a “condition in which things are out of their normal or proper places or relationships”, though the net result may be unpleasant, is difficult to think of the atmosphere-ocean-land system as “out of the its normal order”, it is in fact following definite equations and laws. Such relations and laws are usually well accepted and in general there is a good degree of agreement among scientists on their number and formulation. The big shock that the original Lorenz (1963) result brought was mainly originated in the training that scientists have received based on the idea that nature is smooth, simple and elegant. The sensitivity to initial condition broke with this scenario generating consternation and the impression that the class of equation that include the atmospheric equations were special and mysterious in some sense, therefore the name "strange attractors". The result in a nutshell can be seen as the statement that if one picks arbitrarily a system of ordinary differential equations of more than 3 variables the probability of choosing a system with a strange attractor is virtually one. Strange attractors are not strange at all, but they are everywhere, on the other hand, extremely intricate behavior can be described by powerful laws and equations. The sensitivity of the atmospheric motions to initial condition is therefore a very common occurrence typical of systems of several (greater than three) strongly interacting state variables.

The situation should not be confused with a real stochastic systems, like quantum mechanical systems, where the stochastic component is intrinsic, i.e. an experiment with exactly the same initial and boundary condition would end up in different results. The atmospheric motions are not intrinsically stochastic, because experiment with the same initial condition and boundary condition would result in the same evolution. However, the atmospheric circulations, and even more so the interlocked atmosphere-ocean-land-ecosystem, are largely composed of unobservable state variables that we will probably never be able to observe or specify in a model. This level of uncertainty, coupled to the sensitivity to the initial condition, creates a situation where is conceptually more convenient and more economic to treat the system as if it were an intrinsic stochastic system, exploiting the power of statistical and probabilistic concept into its description.

A single simulation or forecast is therefore going to be scarcely representative of the true behavior of the system, unless in special circumstances. In principle both initial and boundary conditions are going to be affected by unobserved state-variables, but the sensitivity to initial condition has been greatly investigated and it is certainly the most relevant in the classical initial value forecasting problem. Restricting to the initial condition case, we assume that the initial condition is really made up of two parts:

$$x = x_o + x_u \qquad (1)$$

representing the observed part x_0 and the unobserved part x_u that in this formulation contains also the unknown error on the observed variables. The true initial condition corresponds to a certain unknown value of the unobserved variables. Since in a prediction experiment we only have available x_0, the evolution of the forecasts will be dependent on the value of the unobservable variables. A single forecasts will not have a great probability of hitting the right value, to increase our chances will be necessary to perform multiple forecasts, hopefully filling up a hyper dimensional sphere that includes the true value.

Because the effort of the unobserved variables is so important becomes crucial to investigate and understand the dynamics of their evolution. There is a general feeling that this deviations must be small and, though we cannot have any hard proof, the empirical and operational experience indicate that this is indeed the case. The growth of the unobserved variables becomes then equivalent to the growth of small errors. How small errors grow, interact and equilibrate nonlinearly has been much the history of synoptic meteorology in the last 30 years. The following sections are devoted to the major conceptual steps along this line.

8.2 Sensitivity to initial conditions

The previous discussion has indicated that the initial evolution of a system is dominated by the dynamics of the instabilities. It is the number of instabilities triggered, their capability to grow rapidly and the physical character of the instability, that lead to the development of a system rather than another. However, this initial phase, the "linear era", is not bound to last. The rapidly growing disturbances outgrow the assumptions of linear theory and when they reach full amplitude they start to interact nonlinearly with each other and with the "basic flow". At this stage they can modify the basic state. The description in terms of "basic state" and "perturbations" is not useful anymore, the flow has evolved into a fully turbulent state, dominated by the statistics of the fluctuations. The nonlinear interactions in general tend to equilibrate and attenuate the explosive growth of the perturbations, but the boost caused by the initial selection of the set of instabilities will have already hurled the system towards a particular evolution.

The difference in evolution can be highly sensitive to small errors in the initial condition. Fig. 8.1 shows some examples obtained using the Lorenz (1963) system. The Lorenz system is a simple system of three equations that can be obtained from a drastic simplification of the equation for the motion of the atmosphere. They are in fact a very basic model for the atmospheric circulation, assuming that there is only the circumpolar vortex and two longitudinal waves. It was studying the properties of this system that Lorenz made the famous discovery of the sensitivity to initial condition. The legend says that having to suspend a numerical integration for lunch, he was too lazy after the break to reintroduce all the digits of the initial condition and to his amazement the evolution turned out to be completely different. The system is made up of three simple equations that interact nonlinearly with quadratic interactions with each other.

$$
\begin{aligned}
\frac{dx}{dt} &= -\beta x + yz \\
\frac{dy}{dt} &= -\sigma y + \sigma z \\
\frac{dz}{dt} &= -xy + \rho y - z
\end{aligned}
\qquad (2)
$$

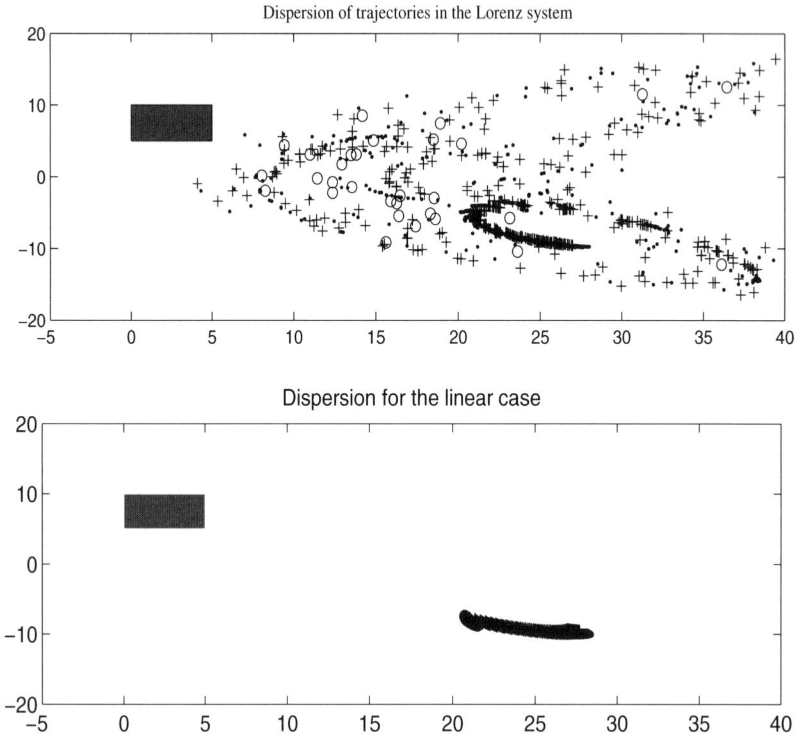

Fig. 8.1 Sensitivity to initial condition. The picture shows the final position of 900 integration of the Lorenz system, projected on the plane formed by two of the Lorenz variables. The symbol indicate the origin of the integration, each symbol corresponds to one quadrant of the rectangle.The top panel shows the case for the Lorenz system with all the interactions activated, the bottom panel represents the position of the 900 simulations in the case that the nonlinear terms are eliminated.

Fig. 8.1 illustrates this point by showing the end location of 900 simulations of the Lorenz system that all started from the rectangle at the top left. The picture shows a particular projection, corresponding to (y,z) for clarity, but the end points are distributed also in the third direction. The location of the trajectories after 2 time unit is all over the set of allowed values for the system (the "attractor set") even if the integrations were all bunched together initially. In fact, the separation between trajectories is such that is impossible to tell that they were selected in any special way. This particular set of initial condition cannot be distinguished from another set generated by randomly selected initial conditions loosely distributed all around the attractor. It was quickly found that the sensitivity does not necessarily scale with the size of the initial error (or initial separation of trajectories), in some cases small errors would tend to grow faster. The parameters, (β,σ,ρ) control the behaviour of the system. The sensitivity can be more or less pronounced, even disappear, depending on the values of the parameters.

If the nonlinear terms are eliminated from the Lorenz system the dramatic spread disappears. The bottom panel of Fig. 8.1 shows the case in which the nonlinear interaction in the Lorenz system are artificially turned off. After the same period of time the same initial condition as in the top panel are all bunched together in a small area rather than being dispersed all over the possible range of values. We have reduced the Lorenz system to a linear system, that does not exhibit a strong sensitivity to initial condition. Linear systems are the simplest example of equations that do not exhibit this dramatic sensitivity, but there are also nonlinear systems that keep neighboring initial conditions close together also at later times.

8.3 The traditional paradigm

In simple terms, the atmospheric circulation appears to be composed by a circumpolar vortex, characterized by the appearance of transient deviations from the zonal symmetry that can have time scales of a few days or months. The first physically and mathematically consistent interpretation of the atmospheric variability was offered by Charney (1947) in a famous paper in which he sketched the interpretative paradigm that has guided science for the following 40 years.

In Charney's vision the atmospheric motions could be seen as made up of the circumpolar current (the basic state) plus deviations that would draw their energy from the basic state itself. The deviations would be created continuously by the intrinsic instability of the basic state and they would grow linearly in the initial low-amplitude phase, but they would then interact with each other and ultimately with the basic state itself in a complex process of nonlinear equilibration. The fundamental process was however the initial growth of the perturbation that would set the stage, fixing the time scales and the energetic of the whole process.

The approach is now well described in textbooks (Pedlosky, 1979), but we can synthesize here in slightly different terms for the sake of our discussion. The equation of motion can be linearized around a time independent basic state,

$$\frac{dx}{dt} + Lx = 0 \tag{3}$$

where the operator L is representing the linearized equation of motion and the vector x is representing the state variable. In finite terms, the equation (2) is a vector equation for the time evolution of the vector x, representing, for instance the grid-point values in a finite-difference model or the spectral harmonics in a spectral model, the finite representation of the operator L is usually a matrix, A, and the entire equation can be written as

$$\frac{dx}{dt} + Ax = 0 \tag{4}$$

The linear equation (3) is valid for small amplitudes of the state vector x and so it describes small departures from the basic state vector. The solution of the problem can be written as

$$x(t) = e^{At}x(0) = S(t)x(0) \tag{5}$$

where the propagator operator $S(t)$ contains all the information on the time evolution of the perturbations. The structure of $S(t)$ is completely specified by the spectrum of A, i.e. by the eigenvector/eigenvalue structure of the equation of motion linearized around the basic state. The solution at time t can be written as a superposition of the eigenvectors ϕ of A, also known as the normal modes,

$$x(t) = \sum_i e^{\lambda_i t}\phi_i(\phi_i, x) \tag{6}$$

weighted by a time-evolving exponential with an e-folding time given by the corresponding eigenvalue. The symbol $(\cdot, \cdot)$ indicates a suitable scalar product.

Depending on the properties of A, and consequently of the basic state, the sign of the real part of λ will be positive or negative. With our conventions, positive real part indicate a growing mode, whereas a negative real part indicates a damping mode, whose amplitude decreases with time. Great importance is usually attached to the mode with the largest positive real part, the most unstable normal mode, since it is the one that asymptotically is going to dominate all the others. The growing rate will indicate the time scale of the process and the normal mode will provide the spatial structure of the modes, making possible detailed analysis of the life-cycles of the perturbation, energy budget, etc. The range of basic states that could be studied using analytical tools were soon exhausted by classical problems (Charney, 1947; Eady, 1949; Green, 1960) and only numerical treatment was then possible, i.e. computing the eigenvalue problem for the matrix **A**. More and more realistic basic states have been analyzed, from purely barotropic, to baroclinic, to time means flows that include large deviations from symmetry (Simmons et al., 1983; Frederiksen, 1983)

The classical paradigm then considers the atmosphere as a gas of evolving normal modes, some of them just starting to grow, others towards the end of their life cycle, strongly interacting with the basic state and contributing to the maintenance of the basic flow itself through the rectified eddy fluxes of heat and momentum. A basic state is said to be asymptotically stable, if there are no growing normal modes. The exponential growth of the normal modes guarantees in fact that a single growing mode, even with a very small growth rate, will eventually dominate.

The success of the classic paradigm lies in the successful estimation of the dominant spatial and time scales of variability of atmospheric motions. Atmospheric synoptic disturbances have typical spatial scales of a few thousand kilometers and typical time scales of a few days and the baroclinic instability theory predicts growth rates and a structure of the eigenmode compatible with the observations.

The classic paradigm is elegantly illustrated in the book by Pedlosky (1979) and some of the landmark papers are collected in the Charney volume (Lindzen, 1990).

The baroclinic instability theory became the instrument of choice to investigate and understand the dynamics of the extratropical cyclones. The theory was expanded to the spherical geometry both for the primitive equations case and the quasigeostrophic case (Simmons and Hoskins, 1976) using numerical techniques. A comprehensive review of the development of the theory is available (Hoskins, 1990).

In the late 80's, sufficient data were becoming available for detailed analysis of atmospheric variability at a global scale. Using both numerical simulations (Lau and Holopainen, 1984) and observations (Blackmon, 1976, 1977) the atmospheric variability was investigated in detail from time scales of a few days to the seasonal scale. The basic components of the classical paradigm however failed to emerge from the extensive diagnostic studies conducted at that time, In particular, the signature of the structure of the normal modes was nowhere to be found in real observations or in model simulations.

8.4 Finite time instabilities

At the end of 80s a series of studies (Lacarra and Talagrand, 1988; Farrell, 1985, 1989, 1990) indicated that the normal mode analysis suffered from severe problems when it was applied to the understanding of a real turbulent flow. The most important finding was that finite time transient growth can be present even if the flow is asymptotically stable, namely that there are no growing normal modes. This intriguing result is linked to the deep structure of the operators that can be obtained from the linearization of the equation of motion. Unless a very special basic state is chosen, e.g. a solid rotation, the operators will have no particular structure.

The eigenvalue/eigenvector analysis correspond to the search for a decomposition of the matrix A,

$$A = V\Lambda V^{-1} \tag{7}$$

such that V is the matrix whose column are the eigenvectors and Λ is a diagonal matrix. The interpretation of the eigenvector is that they form the basis that diagonalized the matrix A. The operation is not always possible and not all matrices are diagonalizable. Special interest have the operators that have a set of eigenvectors that are orthogonal to each other, in that case V is an orthogonal matrix and the eigenvector form an orthogonal basis. Such operators and their matrix representation are called self-adjoint.

Unfortunately the class of self-adjoint operators is rather small and none of the operators arising in meteorological applications belongs to this class. Unless special cases, the linearized meteorological operators, even assuming a very smooth basic state, are not self-adjoint. The orthogonality property of eiegnvectors is therefore lost together with the capability of using them as a basis. If the matrix V is not orthogonal then it cannot be used to project on the eigenvectors.

The complicated situation described demands a modification of the classical paradigm. As we have seen in the preceding section, the assessment of instability is based on the existence of growing normal modes, corresponding to growing amplitudes at all times. A more general requirement is to require that the amplitude grows (in some sense) at a specific time, in more precise terms we are looking for the conditions that allow a norm greater than one at some time (assuming normalized initial perturbation) without requiring exponential shape. The conditions can be expressed mathematically as follows

$$\frac{\|x(t)\|}{\|x(0)\|} = \frac{\|e^{At}x(0)\|}{\|x(0)\|} \leq \frac{\|e^{At}\|\|x(0)\|}{\|x(0)\|} = \|e^{At}\| \geq 1 \tag{8}$$

Equation (8) shows that the maximum amplification at a finite time is given by the norm of the propagator, i.e. the exponential operator e^{At}. This relation is more general than the normal mode equation because does not make any requirement on the form of the time evolution, it is only necessary that the norm is amplifying, but the time evolution may not be exponential. In fact, this is linked to well-known mathematical properties of exponential operators that allow polynomial growth for finite times, even if asymptotically the normal modes are the one that count. The formulation also allows to see clearly that the linear stability analysis can be formulated as a study of the properties of the exponential operator. The strange things that may happen in the calculation of the exponential are discussed in a beautiful paper by Moler and van Loan (1978).

Mathematically, the norm of the propagator can be calculated performing a Singular Value Decomposition (SVD) of the propagator. The SVD is a generalization of the eigenvalue/eigenvector decomposition. A very powerful result guarantees that every matrix can be decomposed as follows (Golub and van Loan, 1989)

$$A = U\Sigma V^T \tag{9}$$

where U and V are orthogonal matrices and Σ is a diagonal matrix of positive numbers, the singular values. A norm of the matrix A is given by the largest singular value. The result is so general that it holds also for general matrices m x n. The generality has been acquired at the cost of allowing two, rather than only one orthogonal basis, so that any matrix can be seen as a rotation, a stretching and another rotation. Another loss is the absence of a time-invariant shape for the growing modes. In general, the shape of the modes will be different at different times, and it is not possible to follow a single mode from the initial time, identifying it as the "most amplifying". At each time there will be a different one.

Another important difference with the eigenvalue/eigenvector case is the fact that because we have expressed the stability condition in terms of a norm, we will also have to specify in which norm we are measuring the growing mode. This is a delicate problem, but choosing a norm that corresponds to the energy of the perturbation usually gives good results.

The loss of the self-adjointness also produces strange effects on the finite time behaviour. The normal modes dominate the asymptotic behaviour of the evolution of a system like eq. (1), but they are not able to dominate the evolution uniformly from the start. This effect is illustrated in Fig. 8.2, which depicts synthetically the differences between long time and short time behaviour. The top two lines describe the eigenvalue distribution (middle) and the distribution of the singular values (top) for the propagator and for the matrix A. The examples shown are constructed with a distribution of singular values, typical of barotropic problems (solid line in the top row), that generates a peculiar distribution of singular values of the propagator (dotted lines) (see Navarra, 1987). The most amplifying mode correspond to the largest value and in general amplifying modes have values larger than one, whereas decaying modes have factor less than one. The top lines show the distribution of eigenvalues in the complex plane. The problems have been stabilized by adding suitable dissipation and so all the real parts are confined to the negative semiplane. By all accounts of the normal mode paradigm this flow is stable (asymptotically stable), since at long times the perturbations are exponentially damped.

The example shows clearly what happens with non self-adjoint operators. At finite times the evolution may be amplifying in the general sense we have defined above, but the long time behaviour is controlled by the eigenvalue spectrum. In the case of the atmosphere it is the characteristic at finite times that are more relevant, since at longer time, non-linear interactions will take over and so it is not at all clear that the modes will have time to evolve linearly into a asymptotic state. In fact the identification of normal modes in models and/or observations has always been difficult (Gall, 1976). Normal mode theory provided realistic estimation of time and space scales and a useful interpretative tool, but the identification of more specific prediction of the theory has been more controversial.

8.5 Ensembles

The sensitivity to initial conditions and the uncertainties in the development of initial errors, including both unobserved variables and real errors, lead to the consequence that a single integration may not really be indicative of the evolution of the atmosphere. It is likely that a single forecast may take a wrong development simply because it did not contain the right combination of uncertainties. In other words a single integration is not an adequate sampling of the possible values of the uncertainties around the nominal initial condition.

It becomes therefore necessary to sample more widely around the initial condition, usually perturbing the initial state in one way or another. The simplest approach would include just random perturbations of the initial conditions, but other techniques have been developed that tend to use "optimal" perturbations, in other words perturbations that excite the errors that grow fast. The finite time insta-

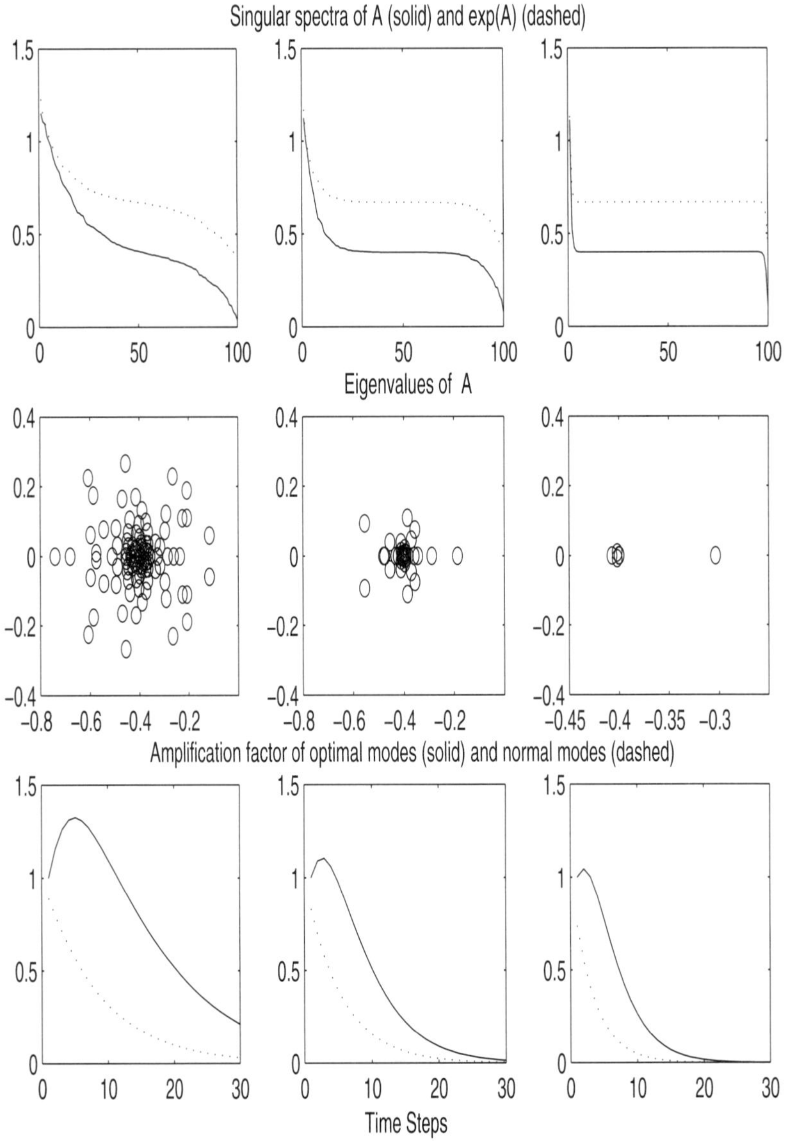

Fig. 8.2 Normal modes and finite time instabilities. The figure illustrates the difference between the asymptotic problem of stability (normal modes of exponential shape in time) and the finite time behaviour (polynomial growth in time). It is also shown that when the operator tends to become self-adjoint the differences disappear. The bottom panel also indicate how basic states that are stable regarding the asymptotic shape are in fact capable of growth over finite times.

bilities by Farrell (1988) are very suited in this case (Molteni et al., 1996) but other choices are possible (see the breeding technique by Toth, 1993). Whatever the technique to generate the perturbations, the net result is that the single forecast is substituted by a bundle of forecasts, the ensemble, that takes its place. Ensembles can give more informations, because they not only provide the deterministic forecast but also probabilistic information about the expected reliability and can give a sense the possible alternative evolutions.

The need for ensembles is also found in longer climate simulations that do not have the goal of practical forecasts. Usually these kinds of simulations have the objective of reproducing the statistical properties of the atmospheric and ocean dynamics, rather than a point by point time evolution. In this regard one might argue that they allow a lower standard, but on the other hand they require a much higher standard on the budget of quantities, like energy, radiation and salinity that result from delicate balances. Small errors in the radiation balance can be tolerated for a model designed for ten-days forecasts, but they are not acceptable in a model designed to simulate for hundreds of years.

The longer simulations have also been crucial to reveal another source of variability and statistical behaviour. These kind of simulations are usually used to simulate the general circulation and the variability of the climate system in general. The objective is here to reproduce as faithfully as possible the statistics of the system, yielding the right seasonal averages and an accurate balance of conserved quantities. In this regard it is expected that initial conditions are not very important because the model will produce its own equilibrium state, in practice initial conditions corresponding to random dates are often used. For instance, for a simulation starting in winter an arbitrary winter day is selected. Coupled models are often used in these experiments, but sometimes atmospheric GCM forced by prescribed Sea Surface Temperatures (SST) are also used.

Very long simulations performed with prescribed SST revealed a dynamical complexity that was somewhat unexpected. The atmospheric variability was found to be sensitive to the external forcing caused by the varying distribution of SST, but the response was not strictly consistent. Simulations that differed only in a choice of initial condition, surely well beyond the deterministic range and therefore forgotten after months and years of simulations, were still producing different variability events. The reason lies in the turbulent structure of the atmosphere. Referring to Fig. 8.3 it is possible to see how the response to an equatorial SST usually consists in upward motion that is strongly correlated with local precipitation. The detailed shape and intensity of the vertical motion is a function of the basic state according to a mechanism first proposed by Gill (1980). The response can be different even if the same SST is used for different realizations. An adequate number of realizations is necessary to identify the part of the response that is obviously linked to the SST. Simple ways include averaging over the ensemble (ensemble mean), separation of variance (Rowell, 1997; Ward and Navarra, 1997), methods based on the decomposition of variance (Wallace and Gutzler, 1981;

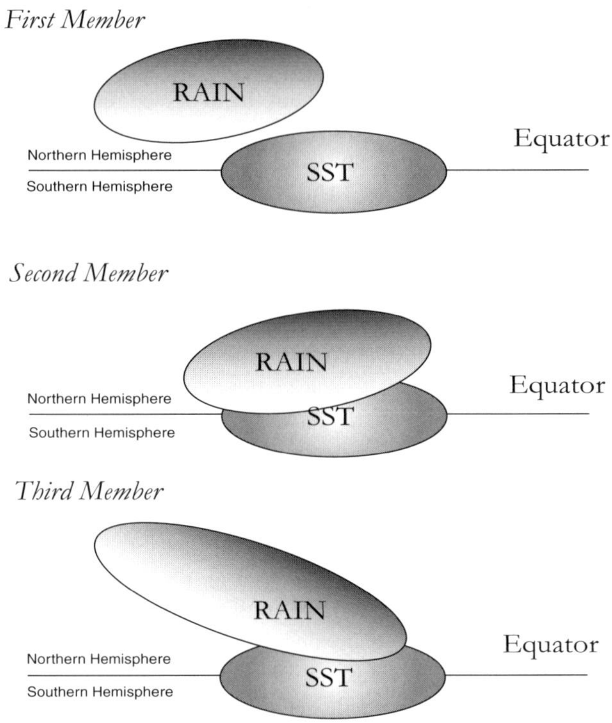

Fig. 8.3 Schematic description of the response of the atmosphere to SST. The usual equatorial response to a warm SST anomalies is precipitation. The intensity and location of the precipitation respect to the SST is however a function also of the atmospheric basic state that is prevalent in that period. The picture illustrates how the same seasonal anomaly can generate different precipitation pattern in different realizations according to the differences in the details of the basic state that are prevalent for that season in each member of the ensemble.

Cheng et al., 1995) and SVD decomposition of the cross-covariance between the atmospheric response and the SST itself (Bretherton, 1992).

For this class of boundary forced problems the predictability problem takes the form of a reproducibility issue, i.e. the potential predictability of a seasonal anomaly is given by how strongly the member of an ensemble will agree on the response. Strongly reproducible responses, i.e. all realizations obtaining very similar responses, are indication of anomalies strongly or totally influenced by SST and therefore predictable if the right SST are given. Fig. 8.4 shows a index that quantifies this concept introduced by Stern and Miyakoda (1995). The index is given by the ratio between the inter ensemble variance and the total variance, averaged over the duration of the simulation. Values larger than 1.0 indicate that the inter ensemble separation (Spread) is larger than the total variance that includes the normal climate variability, whereas values smaller than 1.0 indicate that the realizations are closer to each above the usual clutter of climate. It is possible to see how the reproducibility is space-dependent (and also time dependent, as Winter predictability is different from Summer) with major reproducible areas in the subtropics. Interestingly, small areas of high reproducibility appear in selected portion of the middle latitudes. These regions are those influenced by teleconnection patterns emanating from the tropical SSTs. A clear analysis of the concept of reproducibility can be found in Shukla (1999).

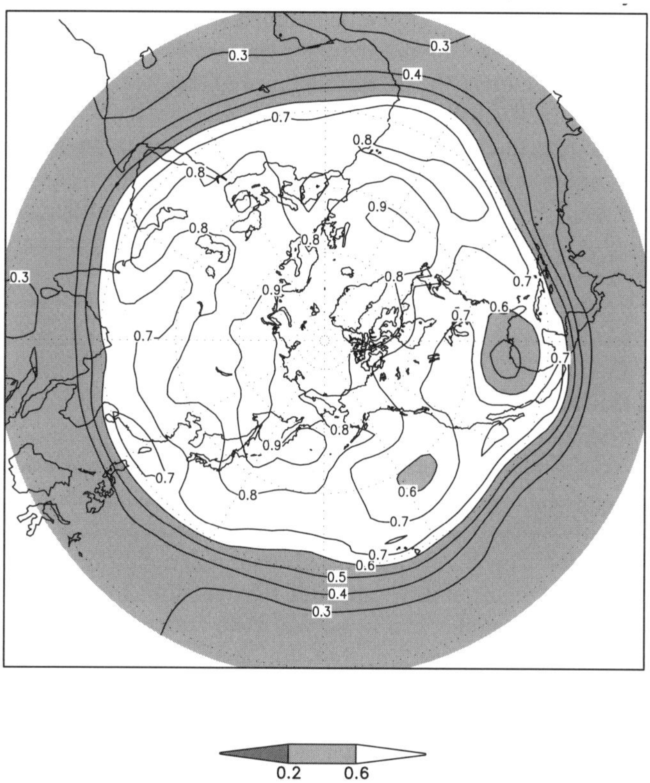

Fig. 8.4 Reproducibility index according Stern and Miyakoda (1995). Values of the index less than 1.0 indicate areas of potential predictability with respect boundary forcing by Sea Surface Temperature. In these regions the signal forced by the external forcing is greater than the internal variability of the atmosphere and therefore the ensemble members tend to reproduce the same seasonal anomalies.

8.6 Forecasting with ensembles

The need for ensembles changed the usual notion of the forecast as a single deterministic realization attempting to reproduce the time march of the atmosphere. Since the beginning of modern weather numerical forecasting in the 50's the development has been driven by a constant attempt to push the limit and the extension of accurate forecasts as further in the future as possible. It appeared that the improvement in the numerical technology would guarantee a smooth progress of the predictability limit. In fact, in the late 60's the possibility of using global models, higher resolution and detailed description of the physical processes description allowed the extension of the predictability limit to the medium range (7 - 10) days. The development of sophisticated techniques of data assimilation to generate good initial condition has strengthened the trend and allowed the routine realization of medium range forecasts. In the 70's a determined attempt was made to extend the range of useful predictions to the monthly time scales, but this time the task appeared to be much more difficult. Results showed that monthly predictions were an erratic business in which exceptional successes were mixed with clamorous failures. Cases could be found of particular situations in which the atmosphere was predictable for more than 2 weeks, but in general the reliability of the predictions was low. The main reason for this difficulty was the inability of the models to follow in a deterministic way the turbulent transitions that the atmosphere is performing on that time scale.

The interpretation of the setback was soon found in the Lorenz discovery. Small errors would amplify and it was impossible to keep them under control for a time longer than a week or so. It appeared that a very gloom future was in store for forecasting and that a cloud of uncertainty was going to obscure forever the future of the climate system. But the situation was not hopeless. The limitation on predictability observed in monthly forecasting was not completely consistent with the results from long climate simulations with prescribed SST that showed a substantial reproducibility of the atmospheric variability at time scales of a season or more. It was necessary to reconcile the predictability limit derived from Lorenz' discovery and the potential predictability implied in the long climate experiments.

In fact the apparent paradox can be solved if we realize that there is no reason why Lorenz' theory, namely the sensitivity to initial conditions, should apply also to long climate simulations with prescribed SST. The unpredictability of forecasts stems from the nature of this problem as an initial value problem, for which one has to worry about the development of unknown components of the initial conditions as we have seen in the preceding sections. Long climate simulation do not feel the effect of the initial conditions, in fact they are mostly independent of the particular initial condition chosen, because they are boundary value problems, in which the statistics is driven by the prescribed SST forcing.

A tight window for the possibility of forecasting at the seasonal time scale then appear, but it is linked to the availability of accurate SST. The atmosphere will respond to an SST distribution in a coherent and reproducible way, opening the

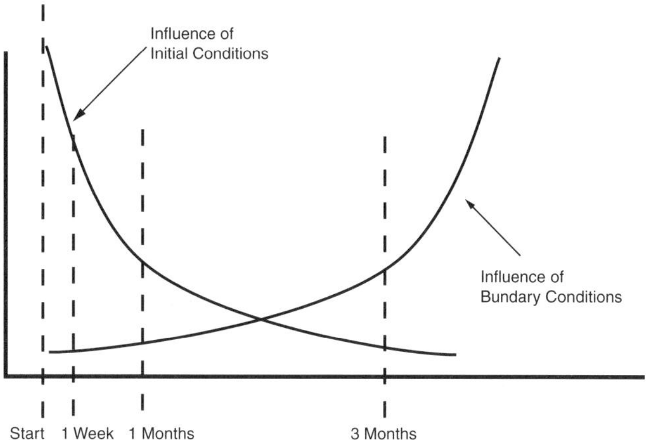

Fig. 8.5 Schematic view of the relation between medium range forecasts, monthly forecasts and seasonal forecasts. Medium range forecasts are dominated by initial conditions and therefore can be improved by better initial conditions and models, seasonal range is dominated by boundary (SST) forcing and so it can be improved by improving ocean initial conditions and models. The middle range, that include monthly forecasts is dominated by the internal turbulence of the atmosphere and it can be improved only a better understanding of the fundamental atmospheric interactions. At present time we do not have a very promising strategy to address this area.

way to practical applications if SST can be forecasted in an accurate and reliable way. Predictions at this range become then a matter of predicting the SST, resulting in an atmospheric boundary problem and in a oceanic initial value problem. The dynamic of the ocean shares the same fundamental nonlinearity of the atmosphere and so we can stay assured that it will exhibit the same sensitivity to initial conditions, but because the characteristic time scales of the ocean are much longer the predictability limit will not be two weeks, but rather 2-3 seasons in advance in the equatorial region and to a season off the equator.

The limit of predictability can therefore be sidestepped by going to a regime in which the climate system is sensitive only to the external forcing. The general vision is summarized in Fig. 8.5, that shows schematically the relative importance of initial conditions and boundary forcing for different time ranges.

Though the prospects of seasonal forecasts may be bright, they are not going to be the usual forecasts. Ensembles will have to be heavily used and the typical product will be a probabilistic information on seasonal anomalies, namely measures of the probability distribution of seasonal anomalies. Fig. 8.6 shows a typical product of a seasonal forecasting system, in this case the forecasts produced at the ECMWF. Shown here are ensembles of seasonal forecasts for NINO3 temperatures in the equatorial Pacific. The thin lines correspond to members of the ensemble and

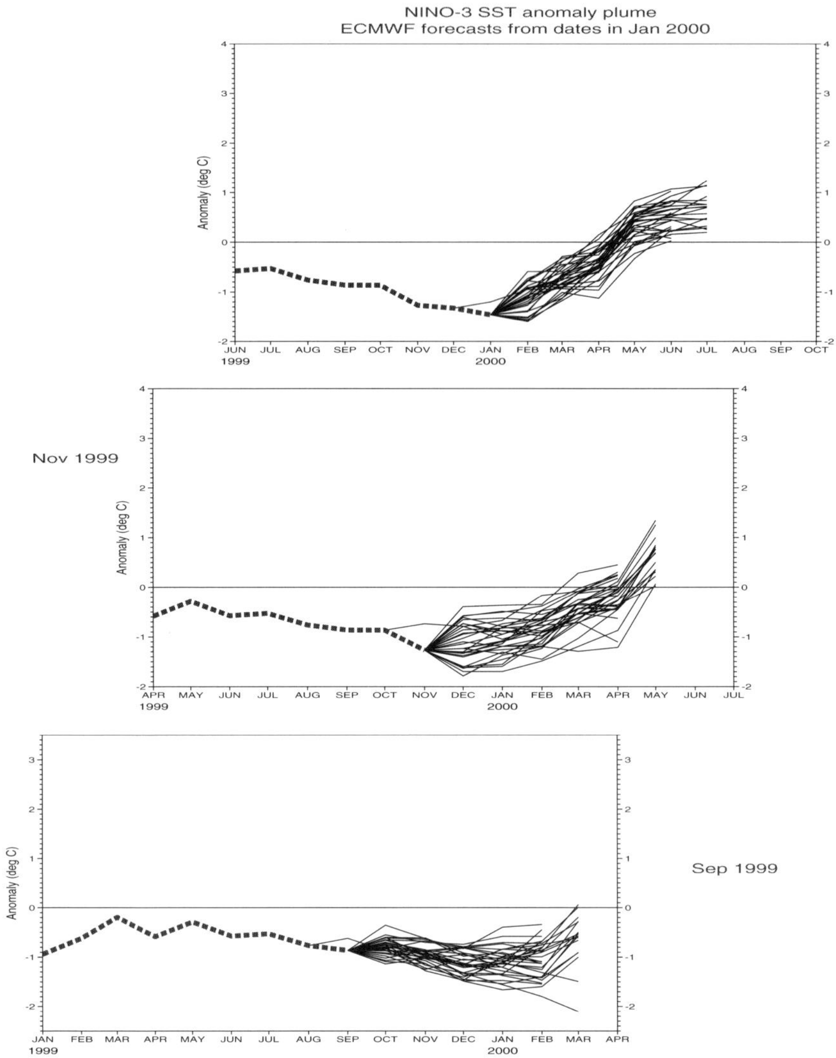

Fig. 8.6 Forecasts "plumes" obtained from the ensembles created by the ECMWF seasonal forecasting system. The picture shows ensembles of forecasts of NINO3 SST (thin lines) generated around different initial condition corresponding to the nominal period indicates in the pictures. The thick dashed line indicate the evolution of the real SST. It is possible to see how certain initial condition give rise to a larger spread in the separation between the different members in the forecasts. (Picture has been redrafted from images obtained from the ECMWF)

the thick line is the observed evolution of the NINO3 SST. The panels correspond to ensembles initialized in January 2000, November 1999 and September 1999. The ensembles include the true evolution of the system and this is encouraging, but the spread of the ensemble is substantial and the members predict, in the January case for instance, from zero anomalies to one degree anomalies. They all agree to a warming trend, but there is a large uncertainty about the magnitude of the warming. Reflecting that then these anomalies must be transformed into atmospheric anomalies by the response of the atmospheric model, it is possible to have a feeling of the long road still laying ahead of us.

8.7 Conclusions

The seasonal forecasts are the first examples of the kind of environmental forecasts we are going to expect in the next century. There is an increasing demand for forecasts of environmental parameters, like ice condition on roads, crop yield results, even price of agricultural and energetic commodities that reflect the influence of climate variability. To respond to this growing demand we will have to evolve our forecasting models from atmospheric and/or ocean models to Earth system models capable of giving estimates of the probability distribution of these various parameters.

As we move steadily away from parameters governed by well-defined mathematical and physical laws there will be a growing need to introduce statistical methods to extract the information from our predictions. We will make a transition from deterministic forecasts to estimation of probability distribution of environmental parameter, using ensemble and MonteCarlo methods. The task is harder and harder, but if we were able to dodge the predictability limit, at the end of the road the reward will be big and exciting.

9 On North Atlantic Intedecadal Variability: A Stochastic View

MOJIB LATIF, AXEL TIMMERMANN, ANSELM GRÖTZNER, CHRISTIAN ECKERT, REINHARD VOSS
Max-Planck-Institute für Meteorologie, Hamburg, Germany

9.1 Introduction

The North Atlantic climate system is characterized by considerable interdecadal variability. We show examples of interdecadal variability in Figs. 9.1 and 9.2. One of the main modes of the atmosphere over the North Atlantic is the North Atlantic Oscillation (NAO) (e.g. van Loon and Rogers (1978), Hurrell (1995)). The NAO is a dipole in sea level pressure (SLP), with centers of action near Island and the Azores (Fig. 9.1b), originally described by Walker (1924) and Walker and Bliss (1932). Hurrell (1995) defined an index of the NAO by the difference of the SLPs measured at Lisbon (Portugal) and Stykkisholmur (Iceland). Its time evolution (Fig. 9.1a) exhibits considerable interdecadal variability, with a maximum during the beginning of this century, a minimum during the 1960s, and strongly increasing values thereafter up to present. Moreover, there are observed fairly regular quasi-decadal [O(10 years)] variations during the most recent decades. The relatively strong upward trend observed during the last 20 years which contributed strongly to the rise in mean Northern Hemisphere surface temperature (Hurrell (1996)) has been the matter of intense scientific debate, since it is not clear as to whether this trend reflects greenhouse warming or is simply an expression of interdecadal variability.

Similar low-frequency variations on interdecadal time scales can be observed in the North Atlantic Ocean (Fig. 9.2). The potential temperature in the Labrador Sea in the depth range 800-2000m (curve labeled c), which is a good index of convection in this region, exhibits a time evolution, which is highly anti-correlated with the NAO index (Dickson et al. (1996)). Colder (warmer) temperatures imply stronger (weaker) convection, so that the cooling trend observed in the Labrador Sea since 1970 can be identified with enhanced convection. Interdecadal changes are also observed in the convective activity in the Greenland Sea (Fig. 9.2d) and in the salinity of the Sargasso Sea (Fig. 9.2e). The behaviours in the Labrador and Greenland Seas during the recent decades can be readily understood by atmospheric forcing. As the NAO strengthened and the Icelandic low intensified during the last few decades, more polar air masses are blown over the Labrador Sea, resulting in colder SSTs and stronger convection. Likewise temperatures in the Greenland Sea rised and convection weakened. This is visualized by Fig. 9.3 which shows the trends in SST and SLP observed over the last few decades. The trend in sea level pressure (Fig. 9.3b) is very similar to the NAO pattern (Fig. 9.1b), and consistent with a change in the NAO, the SSTs in the northern North Atlantic cooled, while they warmed further to the south (Fig. 9.3a). On the one hand, the surface temperature pattern can be explained as the response of the ocean to the surface heat flux changes that result from the change in the NAO. On the other hand, the SST anomaly pattern may well force a change in the NAO, as shown by Rodwell et al. (1999). Thus, one has to consider the interdecadal variability within the framework of large-scale air-sea interactions, as originally suggested by Bjerknes (1964).

We like to note one further interesting feature. As described above, the NAO exhibits quite strong quasi-decadal variations during the recent decades (Fig. 9.la). These fluctuations are not seen in the convection indices of the Labrador and Greenland Seas (Figs. 9.2c and 9.2d), while decadal variations are obvious in the upper ocean salinity time series of the Sargasso Sea (Fig. 9.2e) and the subsurface temperature at ocean weather ship 'C' (52.5°N, 35.5°W) which is shown in Fig. 9.4. Furthermore, the quasi-decadal variations in the NAO and those shown in Figs. 9.2e and Fig. 9.4 are not coherent. This may point towards different mechanisms for the generation of the decadal variabilities.

Different mechanisms were put forward to explain the interdecadal variability in general. On the one hand, external forcing mechanisms have been discussed for a long time. Variations in the incoming solar radiation, for instance, were proposed as one of the major sources of interdecadal variability (e.g. Labitzke (1987), Lean et al. (1995)). Cubasch et al. (1997) show indeed that some climate impact of the sun might exist on time scales of many decades and longer by means of a coupled model experiment prescribing variations in the incoming solar radiation, but it is fairly controversial at present how strong the fluctuations in the solar insolation actually are. The forcing of interdecadal climate variability by volcanos is well established and therefore less controversial, and major volcanic eruptions can be easily seen in regional and globally averaged temperature records (e. g. Robock and Mao (1995)).

On the other hand, interdecadal variability arises from interactions within and between the different climate sub-systems. The two most important climate subsystems are the ocean and the atmosphere. Non-linear interactions between different space and time scales can produce interdecadal variability in both the ocean (e.g. Jiang et al. (1996), Spall (1996)) and the atmosphere (e. g. Lorenz (1963), James and James (1989)) as shown by many modeling studies. More important in the generation of interdecadal variability, however, seem to be the interactions between the ocean and the atmosphere. The stochastic climate model scenario proposed by Hasselmann (1976) is a 'one-way' interaction: The atmospheric 'noise' (the random weather fluctuations) drives low-frequency changes in the ocean, leading to a red spectrum in the ocean's sea surface temperature (SST), for instance. It has been shown that the interannual variability in the midlatitudinal upper oceans is consistent with Hasselmann's (1976) stochastic climate model (e.g. Frankignoul and Hasselmann (1977)). This concept has been generalized recently by Frankignoul et al. (1997) who incorporated the wind-driven ocean gyres into the stochastic climate model concept, which extends the applicability of the stochastic climate

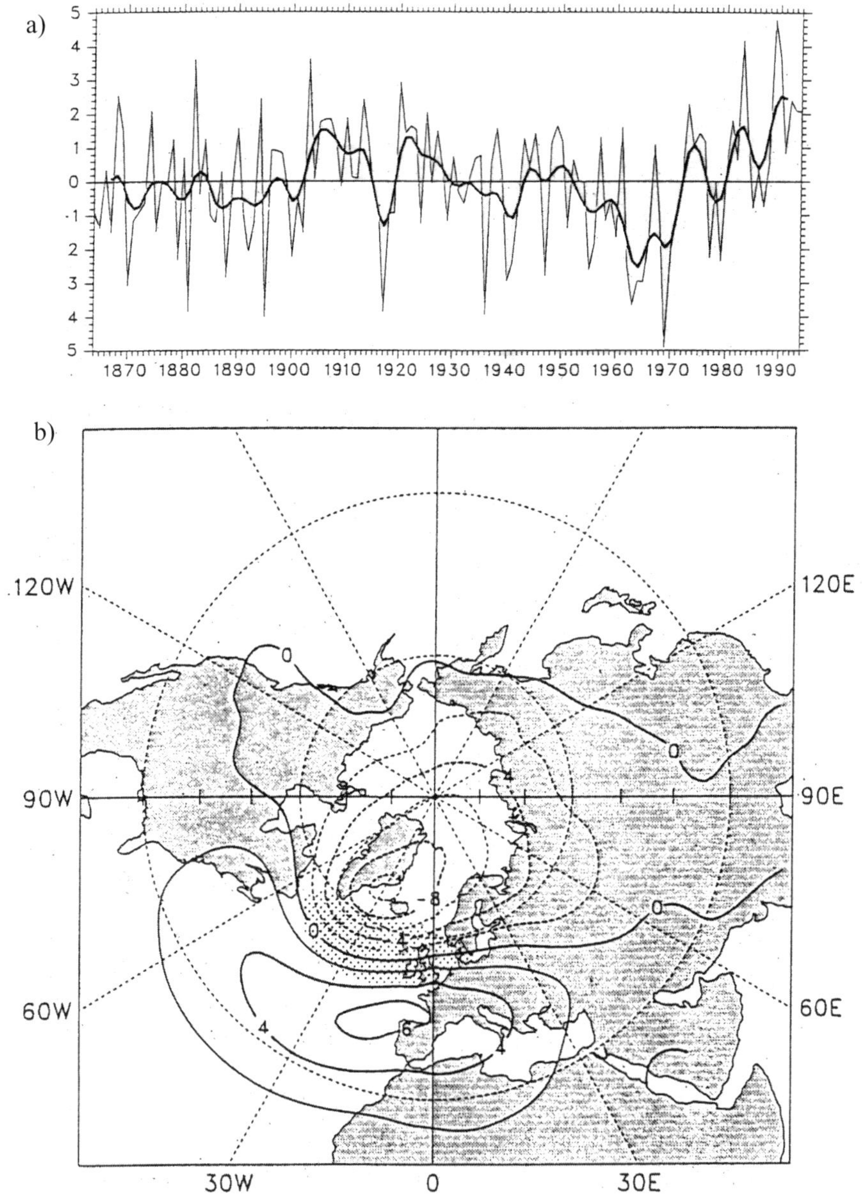

Fig. 9.1 a) Time series of the North Atlantic Oscillation (NAO) index as defined by the normalized SLP difference between Lisbon (Portugal) and Stykkisholmur (Iceland). b) Spatial pattern of the NAO as defined by the difference in SLP between 'high' and 'low NAO index' winters. From Hurrell (1995).

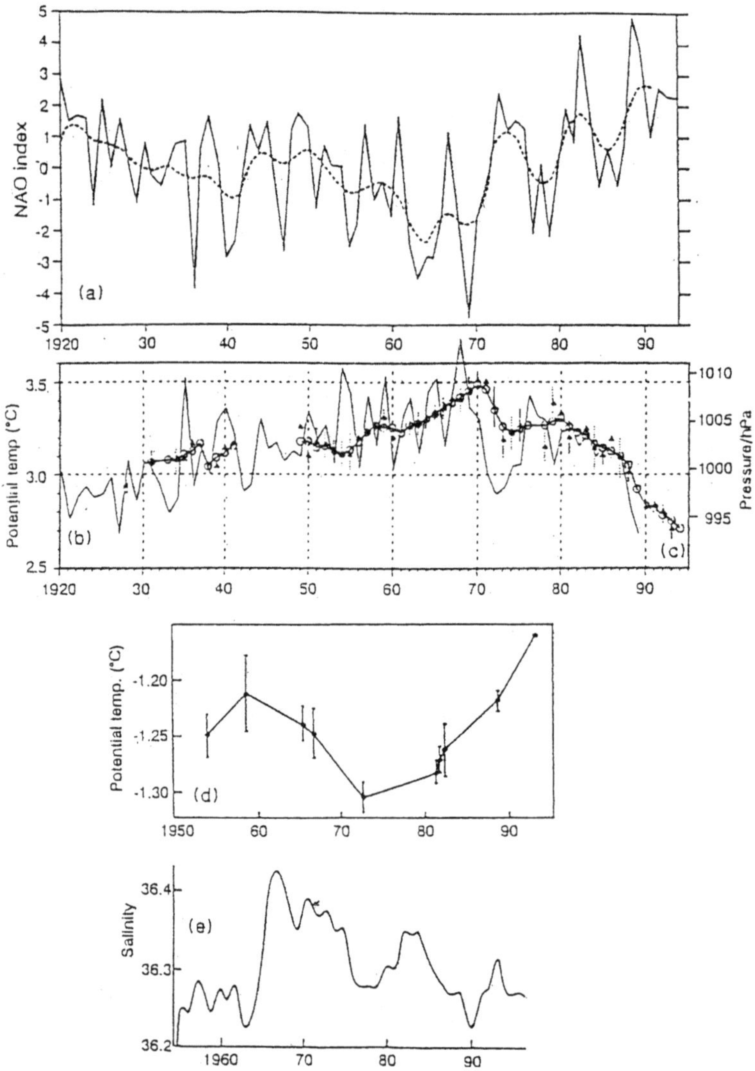

Fig. 9.2 Time series which vizualize the connection of atmospheric and oceanic changes at decadal time scales. a) NAO index (DJFM), b) mean SLP for western Greenland (DJF), c) potential temperature within the layer of minimum potential vorticity in the Labrador Sea, d) mean potential temperature in the Greenland Sea below 2000 m, e) salinity on the density surface $\sigma_\theta = 26.6$ off Bermuda. From Dickson et al. (1996).

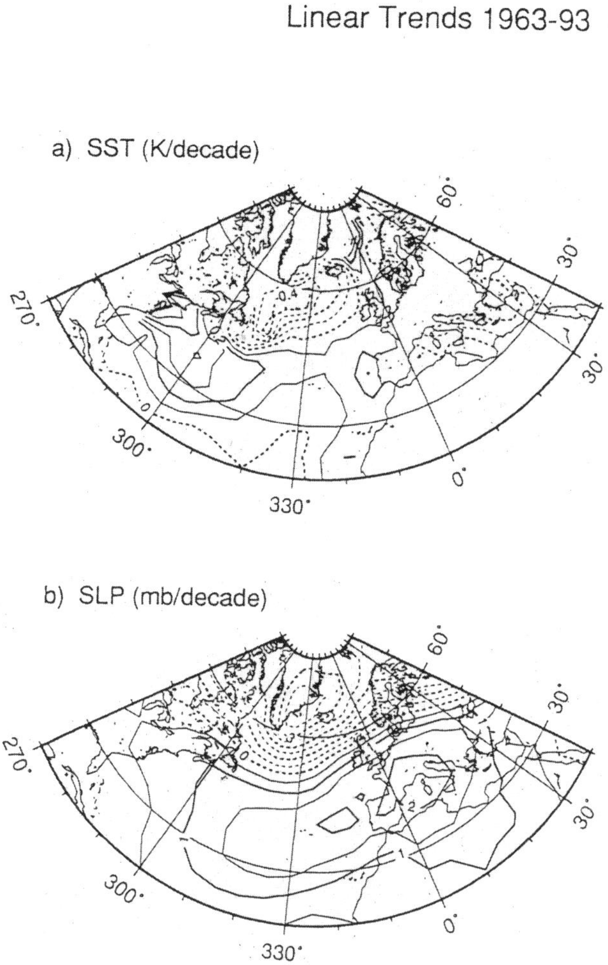

Fig. 9.3 Observed linear trends calculated for winter (DJF) December 1963 - February 1993 in a) SST (°C/decade), and b) SLP (hPa/decade). From Robertson (1998, pers. comm.)

model to interdecadal time scales. The atmospheric noise can excite also damped eigenmodes of the ocean circulation on interdecadal to centennial time scales, as shown, for instance, by the model studies of Mikolajewicz and Maier Reimer (1990), Weisse et al. (1994), and Griffies and Tziperman (1995).

'Two-way' interactions between the ocean and the atmosphere were also proposed to cause interdecadal climate variability (e.g. Latif and Barnett (1994), Gu and Philander (1997), Chang et al. (1997)). Similar to the El Niño/Southern Oscil-

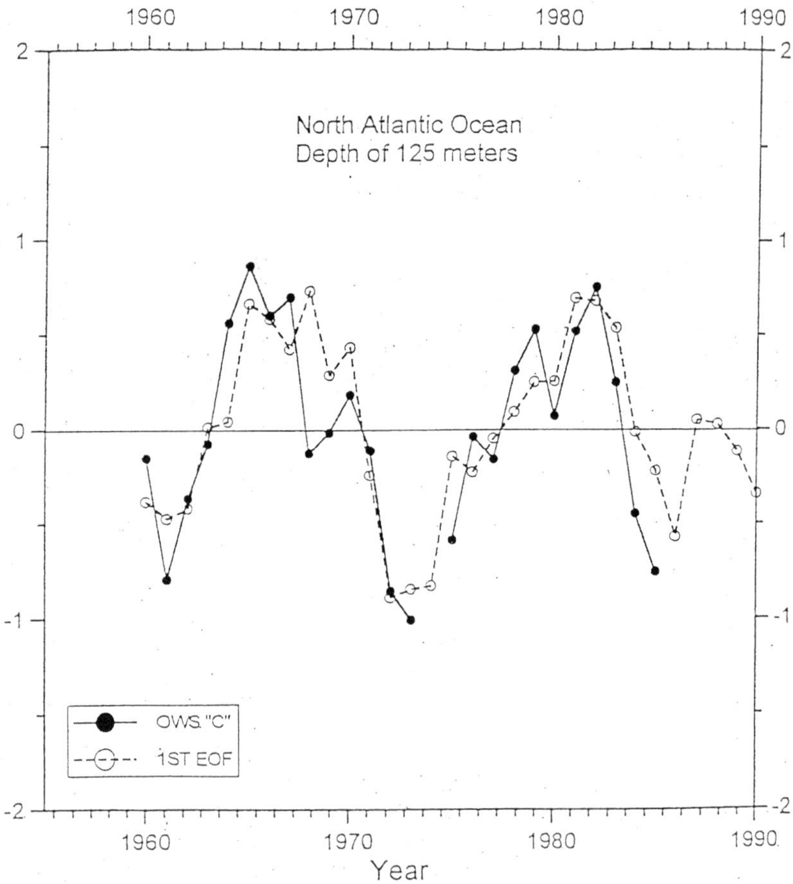

Fig. 9.4 Time series of the first EOF (multiplied by -1) of North Atlantic temperature anomalies and temperature residuals about linear trend for 1960-1985 at ocean weather ship 'C' (52.5°N, 35.5°W) at 125m. From Levitus et al. (1994).

lation (ENSO) phenomenon (e.g. Philander (1990), Neelin et al. (1994)) ocean and atmosphere reinforce each other, so that perturbations can grow to climatological importance. The memory of the coupled system (which resides generally in the ocean) provides delayed negative feedbacks which enable continuous oscillations that are forced by the internal noise within the coupled ocean-atmosphere system. The relative roles of the noise and non-linearities in the context of interdecadal variability, however, needs to be addressed further.

We shall investigate in this paper the interdecadal variability in the North Atlantic region only. There are many other regions which exhibit strong interdecadal variability, such as the tropical and North Pacific regions. Latif and Barnett (1994), for instance, postulate the existence of an interdecadal cycle in the North Pacific, and Gu and Philander (1997) put forward the idea of a coupled tropical Pacific-North Pacific interdecadal cycle. It is beyond the scope of this paper to address all the different aspects of interdecadal variability. The reader is referred to the observational papers of e.g. Folland et al. (1986), Dickson et al. (1988), Mysak et al. (1990), Deser and Blackmon (1993), Kushnir (1994), Trenberth and Hurrell (1994), Mann and Park (1994), Levitus and Antonov (1995), Zhang et al. (1997), Mantua et al. (1997) and references therein for further information on the observational aspects of interdecadal variability. A fairly comprehensive overview of the different aspects of interdecadal variability (including theoretical, modeling, and observational aspects) can be found in the recent book published by Anderson and Willebrand (1996), while an overview of the interdecadal variability simulated in coupled ocean-atmosphere models is given in the review article by Latif (1998).

The paper is organized as follows. We describe briefly the stochastic climate model concept and show some observational results supporting it in section 9.2. Section 9.3 deals with the quasi-decadal variability that arises from variations in the North Atlantic subtropical gyre circulation, as simulated by our coupled ocean-atmosphere model. We present the interdecadal variability associated with the North Atlantic thermohaline circulation of the same model in section 9.4. The predictability of interdecadal changes, as derived by classical predictability experiments, is addressed is section 9.5. The paper is concluded with a discussion in section 9.6.

9.2 The stochastic climate model

In contrast to the atmosphere, which has variations with typical time scales of a few days, the ocean is much more inert. Anomalies of the sea surface temperature (SST), for instance, persist typically for several months, while the characteristic time scale for the deep ocean is of the order of thousand years. Many aspects of the interactions between climate-subsystems with considerably different time scales can be described analogous to the Brownian motion in statistical physics. The picture is that oceanic anomalies (such as SST anomalies) result from the integration of many statistically independent atmospheric events, e.g. the passage of high and low pressure systems. One may view the SST variability, for instance, as the result of the integration of the surface heat flux anomalies within the oceanic mixed layer. This kind of interaction was introduced by Hasselmann (1976) and is referred to as the 'stochastic climate model'. The stochastic climate model scenario is a relatively simple concept which may be regarded as a kind of 'null hypothesis' for the generation of climate variability in general and decadal variability in particular.

The time evolution of a typical oceanic quantity y (such as SST) may be described by a Langevin equation

$$y'(t) = -\lambda y(t) + \xi(t), \tag{1}$$

with λ representing a damping parameter and ξ the atmospheric forcing that is assumed to be white noise. The oceanic response $G(\omega)$ is given within this concept by

$$G(\omega) = \sigma^2/(\lambda^2 + \omega^2), \tag{2}$$

Here, ω denotes the frequency and σ the standard deviation of the white noise forcing. We present in Fig. 9.5 schematically the spectra that result from the simplest version of the stochastic climate model (1). The atmospheric input spectrum (e. g. that of the surface heat flux) is assumed to be white noise, i.e. the spectral density does not depend on frequency. The oceanic response spectrum is red, with a slope ω^{-2} down to a frequency that depends on the damping λ. At frequencies smaller than λ the oceanic spectrum is flat (white). It should be noted that no response of the atmosphere to the variations in the ocean is considered here. It is generally assumed, however, that such a feedback exists, at least at low frequencies, which would explain some of the interdecadal variability observed in the atmosphere (e. g. the NAO, see Fig. 9.1a).

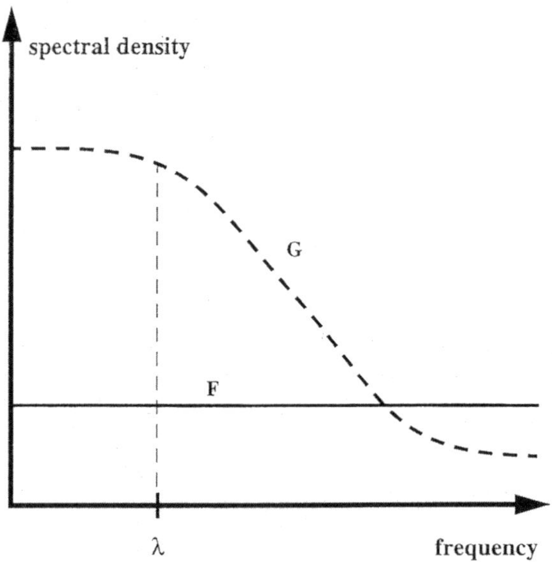

Fig. 9.5 Schematic input spectrum F and response spectrum G which result from the simplest version of the stochastic climate model (1). The atmospheric input spectrum is white, while the oceanic response spectrum is red down to a frequency which is determined by the damping λ.

Spectra computed from atmospheric and oceanic observations taken at ocean weather ship 'I' (60.8°N, 20.6°W) support the stochastic climate model idea (e g. Frankignoul and Hasselmann (1977), Hall and Manabe (1997)). This is visualized by Fig. 9.6 which shows spectra of anomalous SST and sea surface salinity (SSS) taken at weather ship 'I'. By and large the two spectra are consistent with the stochastic climate model, and a perfect red noise spectrum is not an unreasonable fit over the entire frequency range.

We expect according to the stochastic climate model that the spectrum of the SST anomalies flatens at a higher frequency relative to that of the SSS anomalies. The damping for SSS anomalies is considerably weaker than that for SST anomalies, which is due to the fact that the anomalous fresh water flux does not depend directly on the SSS anomalies, while the anomalous surface heat flux does usually depend critically on the SST and tends to damp its anomalies (a counter example, however, will be shown in section 9.3). A spectral analysis from the North Pacific (Hall and Manabe (1997)), for instance, confirms this. Since the decay time scales for SST and SSS anomalies in the North Atlantic are not different (Fig. 9.6) and the variations in SST and SSS coherent beyond frequencies of about 1 yr^{-1} (Hall and Manabe (1997), not shown), one may conclude that non-local processes are also important in generating low-frequency variability in the North Atlantic. Such processes may be associated with changes in the large-scale ocean circulation and resultant changes in the advection of heat and salt, as described below.

In contrast to the spectra shown in Fig. 9.5 which result from the simplest version of the stochastic climate model, some spectra computed from observations in the North Atlantic region show also peaks (e.g. Deser and Blackmon (1993)). These peaks can be also explained by the stochastic climate model concept (e.g. Mikolajewicz and Maier-Reimer (1990)), and resonant interactions between the ocean and the atmosphere play a key role. The ocean or the coupled ocean-atmosphere system may support damped eigenoscillations that are excited by the noise (i.e. stochastic forcing) in the system (like a swing in the wind). A good prototype for such a scenario, according to which damped oscillations are excited by the stochastic forcing inherent to the system, is the stochastically forced harmonic oscillator

$$y''(t) + \lambda y'(t) + \omega_0^2 y(t) = \zeta(t) \tag{3}$$

Here ω_0 is the eigenfrequency, and the meaning of the other symbols is as in (1). The oceanic response G(ω) is given by

$$G(\omega) = \sigma^2/[(\omega^2 - \omega_0^2)^2 + \omega^2\lambda^2] \tag{4}$$

Such a generalized concept yields peaks at resonant frequencies ω_r which depend on the eigenfrequencies ω_0 and the damping λ

$$\omega_r = (\omega_0^2 - \lambda^2/2)^{1/2} \tag{5}$$

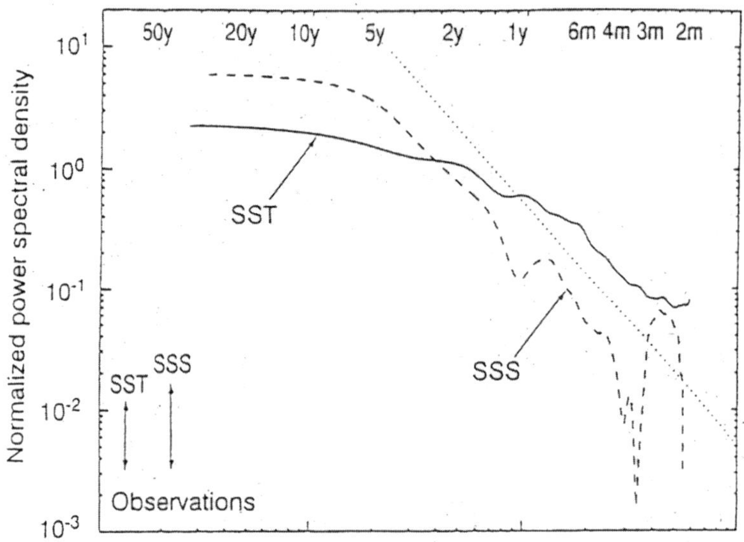

Fig. 9.6 Spectra of anomalous North Atlantic SST and SSS observed at ocean weather ship 'I' (60.8°N, 20.6°W). The dotted line shows the slope ω^{-2}. From Hall and Manabe (1997).

The peaks will be generally superimposed on a red background spectrum which results from the 'pure' integration of the noise according to (1). In the case of an 'ocean-only' mode, in which the feedback from the ocean to the atmosphere is not important to the existence of the mode, the stochastic theory predicts a peak in the oceanic spectrum only, while a typical atmospheric spectrum is white (Fig. 9.7a). Such a situation corresponds to the mode described by Delworth et al. (1993) investigating the results of a coupled model simulation, as discussed in detail by Griffies and Tziperman (1995). In the case of the excitement of a 'coupled ocean-atmosphere' mode by the stochastic forcing, in which ocean and atmosphere are controlled by the boundary conditions of the respective other component, peaks will be found in both atmospheric and oceanic spectra, as it is the case for the El Niño/Southern Oscillation phenomenon (e.g. Philander (1990)). This situation is shown schematically in Fig. 9.7b and may apply to the modes simulated by some coupled models in the North Atlantic, as described by Grötzner et al. (1998) and Timmermann et al. (1998). It is plausible to assume that those eigenmodes will be excited preferably which have surface expressions that match those of the stochastic forcing (this is usually referred to as 'spatially resonant interaction').

We would like to point out another class of stochastically forced variability. The internal atmospheric variability can be decomposed into relatively few spatial patterns such as the Pacific North America (PNA) pattern or the North Atlantic Oscillation. If oceanic advection is included in the stochastic climate model (1), peaks in

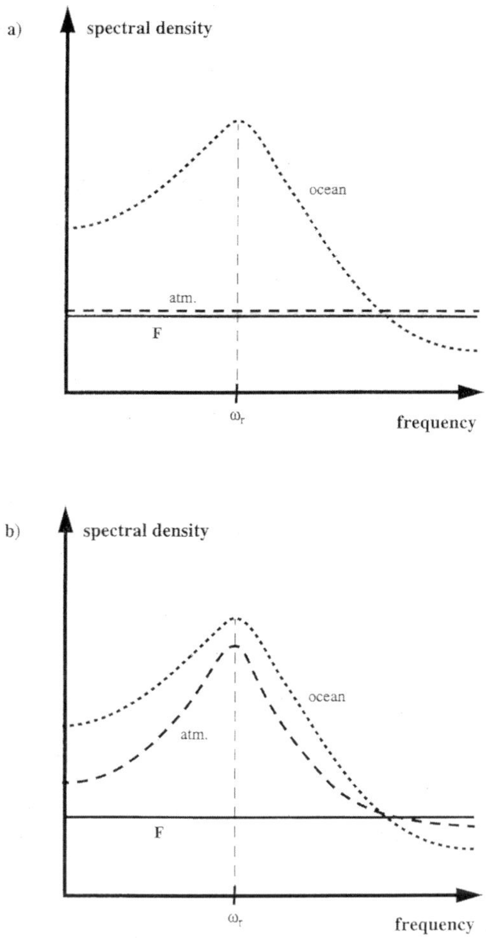

Fig. 9.7 Schematic atmospheric and oceanic spectra which result from the stochastic excitement of a) an 'ocean-only' mode and b) a 'coupled ocean-atmosphere' mode. In the former case, the atmospheric spectrum is white, while the oceanic spectrum shows a peak at the resonance frequency; in the latter case both the atmospheric and the oceanic spectra show a peak at the resonance frequency.

oceanic spectra are possible, although the atmospheric forcing is still white in time. The spatial coherence of the atmospheric forcing in concert with the mean currents are able to produce distinct time scales, and a peak in ocean spectra may be found at the frequency $\omega = k_0 u$, with k_0 the dominant wavenumber of the atmospheric forcing and u the mean current velocity advecting oceanic anomalies. Such type of stochastically forced variability was proposed by Lemke et al. (1980), Frankignoul and Reynolds (1983), Sutton and Allen (1997), Saravanan and McWilliams (1997), and Weisse et al. (1997). The latter attribute the dynamics of the Antarctic Circumpolar Wave (ACW, White and Peterson (1996)) to such a stochastic climate model with advection, while Sutton and Allen (1997) and Saravanan and McWilliams (1997) explain part of the North Atlantic low-frequency variability by this mechanism.

In summary, the stochastic forcing of the climate system by the noise inherent to it is an important mechanism in generating climate variability. The shape of the spectra derived from observations (or model simulations) may provide important indications for the underlying dynamics leading to the variability considered. If the variability is consistent with the simple version (1) of the stochastic climate model, one may assume that ocean dynamics are not important, and that the variability arises from the integration of the atmospheric weather noise by the oceanic mixed layer. If (statistically significant) peaks are found, this points generally towards a prominent role of the ocean dynamics in producing the variability at decadal time scales. We shall show two such examples below which were both derived from a multi-century simulation with a coupled ocean-atmosphere general circulation model.

9.3 Quasi-decadal variability in the North Atlantic

We turn now to the integration with our coupled ocean-atmosphere general circulation model (ECHAM3/LSG). Details of the coupled model and the integration can be found in Voss et al. (1998), Timmermann (1996) and Timmermann et al. (1998). The coupled model was integrated for 2000 years, and the analyses shown below were performed using annualy averaged values. The coupled model simulates reasonably well the low-frequency variability in the North Atlantic, which is shown by the comparison of the standard deviations of annually averaged SSTs as computed from observations and the coupled model simulation (Fig. 9.8).

One can separate conceptually the ocean circulation in the North Atlantic into a 'wind-driven' and a 'thermohaline' part. The former is forced by the surface wind stress and associated with horizontal circulations in the upper ocean (subtropical and subpolar gyres), while the latter is associated with a meridional (north-south) circulation and deep convection in the northern North Atlantic. We shall show that variations in both types of circulation systems can lead to interdecadal variability. Since the adjustment times of the gyre circulations are considerably shorter than that of the thermohaline circulation, variations associated with the gyres have typical time scales of about 10-20 years (quasi-decadal), while those linked to the thermohaline circulation have typical time scales of several decades (interdecadal). Interestingly, both types of variability seem to co-exist in the coupled model simulation, as shown below, and there are some indications from observations that this is the case in the real world too (e.g. Deser and Blackmon (1993)).

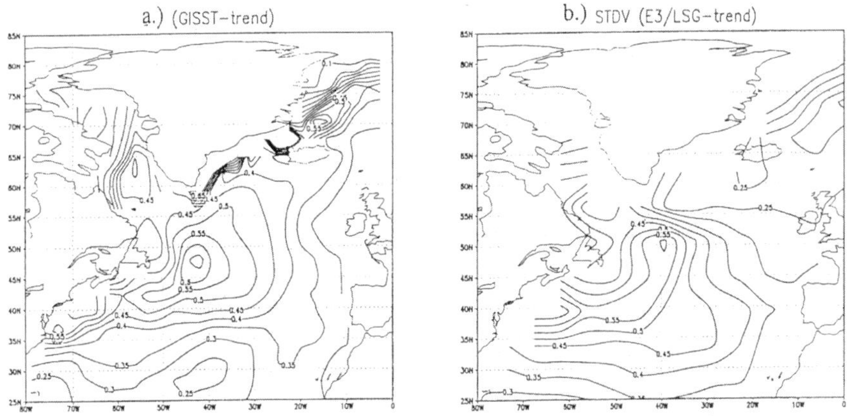

Fig. 9.8 Comparison of the SST standard deviations in the North Atlantic (computed using annually averaged values), as derived from a) the observations (GISST) and b) the coupled model simulation with the ECHAM3/LSG CGCM. From Timmermann et al. (1998).

We present in Figs. 9.9 and 9.10 time series and spectra of model-simulated anomalous North Atlantic SST, meridional overturning (an index of the strength of the thermohaline circulation) and SLP. It is obvious from the time series that low-frequency variations are much more pronounced in the ocean (Figs. 9.9a and 9.9c) than in the atmosphere (Fig. 9.9b). Consistent with the stochastic climate model picture, the spectra of the SST and overturning anomalies are red (Figs. 9.10a and 9.10b), while that of the anomalous SLP is (almost) white (Fig. 9.10c). At periods of about 15 and 35 years, however, both the SST and SLP spectra exhibit peaks, which indicates that two coupled modes may exist. The 35-year peak only is found in the spectrum of the anomalous overturning, which indicates that the interdecadal variability is connected to variations in the models's thermohaline circulation. The significance of the spectral peaks was assessed by testing the spectra against the null hypothesis, that the variability can be described by a first-order autoregressive process (which would result from the simplest version (1) of the stochastic climate model). The spectral peaks are marginally significant, and the overall structures of the spectra indicate that the decadal and interdecadal modes are strongly damped and excited by the stochastic forcing. However, as shown below, the physics of the quasi-decadal variability differ strongly from those of the interdecadal variability (section 9.4). Thus, it makes sense to distinguish these two types of low-frequency variability from a physical point of view.

Our hypothesis for the generation of the quasi-decadal mode is that it arises from air-sea interactions and is associated with variations in the subtropical gyre circulation. In order to describe the spatial structure of the mode, we performed a Canonical Correlation Analysis (CCA) of the North Atlantic SST and SLP anomaly fields.

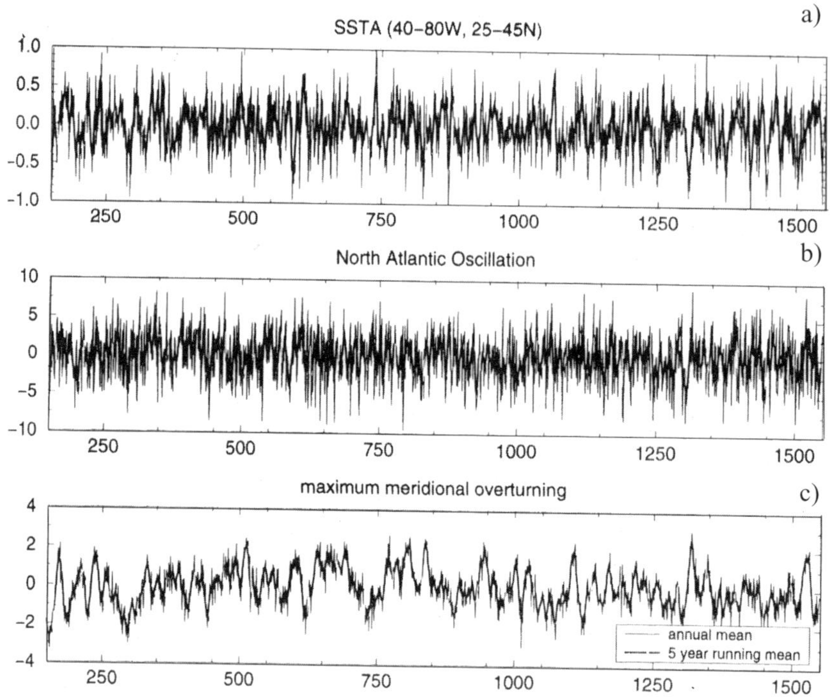

Fig. 9.9 Time series of anomalous North Atlantic a) SST (K), b) NAO, and c) meridional overturning (Sv) simulated by the the coupled model ECHAM3/LSG. From Grötzner (1999, pers. comm.).

The leading CCA mode (canonical correlation = 0.83) is connected to the quasi-decadal mode, which was inferred from the two CCA time series (not shown) which exhibit spectral peaks at a period of 15 years. The leading CCA mode explains 18% and 52% in the SST and SLP anomaly fields, respectively. The resulting SLP anomaly pattern (Fig. 9.11b) is the North Atlantic Oscillation (NAO), with opposite changes equatorward and poleward of about 50°N. The corresponding SST anomaly pattern (Fig. 9.11a) is a tripole with a band of positive anomalies in the region 20°N-40°N and negative anomalies to the north and south. These patterns are consistent with those shown by Deser and Blackmon (1993) and Grötzner et al. (1998), who derived the spatial SST and SLP structures of the quasidecadal variability over the North Atlantic from observations.

The memory of the quasi-decadal mode resides in the ocean. This can be demonstrated by a POP-analysis (Principal Oscillation Pattern analysis) of upper ocean heat content anomalies (see Timmermann (1996) for details). The leading POP mode explaining 21% of the variance has a decadal rotation period, as was revealed by a spectral analysis of the corresponding POP time series (Fig. 9.12). The evolu-

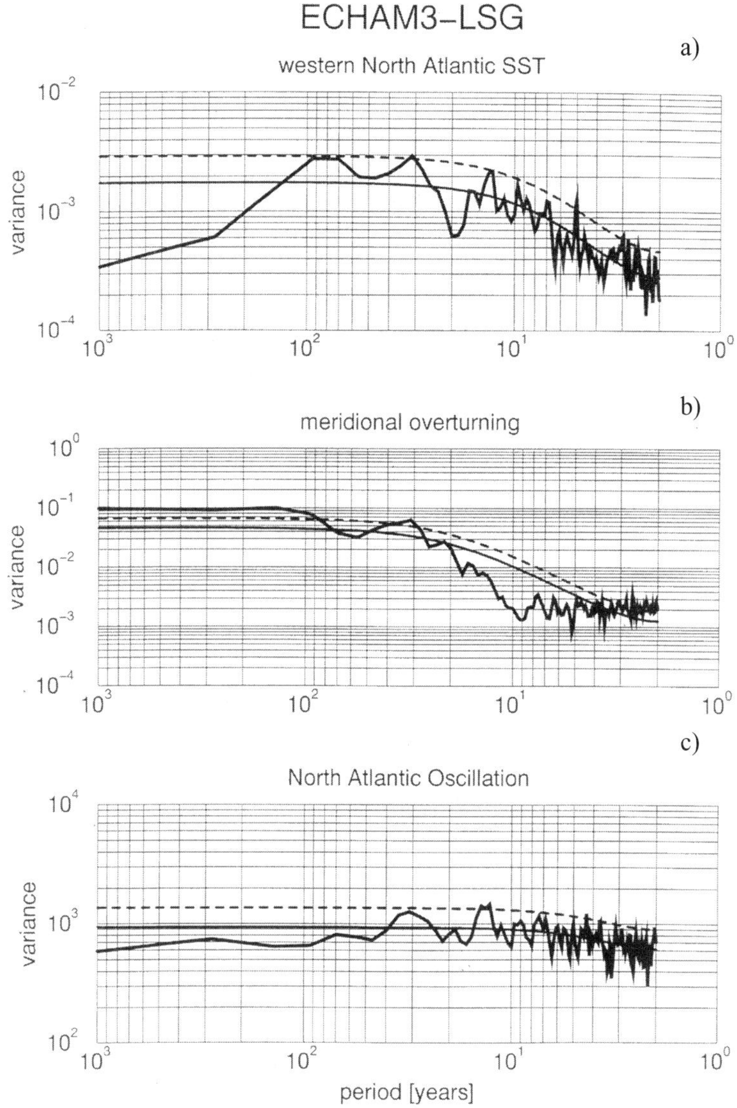

Fig. 9.10 Spectra of the anomalous North Atlantic a) SST, b) meridional overturning, and c) NAO shown in Fig. 9.9. From Grötzner et al. (1999, pers. comm.).

tion of the anomalous upper ocean heat content is dominated by a clockwise rotation around the subtropical gyre (Fig. 9.13). Advection of temperature anomalies by the mean currents and wave processes (in response to changing winds) are likely to contribute both to the evolution in the upper ocean heat content, as argued

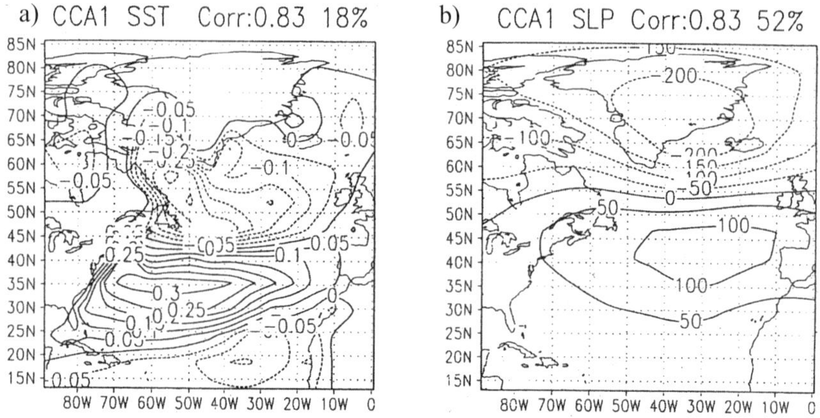

Fig. 9.11 a) North Atlantic SST anomaly and b) SLP anomaly fields as derived from the leading CCA mode. These patterns are characteristic of the quasi-decadal mode. From Timmermann (1996).

by Grötzner et al. (1998) studying the quasi-decadal variability in another coupled model. Thus, the time scale of the oscillation is probably set by a combination of advective and wave time scales, since both time scales are of the same order. The importance of advection in the quasi-decadal variability was also highlighted by Sutton and Allen (1997) who investigated North Atlantic SST observations for the period 1945-1989.

In summary, the quasi-decadal variability in the North Atlantic region may be regarded as an eigenmode of the coupled ocean-atmosphere system, so that the interactions of the ocean and the atmosphere are crucial to the existence of the quasi-decadal mode. Ocean dynamics appear to be crucial in producing the oscillatory behaviour at the quasi-decadal time scale. We would like to point out however, that the quasi-decadal mode is strongly damped (as it is in the observations) and that the bulk of the quasi-decadal variability may well be explained by the stochastic climate model scenario.

9.4 Interdecadal variability in the North Atlantic

On longer time scales, we expect that variations in the thermohaline circulation become more and more important. The adjustment time of the wind-driven gyres is of the order of one to two decades only (depending on the ocean basin considered), while the adjustment time of the thermohaline circulation ranges from decades to many centuries (depending on which portion of the global thermohaline circulation is considered). As can be inferred from Fig. 9.10, enhanced variability in both the

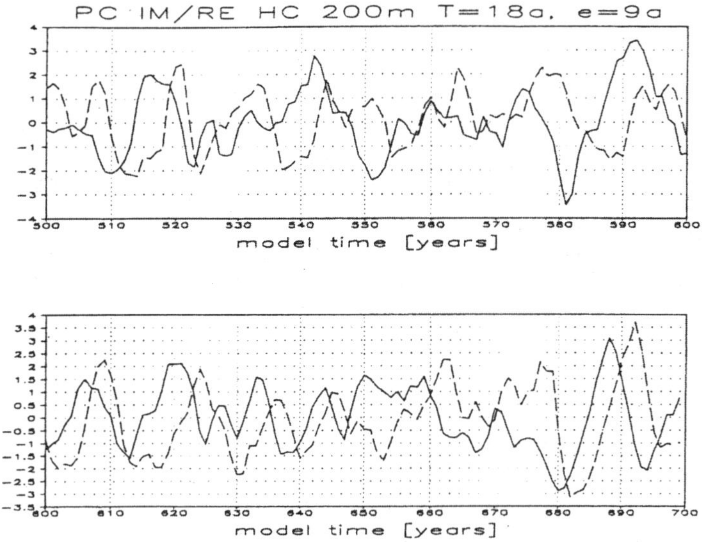

Fig. 9.12 Time series of the leading POP mode of North Atlantic upper ocean heat content (0-200m) anomalies. The POP analysis has been performed using the model years 500-700. The dashed line shows the real part and the solid line the imaginary part. The real part time series is highly coherent with those of the leading CCA mode (with opposite signs). From Timmermann (1996).

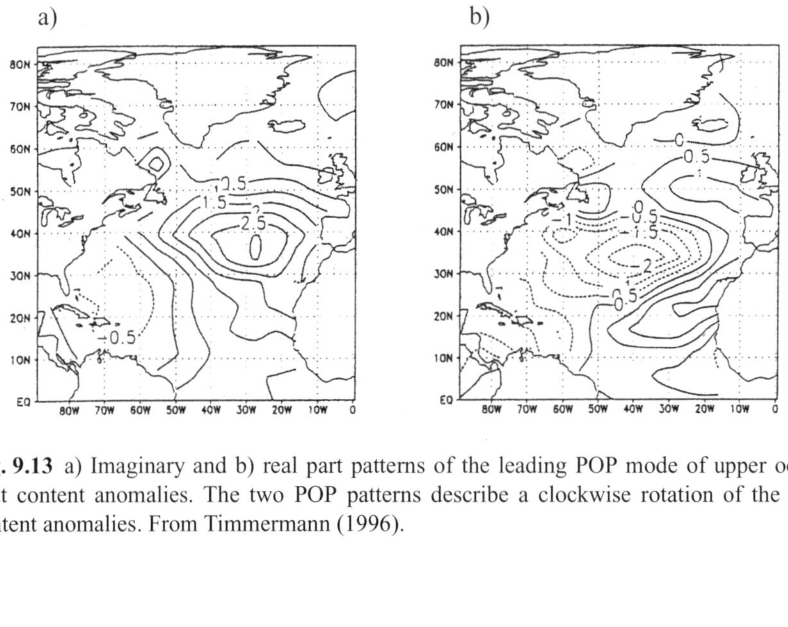

Fig. 9.13 a) Imaginary and b) real part patterns of the leading POP mode of upper ocean heat content anomalies. The two POP patterns describe a clockwise rotation of the heat content anomalies. From Timmermann (1996).

ocean and the atmosphere is simulated at a period of about 35 years. We shall refer this type of interdecadal variability to as the 'interdecadal mode'. Since enhanced variability is simulated in both media at the 35-year time scale, our working hypothesis is (as for the quasi-decadal variability) that the variability arises from ocean-atmosphere interactions (see Timmermann (1996) and Timmermann et al. (1998) for further details).

The 35-year period is seen nicely in the spectrum of the overturning index (Fig. 9.10b). We used therefore the index of the meridional overturning (Fig. 9.9c) to describe the evolution of the interdecadal mode by computing lagged regression patterns of selected quantities (SLP, SST, fresh water flux, SSS, surface currents, and convection). In order to highlight the interdecadal variability, we band-pass filtered the model data prior to the regression analyses retaining variability with time scales of 25-45 years. The anomalous SST pattern ten years prior to the maximum overturning (lag - 10 years) is characterized by negative anomalies which cover the entire North Atlantic Ocean, with maximum cold anomalies near 40°N (Fig. 9.14b). The associated SLP anomaly pattern is the North Atlantic Oscillation (NAO) in its reverse polarity (Fig. 9.14a). Important to the subsequent phase reversal are the positive salinity anomalies which develop primarily east of Newfoundland (Fig. 9.14d). As shown in Timmermann et al. (1998), this positive SSS anomaly is created by anomalous fresh water input (primarily through enhanced evaporation, Fig. 9.14c) and anomalous salt transport by anomalous Ekman currents (Fig. 9.14e). The convection at this time is near normal (Fig. 9.14f). The SSS anomaly has grown and expanded in area five years later (lag -5 years), as shown in Fig. 9.15d. At this time, the SLP and SST anomaly patterns are not well developed and in a kind of 'transition stage' (Figs. 9.15a and 9.15b). The convection, however, shows a clear intensification in the sinking region south of Greenland (Fig. 9.15f), in response to the enhanced density which is due to the increased sea surface salinity in this region. This will lead to a strengthened thermohaline circulation and subsequently to an enhanced poleward heat transport, which will in turn reverse the SST tendendy. This is visualized in Fig. 9.16b which shows positive SST anomalies that cover most of the North Atlantic at the time of the maximum overturning (lag 0). The SLP anomaly pattern has also reversed, with an intensified Icelandic low and Azorian high (Fig. 9.16a). The signals in the anomalous salinity and convection are now relatively weak (Figs. 9.16d and 9.16f). The anomalous SLP and SST pattern will grow further, until half a cycle is completed at lag + 5 years (not shown). A negative SSS anomaly develops through air-sea interactions at lag + 5 years which will eventually reduce the density in the sinking region and weaken the convection. This will lead to a weakend thermohaline circulation, reduced poleward heat transport and a change towards anomalous cold temperatures in the North Atlantic, which completes a full cycle (not shown).

Thus as for the quasi-decadal variability, large-scale air-sea interactions are important in generating the interdecadal variability. The atmospheric response to mid-latitudinal SST anomalies seems to be a key process for the generation of mid-latitudinal low-frequency variability on the time scales of decades. While the

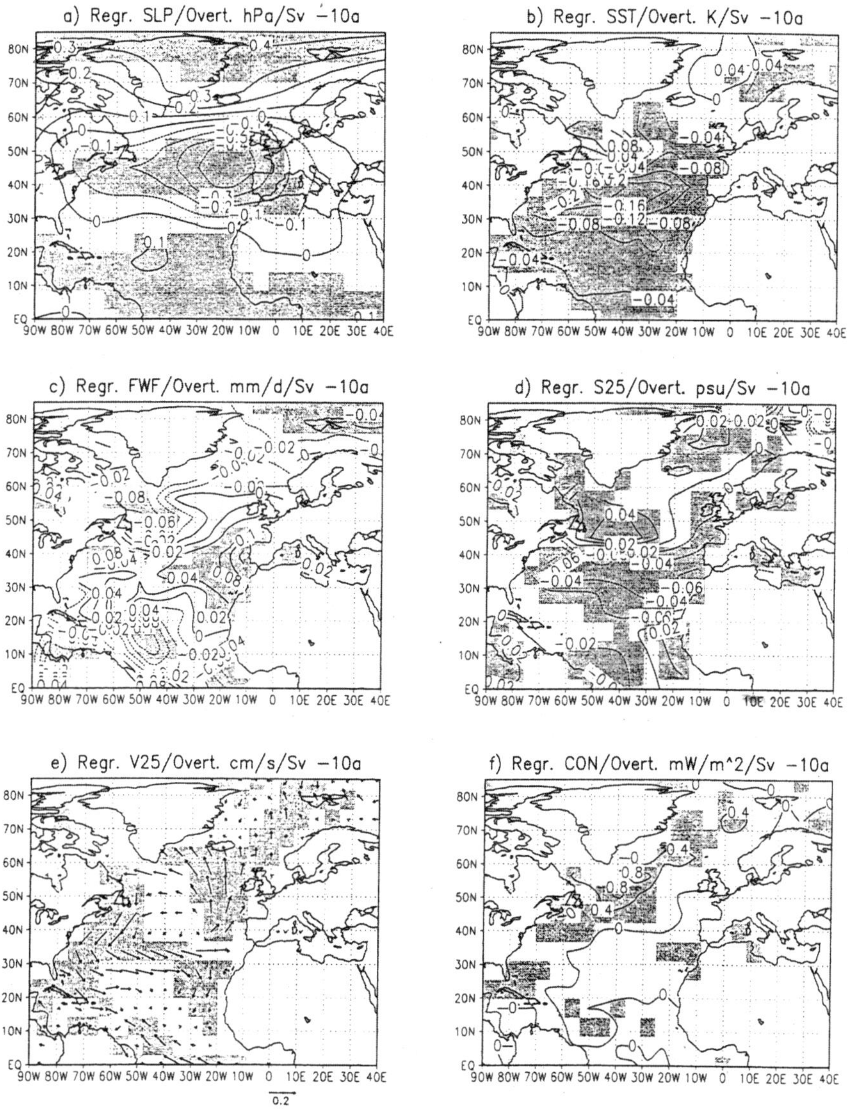

Fig. 9.14 Maps of lagged regression coefficients of selected anomaly fields (lag -10 years, i.e. the fields show the conditions 10 years prior to the maximum overturning). a) SLP (hPa/Sv), b) SST (K/Sv), c) fresh water flux (mm/d/Sv), d) SSS (psu/Sv). e) surface currents ((cm/s)/Sv), and f) loss of potential energy by convection [$(mW/m^2)/Sv$]. The shading denotes regions in which the results are statistically significant at the 95% level. From Timmermann et al. (1998).

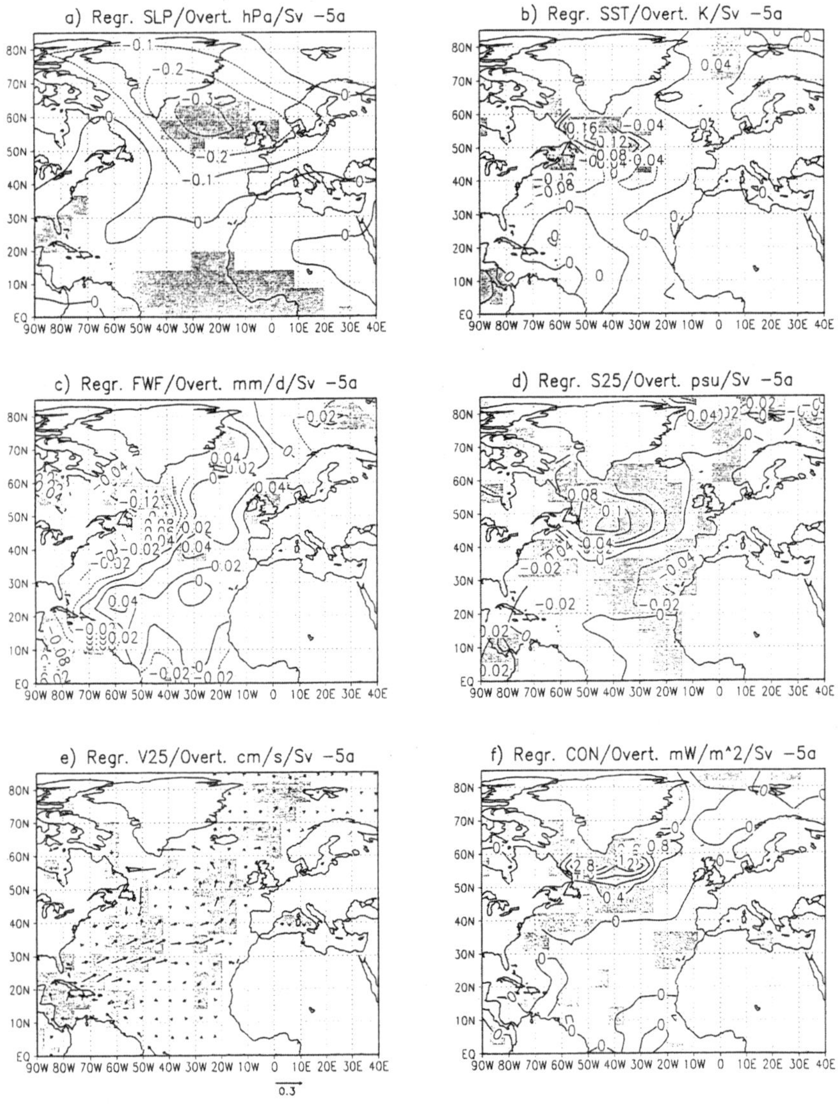

Fig. 9.15 Maps of lagged regression coefficients of selected anomaly fields (lag -5 years, i.e. the fields show the conditions 5 years prior to the maximum overturning). a) SLP (hPa/Sv), b) SST (K/Sv), c) fresh water flux (mm/d/Sv), d) SSS (psu/Sv), e) surface currents ((cm/s)/Sv), and f) loss of potential energy by convection [(mW/m^2)/Sv]. The shading denotes regions in which the results are statistically significant at the 95% level. From Timmermann et al. (1998).

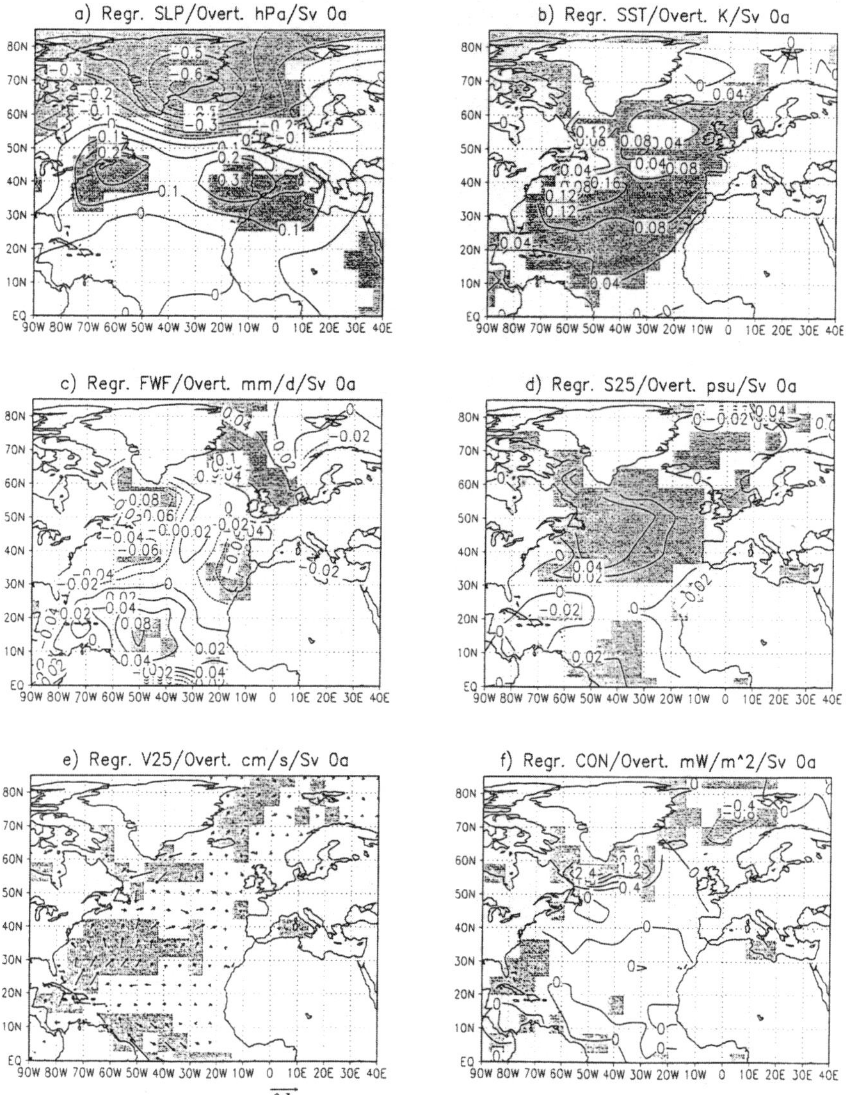

Fig. 9.16 Maps of lagged regression coefficients of selected anomaly fields (lag 0 years, i.e. the fields show the conditions at the time of the maximum overturning). a) SLP (hPa/Sv), b) SST (K/Sv), c) fresh water flux (mm/d/Sv), d) SSS (psu/Sv), e) surface currents ((cm/s)/Sv), and f) loss of potential energy by convection [$(mW/m^2)/Sv$]. The shading denotes regions in which the results are statistically significant at the 95% level. From Timmermann et al. (1998).

anomalous surface stress is important to the generation of the quasi-decadal variability, the anomalous fresh water flux is important to the generation of the interdecadal variability. In both types of variability, however, the air-sea coupling appears to be rather weak, and the resulting modes are rather strongly damped, as can be inferred from the spectral analyses presented above.

9.5 Predictability of interdecadal variability in the North Atlantic

After having discussed some aspects of the dynamics of the decadal and interdecadal variabilities in the North Atlantic, we would like to address now the question of the predictability at decadal time scales. We can distinguish three basic cases from the above discussion (Fig. 9.17). We consider first the 'pure' stochastic climate model. In this case, the predictability limit of oceanic quantities is given by their persistence (autocorrelation), which ranges typically from a few months (for surface quantities) to many years (for subsurface quantities). The predictability of the atmosphere at time scales beyond the predictability limit of individual weather phenomena, which is of the order of about fourteen days, is not affected, since no feedback from the ocean to the atmosphere is considered (Fig. 9.17a). The second case is the stochastically forced 'ocean-only' mode. Enhanced predictability relative to persistence is found for oceanic quantities around the resonance frequency, while atmospheric quantities cannot be predicted with higher skill relative to the 'pure' stochastic climate model (Fig. 9.17b). Thus, if an oceanic mode exists with a decadal period, the variations in the ocean associated with this mode will be predictable at decadal time scales. Finally, third, we consider the stochastically forced coupled ocean-atmosphere mode. Both atmospheric and oceanic quantities exhibit enhanced predictability (again relative to persistence) at the resonance frequency (Fig. 9.17c). An example of this latter case is the ENSO phenomenon, for which both oceanic (e.g. eastern equatorial SST anomalies) and atmospheric quantities (e. g. the Southern Oscillation Index) are predictable at lead times of about one year (for a recent review on ENSO predictability see e.g. Latif et al. (1998)). If coupled modes with decadal periods exist in the North Atlantic, both the atmosphere and the ocean will exhibit enhanced predictability at decadal time scales in the North Atlantic region.

We have described two such modes from our coupled model simulation, and it will depend critically on the degree of damping, whether the coupled nature of the mode will affect significantly the predictability of the atmosphere at decadal time scales.

Thus, the predictability characteristics of the low-frequency variability in the North Atlantic can provide important informations about its underlying dynamics. We conducted therefore an ensemble of predictability experiments with our coupled model. Before we describe the results of the predictability experiments, we discuss briefly the spectral characteristics of the three key quantities (SST, over-

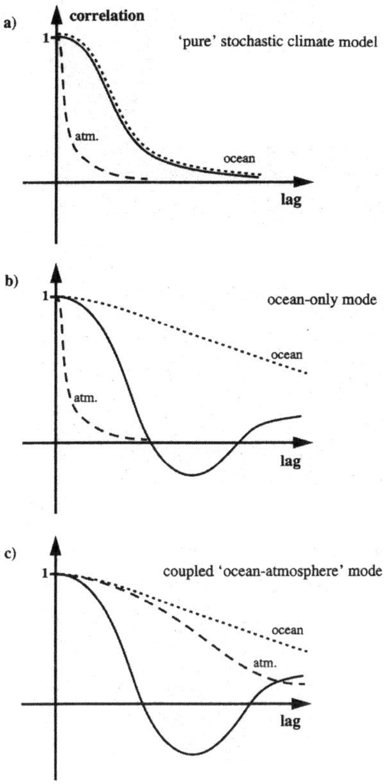

Fig. 9.17 Schematic atmospheric and oceanic skills expected from three scenarios. a) The 'pure' stochastic climate model, b) the stochastic excitement of an 'ocean-only', c) the stochastic excitement of a 'coupled ocean-atmosphere' mode. The persistence of a typical oceanic quantity (such as the anomalous SST) is given by the full lines. The typical oceanic skills are given by the dotted lines, while those of typical atmospheric quantitities are given by the dashed lines.

turning, NAO) shown in Fig. 9.10. All three spectra show peaks at a period of 35 years. Most striking, however, is the difference in the amount of the high-frequency variability. While the overturning index exhibits relatively weak variability up to time scales of about one decade, the SST and NAO are characterized by much more high-frequency variability. Thus, the deep ocean acts as a kind of low-pass filter, and we expect that subsurface quantaties exhibit much more predictability at decadal time scales than surface or atmospheric quantities. In particular, it is likely that the interdecadal mode will be predictable in the subsurface ocean only, although all three quantities shown exhibit peaks at the period of 35 years. The amount of high-frequency noise in the SST and NAO is simply too high and destroys the predictability at relatively long time scales. Another factor limiting the predictability of the two modes discussed here is the width of the spectral peaks, implying a relatively strong damping of the modes.

We have applied our coupled ocean-atmosphere circulation model (which was also used to study the dynamics of the quasi-decadal and interdecadal variabilities; see sections 9.3 and 9.4) to assess systematically the predictability of climate variations in the North Atlantic region at decadal time scales (further details can be found in Grötzner et al. (1999)). Classical predictability experiments were conducted in ensemble mode. We have chosen four states from the control integration, from which the coupled model was restarted. While the oceanic initial conditions remained unchanged, the atmospheric conditions were perturbed. Each predictability ensemble comprises of ten individual members, i.e. for each of the four oceanic states ten experiments were performed with different atmospheric initlal conditions. Each predictability experiment has a duration of twenty years (the duration in the last forecast ensemble amounts to only ten years). The divergence of the trajectories within an ensemble provides an indication of the predictability at decadal time scales. Our experimental setup yields an upper limit of decadal predictability, since we assume a perfect knowledge of the oceanic initial conditions, an assumption which will be never fulfilled in a real forecast situation.

We present in Figs. 9.18, 9.19, and 9.20 the results from our forecast ensembles. We present the results in terms of the leading EOFs of the anomalous meridional overturning, SST, and 500 hPa heights. The individual forecast trajectories show relatively little spread, when the meridional overturning is considered (Fig. 9.18), which is visualized by the ensemble means and normalized ensemble variances. The variance was normalized using the variance computed from the control integration. When the normalized variance reaches unity, the variations between the individual forecast members are as large as typical variations in the control run, which defines a predictability limit. The North Atlantic meridional overturning is predictable for about one to two decades, depending on the forecast ensemble considered. The anomalous North Atlantic SST, however, appears not to be predictable beyond time lags of a few years (Fig. 9.19), which is also true for the leading EOF of the anomalous Northern Hemisphere 500 hPa heights (Fig. 9.20). This latter result, however, depends somewhat on the forecast ensemble considered: when predictability experiments are started from states characterized by strong SST anomalies, the predictability of anomalous SST and 500 hPa heights seems to be enhanced.

We anticipated these results from the spectral analyses shown in Fig. 9.10. Thus, our predictability experiments highlight the ocean's role in generating the interdecadal variability. Although the coupling between the ocean and the atmosphere seems also to play a key role in the generation of the quasi-decadal and interdecadal modes, the coupling appears to be too weak (or equivalently the damping too strong) to influence the predictability of surface quantities at decadal time scales in the presence of the internal high-frequency noise. The results of our predictability experiments are consistent with those shown by Griffies and Bryan (1997), who

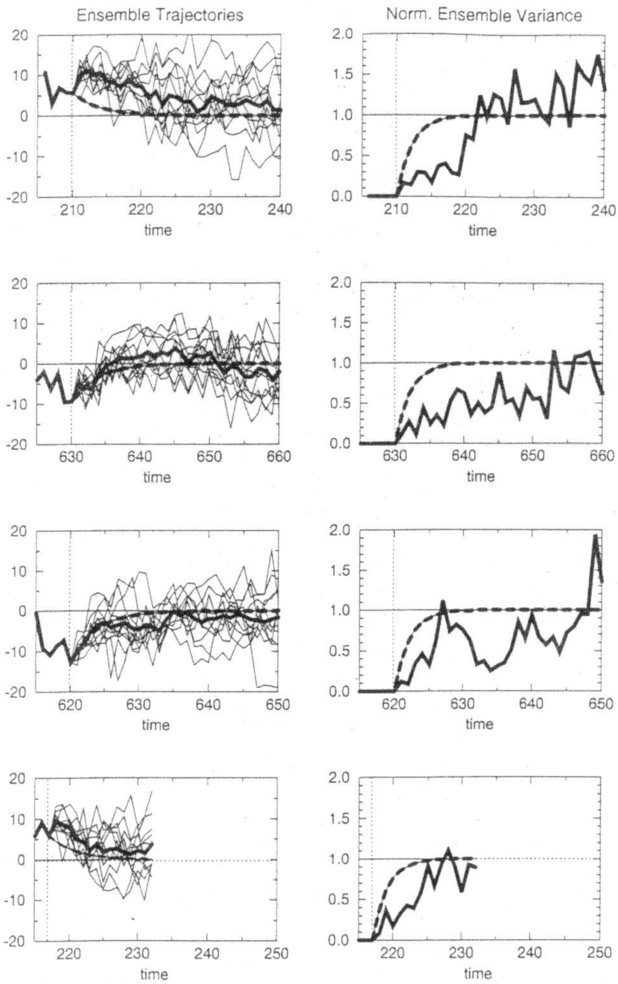

Fig. 9.18 Individual ensemble trajectories and ensemble mean (thick line) of projections onto the first EOF of the anomalous North Atlantic meridional overturning for the four different forecast ensembles (left) and time series of the respective normalised ensemble variances (right). The thick dashed lines show the temporal behavior of fitted first order autoregressive processes. From Grötzner et al. (1999).

were the first to apply a coupled ocean-atmosphere model to investigate the decadal predictability in the North Atlantic region.

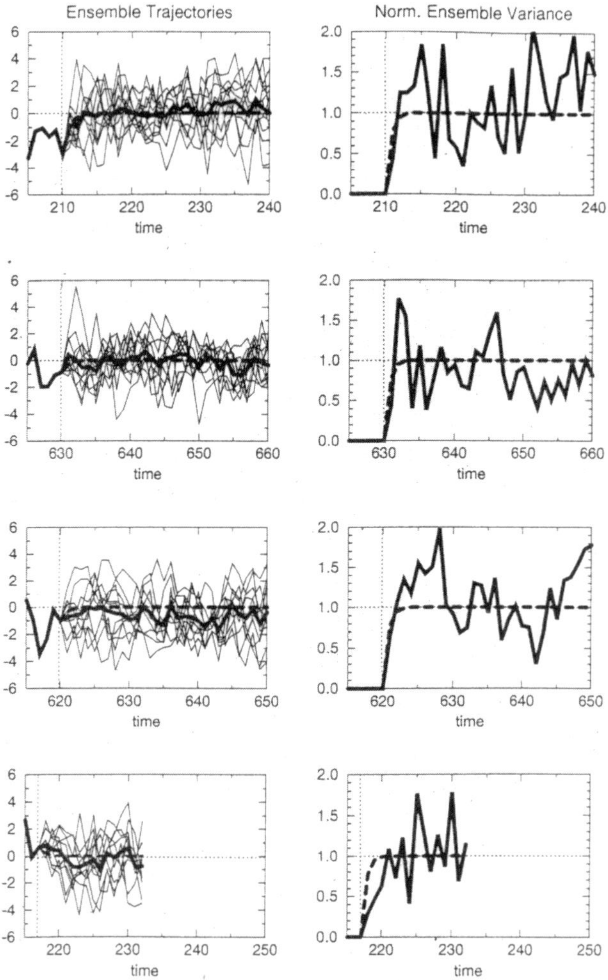

Fig. 9.19 Individual ensemble trajectories and ensemble mean (thick line) of projections onto the first EOF of the North Atlantic SST anomalies for the four different forecast ensembles (left) and time series of the respective normalised ensemble variances (right). The thick dashed lines show the temporal behavior of fitted first order autoregressive processes. From Grötzner et al. (1999).

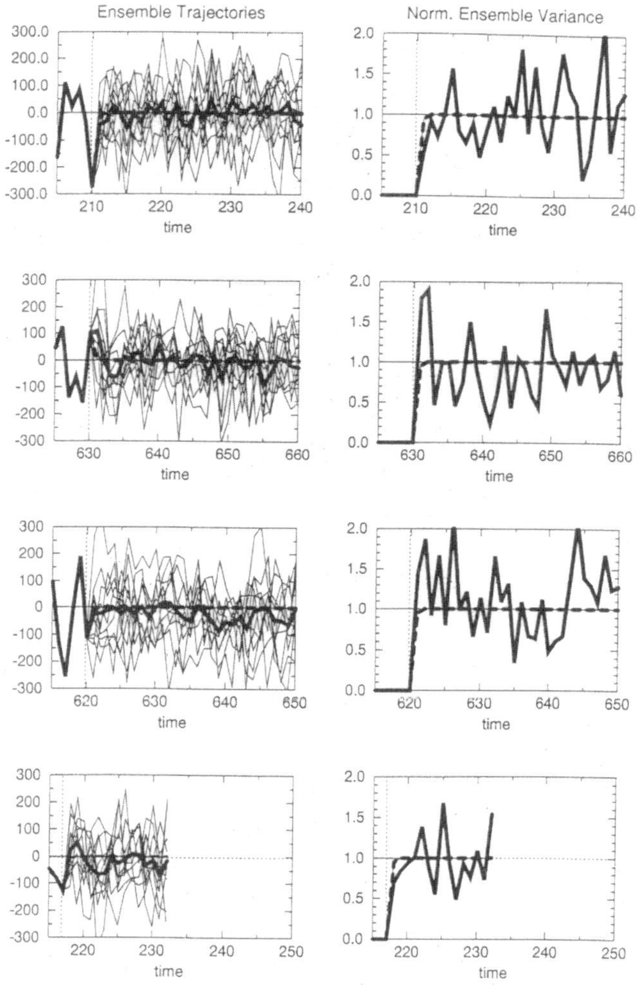

Fig. 9.20 Individual ensemble trajectories and ensemble mean (thick line) of projections onto the first EOF of the North Atlantic 500 hPa geopotential height anomalies for the four different forecast ensembles (left) and time series of the respective normalised ensemble variances (right). The thick dashed lines show the temporal behavior of fitted first order autoregressive processes. From Grötzner et al. (1999).

9.6 Discussion

We have shown that the North Atlantic climate variability at decadal time scales is largely consistent with the picture that the noise immanent to the coupled ocean-atmosphere system drives its low-frequency variability. This stochastic concept is supported by observations, simulations and predictability experiments with a sophisticated coupled ocean-atmosphere general circulation model. Large-scale ocean-atmosphere interactions play an important role in the generation of the inter-decadal variability, and eigenmodes of the coupled ocean-atmosphere may exist in the North Atlantic region. It is likely, however, that these eigenmodes are strongly damped.

In this article the stochastic climate model scenario served as a paradigm for the description of climate variations. Many observations and modeling results agree with its predictions. There are, however, alternative paradigms which as well capture essential features of climate variability. Imagine, for example, the NAO. One may hypothesize that the dynamics of this climate fluctuation may be represented by a low-dimensional non-linear system. Its attractor in phase space may consist of regions where the system stays for a relatively long time. These are the minima and maxima of the observed time series. Transitions between states occur quasi-regularly (see Lorenz (1963), Palmer (1993), and references therein for further details). The observational records do not permit a distinction between these two paradigms, since they are too short. In the near future long simulations with coupled ocean atmosphere GCMs could allow such inferences.

Our studies raise some important scientific questions that need to be addressed in more detail in the future. One of the most important challenges is to understand the atmospheric response to mid-latitudinal SST anomalies. This response is likely to be highly non-linear, since it involves changes in the statistics of the transient eddies (e. g. Palmer and Sun (1985)) which feed back onto the mean flow. We have reasons to believe that the atmosphere is indeed sensitive to mid-latitudinal SST anomalies, as argued by Latif and Barnett (1994). The spectral peaks in the model's NAO, the pan-oceanic connections between the North Atlantic and the North Pacific described by Timmermann et al. (1998), and additional atmospheric response experiments with prescribed SST anomalies (not shown), all of these indicate that the atmosphere feels the slowly changing North Atlantic SSTs. However, the internal atmospheric variability is relatively strong, and it remains unclear as to whether the atmosphere is affected by the slowly varying boundary conditions to a degree that enables useful predictions at decadal time scales.

The oceanic processes that lead to the interdecadal variations in the North Atlantic SST need also to be analysed in more detail. We need to quantify the relative contributions of mixed layer processes and ocean dynamics to the SST changes. This can be achieved, for instance, by conducting an experiment, in which the atmosphere is coupled to a fixed-depth mixed layer (slab) model. Such an experiment is underway, and its outcome will be discussed in a forthcoming paper. A series of experiments was performed by Manabe et al. (1996), who compare three thousand-year integrations: A fully coupled run with an ocean-atmosphere general circulation model, an atmosphere-slab run, and an atmosphere-only run with fixed climatogical SSTs. Our results presented in this paper are mostly consistent with the results of Manabe et al. (1996). In particular, Manabe et al. (1996) conclude that the stochastic forcing plays a major role in generating low-frequency variability.

Moreover, it would be desirable to understand the relative roles of changes in the wind-driven and thermohaline circulations in changing the North Atlantic SST. Sensitivity experiments with our coupled ocean-atmosphere model would help to get further insight into this matter. One could, for instance, inhibit certain atmospheric feedbacks in coupled integrations, such as the fresh water feedback or the wind stress feedback and compare the level of SST variability with that in the control run. Such experiments will provide important informations about the mechanisms that lead to the decadal and interdecadal variations in the SST. These experiments are underway, and we shall report on the their outcomes in forthcoming papers.

Finally, model verification is still an important issue. We have used so far relatively coarse-resolution models to investigate the dynamics and predictability of decadal and interdecadal variability. Such models have considerable problems in simulating important aspects of the global climate and its variability. Relatively large flux corrections, are applied to inhibit the coupled models to drift into unrealistic climates. It is unknown how these corrections affect the variability of complex non-linear systems such as the climate system. Furthermore, we have no good estimate of the space-time structure of the low-frequency variability from observations. Paleoclimatic datasets will become more and more important to validate the coupled models. An example of such a dataset, for instance, is the multi-century reconstruction of the North Atlantic Oscillation from tree rings (Cook et al. (1997)).

Acknowlegements

The authors would like to thank Drs. G. Lohmann, E. Maier-Reimer, and U. Mikolajewicz for many fruitful discussions. This work was supported by the German government by its "Decadal Predictability" and "Ocean-CLIVAR" programmes and the European Union through its "DICE" and "SINTEX" projects. The model runs were conducted at the Deutsches Klimarechenzentrum in Hamburg.

10 Strategy for Regional Seasonal Forecasts

KIKURO MIYAKODA
George Mason University, COLA, Maryland, USA

10.1 Numerical predictions in general

The topic is numerical weather prediction, in particular air-sea coupled forecasts over the Mediterranean Sea and the surrounding areas. At this stage, it is not certain whether or not this venture will be successful. The key questions are: (Question 1) **Are three month forecasts over the Mediterranean Sea possible ?** and (Question 2) **Are forecast studies of the ocean state or ecological conditions useful ?**

Global-to-mesoscale

The first issue is how small a scale is predictable for a season. In the 1970's, when mesoscale weather forecasts became a focus of research, various definitions were proposed for mesoscale disturbances.

Orlanski's (1975) nomenclature for the scale of atmospheric disturbances is as follows:

1) macro α (L>10^4 km);
2) macro β (10^4 km>L>2 10^3 km)
3) meso α (2 10^3 km>L>2 10^2 km)
4) meso β (200 km>L>20 km)
5) meso γ (20 km>L>2 km)
6) micro α
7) micro β

where L is the disturbance wavelength.

This list may lead one to speculate that for seasonal time scales, meso β scales may not be predictable, and perhaps even the meso α scales are questionable. But this question itself is a major research issue.

For the ocean the dominant time scales are much longer than those for the atmosphere because of the greater thermal and kinetic inertia. In addition, the horizontal length scales of the relevant motions in the ocean are approximately one tenth of those in the atmosphere.

One measure of the relevant horizontal length scale is the radius of deformation, L defined as the ratio of the phase speed of gravity waves, C_g to the Coriols parameter, f:

$$L = C_g/f \tag{1}$$

In the atmosphere C_g= 100ms^{-1}, so that for f = 10^{-4} s^{-1}, L=1000km, whereas in the ocean, C_g= 10ms^{-1}, and L is of the order of 100 km. Because of the smaller radius of deformation in the ocean, the horizontal grid size of an ocean model needs to be much smaller than that for an atmospheric model. Near the equator the ocean model requires even finer spatial resolution, i.e., 25km, in order to resolve the oceanic Kelvin and Rossby waves.

The second issue is which type of model should be used, global or regional. A brute force approach is simply to take a global model of homogeneously small grid size, say 10 km. This approach is too expensive for typical research institutions and universities, and perhaps can only be pursued by two or three major forecasting centers. However, whether or not the major world meteorological centers can keep a constant and careful watch on global weather is a good question. Of course, Europeans are more intimately monitoring the results over the European sector than other parts of the world.

10.2 Atmospheric and oceanic general circulation models (GCMs)

10.2.1 Basic questions

The governing equations for the atmosphere and the ocean are written below.

In general, atmospheric model computations are numerically more complicated than the oceanic calculations. This may be due to the far larger Reynolds number for the atmosphere than for the ocean, and the large effect of condensation in the atmosphere. The compressibility of the atmosphere is another complicating factor. Since seawater is nearly incompressible, the equation of continuity for the ocean is simplified and the Boussinesq approximation is used.

For simplicity, water vapor in the atmosphere and salinity in the ocean are omitted from the following equations.

Atmosphere **Ocean**

Eqs. of motion

$$\frac{\partial u}{\partial t} + \vec{V} \cdot \nabla u + w\frac{\partial u}{\partial z} - fv = -\frac{1}{\rho}\frac{\partial p}{\partial x} \qquad \text{the same} \tag{2}$$

$$\frac{\partial v}{\partial t} + \vec{V} \cdot \nabla v + w\frac{\partial v}{\partial z} + fu = -\frac{1}{\rho}\frac{\partial p}{\partial y} \qquad \text{the same} \tag{3}$$

Hydrostatic assumption

$$0 = -\frac{1}{\rho}\frac{\partial p}{\partial z} - g \qquad \text{the same} \tag{4}$$

Continuity

$$\frac{\partial \rho}{\partial t} = -\nabla \cdot (\rho \vec{V}) - \frac{\partial}{\partial z}(\rho w); \qquad \nabla \cdot \vec{V} + \partial_z w = 0 , \tag{5}$$

Thermodynamic equation

$$\frac{d}{dt}(c_p \ln \theta) = \frac{Q}{T} \qquad \frac{dT}{dt} = F, \tag{6}$$

Equation of state

$$p = \rho R T \qquad \rho \cong f(T) \approx (-\alpha(T - T_o)) \tag{7}$$

where $\vec{V} = (u, v)$ is a horizontal velocity vector for x and y, w the vertical component, T temperature, ρ density, p pressure, and θ the potential temperature, g is the gravitational acceleration, C_p is the specific heat at constant pressure, and R is the gas constant, and Q and F are the heat given to the atmosphere and to the ocean, respectively.

10.2.2 Vertical boundary conditions

Atmosphere

(a) Lower boundary conditions:

The lower boundary condition at the Earth's surface is

$$w = \vec{V} \cdot \nabla H \tag{8}$$

where $H = H(x,y)$ is the height of the mountains. Eq. (4) may be rewritten using σ – coordinates ($\sigma = p/\pi$ and $\pi = p_s$, p_s being the surface pressure which includes the effect of H, Phillips 1957) as,

$$\dot{\sigma} = \frac{d\sigma}{dt} = 0 \tag{9}$$

at the surface.

(b) Upper boundary conditions:

In the atmospheric model with p-coordinates, it is practical to assume, that

$$w = 0 \qquad p = p_T \tag{10}$$

or the condition $\dot{\sigma} = 0$ at $p = p_T$ is used in σ-coordmate.

This is called the "top-lid" condition, which leads to distorted solutions. However, a similar effect to the radiation condition can be applied, if viscous dissipation is specified, *i.e.*, a *sponge layer.* The solutions near the upper boundary are still considerably distorted, but are close to those using the radiation condition, except at the top several levels.

Ocean

Surface boundary conditions

a-1) Rigid-lid condition: w=0 at z=0 (11a)

a-2) Free surface condition:

The surface pressure is related to the displacement, η , of the free surface from mean sea level (z=0), i.e., $p_s = \rho_0 g\eta(x,y,t)$. The total pressure p vanishes at $z = \eta$, neglecting the pressure due to the overlying atmosphere

$$w = \frac{d\eta}{dt} \qquad \text{at } z=0 \tag{11b}$$

The equation for η will be shown later, see (24).

10.2.3 Numerical aspects of atmospheric GCMs

i)**Vertical representation**

σ-coordinates or σ-p hybrid coordinates are often used more often than z-coordinates.

More recently, a σ-θ coordinate has been developed (Bleck 1984; Hsu and Arakawa 1990; Zhu et al. 1992). The solutions of the GCM are very satisfactory.

ii)**Spherical harmonic expansion**

For numerical predictions, the spherical harmonic function S has been used extensively (Bourke 1972; 1974). This is because there is no horizontal discontinuity except at the mountain and the mathematics is precise, so energy and enstrophy conservation is guaranteed. For example, an atmospheric variable, ψ, is expanded in terms of these functions as

$$\psi(\lambda, \mu) = \sum_m \sum_n \psi_{mn} \cdot Y_{mn}(\lambda, \mu) \tag{12}$$

where Y_{mn} are the spherical harmonic functions, λ and μ are spherical coordinates ($\mu = cos\varphi$), and ψ_{mn} are the coefficients.

For numerical efficiency, the "transform method" has been used, as opposed to the "coefficient interactive method." The former method was developed by Orzag (1970) and Eliasen et al. (1971). In this scheme the calculation of non-linear products is simplified by (i) transforming spectral fields to a spatial grid; (ii) evaluating the non-linear terms on these grid-points (iii) inversely transforming these terms back to the spectral domain, consistently, and (iv) besides, evaluating the physics through the grid-points.

For the λ direction, a truncated Fourier series is used. The transform calculation is performed efficiently by using the FFT (Fast Fourier Transform) technique. For the φ direction, the integral can be computed exactly by the trapezoidal quadrature formulae using a sufficient number of quadrature points. On the other hand, for the φ direction, the Gaussian quadrature formula is used, which gives an exact inverse relation by using the so-called Gaussian grid points.

iii) **Semi-Lagrangian method**

More recently a new approach was developed by Robert (1982), Bates and McDonald (1982), Bates (1984) and Robert et al. (1985), based on the semi-Lagrangian method. This method permits treating advective terms without any limit; there is no computational instability, i.e., no C-F-L. (Courant-Friedrich-Levy) constraint. Therefore, greater computational efficiency can be obtained by using larger time step.

Using this technique a multi-level primitive equation model can be run with a time steps 25 times larger than those allowed by the C-F-L criterion in an explicit scheme. Combining it with the semi-implicit algorithm for the gravity waves, the semi-Lagrangian method is very effective.

In a Lagrangian scheme, a displacement of a particle, a, b and c in three dimensional space is determined by the advection velocity, (u,v,ω), where $\omega \equiv \frac{dp}{dt}$. The displacement is calculated by an iterative process. Upstream values of u,v,ω and the dependent variable Q are estimated by interpolation, using an x-y-σ grid. In this sense, this method is called "semi-Lagrangian", as opposed to real "Lagrangian" referring to a-b-c coordinate.

Advantages (i) Moisture and tracer transports are positive definite (ii) The same advection calculation is applied three-dimensionally, as opposed to two-dimensionally in the "spectral model", (iii) Substantial reduction of gridpoints near the poles, because of no constraints due to Gaussan grids or alias free calculation, and (iv) Calculation is economical due to no computational instability.

Disadvantages: (i) No guarantee of conservation of kinetic energy, though in practice, good conservation is obtained with high spatial resolution, and (ii) Substantial amount of overhead for the preparation.

10.2.4 Surface boundary conditions of oceanic GCM (z-coordinate)

The model was developed by Bryan (1969), subsequently, by Semtner (1974) and Cox (1984). Pacanowski et al. (1993) have recently generalized the computer codes, MOM (Modular Ocean Model), which is being extensively used throughout the modeling comunity.

a. Free or Rigid surface condition

i) Rigid lid version

No vertical motion is allowed at the top. This boundary condition automatically eliminates high frequency motions, allowing a larger time step (in the explicit scheme).

The velocity is divided into two components, i.e., the barotropic and the baroclinic components, i.e.,

$$\vec{v} = \vec{V}_b + \vec{v}' \quad (13)$$

where, $\vec{V}_b = \frac{1}{H}\int_{-H}^{0}(u, v)dz$ is the barotropic component, and it satisfies

$$\nabla \cdot (H\vec{V}_b/m) = 0\,. \quad (14)$$

Applying $\int_{-H}^{0} dz/H$ to eq. (7) and (8), the equation for the barotropic component becomes,

$$\partial_t \vec{V}_b + \frac{1}{H}\int_{-H}^{0} L(\vec{V})dz + f\vec{k} \times \vec{V}_b = -\frac{1}{\rho_0}\nabla p_s - \frac{g}{H\rho_0}\int_{-H}^{0}\nabla\int_z^0 \rho dz' dz \quad (15)$$

where $p = p_s(x, y, t) + p_H$, $\quad p_H = \int_z^0 \rho g dz$ and ρ_0 is a reference value of density, which is a function of depth z. The advection operator ,L, and del-operator, ∇, are represented by

$$L(\vec{V}) = \frac{m}{a}[\partial_\lambda(u\vec{V}) + \partial_\varphi((u\vec{V})/m)] + \partial_z(w\vec{V}), \qquad \nabla = (m/a\partial_\lambda 1/a\partial_\varphi) \quad (16)$$

.

a-1) _Streamfunction formulation_

The assumption of incompressibility and the rigid lid condition, (14), gives a stream-function ψ,

$$HU_b = -\partial_\varphi\psi, \qquad HV_b/m = \partial_\lambda\psi \quad (17)$$

Applying *1/m*$\nabla\times$ to eq. (15), one obtains,

$$\partial_\lambda\left(\frac{m}{H}\partial_\lambda\psi_t\right) + \partial_\varphi\left(\frac{1}{mH}\partial_\varphi\psi_t\right) = \partial_\varphi\left(\frac{aG_u}{mH}\right) + \partial_\lambda\left(\frac{aG_v}{H}\right) \quad (18)$$

where

$$G_u = \int_{-H}^{0} L(u)dz - fHV_b + \frac{gm}{a\rho_0}\int_{-H}^{0}\partial_\lambda\int_z^0 \rho dz\text{dz} \quad (19)$$

$$G_v = \int_{-H}^{0} L(v)dz - fHU_b + \frac{g}{a\rho_0}\int_{-H}^{0}\partial_\varphi\int_z^0 \rho dz\text{dz} \quad (19')$$

In order to solve eq. (18), boundary conditions on the streamfunction must be specified. The conditions are that the normal components of the barotropic velocities vanish along the coastlines, and, therefore, the streamfunction must be constant along coastlines. In multiply-connected basins, the streamfuntions have different values for each island, which change with time. The value of ψ around each island can be obtained by performing line integrals of eq. (18) around the coasts. This leads to a set of equations for all islands that must be solved simultaneously at internal ocean points.

By integrating (18) over λ and φ around each island, the following equation is obtained:

$$\int\left(-\frac{\cos\varphi}{H}\partial_{\varphi}\psi_t d\lambda + \frac{1}{H\cos\varphi}\partial_{\lambda}\psi_t d\varphi\right) = funct.G_u \& G_v \tag{20}$$

ψ is the streamfunction at the the periphery of each island, and ψ_t is the time derivative.

b-2) *Surface pressure formulation*

In order to simplify the procedure for the muti-island case, a different method was proposed by Dukowicz and Smith (1994). Applying *1/m* $\vec{\nabla}$ to eq. (15), one obtains

$$\frac{1}{a\rho_0}\left[\partial_{\lambda}(Hm\partial_{\lambda}p_s) + \partial_{\varphi}\left(\frac{H}{m}\partial_{\varphi}p_s\right)\right] = -\partial_{\lambda}G_u + \partial_{\varphi}G_v \tag{21}$$

The crucial difference between (18) and (21) is the fact that the depth H is included in the denominator in (18) or numerator in (21). Since H=0 along the coastlines, no boundary conditions for the pressure are imposed or required for (21). A theory suggests that the latter approach, (21), is appropriate when Neumann boundary conditions are intended, provided the topography being modeled satisfies $(x/H)\partial H/\partial x > 1$ on the coastline (where x is the distance from it). This modification has perhaps been the largest change since the original Bryan-Cox formulation of 1968 (Bryan and Cox, 1968).

ii) Free surface version

Dukowitz and Smith (1994) developed a method for a system allowing movement of the free surface.

The continuity equation, (5), for the ocean is then integrated vertically from -H to 0, and using (11b), obtaining the equation for the barotropic component:

$$\partial_t\eta + \nabla\cdot(H\vec{V}_b/m) = 0 \tag{22}$$

instead of (17) where V_b is given by (13). Using the relation,

$$\frac{1}{H}\int_{-H}^{0}\nabla p dz = \nabla p_s + \frac{1}{H}\int_{-H}\nabla p_H dz \tag{23}$$

and $p_s = \rho_0 g\eta$, eq. (14) is re-written by replacing term ∇p_s by $g\nabla\eta$.

There are three equations which determine the barotropic components, *i.e.*, two of (15) and (22). These equations are discretized implicitly, using the same time step as is used for the baroclinic equations. The computation for the free surface version can be efficently programmed for "Massively Parallel" computers at Los Alamos (US).

10.3 Nested techniques

10.3.1 Eta-coordinate

Ji and Vernekar (1997) developed a nested climate model, which will be described later. They used an eta-coordinate model. One of the disadvantages of the σ-coordinate is that the surface is not horizontal, so that horizontal diffusion is not correct if diffusion is computed on a sigma surface, (though Blumberg and Mellor (1987) included a correction of this effect in their σ coordinate model). In addition there is some difficulty in utilizing the observations for the initial conditions (conversion from p to σ surface). For this reason, the η-coordinate was introduced by Mesinger (1984). η is defined by

$$\eta = \frac{p}{p_s}\eta_s \quad , \quad \eta_s = \frac{p_{rf}(Z_s)}{p_{rf}(0)} \tag{24}$$

where p_{rf} (Z_s) is a suitably defined reference pressure as a function of mountain height, Z_s. They are permitted to take only a discrete set of values, chosen so that mountains are constructed from the three-dimensional grid boxes in the model, (see Fig. 10.1).

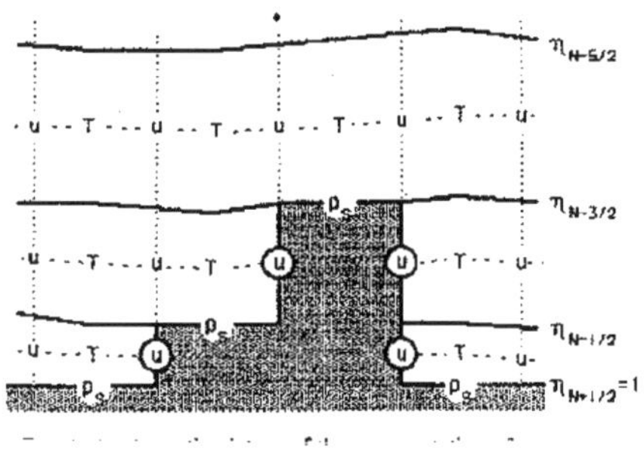

Fig. 10.1 A schematic picture of representing mountains using the eta coordinate (24)

10.3.2 Horizontal grids

One of the characteristics in the Mediterranean coupled system is that forecasts can utilize the effect of the small-scale ocean eddies, which have long time scales (more than several months). One of the GCMs applied to the Mediterranean Sea was "the curvilinear orthogonal grid" model (Mellor 1991; Zavatarelli and Mellor 1995).

As is seen in Fig. 10.2, the oceanic grid can be irregular, while the atmospheric grid should be quite regular, because of the reasons mentioned in section 10.2. In order to couple the oceanic model of the curvilinear grid with the atmospheric model of the regular grid, care is needed. Examples of the smooth and soft interaction technique are the "delocalized" method of Vintzileos and Sadourney (1997) and the "Flux Coupler" by Bryan et al. (1997).

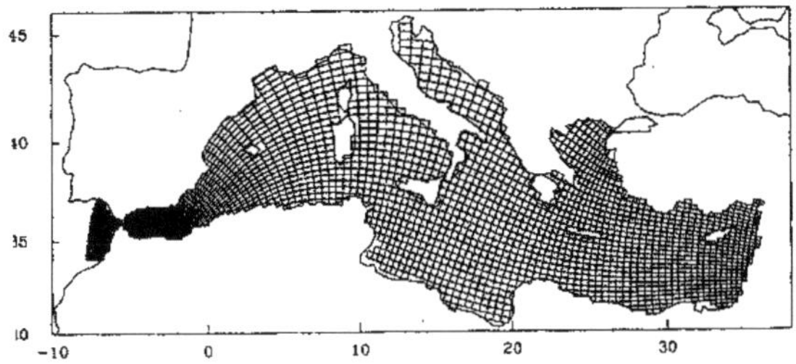

Fig. 10.2 The curvilinear grid (Zavatarelli and Mellor, 1995)

10.3.3 Potential difficulty in mesh nesting

Regional models have finer grid sizes and sometimes more elaborate physics, while global GCM have coarser grid sizes. Finite difference calculations of different resolutions produce a discontinuity at the boundary, because the different grid sizes produce different numerical solutions. For example, the phase speeds are different, while the original solutions have the same speed. Another problem is that the centered difference generates an additional solution, which is referred to as the "computational mode", as opposed to the "physical mode". The computational modes sometimes emerge as irregular and noisy solutions.

Vichnevetsky (1981) presented evidence of the discontinuity, using an example of a wave equation, i.e.,

$$\frac{\partial u}{\partial t} + c\frac{\partial u}{\partial x} = 0 \tag{25}$$

The finite difference approximation is applied to (25) by using the grid index, j, for the x-coordinate and the grid size, Δx:

$$\frac{\partial u_j}{\partial t} = -c\frac{u_{j+1} - u_{j-1}}{2 \cdot \Delta x} \tag{26}$$

Applying the Fourier transform to (26), e.g.,

$$\hat{u}_j(\omega) = \int_{-\infty}^{\infty} u_j(t) \cdot e^{-i\omega t} dt \tag{27}$$

where ω is the non-dimensional phase speed, the following conclusion is obtained. There are two fundamental solutions, *i.e.*, p_i - forward propagating waves, and q_i - backward propagating waves. If the set of functions of $\{u\}$ are a solution for the finite difference equation, there is a relation as

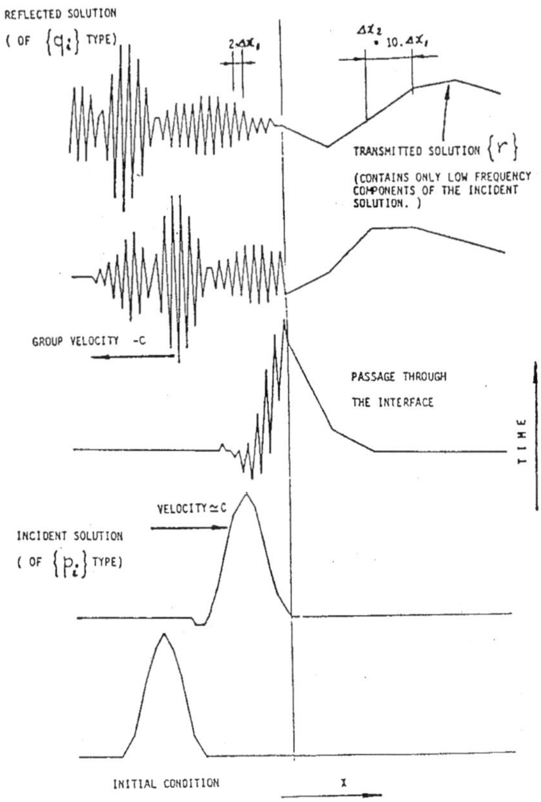

Fig. 10.3 Passage of a smooth function through the interface point x=0 with different grids. A wave comes from left. Time goes from bottom to upward. Finer grid on the left, corser on the rigth.

$$\{u_j\} = \{p_j\} + \{q_j\} \tag{28}$$

where $\{p_j\}$ is the incident wave, and $\{q_j\}$is the reflected wave. The computational group velocities (positive) of the incident waves are calculated as $C_{1g} = c\sqrt{1-(\omega\cdot\Delta x/c)^2}$, and computational group velocities (negative) of the reflected waves are as $C_{2g} = -c\sqrt{1-(\omega\cdot\Delta x/c)^2}$. Fig. 10.3 shows the time evolution of the numerical solution for the case $\Delta x_2 = 10\times\Delta x_1$; x_1 is on right, and x_2 is on left. This case may represent an exaggerated example. In practice, the ratio between Δx_1 and Δx_2 should be less than 3. Note that these discussions do not apply to nested spectral models (Tatsumi 1986; Chen and Kuo 1992; Juang and Kanamitsu 1994).

10.3.4 Lateral mesh nesting

A fine mesh (FM) is embedded in a larger coarse mesh (CM). There are two versions of FM-CM connection, i.e., one-way nesting and two-way nesting. Often there is an overlapping (transitional) zone. In general, the lateral boundary treatment should serve two purposes. It should be capable of transmitting the large scale flow resolved by the external boundary data smoothly into and out of the limited domain, and it should also adequately represent the outgoing gravity waves and fine-mesh-scale flow (see the review by Warner et al. 1997).

For the FM-CM connection, three additional considerations are:

1) An overlapping zone is normally required in both one-way and two-way nesting schemes. 2) In two-way nesting, the conservation of fluxes is essential for first order quantities, such as momentum and vertical velocity, and desirable for the second order quantities, such as kinetic energy and/or enstrophy, though this may not be required in one-way connection. 3) For open boundary conditions, some schemes treat the flow differently dependent on its direction, inflow to or outflow from FM. See Kurihara, Tripoli and Bender (1979).

In order to satisfy these requirements, different methods have been developed:

a. radiation condition: representative phase velocity is estimated numerically, see Orlansky (1976).

b. wind direction dependent scheme: gravity waves should be sufficiently suppressed, and fields smooth near the transition zone. See Kurihara, Kerr and Bender (1989), and the next section.

c. boundary zone: Substantial damping is applied to reference fields in broad overlapping zone: Davies (1976), or tendency weighting (predicted and specified), Perkey and Kreitzberg, (1976).

10.4 Nested predictions

Two examples of nested prediction are given in this section. One is a two-way nesting for hurricanes forecast (Kurihara et al. 1997), and the other is the one-way nesting for short-range weather forecast at NCEP (National Centre for Environmental Prediction, USA (Mesinger et al., 1988)). For seasonal prediction, the latter approach will be used (Ji and Verneckar, 1997). However, the first example will be shown here because of its remarkable achievement as a nested calculation.

10.4.1 Hurricane prediction

Forecast from the NCEP global model are used to specify the time-dependent lateral boundary conditions for the MMM model. Mesh C is a one-way from the global model, while meshes M and F are two-way nested in meshes C and M, respectively.

Mesh movement: The interior meshes F and M are relocated so that their centers are near the forecast storm center at all times. Mesh F moves in the zonal and

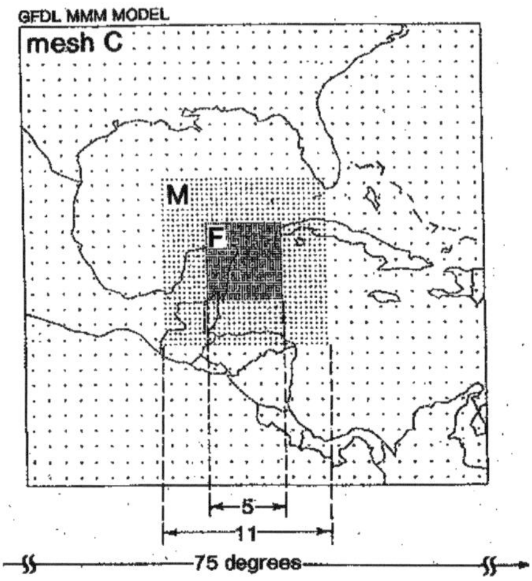

Fig. 10.4 Multiply-nested-movable mesh

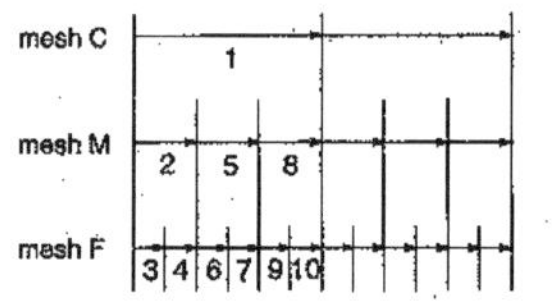

Fig. 10.5. Time integration order of the MMM model

meridional direction if the mesh center becomes separated from the forecast storm center by more than one mesh M resolution.

Construction of initial comditions: Initial condition for the MMM model rely on both the global analysis performed at NCEP and the information on the tropical cyclone structure prepared at National Hurricane Center, USA. The latter is utilized in generation of the tropical cyclone vortex. A tropical cyclone analyzed in the NCEP global analysis tends to be too large and too weak. Therefore, the model initialization method is formulated using a vortex replacement strategy.

Controlled spin-up of a vortex:
In course of time integration which starts from a calm condition, the tangential wind in the free atmosphere is nudged toward a time dependent reference wind field. This wind field gradually approaches the target field in a long time scale of vortex evolution, e.g., 60 hours.

Time integration scheme: At two-step iterative time integration scheme is used,

$$\frac{\partial}{\partial t} p_* h = LF + HF + DF + AD \qquad (29)$$

where h represents a time dependent variable: $h=u$, v, T, r, (r: the mixing ratio of water vapor), p_* is the pressure at the surface, and LF and HF are the low and high frequency terms, respectively, DF, the diffusion term and AD the forcing term (condensation, convection, radiation). To advance a variable h by a time increment, Δt, from a time level τ to another time level τ+l, a tentative value h_* is calculated in the predictor step:

$$h_* = h^\tau + (LF^\tau + HF^\tau + DF^\tau) \cdot \Delta t \tag{30}$$

Let LF^* denote these terms at the time level, τ, computed from the tentative values h_*. In the corrector step, $h^{\tau+1}$, is obtained by first computing

$$h^{\tau+1} = h^\tau + [(1-\alpha) \cdot LF^\tau + \alpha \cdot LF^*] + [(1-\beta) \cdot HF^\tau + \beta \cdot HF^*] \tag{31}$$

and next adding the forcing effect $AD^{\tau+1}$ to the above. Values assigned to the weight parameters α and β are 0.506 and 2.5, respectively. Using these weights, the low frequency waves are preserved in time marching, whereas the amplitudes of high frequency waves are strongly suppressed[1].

This Hurricane Prediction System (GFDL) was adopted in 1995 by the US NWS (National Weather Service) as an official tropical cyclone prediction technique. In both the Atlantic and eastern Pacific basins, storm track predictions in 1995 were substantially improved compared with predictions from several other models, in particular for the period beyond 36 hours. For example, improvements in the track forecasts in the Atlantic at 36, 48 and 72 hours are 14, 19 and 25 %, respectively, relative to the next best NWS guidance at each forecast time.

10.4.2 Eta-model prediction

The dynamical part of the Eta-model is designed following the approach of Arakawa for maintaining a variety of chosen integral constraints of the continuous atmosphere (e.g. Janic 1984, Mesinger et al. 1988). Differencing schemes are of second or approximately second order accuracy.

The current operational program of NCEP's Environmental Modeling Center uses three models; Eta 48 km, Eta 29 km, and Eta 10 km models. These models are identical except for the grid size and a single constant, the "horizontal diffusion" coefficient.

The lateral boundary conditions are provided by a global spectral GCM of resolution T126L28; triangular truncation at 126 wave numbers and 28 vertical levels (The Medium Range Forecasting Model, MRF, see Kanamitsu et al. 1991). This run at an earlier time and is referred to as the "Aviation model". The Eta-model is initialized with a static analysis using the global analysis as the first guess, and other nested grid model used at NCEP is the NGM (Nested Grid Model), which has 80 km/L18 inner grid (Philips 1979).

1. $HF = h\dfrac{\partial p_*}{\partial t}$ + (Coriolis force, metric force, pressure gradient force for $h=u,v$, energy conversion terms for $h=T$)

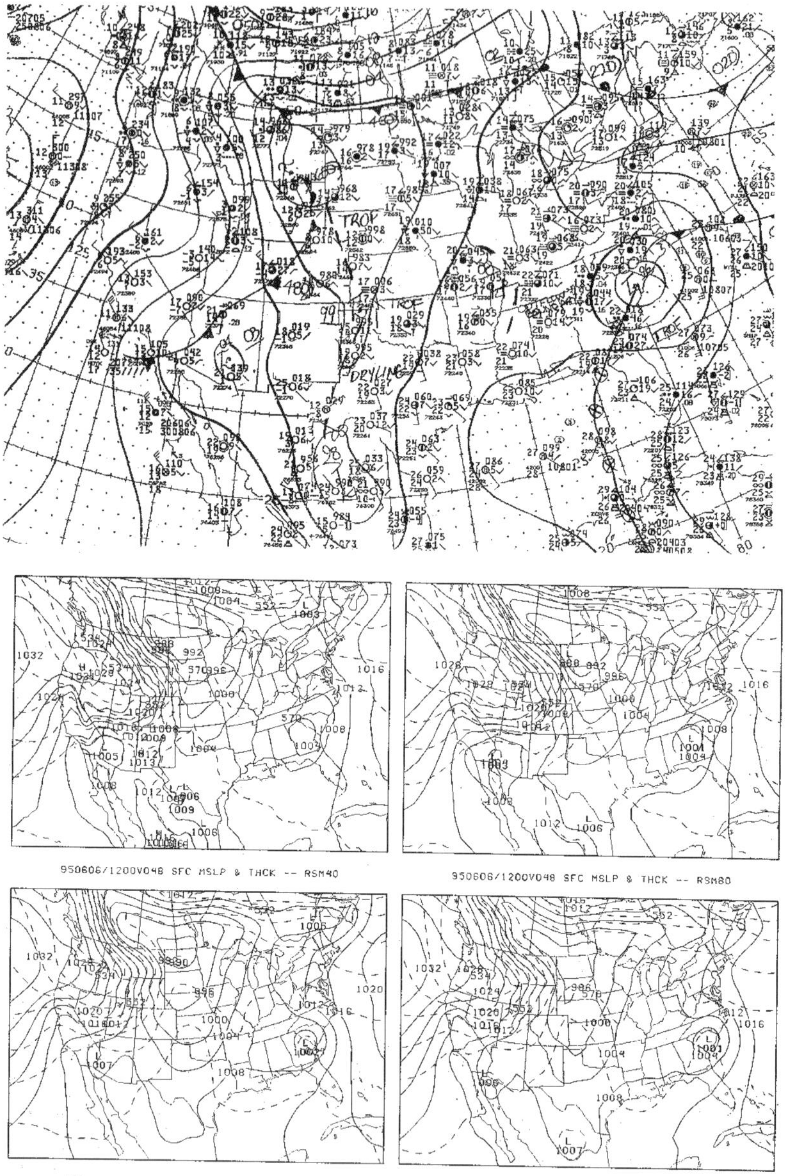

Fig. 10.6 Sections of the sea level pressure and 1000-500 mb thickness forecasts by the NGM (middle left), the Aviation model (middle right), the 80-km eta-model (bottom left), and 48-km eta-model (bottom right). The observation (top) is valid at 1200 UTC 6 June 1995. Isobars are drawn at interval of 14 mb.

Fig. 10.6 shows a comparison of results from the four models. It appears that the 48 km Eta-model performed best compared with the observations (top). The superior performance of the 48 km Eta model is particularly evident in the precipitation forecasts (not shown). A one-year investigation was conducted on the precipitation forecasts, using the so-called "threat score", in which the rates of precipitation are divided into 8 categories. The results show that the 48 km Eta-model was the best in all categories.

10.5 Long-range prediction

10.5.1 Current state of monthly forecasts in general

Here we utilize monthly forecasts of the Japan Meteorological Agency (courtesy of Drs. K. Takano and T. Tokioka). Their results show that, for forecasts longer than about five days, ensemble forecasts are needed and are indeed indispensable. For forecasts up to one month, it is important to select the most appropriate and most efficient ensemble of the initial conditions (about 10 members according to the Japanese Meteorological Agency). The forecast skill depends on the quality of members in the ensemble initial condition. To achieve maximum skill, it is necessary to utilize advanced knowledge of the linear dynamics (Singular Vector Analysis) as to the rate of growth of disturbances.

In contrast, 3 months forecasts may not need this advanced knowledge, because linear dynamics will affect only the first month. Thus, a high level of knowledge of the applied mathematics relevant to this issue may not be required for seasonal forecasts, but multiple forecasts (an ensemble) is absolutely needed (Brankovic and Palmer 1997).

10.5.2 Teleconnections from the tropics to the Mediterranean Sea

Although Europe is somewhat remote from the ENSO region, there is still some ENSO influence on Europe. Ward (1996) showed observational evidence for significant correlations of Sahel rainfall and meteorological variables over the eastern Mediterranean Sea with the Indian monsoon rainfall. This teleconnection can be explained by linear dynamics (Rodwell and Hoskins 1996). It is hoped that Atlantic SST anomalies also influence climate of the Mediterranean Sea in a systematic way.

10.6 Regional long-range prediction

10.6.1 Limited domain 30 day predictions over complex terrain

Giorgi and Bates (1989) made 30 day simulations, for Jan 1979 over the western US. Their model was not nested in another, but the initial conditions and lateral boundary conditions were given by the ECMWF large scale analysis of the obser-

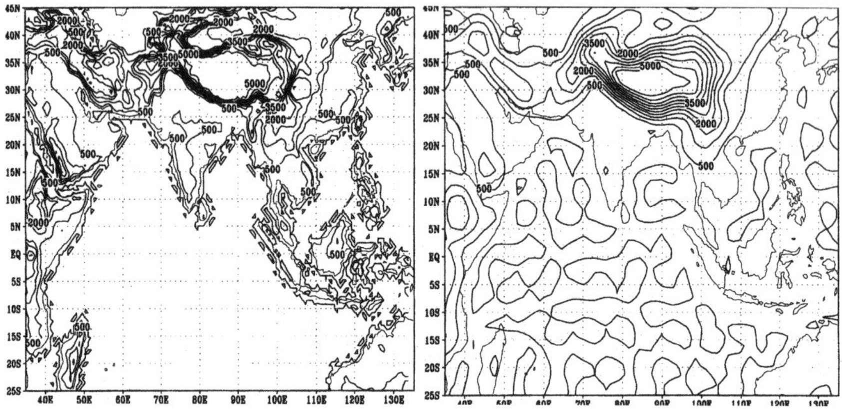

Fig. 10.7 The domain of the nested eta-model. Step mountain topography in the Eta-model. (left). Topography in the GCM over the monsoon region (right). Contour interval is 500 m

vations. The regional model used is the NCAR (National Center for Atmospheric Research) Community Climate Model with 60 km grid point spacing. The domain is centered over the Great Basin of North America, at 38°N and 116°W and is 3000 x 3000 km^2 in size.

Their experiments included 9 major precipitation events over the region. Precipitation verification was performed using data from 140 stations distributed throughout the western US. The model was successful in reproducing most of the regional features of the precipitation over the Cascade Range, Sierra Nevada, and coastal ranges, the dry conditions over the Great Basin. It is interesting to note their conclusion that, "Given good initial and lateral boundary conditions, the model is quite accurate in producing regional precipitation average." The largest difference was found in the boundary layer.

They also concluded that "In the simulated boundary layer internal model physics are more important than the lateral boundary conditions." Overall the inclusion of a new physics package (BATS--Biosphere Atmosphere Transfer Scheme) did not have a major impact on the simulated circulations.

10.6.2 Indian monsoon forecasts

The regional model used is the 80 km / L38 Eta model, which is nested in a global of R4OL18(rhomboidal truncation at wave number 40) GCM(COLA-Center for Ocean-Land-Atmosphere Studies). The nested domain is fairly large, e.g., 12000km x 8000 km. (see Fig. 10.7). In this region, it appears quite essential to use a large domain. The topographic features in these models are, at least, quite different (Fig. 10.7).

A unique aspect of Ji and Vernekar's forecasts (Ji and Vernekar,1997) is that not only the lateral boundary condition but also the initial condition are given by the driver model, i.e., the GCM. In this procedure, the regional model restarts every 48 hours with the GCM output as its new initial conditions, and the results at the beginning of each integration period are largely influenced by the spin-up of the regional model.

In order to show the merit of the nested model runs., the observations, the nested model simulations, and the original GCM simulations (top, ndddle, bottom) for the JJA precipitation are compared (Fig. 10.8).

This procedure is quite different from the method of Giorgi and Bates (1989). However, it is not certain whether this approach is better or not (see the final comment in section 10.7). Perhaps the most important point in their study is that a sufficiently large area is adopted. It is interesting to note, however, that, although the eastern Pacific was not included in the regional models domain, the impact from this region is transmitted through the lateral boundary conditions.

In order to investigate the interannual variability of the monsoon, JJA of 1987 and JJA 1988 were compared from integration initialized in the mid-April. 3 runs were done for each year, starting at 14, 15 and 16 April of each year. The nested model did a much better job of simulating the interannual variability in the monsoon between 1987 and 1988 than did the GCM alone.

10.7 Conclusions and commcnts

For **Question 1**, a larger region than the Mediterranean is required, such as 25°S-70°N, 120°W-0°-130°E. The effect of the Indian monsoon as well as the influence of the Atlantic Ocean and ENSO should then be included through the lateral boundary conditions.

Comments What is the best approach to the nesting? Is it that of Ji-Vernekar (1997) or that of Giorgi-Bates (1991)? According to Fennessy (COLA-personal communication), updating the interior of the regional model every 2 days is not necessary with the latest versions of the Eta model. In other words, the initial condition is specified only at t=0.

Recent results relevant to the limited domain seasonal forecasts have come from BALTEX (Bengtsson and Wang 1998). A moisture budget was calculated using a regional model forced by observations of the in/out-fluxes of moisture along the lateral boundaries over the Baltic Sea. The experiment was set up by using continuous boundary values from Deutsche Wetter Dienst 6 hourly operational analysis. The model was run for 3 months (August 1 - October 31,1995) without restart. Resolution was 24 levels and 1/6° lat./lon.. The simulated precipitation agrees well with the observed rainfall in that region. Thus there are many examples that the regional seasonal forecasts are possible.

However, as a final remark , it must be kept in mind that, a spectral model may be another choice (see Juang et al. 1994; 1997).

For **Question 2,** a smaller region can be taken, for research, such as Equator-60°N, 10°W-60°E. However, this region is so small that realistic forecasts cannot

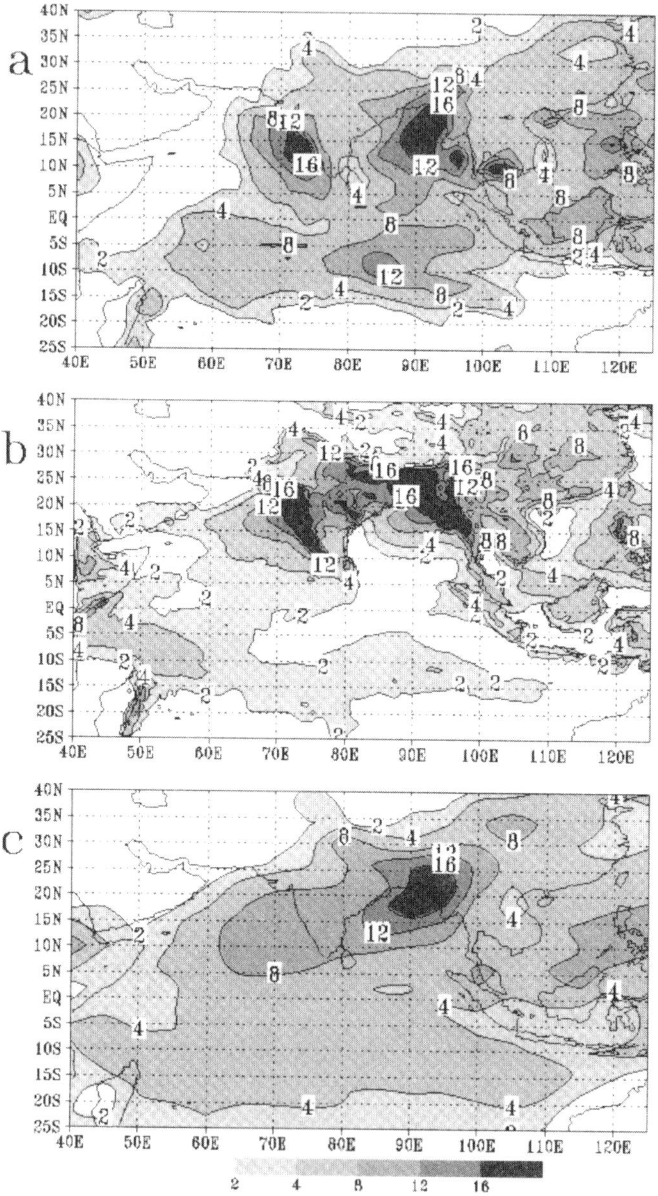

Fig. 10.8 Mean precipitation for JJA 1988. Contours are 2,4,8,12 and 16 in units of $\mathrm{mm\ d^{-1}}$. The observations (top), tihe regional model (middle), and the global GCM (bottom).

be expected. But it is large enough for basic studies in ecology , bio-hydrology, and preliminary studies of air-sea coupling over the Mediterranean Sea eddies.

Both land surface and ecological process may not have profound impact on the atmosphere or ocean (though there are some coupling effects with the former), but the "coupled calculations" with the atmospheric processes or the oceanic processes are necessary in the "real time", because the ecological and soil models require all the information that the atmospheric or oceanic model produces temporally as well as three-dimensionally. Archiving all this information is not practical due to its immense volume.

Acknowledgement

The author would like to thank Drs. L. Bentsson., T. Hollingsworth , V. Ji, Y. Kurihara, S. Lord, F. Mesinger, M. Miller, A. Navarra, N. Pinardi, K. Takano, T. Tokioka, N. Ward, A.D. Vernekar, and M. Zavatarelli. for their contributions and help. This paper was completed at George Mason University, COLA in courtesy of Drs. J. Shukla, J. Kinter, and Mr. M. Fennessy.

11 Rapid Assessment of the Coastal Ocean Environment

ALLAN R. ROBINSON
Harvard University

JURGEN SELLSCHOPP
NATO SACLANT Undersea Research Centre

11.1 Introduction

The concept of Rapid Environmental Assessment (REA) is to provide environmental nowcasts and forecasts accurate and efficient enough to support operational activity in any arbitrary region of the global coastal ocean, and to respond to operational assessment requests effectively on very short notice. Ocean science and technology today are rapidly evolving and recognized as generally involving interdisciplinary processes and interactions on multiple scales in space and time (Robinson et al., 1999a). To deal with the complex ocean forecasts and simulations required both for research and for operations and management, the concept of an Ocean Observing and Prediction System (OOPS) has emerged for assimilating data from the system's observational network into the system's forecasts models to generate requisite field estimates (Mooers, 1999; Robinson and the LOOPS Group, 1999). REA requires a rapidly deployable, generic, portable OOPS.

REA has been articulated in the context of NATO naval operations (Pouliquen et al., 1997) as "a new approach to environmental support, designed specifically to provide ... tactically useful environmental data in a tactically useable time scale" (Hammond, 1997) and US requirements have recently been articulated by Curtin (1999). The Harvard and SACLANTCEN oceanography groups have collaborated in recent years (Sellschopp and Robinson, 1997) in providing real time nowcasts and forecasts for a number of NATO REA exercises (Sellschopp, 1998; 1999). Fig. 11.1, taken from an exercise operational web site, illustrates multiscale interdisciplinary forecast products for the Gulf of Cadiz for 21 March 1998 obtained by assimilating multiple streams of *in situ* and remotely sensed data. It will be discussed further below.

Section 11.2 discusses physical forecasting and section 11.3 the use of the physical forecast. The concept of an REA OOPS is introduced in section 11.4 and validation issues are presented in section 11.5. Section 11.6 overviews and evaluates experience gained in NATO Rapid Response REA exercises in 1996-1998. Sections 11.7 and 11.8 summarize and discuss prospects for improved REA respectively from an operational and research viewpoint.

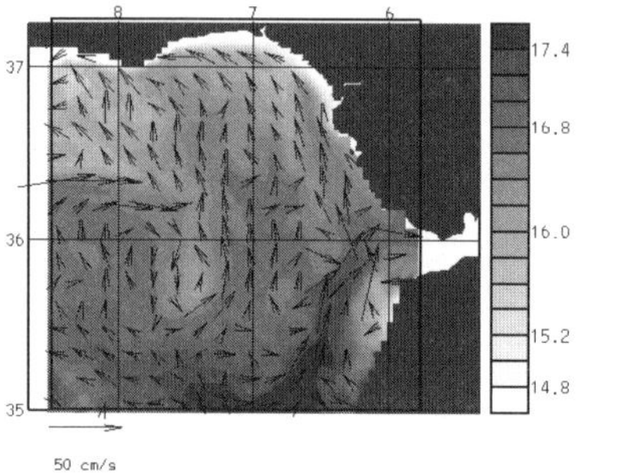

(a) Temperature Forecast 21 Mar. 1998

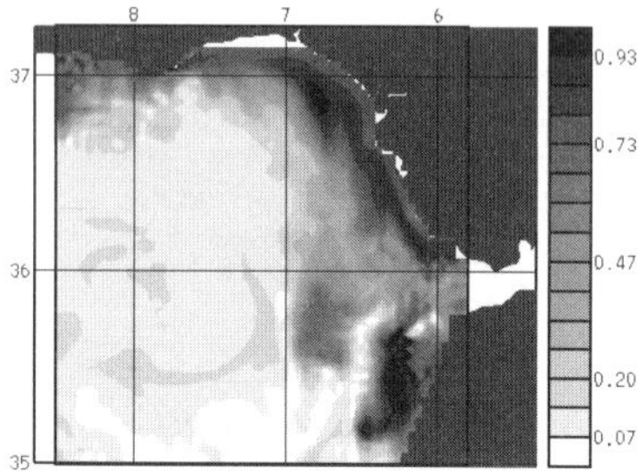

(b) Chlorophyll Forecast 21 Mar. 1998

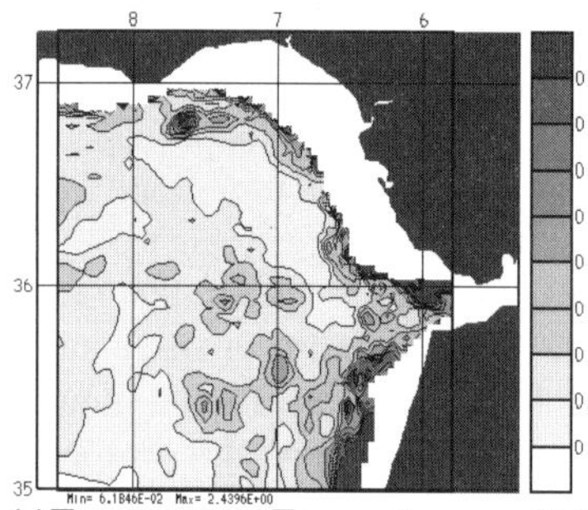

(c)Temperature Forecast error (100m)

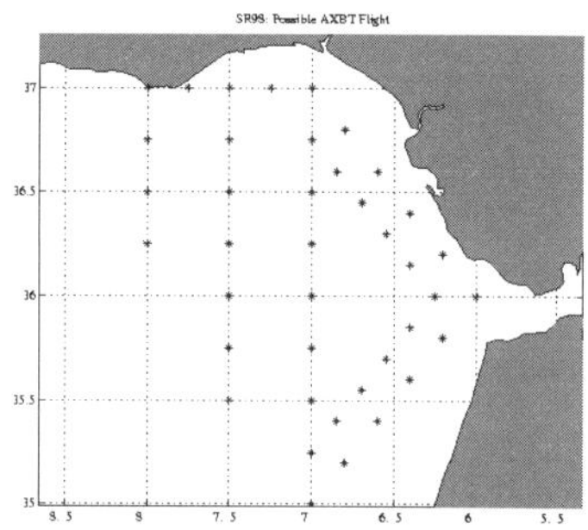

(d) Adaptively designed sampling

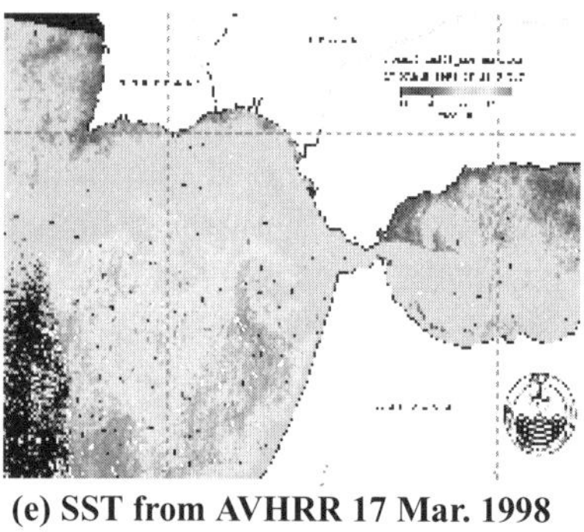

(e) SST from AVHRR 17 Mar. 1998

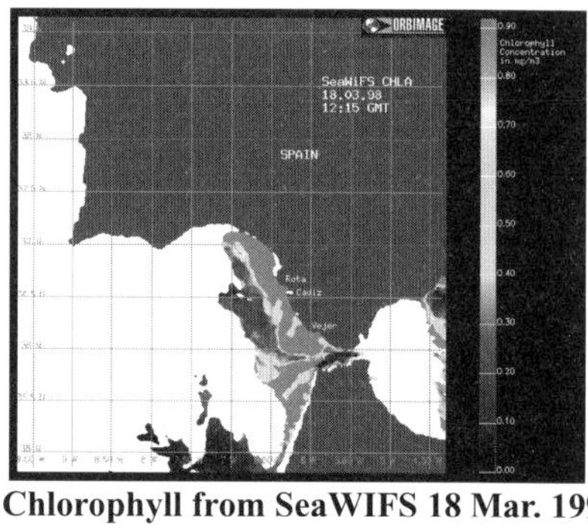

(f) Chlorophyll from SeaWIFS 18 Mar. 1998

Fig. 11.1 Fig. 11.1a shows the melded estimate of forecast temperature from a real time multi-disciplinary forecast for the Gulf of Cadiz and Strait of Gibraltar with data assimilation, Fig. 11.1b shows the melded estimate of forecast chlorophyll with data assimilation, Fig. 11.1c maps the forecasted error associated with the estimated field of Fig. 11.1a carried out by ESSE methodology (Lermusiaux and Robinson, 1999; Lermusiaux, 1999), Fig. 11.1d depicts the sampling track adaptively designed from the forecast and forecast errors, Fig. 11.1e portrays a remotely sensed sea surface temperature field, and Fig. 11.1f illustrates a remotely sensed (SeaWiFS) chlorophyll field. The data from 11.1e and 11.1f were both assimilated into the multi- disciplinary real time forecast fields (Figs. 11.1a, b, c). Figs. 11.1a, b, c, d were displayed on the RR97 exercise web site.

This chapter is addressed to both ocean forecast scientific researchers and to applied REA managers and naval officers. Thus some material will be of more or less interest to each group.

11.1.1 Nowcasting and forecasting for REA

Mankind has learned to intuitively assess weather and sea conditions and to predict their likely development from observations. In the first place the experience of shepherds and fishermen is related to their home area and can only be relocated if the same physical processes are dominant in another area. Universal environmental assessment systems must include all physics of potential relevance. The theory of tides is an excellent example of a relocatable assessment system with long range forecast capability. Prediction of ocean features that are less deterministically forced, suffers from spatial scales being an order of magnitude smaller than in meteorology. While in meteorology a worldwide observational network has constantly been maintained, a comparable oceanographic eddy resolving observation system would still be too expensive. In oceanography detailed environmental information, though technically always possible, is only assembled in limited areas and for limited time periods.

Operational oceanography that sustains the comparison with atmospheric weather analysis and forecast must be based on a relocatable ocean assessment system that yields results within short time (order of a day) and, besides a description of the present status, predicts changes in the ocean. As noted above, an ocean assessment campaign that meets these requirements is called Rapid Environmental Assessment (REA).

An REA operation is required whenever necessary information cannot be obtained from available sources. Pre-existing information usually provides good estimates of ocean depth and climatological means of parameter fields such as temperature, salinity or nutrients. Even though for many applications this information might already be sufficient, there are cases where small discrepancies between climatology and actual fields create large errors in a derived quantity. The requirements for parameter accuracy and spatial or temporal resolution depend on the final products that are desired. They control the measurement strategy and the numerical assessment and prediction tools to be applied.

11.1.2 Naval REA

Navy vessels are more affected by environmental conditions than merchant ships. Currents and waves would impact navy ships or soldiers the same as in the civilian world. However, certain specific environmental parameters have a meaning only in a military scenario. Sonar range, e.g., confines the ability to fight or protect against an invisible enemy. Navy ships do have limited capability for environmental *in situ* measurements that can be combined with *a priori* knowledge. Single point measurements however do not give full insight into the environmental situation. On the contrary, the melding of a point measurement with climatology can in some conditions be worse than the climatological database alone.

In military operations, there is room for improvement by REA (Sellschopp, 1999). In many cases, overt REA operations may be carried out without antagonist interference. To give an example: In an oceanographically active region, the safest route may be found by REA to bring a convoy across. Survey results will be kept secret for a few weeks, positions of survey deployments not. In a not totally unrealistic scenario, navy units are sent to an area that is close to a focus of mischief. Good knowledge of the littoral environment is eminent and can be obtained via REA, the results of which need not be classified.

In denied areas one would not deploy survey ships. Remote sensing from satellites and aircraft is primarily useful for the investigation of surface features. Except for visible light, electromagnetic waves are not appropriate for illumination below the sea surface. Surface signatures are, however, often caused by subsurface features, such as the transformation of surface waves by currents or bathymetry.

Air deployed autonomous systems are an alternative for ship measurements. Buoys may be moored or drift through a denied area and from time to time transmit data to an aircraft or satellite. For several weeks of continuous operation, a bottom deployed trawl-safe system can collect water column data; it can use a small pop-up buoy with an antenna to send the data at programmed intervals. Autonomous underwater vehicles are becoming increasingly important for underwater research and surveying. Their appropriateness for denied areas is coupled with high expenses for bridging the range from the deployment position. In summary, REA is aggravated but still feasible in a hostile military environment.

11.1.3 Civilian REA

At first glance it might seem that rapidity is not an issue in civilian environmental assessment except for standard products such as storm surge prediction. But there are scenarios, that would compel REA operations. An appropriate reaction to natural disasters or accidents might be impeded by lack of knowledge about environmental parameters such as mixing rates or time variable current fields. With an REA capability in place, it could be possible to close the knowledge gap in a timely manner. REA methodology should also be applied when management plans at-sea events such as byproduct dumping. Once an ocean area is sufficiently precisely modeled, variations of forcing fields will cost effectively yield the potential impact of the management event on the environment.

11.2 Nowcasting and Forecasting the Physics

The water column physical state variables associated with the circulation and its variabilities play a central role in the prediction of many REA parameters of interest. The fundamental physical state variables (velocity components, pressure, temperature, salinity, density) evolve in time according to the coupled conservation laws for momentum, mass, heat and salt, and the equation of state. We will treat those scales and phenomena whose dynamics are adequately approximated explicitly by the almost incompressible and hydrostatic primitive equations (Cushman-Roisin, 1994; Robinson, 1996). Spatial scales are lower bounded by several hundred meters and temporal scales, by a few hours. Phenomena include, e.g., currents, meanders, eddies, fronts, jet filaments, topographic and coastline effects, surface and bottom boundary layers, upwelling, coastally trapped waves, barotropic and baroclinic tides. The effects of smaller scale and higher frequency physical processes (e.g. turbulence, tidal mixing) are parameterized in the primitive equations and referred to as sub-grid scale processes in accord with the numerical representation of the dynamical model.

The ocean evolves in time, both as a direct response to external, surface and body forces, and also via internal dynamical processes. The former include winds and surface fluxes of heat and fresh water. Where air-sea interactions are important, an accurate meteorological forecast is needed for the ocean forecast. Oceanic internal instabilities and resonances, which include the meanders of currents, frontogenesis, eddying and wave propagation, are generally analogous to atmospheric weather phenomena and are called the internal weather of the sea. The spatial scales of important internal ocean weather phenomena range from O(10 km) (sub-mesoscale or synoptical dynamical event scale) to O(100 km) (mesoscale or evolutionary scale). These relatively short scales require ocean forecasts generally to be carried out regionally rather than globally. The regional forecast problem then has additional forces appearing as fluxes through horizontal boundaries, representing both larger scales of direct forcing, remote internal dynamical events and land-sea interactions in the littoral zone. The development of a regional forecast system and capability depends both upon the scales and processes of direct interest and the scales and processes that are dominant in the operational region and its surroundings. The forecast region or region of influence is often necessarily larger than the region of operational interest.

The ocean is intermittent, eventful, and episodic, and ocean circulation is characterized by very many dynamical processes occurring over a broad range of nonlinearly interactive scales in space and time. Intermittencies and multiscales have led to the concepts of adaptive sampling and nesting in ocean forecasting. Observations are used to initialize dynamical forecast models, and further observations are continually assimilated into the models as the forecasts advance in time. Such observations are generally difficult, costly and sparse. If a region of the ocean were to be sampled uniformly over a predetermined space-time grid, adequate to resolve scales of interest, only a small subset of those observations would have significant impact on the accuracy of the forecasts. The impact subset is related to intermittent energetic synoptic dynamical events. For most of the energetic variability in the ocean, the location and timing of such events is irregular and not *a priori* known. However, a usefully accurate forecast targets such events and forms the basis for the design of a sampling scheme tailored to the ocean state to be observed. Sampling schemes can be determined subjectively by experience or objectively by minimizing a selected forecast error metric. Adaptive sampling is efficient, can drastically reduce observational requirements, and is essential for effective REA.

The smallest scales of operational forecast interest generally require compatible observational and modelling grids of too high a resolution to be practical for the entire forecast domain. Thus special interest regions are nested. Two-way modelling nests designed to run in parallel with equal elapsed computing times are efficient. Steep topographic slopes require careful treatment for accurate forecasts (Haidvogel and Beckman, 1998). Data assimilation (Malanotte-Rizzoli, 1996; Robinson et al., 1998) is essential to control phase and loss of predictability errors and to optimize forecast accuracies. A field estimate made by melding data and dynamics by a short dynamical adjustment model run, after assimilating data, is called a nowcast. The coupled system required for ocean forecasting composed of: an observational network; numerical dynamical models; and data assimilation, analysis and management schemes, is called an Ocean Observing and Prediction System (OOPS) (Robinson, 1999; Robinson and LOOPS Group, 1999). The OOPS concept is developed in the REA context in section 11.4.

11.3 Implications and Applications of the Physics

11.3.1 Interdisciplinary Processes

Ocean science and marine technology are inherently interdisciplinary subjects and physical forcing plays an important or dominant role in many aspects of, e.g., acoustical, biological and sedimentological dynamics in the sea. Thus there are a variety of scientific and practical applications of the forecast physical fields and many situations occur where coupled interdisciplinary simulations and forecasts are necessary. Recent rapid progress in understanding physical processes and in achieving realistic physical field estimation now makes feasible novel interdisciplinary prediction relevant both to climate, biogeochemical cycles, and ecosystem dynamics (Robinson et al., 1999a), and to the management of, and operations in, coastal oceans and multiuse Exclusive Economic Zones. Interactions and feedbacks occur among the physical-acoustical-optical-biological-chemical-sedimentological fields. Examples include living and other particulate control of incoming solar radiation, biological and chemical reactions during pollutant dispersion, motion induced bioluminescence, resuspension processes, the effects of temperature gradients on acoustic propagation and, inversely, the use of acoustic travel times to estimate temperature gradients.

11.3.2 Naval Applications

Navy interest in the ocean is due to the impact on system performance under different environmental conditions. Navy operations are usually categorized under warfare disciplines that, with some overlap, also reflect the water depth of the operations. There are three attack-defense pairs: Submarine operations and anti submarine warfare (ASW), mine warfare (MW) and mine counter-measures (MCM), amphibious warfare (AW) and coast protection. Both facets should be considered, when for simplicity only ASW, MW and AW are mentioned. Environmental conditions that favor the aggressor must be known to the defender and vice versa.

In an ASW scenario, the structure of the sound velocity field and its variability are most important. Sound radiated or reflected by a subsurface structure is nearly the only means for the detection of a submerged submarine. Under favorable conditions, sound can reliably be detected after thousands of kilometers. On the other hand, the noise produced by a modern coastal submarine often drops below the ambient noise level after only a couple of kilometers, and a sonar echo from its hull may not exceed the reverberation from the sea surface or bottom. The non-linearity of the relation between the depth dependent sound velocity and the detection range does not allow for a simple statement about the required accuracy of a modeled sound velocity field. An exemplary critical parameter is the sign of the sound velocity gradient in the upper layer.

Acoustical methods are also used for mine detection. Because of the limited range of detection systems, the necessary information on the sound velocity structure is fairly easily obtained by standard tools. In mine warfare, littoral ocean currents and their prediction play a dominant role. On a sandy bottom in the presence of a sufficiently strong current, mines are buried by scouring. Mine divers are limited by a maximum current strength that obstructs swimming, and by a sediment load that reduces visibility.

Amphibious operations depend most critically on surf conditions. Together with the information on approaching deep water wave fields, accurate bathymetry is required for surf predictions. Near shore bathymetry can be modified by relocation of sand in the currents that are generated by a storm. REA should provide an update on coastal bathymetry, either by direct inspection or by calculation of the sand transport in a numerical model.

In a military REA operation, potential data sources reach from public databases over results from dedicated survey units down to sporadic measurements from navy ships. The information is merged at a data fusion center and made available for assimilation into models and for REA product generation (Sellschopp, 1998).

Oceanographic advisors and naval operations planners have environmental support systems and tactical decision aids at hand, that use environmental information for the benefit of the military task. Without REA, these tools have to rely on climatology or single measurements extrapolated over an area. REA for the first time offers the opportunity to use realistic physical fields in coupled tactical decision making models.

11.3.3 Fisheries and other applications

Advanced contemporary fisheries operations and management provides another example of environmental forecast fields coupled to strategic and tactical decision making models. Both governmental regulation of fisheries and commercial fishing are involved and subjective models are in the process of being augmented by quantitative numerical models. Fish spawning, larval survival and metamorphosis to adults (recruitment) affect inter-annual and longer variations in fishing stocks, and the migration of fish stocks over one or two weeks is of great importance to the fishing fleet. It is the latter forecasting problem which overlaps significantly water column REA forecasting requirements. Currents transport and entrap nutrients, phytoplankton, zooplankton and larvae, and the concentration distribution (mass abundance) of adult fish shifts to seek food (e.g., plankton, larvae) to remain in a comfortable environment (e.g., temperature, salinity), and to avoid predators.

This is only one example of many uses of REA involved in management and operations in a multi-use coastal ocean and extended EEZs. Planned operations, e.g. dredging, and sudden crises, e.g. oil spills, ship and aircraft crashes, represent events requiring REA. The ocean science and technology community is becoming increasingly aware of the advantages of dual use research and the development of dual use methodologies (military and civilian).

11.4 The REA OOPS (Ocean Observing and Prediction System)

11.4.1 The overall system and components

Purely deterministic processes such as the orbital motion of the planets can be predicted by integration of the governing physical laws, starting from precise initial conditions at a certain time. In deterministic systems, inaccuracies of the initial conditions can in principle be decreased below any given value by repeated observations and backward modeling. Systems that are subject to stochastic influences, either implicitly through physical processes or by random forcing, can be predicted only with limited accuracy. In the model of a linear system, through its dependence on initialization and parameterization of the physics, the prediction error increases gradually with time. In a nonlinear system, predictions can degrade rapidly when nonlinear terms, including nonlinear transfer of error scales, amplify differences between predictions and reality.

The physical ocean is a nonlinear system with inherent stochastic processes and stochastic forcing. Physical laws are represented by the set of hydrodynamic and thermodynamic equations, which in general have no closed form solution. Numerical ocean models integrate simplified versions of the dynamical equations, that neglect physical processes having minor impact on the problem under consideration. The model formulation for an ocean circulation model, for example, would include gravity and friction forces and would omit density fluctuations that are responsible for sound propagation and vice versa.

A valid model, initialized with reasonable climatological fields and forced with accurate time dependent boundary conditions, can be expected to converge towards reality after a sufficiently extended spin-up period. Numerical calculations of this kind have been extensively used for studies of physical processes in the ocean, for the explanation of inter-annual changes and for climate predictions. Internal dynamics will produce meso-scale ocean features in an eddy-resolving model, if the resolution of the forcing fields in space and time is not too coarse. In this case, the statistical properties of meso-scale features may be correct, but their occurrence must not be regarded as a realistic synoptic prediction.

In an alternative and more realistic approach, real-time ocean assessment is initially based on observations, which are presumed to be essentially coincident in time, and with spatial resolution high enough to resolve the processes of interest. Ocean models initialized with the observations are then used to produce parameter fields that are consistent with the observations. This process, often called "nowcast", balances the fields in a way that they obey the laws of physics and therefore are more than just interpolations between observations.

Starting from the nowcast, the ocean model is integrated over a certain number of time steps, driven by its own internal dynamics and by predictions of the external forcing fields. Inevitably, the period for which reliable forecasts can be produced by this technique is limited by three factors: a) the quality and density of observations, b) the appropriateness of physical approximations in the forecast model and c) the quality and resolution of forecasts for the forcing fields. Before model results would degrade too much, model fields must be adjusted by new observations. The melding of new data into a model run is a non-trivial scientific issue, it is the third component of the ocean forecast system. An oceanographic observation and prediction system (OOPS) for REA thus consists of 1st an observational network, 2nd a numerical prediction model and 3rd a data assimilation scheme.

11.4.2 Data Acquisition and Management

In an REA scenario, very detailed results may be required for a small ocean area. Since the area of interest cannot be isolated from the surrounding ocean, the influences from outside must be taken into account by appropriate boundary conditions. A request for continuous *in situ* measurements along the boundary would be unreasonable. Boundary conditions can instead be obtained from a larger scale model running on a coarser grid and either assimilated from time to time into the small scale model as if it were measured data, or the different scale models can be synchronized leading to frequent adjustments at the boundary of the small model domain. Synchronized models usually have the same number and vertical position of layers. Their horizontal grid spacing differs by a factor of 3 in order to obtain matching points on the staggered computation grid. Model nesting (recall section 11.2) may be carried out also with two-way data flow, thus improving the fields of the larger scale model with results from the more accurate small scale. The observational network required for nested models, includes an appropriate number of measurement stations in the outer domain in order to avoid discrepancies between water mass properties that might otherwise occur especially at depth. If climatological data is also used it must be checked and adjusted as necessary for compatibility with the new observations in both domains. Simple adoption of spun-up climatological data could result in unrealistic density driven currents.

Data acquired for assimilation into a model need not satisfy the conditions for sampling frequency in space and time that must be observed for initial conditions and forcing fields. Since they are only required for the compensation of developing model deviations, they are most effective in those parts of the parameter space that are most difficult to predict. The value of sparse observations may be extended by applying models that describe typical ocean features.

Data managers in an REA system are responsible for timely data transfers between the observational network, the assimilation and modeling effort and the users of observational and forecast products. Data formats should be pre-defined, but in the haste of real time operations, data cannot be guaranteed to be fully quality controlled. Data users, especially modelers, will check again. A single centralized authority for REA data management could slow down the final delivery. Collaborators should rather be connected in a network, preferably the Internet with data access restricted to authorized sites. Data originators or sub-nodes to which they report can keep the data on their computers and allow others to share the information. There should be a site that acts as an REA data fusion center. It has the responsibility to keep track of all information offered by participants, to maintain inventories and to establish links. A distributed system for data management is more flexible and potentially faster than a centralized solution.

The number of permanent platforms such as weather ships, tide gauges or wave poles, is very limited, and a sufficient number can hardly be expected in an area designated for REA. Quasi-permanent platforms are available through satellites that regularly look at the sea surface. Time intervals between repeated coverage of the same area ranges from hours for sea color and surface temperature to weeks for synthetic aperture radar imaging and altimetry. REA can take advantage of satellite data, but for littoral areas substantial information must come from dedicated smaller scale surveys.

Platforms used on demand are moored or drifting buoys, autonomous underwater vehicles, ships and aircraft. The most important *in situ* sensor in the assessment of the physical ocean measures temperature, electrical conductivity and pressure in the water column, which through the equations of thermodynamics also results in salinity, density and sound velocity. Conductivity-Temperature-Depth (CTD) sensors are lowered from a stationary ship, are towed in a yo-yo-ing underwater body or in a wide aperture multi-sensor chain. As expendables (XCTD), they are dropped from a ship or aircraft. Because of the costs, XCTDs are mostly replaced by expendable thermometers (XBT) and the data is assimilated without salinity or with an assumed relation between temperature and salinity, which is usually a good assumption in the open ocean but can be misleading in a littoral environment.

11.5 Forecast System and Predictive Skill Evaluations

11.5.1 Regional System Validation, Calibration and Verification

In the development of any regional forecast system, evaluation of the integrated system, the system components and the forecast products is of course essential. The development of a regional system is usefully conceptualized as having three phases, a *descriptive* phase, a *dynamical* phase, and a *predictive* phase (Robinson et al., 1996). In the descriptive phase the relevant circulation structures, their time and space scales and their variabilities need to be identified. Concomitantly, the forecast models must be *validated*, i.e., shown to be applicable to the relevant structures (e.g., a barotropic model is not appropriate for baroclinic structures) and observational requirements must be established. The extension to interdisciplinary phenomena is direct. In the dynamical phase, the specific dynamical processes responsible for i) the evolution of the circulation and its variabilities, and ii) for synoptical dynamical events must be determined. In this phase the models are *calibrated*, i.e., physical, domain and computational parameters tuned to the region and its phenomena by sensitivity analyses (Lermusiaux et al., 2000) and measurement requirements refined. Although forecasts are carried out in every phase, the predictive phase is devoted to forecast model and assimilation scheme *verification*. Real time forecasts with high quality data sets for initialization, assimilation and verification, obtained by oversampling, are required (e.g., Robinson et al., 1996). These data sets can then be used to design an optimally efficient observational network component for the regional forecast system.

The *predictive capability* of the regional system needs to be evaluated both qualitatively and quantitatively. Regional dominant variabilities need to be defined and used to characterize the forecast. Examples include: the number of branches present in a coastal current that may bifurcate or trifurcate; the existence or not of transient vortices; the location and shape of a permanent meander. Additionally, appropriate quantitative skill metrics need to be defined such as root mean square errors and pattern correlations (Miller et al., 1995; Robinson et al., 1996). The application of such metrics must generally allow for phase errors, i.e., errors in the location or timing of events. Predictive capability without data assimilation is ultimately limited by loss of predictability (Ehrendorfer, 1997; Goswami and Shukla, 1991; Houghton, 1991; Latif et al., 1998). Small differences in initial conditions after some time lead to reasonable but completely different synoptic states. In geophysical fluid dynamical systems, small-scale initial errors are non-linearly transferred to scales of operational interest. Error attribution (initial or boundary conditions, atmospheric forcing, model deficiencies) is important. It is desirable to forecast error fields as well as state variables, as is done, for example, in the ESSE method introduced in section 11.6.2 (Lermusiaux and Robinson, 1999). Automated objective adaptive sampling error minimization metrics can then be related to the quantitative forecast skill metrics (Robinson and Glenn, 1999). The topics discussed in this paragraph require extensive research efforts.

11.5.2 REA System Validation Issues

The sum of differences between validation measurements and model predictions has only limited value as a quantitative measure for predictive capability for operational applications. Capability assessment should comply with the requirements of the end user of the prediction, who is affected by certain deviations much more than by others. The end user cannot be convinced by a small rms error if erroneous predictions in a parameter subspace have high impact on operations. Depending on what the operations are, predictive capability would be estimated differently, obviating a general reliability assessment scheme. As a compromise, one might investigate the change of patterns and features and compare with prediction. At least the tendency, more pleasingly also the magnitude of changes, should match (Sellschopp and Robinson, 1997).

Processes with length scales comparable to or smaller than the cells in the numerical scheme cannot be forecasted. But also time and space resolution of the forcing fields, which is coarser than the grid of the ocean model, limit predictability. The reliability of forecasted atmospheric forcing fields can be of minor importance if internal dynamics exceed external forces, otherwise it has immediate influence on ocean prediction quality. If ocean internal dynamics is dominant, the somewhat longer ocean predictability time is then limiting.

Validation, calibration and verification of a generic REA system rapidly deployable in any region of the global coastal ocean is a challenging and demanding task. Curtin et al. (1993) advocate a series of Coastal Ocean Predictive Skill Experiments (CPSE), which are the essential elements of the predictive phase of regional forecast system development discussed above. Since REA predictive skill experiments must be designed to determine forecast skill on the basis of minimal and covertly attainable observations, they may be most efficiently carried out in the context of the definitive over-sampling provided by a series of regional CPSEs encompassing a broad range of coastal processes.

11.6 Illustrations from Rapid Response Exercises 1996, 1997, 1998

In three consecutive years from 1996 to 1998, REA exercises have taken place in, and at the entrance to, the Mediterranean Sea. They were connected with NATO naval exercises that succeeded the REA survey in the same areas. Up to 8 ships, 6 patrol aircraft and numerous institutions have been active in REA. For support of warfare commanders, data were collected, only part of which was oceanographic in a narrow sense. The rest was for sea bottom and beach assessment. Measurements for real-time analysis of the physical ocean consisted of CTD stations, XBT and XCTD casts from ships and aircraft, shipborne ADCP velocities, surface drifter deployments and satellite images. Self-recording current meters were used for *a posteriori* validation.

Nowcast and forecast results from ocean modeling were delivered in a format suitable for the customer, who was most interested in surface currents and sound velocity profiles. Maps and vertical sections were produced as image files for electronic distribution. Profiles were reduced to inflection points coded in a special format for automatic insertion into navy environmental systems.

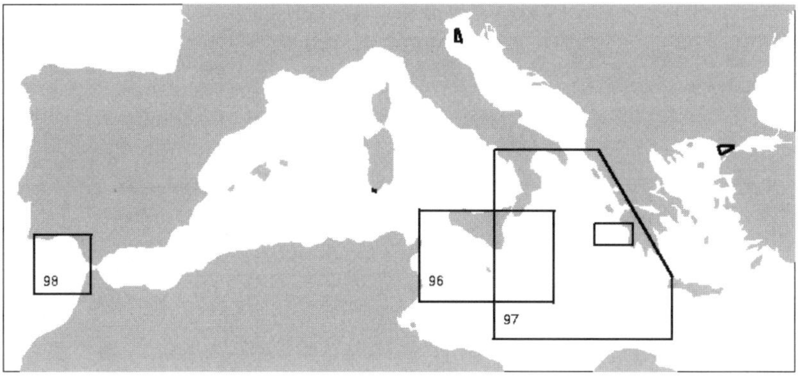

Fig. 11.2 Rapid Response exercise domains of the years 1996 to 98. Ocean models were set up for the three large domains and a nested subdomain.

11.6.1 Location and Characterization of Operational Regions

Rapid Response 1996 and its subsequent naval exercise (both referred to hereafter simply as RR96) took place in the Strait of Sicily and the adjoining western Ionian Sea from mid-August to early October. RR97 spanned the entire Ionian also from mid-August to early October a year later, followed by RR98 in the Gulf of Cadiz just west of the Strait of Gibraltar, only four months later in February and March. Nested modeling domains were utilized in all cases as appropriate for anti-submarine warfare (ASW), mine warfare (MW) and amphibious warfare (AW). Fig. 11.2 depicts the largest ASW domain for each exercise and also the MW domain off the western coast of Greece for RR97.

With respect to the phases of development concept for regional forecast systems these regions differed. During the 1980s Physical Oceanography of the Eastern Mediterranean (POEM) program (Robinson and Malanotte-Rizzoli, 1993) quasi-synoptic basin scale surveys of the Ionian Sea were carried out with the sub-basin scale and coarse mesoscale resolution, and the circulation and water masses have been synthesized by Malanotte-Rizzoli et al (1997). Thus the 1997 exercise served as a predictive phase forecast verification experiment. Prior to RR96 our Harvard and SACLANTCEN research groups were collaborating in Strait of Sicily research and had carried out research cruises there with real-time at-sea forecasting and data assimilation in 1994 and 1995 (Robinson et al, 1999b). A dynamical phase cruise was already planned for 1996 before RR96 was scheduled which was subsequently carried out in conjunction. Thus excellent synoptic flow and water mass data was available for the region with mesoscale and sub-mesoscale resolution (Robinson et al, 1999b; Warn-Varnas et al, 1999) and we regard RR96 as a combined dynamical/predictive phase exercise. The Gulf of Cadiz presented a complex and generally less well understood region for which we were not able to locate a prior synoptic

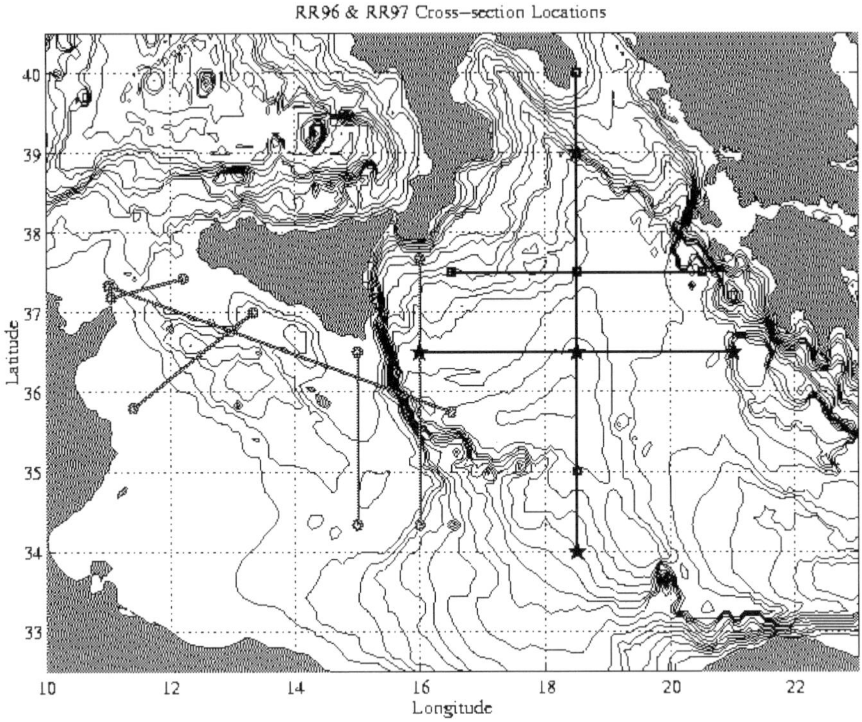

Fig. 11.3 Bottom topography for the RR96 and RR97 regions is shown. Contours are 250m. The red and blue lines indicate the positions of standard vertical sections for the real time forecast products.

mesoscale resolution regional data set. For our team, RR98 represented a challenging combined descriptive/dynamical/predictive phase effort, with little time for preparation. The full scope of phases of these three exercises provided a valuable range of REA experiences.

The Strait of Sicily and the Ionian Sea - The topography of this region is depicted on Fig. 11.3. The choke point of the strait lies in the west to the northeast of 37°N, 11°W. There is a deep central trench in the strait and a narrow (broad shallow) coastal region off the Sicilian (North African) coast. The extremely steep and narrow Ionian Shelfbreak off the eastern coast of Sicily extends southward and fans out and broadens off the coast of Libya. The central Ionian basin and the Cretan passage to the south of Greece have gentle slopes and lie between about 2500 and 3500m; there are steep narrow slopes off the west coast of Greece.

Fig. 11.4 presents a schematic of the near surface circulation and variabilities of the Strait of Sicily which was developed in real time during RR96. Fig. 11.5a is a

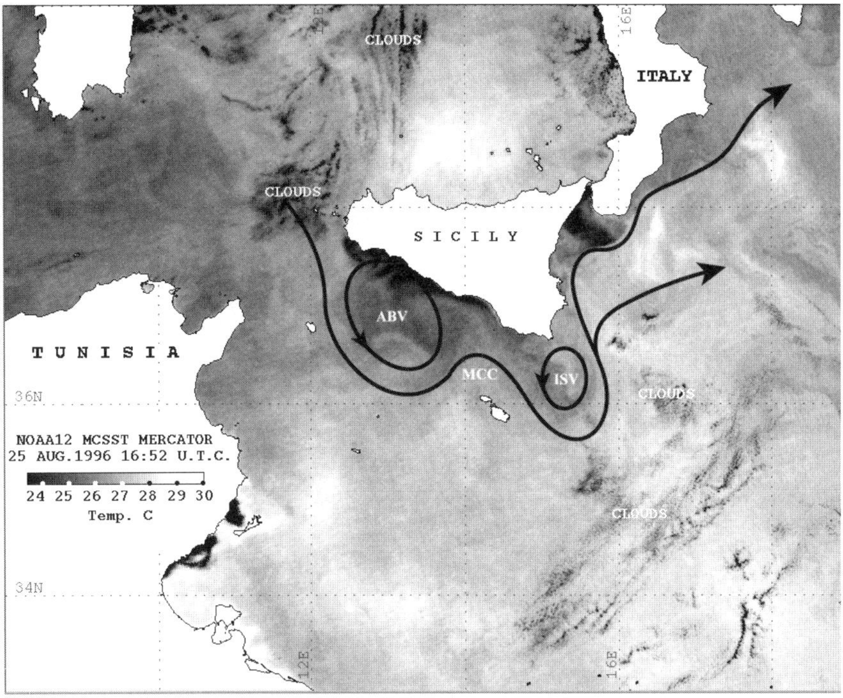

Fig. 11.4 A schematic of the near surface circulation and variabilities of the Strait of Sicily which was developed in real time during RR96 superimposed over an AVHRR image of the area for 25 August 1996. Features noted are: ABV - Adventure Bank Vortex; MCC - Maltese Channel Crest; and, ISV - Ionian Shelfbreak Vortex.

similar schematic for the entire region based upon the POEM August-September 1986 data; Fig. 11.5b identifies the system of cyclonic and anticyclonic eddies and gyres present during RR97 in the Ionian.

In the Strait and the Ionian, the fresh modified Atlantic water flowing generally eastward and salty Levantine water flowing generally westward constitutes a two-current system of the general circulation of the Mediterranean. The Levantine outflow is located at depth and the Atlantic inflow is in the upper ocean. The Atlantic inflow, the Atlantic-Ionian Stream (AIS) is an internal free jet controlled by topography and internal dynamics exhibiting variabilities. At the Ionian Shelfbreak, the AIS bifurcates (or trifurcates) into a branch that turns north and flows along the bathymetry before exiting into the Ionian, and one that proceeds directly east into the Ionian Sea. The strength and volume transports of the branches change considerably due to seasonal and ambient conditions. Mesoscale and sub-mesoscale eddies are always present in the Sicilian Channel. The mesoscale variability has been studied statistically by Lermusiaux (1999). In addition to the general circula-

a)

b)

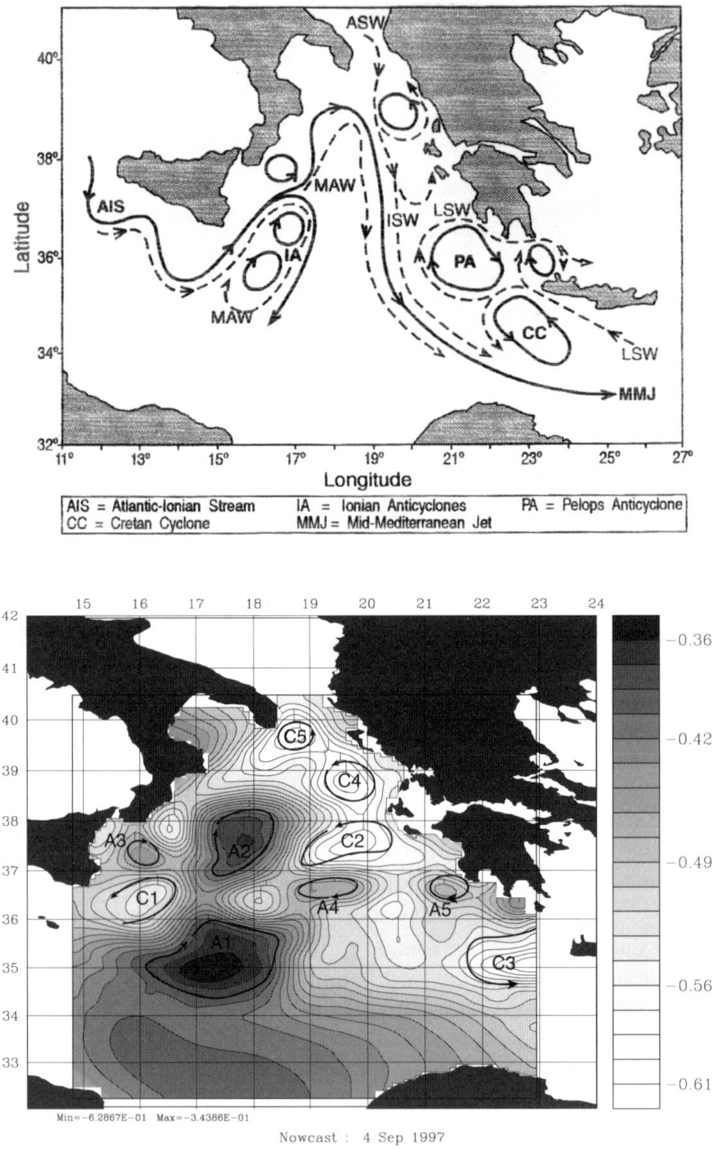

Fig. 11.5 a) A schematic of near surface (solid lines) and mid-depth (dashed lines) circulation for the Strait of Sicily and Ionian Sea based upon the POEM 1986 data; acronyms not identified in the figure legend include: MAW - Modified Atlantic Water, ASW - Adriatic Surface Water and LSW - Levantine Surface Water; b) system of cyclonic (C) and anticyclonic (A) eddies and gyres present during RR97 in the Ionian superimposed upon an analysis of dynamic height (5m/800m) for 4 September 1997.

tion, with its mesoscale variability, there are the wind-driven currents on the from local and remote storms (including the Sicilian coastal current) and upwelling off Sicily. Tides, inertial, gravity, surface, and continental shelf waves occur. This region contains active water mass modification processes between the fresher and warmer Atlantic origin water mass and the saltier and colder Levantine water mass. The free jet filaments of the AIS in the Ionian Sea exhibit considerable variability in occurrence and location as they thread their way to the Cretan passage. In addition a number of (semi-) permanent and transient sub-basin scale to mesoscale cyclonic and anticyclonic vortices occur which are interactive with the jets and each other. The variability of the circulation structures and sub-basin scale features has been studied by Malanotte-Rizzoli et al (1999). Figs. 11.4 and 11.5 will be discussed in more detail below.

The Gulf of Cadiz - Fig. 11.1, introduced earlier, and Fig. 11.6 present aspects of the circulation and topography of the Gulf of Cadiz. For our purpose we define the Gulf of Cadiz as the area east of 9°W and north of 35°N. To the north and east it is bounded by the Portuguese, Spanish and African coast. The western and southern boundaries are open; another narrow open boundary exists in the east at the Strait of Gibraltar. The west-east extension is about 270 km, in north-south direction it extends over about 250 km. Maximum water depth is about 3000m in the southwest corner. Interior to the Gulf, topographic slopes are gentle. The Gulf of Cadiz is incorporated in the eastern recirculation regime of the North Atlantic subtropical gyre (Siedler and Onken, 1996). It is located in the transition area between the Portugal Current and the Canary Current. A small fraction of the North Atlantic thermocline transport flows into the Strait of Gibraltar. Towards the southeast, the area is connected to the northwest African upwelling regime. The sill depth of the Strait of Gibraltar is about 300m at the Camarinal Sill (5°45' W). The circulation in the strait is characterized by a near surface inflow of low salinity Atlantic Water into the Mediterranean and an outflow of salty Mediterranean Water (MW) below. Mean transport rates of both flow components are on the order of 1 Sv, mean velocities are on the order of 1 m/s. After having passed the Camarinal sill, the high density MW slides down the continental slope and progresses further to the west while leaning against the Spanish/Portuguese continental slope. During these stages the MW is subject to strong mixing. The main spreading paths are confined to two warm salinity 'veins' at about 600m and 1100m depth, typical velocities within these veins are 40 cm/s or even more. The Gulf of Cadiz circulation is influenced by wind forcing and surface heat and water fluxes. In wintertime, the area lies close to the main storm tracks of the northern hemisphere, hence wind forcing may play a dominant role. The mean inflow/outflow situation in the Strait of Gibraltar is superimposed by strong tidal currents, the magnitude of which is on the same order as the mean flow. Hence, within a tidal cycle, inflow/outflow may be reversed. Special features of the internal dynamics of the region involve the external effects of internal waves and hydraulic jumps occurring in the Strait of Gibraltar within a tidal cycle, the slope convection of MW west of the Camarinal Sill and the instability of the MW veins causing generation and detachment of meddies (Onken, 1998, personal communication).

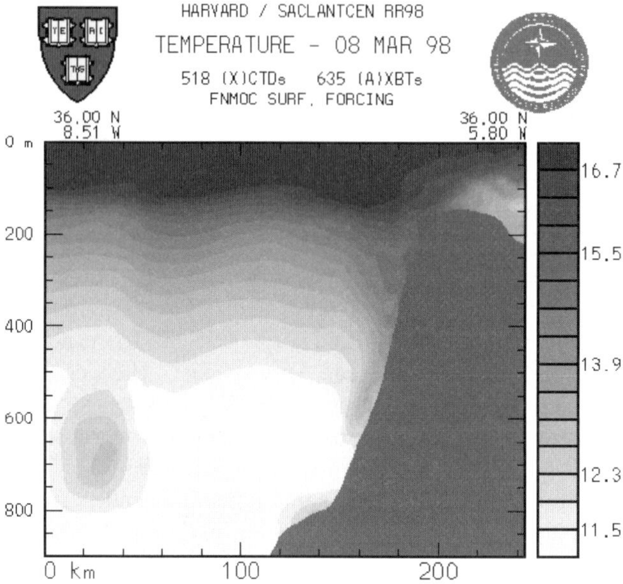

Fig. 11.6 A vertical section of temperature along 36N as taken from the RR98 operational web site. Data noted in the figure is from the RR98 exercise. The figure shows the outflow of Mediterranean water through the Strait of Gibraltar.

11.6.2 Methodologies: Initialization, Assimilation and Sampling

The Harvard Ocean Prediction System (HOPS) is a flexible, portable and generic system for inter-disciplinary nowcasting, forecasting and simulations. HOPS can rapidly be deployed to any region of the world's oceans, including the coastal and deep oceans and across the shelfbreak with open, partially open or closed boundaries. Physical, and some acoustical, real time and at sea forecasts have been carried out for more than fifteen years at numerous sites (Robinson, 1999) and coupled at-sea biological forecasts were initiated in 1997. HOPS has been utilized during the Rapid Response exercises and the operational applications. The present system is applicable from 10m to several thousand meters and the heart of the system for most applications is a primitive equation physical dynamical model.

HOPS is used to build and maintain a regional synoptic description through nowcasting, forecasting and assimilation. The domain is initialized with the best possible estimate from historical synoptic data, climatology and feature models. Dynamical adjustment of synoptic conditions, not a spin-up from climatology, enables a rapid achievement of current conditions. Historical data can be thermodynamically and dynamically adjusted to match the present state. *In situ* or remotely sensed data, acquired via adaptive sampling designed to have optimal impact and control predictability is assimilated.

Data assimilation methods used by HOPS include a robust Optimal Interpolation (OI) scheme with weights set by simple engineering-type assumptions and a quasi-optimal scheme, Error Subspace Statistical Estimation (ESSE). The advanced ESSE method determines the nonlinear evolution of the oceanic state and its uncertainties by minimizing the most energetic errors under the constraints of the dynamical and measurement models and their errors (Lermusiaux and Robinson, 1999; Lermusiaux, 1999). Measurement models relate state variables to sensor data. Substantial real time efficiency is achieved by reducing the error covariance to its dominant eigendecomposition. Error propagation is estimated via an ensemble forecast using the full nonlinear model. The evolving error subspace is characterized by singular error vectors and values, i.e., time evolving three dimensional error empirical orthogonal functions (EOFs). Melding weights for assimilation are determined using a minimum error variance criterion. Importantly, melding occurs in the error subspace and is thus much less costly than a classical analysis with the full error covariances. The error subspace is updated at the melding step by combining the forecast principal errors, i.e., errors arising from the dynamical model and the loss of predictability, with the error covariances of the measurements. ESSE was utilized in real time for the first time during Rapid Response 96. The influence of historical synoptic data is reduced as data is gathered and eliminated when and if sufficient data is acquired.

During RR96, dynamically adjusted (heated while in dynamical balance) historical synoptic data from 1995 was used in combination with the Mediterranean Ocean DataBase (MODB) climatology. In RR97, quasi-synoptic data from the Physical Oceanography of the Eastern Mediterranean (POEM) cruise of 1986, along with MODB data, was used to formulate initial conditions. In the Gulf of Cadiz region (RR98), the MODB MED4 database, supplemented with data from the UK Hydrographic Office, the NODC CD-ROM and the SACLANTCEN web site, provided the basis of the initial conditions.

The Rapid Response exercises have generally provided extensive sampling during their duration and provide extremely rich data sets as a result. Fig. 11.7 shows the sampling patterns during RR97. Three panels (top left and right and bottom left) represent the data collected during the exercise; 5 ships and 17 aircraft flights acquired 416 CTDs and XCTDs and 975 XBTs and AXBTs during the period 18 August - 4 October 1997. The sampling patterns were designed for maximum spatial coverage with constraints imposed by platform capabilities and mission needs. Some patterns were pre-planned while others were adaptively planned as forecasts and conditions dictated. The bottom right panel of Fig. 11.7 shows the data positions for the POEM 1987 data which formed the initial conditions for the RR97 forecasting.

Fig. 11.8 illustrates the effect of, and need for adaptive sampling. The left panel shows the expected error of an objectively analyzed data field after an AXBT flight. The path of the aircraft is obvious and was adaptively designed to reduce

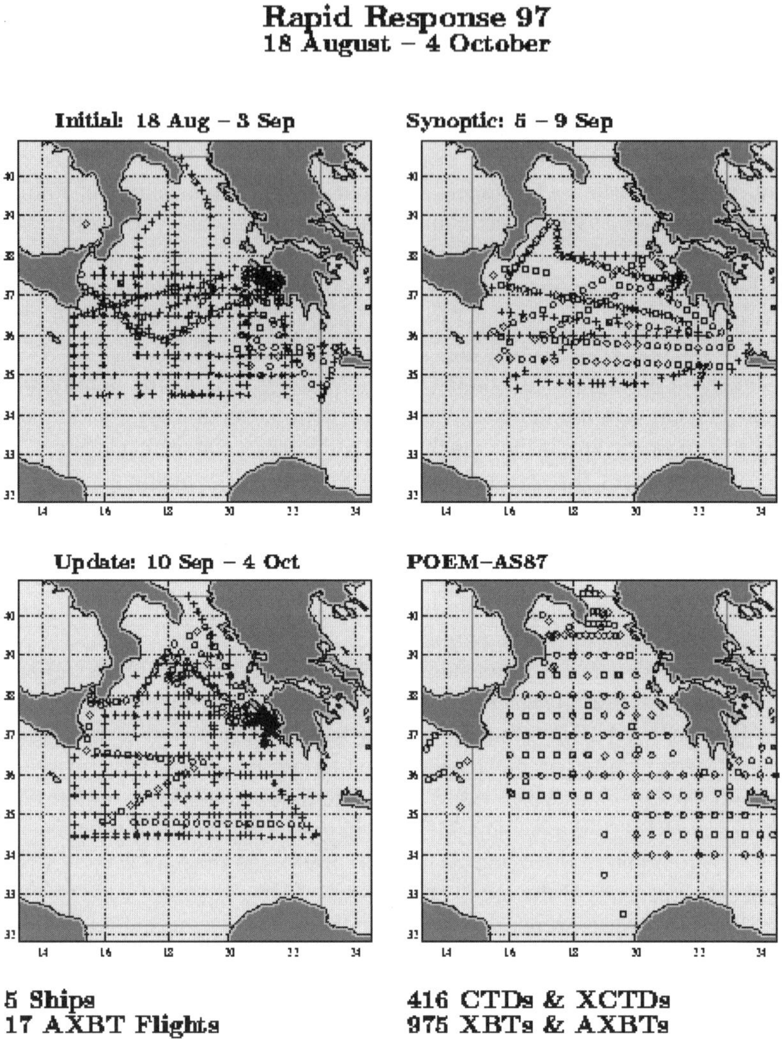

Fig. 11.7 Shows the sampling patterns during RR97. Three panels (top left and right and bottom left) represent the data collected during the exercise; 5 ships and 17 aircraft flights acquired 416 CTDs and XCTDs and 975 XBTs and AXBTs during the period 18 August - 4 October 1997. The sampling patterns were designed for maximum spatial coverage with constraints imposed by platform capabilities and mission needs. Some patterns were pre-planned while others were adaptively planned as forecasts and conditions dictated. The bottom right panel shows the data positions for the POEM 1986 data which formed the initial conditions for the RR97 forecasting. + signs indicate the position of XBTs and AXBTs. The small circles indicate the position of CTDs and XCTDs.

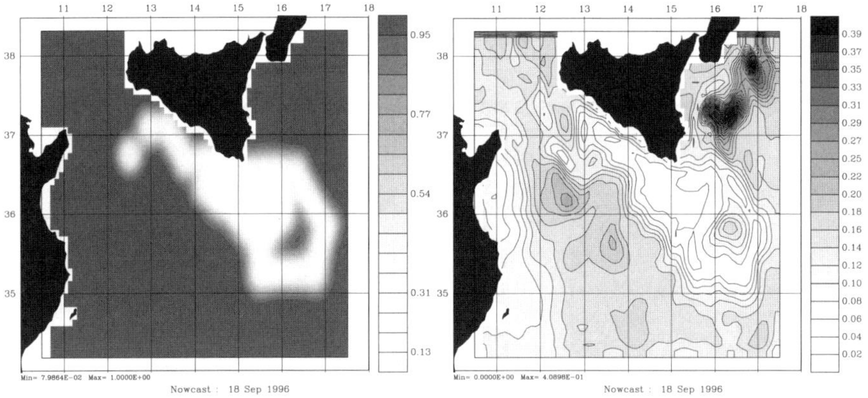

Fig. 11.8 The left panel shows the expected error of an objectively analyzed data field after an AXBT flight. The path of the aircraft is obvious and was adaptively designed to reduce model forecast error of the meandering stream. The right panel is the model forecast error (contours of 0.02) after the data collected was assimilated into the model. The reduction of model error in the region of sampling is indicated and the higher errors in other regions indicate the need for sampling at a future time.

model forecast error of the meandering stream. The right panel is the model forecast error after the data collected was assimilated into the model. The reduction of model error in the region of sampling is indicated and the higher errors in other regions indicate the need for sampling at a future time.

11.6.3 Products

Operational products provided to naval operators have evolved during the Rapid Response exercises, as indicated by the needs of the operators. Primary products have included maps of surface temperature, salinity, mixed layer depth and velocity, maps of bottom current velocity (for mine warfare), vertical sections of temperature along chosen paths, temperature profiles in standard naval JJXX format, maps of forecast error, tidal velocities and maps of surface chlorophyll (as a proxy for bioluminescence). The use of a central web site for product distribution has allowed for the release of products within a few hours after the completion of a forecast. Forecasts were issued, on average, at a 4-7 data interval. Fig. 11.9 represents a sample of the products issued during RR97. Shown are maps of surface temperature and mixed layer depth for 30 September 1997 in the large forecast domain, a south-north vertical section of temperature along 18.5°E for 30 September 1997 and a map of bottom velocity in the mine warfare (Kiparissia Bay) domain for 20 September 1997.

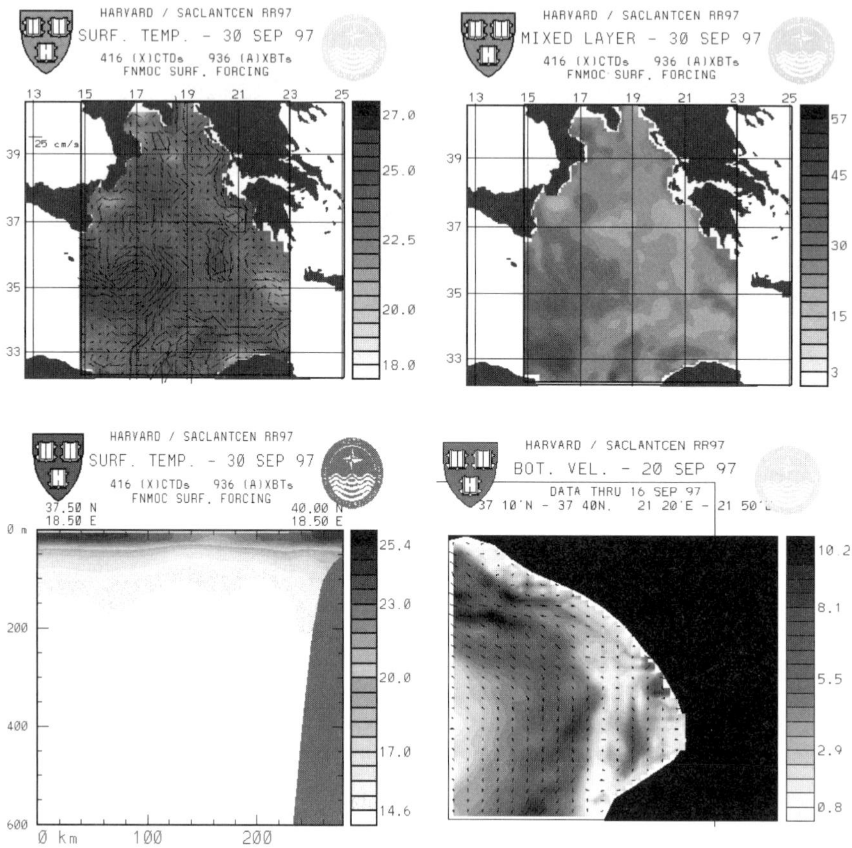

Fig. 11.9 Sampling of the products issued during RR97. Shown are maps of surface temperature and mixed layer depth for 30 September 1997 in the large forecast domain, a south-north vertical section of temperature along 18.5°E for 30 September 1997 and a map of bottom velocity in the mine warfare (Kiparissia Bay) domain for 20 September 1997.

11.6.4 Forecast evaluations

Subjective evaluations of the real-time forecasts during RR96 and RR97 are presented below. These rich data sets also provide the possibilities for quantitative hindcast studies and REA Observational System Simulation Experiments (OSSEs), some of which have been initiated. Quantitative skill metrics for objective evaluations of the real-time forecasts are under development.

A quick-look verification of predicted variations of features in the Sicilian Channel was performed for four dates during RR96 as the exercise proceeded (Sellschopp and Robinson, 1997). The satellite sea surface temperature images and the HOPS nowcasts and forecasts were selected and the position and shape of three dominant features were identified, as well as their development in time. These

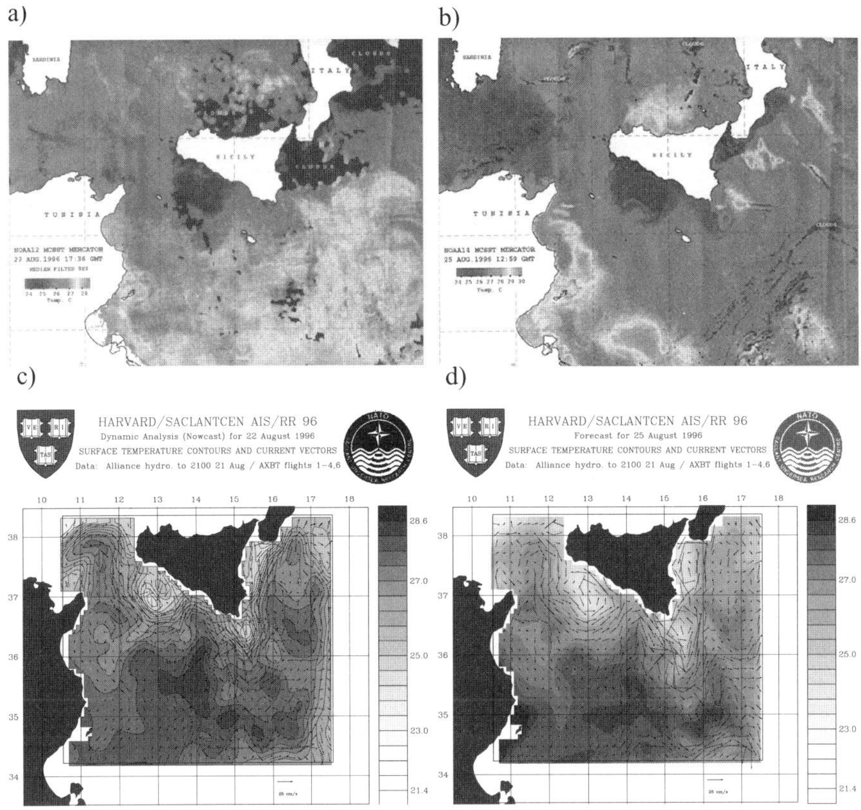

Fig. 11.10 Panels (a-b) show the satellite SST distributions for August 23 and August 25, 1996, respectively. Panel (c) shows the HOPS nowcast of the surface temperature for August 22, overlaid with surface velocity vectors (scale arrow is 25 cm/s). Panel (d) is as Panel (c), but for the forecast fields for August 25.

dominant features are presented in Fig. 11.4: ABV - Adventure Bank Vortex; MCC - Maltese Channel Crest; and, ISV - Ionian Shelfbreak Vortex. The satellite images were chosen with regard to percentage cloud cover. The quick-look comparison of actual development versus predicted development indicated that the HOPS forecasts predicted conditions at a success rate of approximately 70%. Fig. 11.10 depicts the type of comparisons made for the quick-look comparison. A subjective comparison of the SST distributions from AVHRR show excellent agreement with the HOPS nowcast and forecast fields shown below them.

Dr. K. Nittis, from the National Center for Marine Research (NCMR) of Athens, Greece, joined the Harvard/SACLANTCEN group for RR97 to provide expert knowledge of the circulation and processes present in the Ionian Sea. We summarize here his description and subjective evaluation of the RR97 circulation and forecasts. Fig. 11.5b illustrates the upper thermocline dominant circulation mesoscale and sub-basin scale features present during RR97. The most energetic features are the two anticyclonic eddies located in the central Ionian (A1 and A2). They can be easily observed in both altimeter and CTD (dynamic height) data. Between these two anticyclones we observe two cyclonic eddies, C1 and C2. The two eddies C1 and C2 occasionally appear as part of a larger cyclonic system that separates A1 and A2. Two smaller anticyclonic gyres are present in the same area, without being visible in most of the SSH images (probably due to their size). The first one (A3) is located south of the Gulf of Messina. At surface and subsurface layers (0-100m) it appears to be connected with A2. The second anticyclone (A4) is located east of A2 and C2. All four anticyclones (A1-A4) seem to be part of the same system, the Ionian Anticyclone. In the eastern Ionian, along the coasts of Greece we observe the Cretan cyclone (C3), the Pelops Anticyclone (A5), and two cyclonic gyres (C4 and C5). The Cretan Cyclone is a permanent sub-basin scale feature of the East Mediterranean upper thermocline.

The model (Fig. 11.9)seems to reproduce the three phases of the Ionian Anticyclone: the 1st phase when the large scale circulation dominates and the anticyclones are part of the same system; the 2nd phase with i) the intensification of cyclones C1 and C3, ii) the weakening of the large scale circulation, iii) the intensification of A2; and the 3rd phase when A1 and A2 seem to merge again in large scale circulation. Finally, it is interesting to note that the model simulates the cyclonic circulation in the NE Ionian (C4 - C5) but the Cretan cyclone (C3) in the SE Ionian is not well represented at least during the first half of the period (up to Sept. 8).

11.6.5 Accomplishments and implications

The time evolution and thus the predictive capability (recall section 11.5) of regional circulation systems is driven by internal dynamical processes, inflow and outflow boundary conditions and surface atmospheric fluxes. Our knowledge of internal dynamical processes and boundary conditions is affected by the type, density, accuracy and collection strategy of environmental data and the calibration and tuning of the physical dynamical model. Rapid Response forecasts have proven to be usefully accurate. The predictability of the individual regions is limited: in weather prediction, to a few days by atmospheric internal non-linear dynamics, and, within the ocean, to a week or two by similar processes.

Each of the Rapid Response exercises provided new and unique opportunities. While many aspects of the REA process remained consistent between exercises, the variation in domains and dynamical processes contributed to differences between each operation. RR96 was our first experience in REA. Dialogs were initiated with naval officers and products tailored for specific warfare activities. The NRV Alliance served as a shipboard forecast center. HOPS utilized a novel synoptic initialization from warmed historical synoptic data. Aircraft adaptive sampling was initiated both subjectively and based on the first real time ESSE error forecasts.

In RR97, very efficient communications and data management procedures were established. A command ship (R/V Alliance) collected data from the survey fleet (vid Fig. 11.7) which was transmitted to a fusion center (SACLANTCEN) and a remote institutional forecast center (Harvard). An efficient website was central to the almost flawless communication of data and products. This exercise involved a unique preliminary forecast phase with historical synoptic data from the same season in 1987 (POEM) and data acquired from satellite and ships transiting to the preliminary briefing in Crete. New experience was gained in adaptive sampling from many simultaneous platforms and the assimilation of SSH. Nesting of operational domains was initiated and special products for the MW region developed. RR97 was a sustained exercise in a large region with known dynamics which allowed for attention to forecasting and expert qualitative evaluation.

RR98 proved especially demanding due to minimal preparation time in an unknown region with demanding physics. The physics included challenging forcing through the Strait of Gibraltar as well as from the NE Atlantic. Many new initiatives were brought forward during RR98, including: the coupling of physical and biological modeling, assimilation of sea surface color (SSC), the first prototype bioluminescence forecast and the first inclusion of tidal products to the forecast center.

The experience of these three exercises and the lessons learned provide the substantive basis for sections 11.7 and 11.8.

11.7 Organization, logistics and resources

As for any oceanographic field activity, the establishment of an ocean observation and prediction system (OOPS) for rapid environmental assessment has to be arranged within the constraints of limited resources and a given suite of measurement systems and platforms. A single laboratory is hardly able to provide all components of an OOPS by itself. In a cooperative program, laboratories should be tied together to provide the required mix of expertise, instrumentation and platforms.

11.7.1 Platforms and resources

In an early phase of military survey planning, it must be defined whether the process of data collection ought to be covert or overt. Remote sensing of the sea surface is the most obvious method for covert ocean measurements. It can provide the surface parameters: elevation, sea state, roughness, temperature and color, which to a certain extent allow conclusions to be driven for subsurface conditions also. With visible laser light and translucent water, also shallow water bathymetry can be obtained from distance. Currents can be measured by their impact on the velocity of surface waves and the Doppler shift of Bragg-scattered radar signals. Platforms for remote sensing of the sea surface are satellites, aircraft and radar towers.

Covert *in situ* measurements require a submerged platform to be deployed in the area of interest. The range of possible platform size is from a swimmer to a submarine. Autonomous underwater vehicles (AUV) will gain more importance even for overt surveys in future when the maximum range of small affordable units increases to operationally meaningful distances. For time series at fixed locations, bottom moored systems are appropriate, deployed from aircraft, ship or AUV. Following a pre-programmed sequence, they release a surface float with an antenna that transmits the most recent data via satellite to the laboratory (Tyce et al, 1998). Moorings of this kind are required also for non-denied observations because traditional moored instruments would store data during the full deployment period. Data are read out only after recovery and thus not available in time for ocean prediction.

Open military environmental investigations like any civilian ocean research activity mostly rely on ships as the main platforms. The expenses for ship time are easily compensated by the advantage of hands-on interaction with measurement systems. On a ship it is easier to meet the requirements for instrumentation, which need not be as complex and sophisticated as in remotely operated systems.

For the analysis of the physical ocean, the direct measurement of those quantities is most valuable that are represented in the fields of numerical models, namely temperature, salinity and water velocity. Temperature and conductivity below the sea surface require direct probing. The limitation to one-dimensional measurements when *in situ* probes are lowered from a ship, is overcome by towing undulating bodies or multi-sensor chains. Traditional current measurements at single fixed positions have been widely replaced by acoustic profiling current meters.

An optimal ocean observing and prediction system would continuously have access to real time data. Many measurement systems and platforms that are used for an REA campaign, are also suitable for permanent monitoring, and it is only the costs for maintenance and periodic replacement that prevents constant deployment. Sensor fouling can, however, severely degrade data such as electrical conductivity, with the consequence that the precision required for ocean modeling is not guaranteed after weeks of exposure to a shallow littoral environment.

11.7.2 Adaptive sampling and assimilation

In an REA operation, it is not totally predictable when and from what location updating measurements will arrive. This might put some strain on data assimilation procedures as compared with assimilation from fixed station networks. On the other hand, smart assimilation of data from changing locations has the potential for detection and refinement of structures that might fall through the meshes of an immutable observational scheme.

Free decision on the locations for the next day's measurements offers the opportunity for optimal track design. Ocean areas that are suspected or estimated to be incorrectly represented in the model results will be covered together with areas that have not been visited for a longer period. Adaptive sampling significantly reduces the demands for resources and the number of platforms, which otherwise would have to provide full area coverage with acceptable forecast degradation everywhere. Automated adaptive sampling, which is in its infancy, can be expected to impact the efficiency and effectiveness of REA substantially (Robinson and Glenn, 1999).

11.7.3 Data fusion and communication

For successful cooperative adaptive sampling it is necessary that modelers are prepared to timely articulate their preferences for the next sampling locations and data types, and that reliable communication channels exist between modelers, survey planners and data providers. In a small experiment, when cruise management, data acquisition and processing, and ocean modeling is united on a single research ship, timely communication is purely a problem of persons working together. In a bigger scenario with several ships, aircraft and land based laboratories, technical installations for communication require major attention. Traditional ship to ship and ship to shore communication is at the best appropriate for short conversations. For data transfer from ships, high bandwidth satellite communication is preferable. Near to shore also cellular telephones with digital data capability are appropriate, since in most cases a data rate close to ten thousand bit per second will suffice.

Data at high rates can be exchanged between ships or with a land station in sight (15 km) using modern off the shelf spread spectrum radios. They are not standard on ships. Their installation requires a few months preparation time. Standard data transfer protocols are adapted from local area networks. Connections via spread spectrum radio can be virtually transparent for the user, whose computer becomes part of a larger network spread over several platforms.

On land it can be presumed that everybody involved in an REA operation possesses a connection to the Internet and the appropriate technology. Internet standard is appropriate for data exchange even if the data must not be shared with the public. Access control and encryption are in place that would keep unwanted recipients out. REA operations take advantage of the three main services on the Internet. Electronic mail is used for coordination, file transfer for data delivery and the web for distribution of results.

Relevant information for REA consists of databases, archived and actual data, that are likely spread over several agencies and data providers. To make the information usable for the development of derived products, format conversions and data transformations must be carried out for older data. For the expected real-time data, administration programs must be in place prior to the exercise. Paper products are a challenge for appropriate handling and therefore sometimes ignored. Fast delivery and processing requires information in electronic format.

Once the existing data have been determined, everything needed for forecasting is gathered by modelers. Entire databases need not, and the responsibility for data bases should not be transferred to the groups that perform the modeling. In the web structure of the Internet, data can be shared when they reside at different places. For atmospheric forcing fields, doubtlessly the atmospheric center takes responsibility for data presentation and administration. Also the responsibility for oceanographic input data should be with the organization that is closest to the originators.

A distributed data system for REA has to be better organized than ordinary information on the Internet. It is unacceptable for participants in an REA effort to regularly search through the offers of potential data providers and find out whether new information has shown up. There must be an element in the REA organization, called the Fusion Center, that either receives an REA data set as soon as it is generated, or is informed about its existence and the location from where it can be downloaded. The Fusion Center will set up an inventory of all existing data with pointers to the respective files. It generates maps, designs web pages and provides appropriate data search algorithms. Occasionally the Fusion Center can adjust data formats to a common standard. Data originators remain however responsible for data integrity. The Fusion Center provides service to participating groups who for product generation rely on other people's data, it creates and maintains an archive of contributions, and it finally places results to be browsed by customers.

11.7.4 Simple and complex systems for special purposes

The complexity of the distributed data management system and fusion center depends on the size of the REA operation. In the most simple case of a single platform for measurements and modeling, data collection on a computer with a clear directory structure and a few pages written in hypertext markup language (html) for descriptions and cross-links between data types would do. In a complex REA survey, data are physically organized relative to the different data providers. They keep their contributions together, either on an own Internet server or in a directory at the fusion center. The fusion center then creates tables that would combine links to a certain data type regardless of the originator. The distributed data archive thus gets a double structure: the physical placement of files and the menu-guided access to data can differ significantly.

In an REA operation, it is often required to establish a back-up for the data server or to save the complete server contents on a CD-ROM. This should be facilitated by a simple information format and server structure. Some of the modern techniques, that are widely used on Internet servers such as java scripts and database queries, are counterproductive in that respect.

Common formats for all data delivered for REA, though desirable, are unlikely accomplished. But care must be taken that data files are compatible between computer systems. Standard tools under the most frequently used operation systems, at least MS Windows and Unix, must be able to digest all files. Plain ASCII files with a simple structure are preferable. If binary data are more appropriate, platform independent formats such as Netcdf or Matlab save sets are recommended. For text and image data, standard file types of the world wide web (Internet) are preferred including the platform-independent document format (pdf) for vector graphics and formatted documents. Postscript (ps) format is also acceptable for documents that are anticipated for high quality printouts rather than for screen display.

Files generated with a typical office tool such as a spread sheet, presentation graphic or document writer, would require compatible software releases for ingestion, that are unlikely available under Unix and may not be present under Windows. Office file formats are therefore inappropriate for data exchange. By careful choice of the format of electronic products, especially of raster graphic images, the file size can often be remarkably decreased.

REA relies on state of the art communications technology and computer network software. The data flow from acquisition and primary data processing to the generation of derived products and further to appropriate presentation to the end user, cannot be expected to be defined in standard procedures yet. The construction of the data collection and distribution system requires expert knowledge of modern technology. At the same time, in-depth knowledge is required of the essence of the disseminated information. To make an REA communications system functioning, computer specialists, network managers and scientists work closely together. In a fully satisfying REA communications system, scientists have experience in the generation of electronic products, and network managers have a scientific background and a clear idea of the needs of the end users.

Feedback and continuing discussion guarantee for optimum results at the end of an REA communications chain. Data originators continuously monitor the product that is delivered to the end user and verify that the product is presented correctly without degradation by inappropriate integration into the context of other expositions. Data managers give advice on data formats that would comply with a general standard, increase readability or diminish the requirements for transmission bandwidth.

11.7.5 Distribution and impact of information

The final customer must be seen as part of the information chain. It is highly desireable that the customer takes part in the definition of data products and its presentation. Different from usual publications, which will sooner or later find the intended audience, REA products are available to a limited number of users who have the proper network connection. The suitable complexity of REA products very much depends on who the customer is and what capability he has for higher-volume digestion.

A major navy command being the customer of REA products likely has good communication channels and a staff of meteorology and oceanography experts who would be happy to receive detailed and complex environmental descriptions and predictions, even contradictory assessments from different origins. Smaller units and sub-commands that cannot spend time on environmental assessment need concise filtered products.

Security issues complicate data fusion and dissemination, if part of the information is to be released only to a subset of participants. Special password protection can be used to limit access only to authorized persons. By identity checks at login time, it can be decided by the data server which part of information is offered to a customer, who would not even get knowledge about the existence of denied products.

Data protection and restricted accessibility by selected members of an REA effort might be necessary in exceptional cases because of industrial proprietary issues. There is, however, no need for software protection. Essentially all software required for the management of a distributed network can be obtained from the public domain. The same is true for browsers, image viewers, document readers and geographic information system client software. As a precondition, originators of REA products must avoid the delivery of files that would require not freely available proprietary software for ingestion.

It is not suggested to publicly disseminate the entirety of information that is exchanged between participants of an REA operation because of the preliminary and provisional character of partial results. It is however a good idea to prepare a few pages with selected and possibly updated results on the unrestricted Internet for mitigation of the public that might be interested or have environmental concerns.

A visitors page on the Internet is also apt to gain credibility of the scientific community. Regular presentations of results demonstrate that REA is more than a vision. Scientific and logistic state of the art includes the competence for rapid assessment of the ocean environment.

11.8 Summary, conclusions and prospects

In this chapter we have presented the concept of rapid assessment of coastal ocean regions for the purpose of providing detailed synoptic environmental nowcasts and forecasts in support of a variety of naval and other operations. The general concept was illustrated and specific research and operational issues were identified and explored in the context of the NATO REA Rapid Response exercises 1996-7-8. The REA Ocean Observing and Prediction System requirements include importantly portability, flexibility, rapid relocateability, inclusivity with respect to coastal processes, and scalability with respect to hardware and software. Operational and system integration issues with respect to data acquisition and management and product design and dissemination have been discussed.

Scientific research areas which impact REA and provide the fundamental bases for advancement of the concept and capability include interdisciplinary coastal ocean science, ocean forecast science, ocean engineering sciences and information and complex system sciences. All of these sciences are evolving and growing rapidly and novel interdisciplinary interactions among them are being identified and researched. Rapid progress in rapid assessment of all regions of the global coastal ocean must be anticipated.

Acknowledgements

We are grateful to Mr. Wayne G. Leslie for essential input to the preparation of this manuscript and thank Ms. Gioia Sweetland for help in its production. The Rapid Response forecasts were carried out by a dedicated group of collaborative scientists, including Dr. P.J. Haley, Jr., Dr. P. F.J. Lermusiaux, Mr. W.G. Leslie, Dr. C.J. Lozano, Dr. J.A. Dusenberry, all from Harvard, Dr. R. Onken from SACLANT Undersea Research Centre, and Dr. K. Nittis from NCMR Athens. We are grateful to the captain and crew of the NATO Research Vessel Alliance and the technical staff of SACLANT Undersea Research Centre and specifically to E. Nacini, G. Baldasserini, P. Zanasca and M. Zahorodny. The Harvard research and operational effort was supported by grants N00014-95-1-0371 and N00014-97-1-0239 from the Office of Naval Research. The SACLANTCEN effort was due to the REA project 011-1 in the Centre's Programme of Work.

12 Forecasting of Sea-level, Currents and Sea Ice in the Baltic Sea

LENNART FUNKQUIST AND HANS DAHLIN
Swedish Meteorological and Hydrological Institute

12.1 Introduction

The development of operational oceanography is, and should be, driven by the users. To satisfy user requirements scientists and maybe more often consultants are producing the tools or systems for an operational service. Even if the needs expressed by the users seem to be impossible to satisfy, rather modest products can often significantly improve the knowledge on which the users base their decisions.

In many cases meteorology is serving as a model for the development of operational ocanography and existing oceanographical services are in general produced by or in close co-operation with meteorological agencies. The existing services today are accordingly mainly related to the interface between meteorology and oceanography, the sea surface. Forecasts of surface temperature, sea ice, sea level, surface circulation and surface drift have existed for several sea areas for a decade or more. Several oceanographic services based on ocean forecasting exist today also for users in the Baltic Sea area. Most of them have been developed on request and in co-operation with the users.

The Baltic Sea, the world's largest brackish water area, is almost totally enclosed. Only three narrow straits around the Danish Islands connect the Baltic Proper with the Kattegat and the World Ocean. A large drainage area gives a massive fresh water input, which in balance with the water exchange with the North Sea creates a horizontal salinity gradient from zero to ocean levels distributed over about 3000 km. The Baltic is almost tideless, but due to changes in air pressure and wind forcing sea level changes of up to plus or minus one meter from mean sea level are observed. The changing sea level causes barotropic currents, which in the narrow straits at the entrance area can reach values of several knots.

From southern Sweden and northwards, the shore is subject to continuous geological uplift with up to 8 mm/year. Over the centuries, this has created new islands, the change in location of many centres or whole coastal towns and the construction of new harbours. The seasonal and inter-annual variations of sea-level are also large.

The distribution of sea ice can vary from almost total ice coverage during very strong winters to just covering inner parts of the large gulfs in mild winters. In summer, the surface temperature reaches about 20°C.

Almost 85 million people are living in the drainage area of the Baltic Sea belonging to 15 countries of which nine are bordering the sea. The countries have in general high-developed industry, agriculture and forestry. At least Finland and Sweden can, from the view-point of transportation, be regarded as islands and for these Baltic countries in particular, but also for the others, shipping all the year around is of major importance for their export and import.

This paper focuses on three areas of operational oceanographic forecasting put into practice on the Baltic Sea. They are

- Ice,
- Circulation including currents, sea-level, temperature and salinity and
- Surface waves

They will all be explained in the following parts of this chapter.

12.2 Presently used forecasting models at SMHI

With the inclusion of more operational models which are more or less dependent on each other and even two-way coupled, the requirement of a robust and secure operational system has increased dramatically. Once starting from one operational atmospheric model, the operational environment now includes a whole range of forecast models and routines for handling observational data covering all three major disciplines at SMHI, Swedish Meteorological and Hydrological Institute (meteorology, oceanography and hydrology). The possibility that e.g. observations don't get into the assimilation process in time or that computer problems prevent the forecast model to be run, has to be taken care of. This is usually done by prescribing a deadline or in the latter case by giving access to forecast fields produced by other countries. As the computer capacity is often used at maximum, restrictions to the possibility of rerunning single forecast modules exist. This, together with the fact that the distribution of forecast products are also associated with deadlines, makes the whole system very sensitive to any disturbance. A simplified picture of SMHI's operational system as related to oceanography is depicted in Fig. 12.1.

The model system for operational oceanography is built up of three different and more or less coupled modules. The basic component is HIROMB (Funkquist and Kleine, 2000), a 3-dimensional baroclinic circulation model including a fully thermodynamic coupling with the atmosphere. Wind generated waves are calculated with a nested version of a second generation wave model (HYPAS, Günther et al.,1983, 1984). Although a complete ice model is included in HIROMB, BOBA, a uniquely designed ice model for the Baltic Sea, is used for forecasts of ice (Omstedt et al, 1994, 1995). All oceanographic models are forced by forecast fields from the HIRLAM atmospheric model with a resolution of 22 and 44 km. At present, HIROMB is running on a Cray T3E, while the HYPAS and BOBA models are running on local operational workstations.

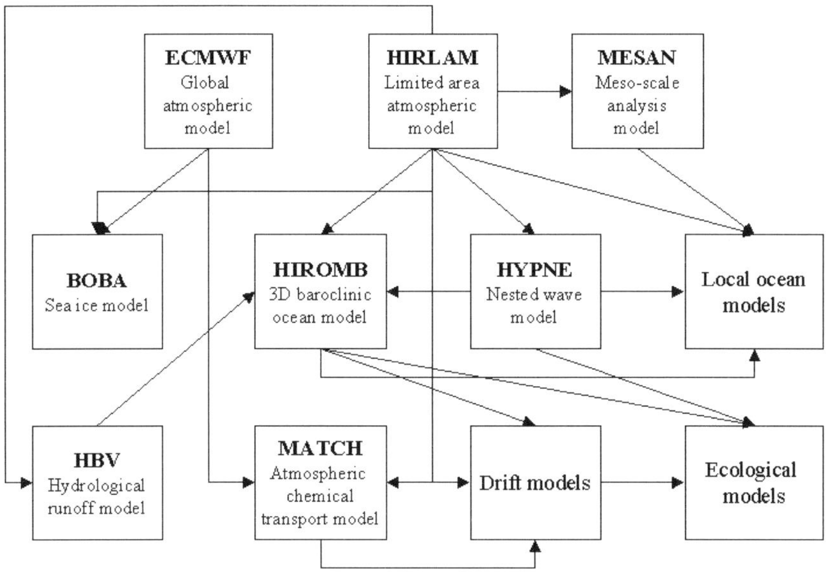

Fig. 12.1 Overview of the operational forecast system at SMHI related to oceanography.

12.3 Observations and data collection

The oceanographic forecasting system has to be supported by a data integration system. At present, ocean data are primarily used for initialisation and validation of model results. The rarity of real-time three-dimensional data does not allow for a daily assimilation but is useful in creating monthly averages, which then can be used to re-initialise the density stratification. This procedure is normally done within a one or two years' interval in order to remove the effect of numeric diffusion, which has a tendency to smear out vertical gradients. The on-going development in data assimilation concentrates on assimilating two-dimensional parameters like SST and ice coverage. The development in remote sensing and the combination of satellite data and in situ observations has now made it possible to produce gridded data with high accuracy.

Today, an extensive monitoring program is run in order to describe sea-level changes, waves, ice situation, SST, hydrography, and environmental status, but the existing observational network in the Baltic is not rigorously designed to support models. The traditional use of these data has been to create statistics, and to construct maps and trend curves. However, an on-going revision of the observational networks is partly changing the aim of the monitoring from the descriptive role to the role of supporting and validating ocean forecasting. An increased exchange of real time data between the Baltic countries and the installation of online automatic fixed stations or moored buoys is planned and will definitely increase the support of ocean forecasting.

12.4 Sea ice

To continue shipping and to keep the harbours open also during severe winters, sea ice forecasting is a task of great economic importance for the countries around the Baltic Sea. Since 1971, Finland and Sweden have, due to strong icebreakers and good ice surveillance, been able to keep the traffic going in spite of some very severe winters. To increase the efficiency and keep the costs down, the ice-breaking services rely upon the national ice information services, which produce diagnostic ice-maps and forecasts. The ice-breaking services are continuously requesting more and better support for their decisions and the responsible agencies in Finland and Sweden have funded a Winter Navigation Research Board, which oversees the development of surveillance methods and forecast models by funding relevant research.

The thickness of ice and its horizontal distribution together with sea surface temperature are routinely analysed every third day in a uniquely designed system (ICEMAP) with the help of in situ measurements and satellite data. However, as the data coverage is still too coarse to automatically create gridded data, a manual subjective analysis is needed for the final product.

Sea ice is modelled by the BOBA model, which covers the Baltic Sea by a 10 nautical miles grid and is composed of three different sub-models, a shallow water model for the interface stress between the ice and water, a one-dimensional model for the vertical salinity and temperature profile applied to 31 quasi-homogeneous regions and a Hibler-type (Hibler, 1979) ice model for the ice dynamics. The sea surface temperature in the model is updated regularly by a simplified nudging procedure and the model is also coupled to the atmospheric model by introducing the latest ice forecast into the HIRLAM assimilation procedure (Gustafsson et al., 1997). Fig. 12.2 shows an overview of the coupling between the active modules. The PROBE model is the one-dimensional part of BOBA and divides the Baltic Sea and the Kattegat into 13 sub-basins, which are coupled through horizontal flows. The model runs for the whole year and supplies the BOBA model with initial vertical stratification when the ice-season starts. However, when forced by climatology, it is also a valuable tool for long-term forecast of the risk for ice formation. This is an important application for an efficient use of the ice-breaking service, because the later the ice-breakers have to be manned and put into operation, the less is their stand-by cost.

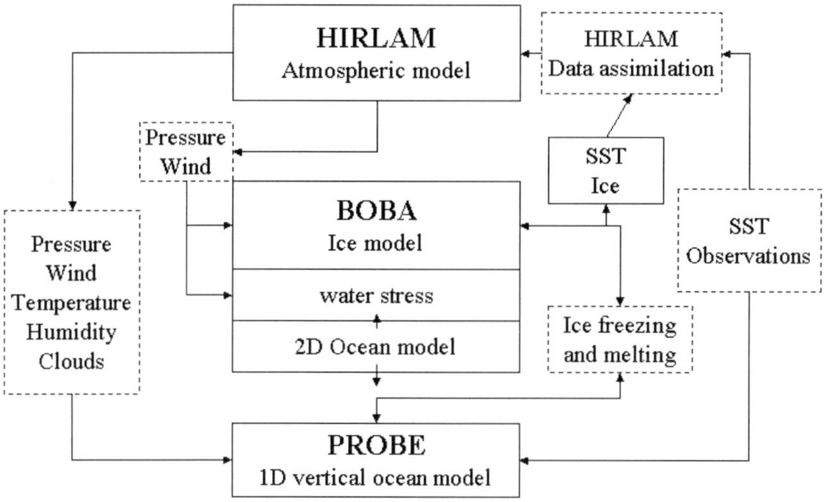

Fig. 12.2 Overview of weak coupling between the atmospheric model and ice model. The results from the ice model is used in the data assimilation for the atmospheric model.

12.5 Sea-level, currents, temperature and salinity

The development of circulation models for the Baltic Sea has during the last two decades been pushed by authorities, which are responsible for oil spills monitoring and mitigation, mainly through the Helsinki Commission. The only allowed method in the Baltic for mitigating oil-slick effects is to collect it. This requires substantial resources, ships, collectors, booms and tanks. To concentrate ships and equipment in the right place and to join forces in the emergency events, a forecast of the oil drift is needed. This forecast is produced by a drift model, which is forced by wind and ocean circulation forecasts. An additional demand to develop drift forecasts also for objects and substances floating below the upper surface layer, called for the development of a fully three-dimensional circulation model in the early nineties. The work resulted in the High Resolution Operational Model for the Baltic Sea, HIROMB. This circulation model is now also used for other forecasts; support for rescue operations, boundary conditions for local models, forcing of ecological models, assessment of the physical marine environment, sea level forecasts and ice forecasts. HIROMB stands for a whole system involving almost all countries around the Baltic Sea. The idea is with common efforts to run a state of the art operational oceanographic model for the Baltic Sea. The kernel of the project is the ocean model and other ingredients are real-time exchange of data for assimilation and validation, a post-processing system including presentation on a WEB-page and regular workshop meetings.

Although tides have no significance in the Baltic area, sea level forecasts are required for flood warnings, to maximize the loading of ships and for simplified forecast models of currents in straits. Flood warning is a governmental service to protect people and property, while the other forecasts have shown to increase the profit for cargo ships and ferries. There is an increasing need of forecasts and information of currents, mainly in connection with offshore construction but also for regular shipping. The surface currents also form the basic module in the oil-drift model, which is also used for sea-rescue operations.

The basic oceanographic model, HIROMB, is a modified version of the BSH (Bundesamt für Seeschiffahrt und Hydrographie) operational model (Kleine, 1994) with identical boundaries between the Atlantic and the North Sea. It is a primitive equation model derived using the hydrostatic and Boussinesq approximations. The model solves for water level, currents, salinity, temperature, ice-coverage and ice concentration. The first operational runs started in mid 1995 and since then the model has been run operationally.

The resolution of the model is mainly chosen as to fit into available computer memory and CPU-time. This has resulted in a 12 nautical miles grid, covering the whole North Sea and Baltic Sea region. East of 6° E, a 3 nautical miles grid (see Fig. 12.3) for the Skagerrak, Kattegat, Belt Sea and Baltic Sea is nested in the larger grid. In the vertical, there is a variable resolution starting at 4 m for the mixed layer and gradually increasing to 60 m for the deeper layers. The model code has recently been parallelized to fit into a distributed memory computer (Wilhelmsson and Schüle, 1998) and the model is presently run in parallel on a CRAY T3E computer. In connection with the transfer of the model to the more powerful computer the horizontal resolution has increased to 1 nautical mile for the area east of 9° E.

The model produces up to 48 hours forecast each day and is forced by the atmospheric model (HIRLAM), which gives 10 m wind, sea level pressure, 2 m temperature, 2 m humidity and total cloudiness. There is a further forcing applied to the surface layer mixing and currents (Stoke's drift) from the wave model. Hydrological forcing is applied at about 80 river outlets as daily forecasted river runoff. The forcing at the open boundary to the North Atlantic is accounted for by specifying 17 tidal constituents and sea level from a North Atlantic storm surge model. A further condition at the open boundary consists of monthly averages of salinity and temperature profiles.

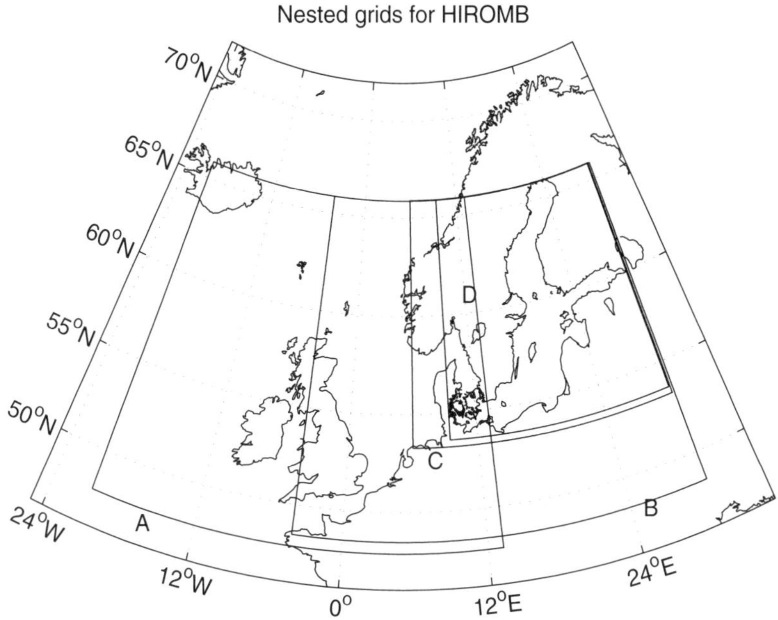

Fig. 12.3 The figure shows the different grids used for the operational ocean model. Grid A is the storm surge model for the NE Atlantic with a resolution of 24 nautical miles (n.m.). Grids B,C and D are for the 3D model which has the resolution of 12, 3 and 1 n.m. respectively.

12.6 Wind waves

Normally, the significant wave height in the Baltic Sea seldom reaches values above 6 m and the highest individual waves recorded are between 12 and 13 m. Still, wave forecasts are of great value both for regular cargo and passenger traffic. Even a rather modest sea state may be critical for special transport where the wave forecast may decide whether the transport will take place or not.

A second-generation wave model, HYPAS has shown to give equal quality as today's third-generation type models and is used operationally at SMHI since 1986. Both the area coverage and resolution have continuously been improved and from 1996 the model system consists of four nested grids (see Fig. 12.4) covering an area from approximately 10° W to 31° E in longitude and 52° N to 71° N in latitude. The resolution is gradually increasing from 22 km for the Atlantic and North Sea to 5 km for the Kattegat and Arkona region. A further refinement of 500 m is used for the Sound, which is of particular interest because of heavy traffic and the construction of the fixed link between Sweden and Denmark.

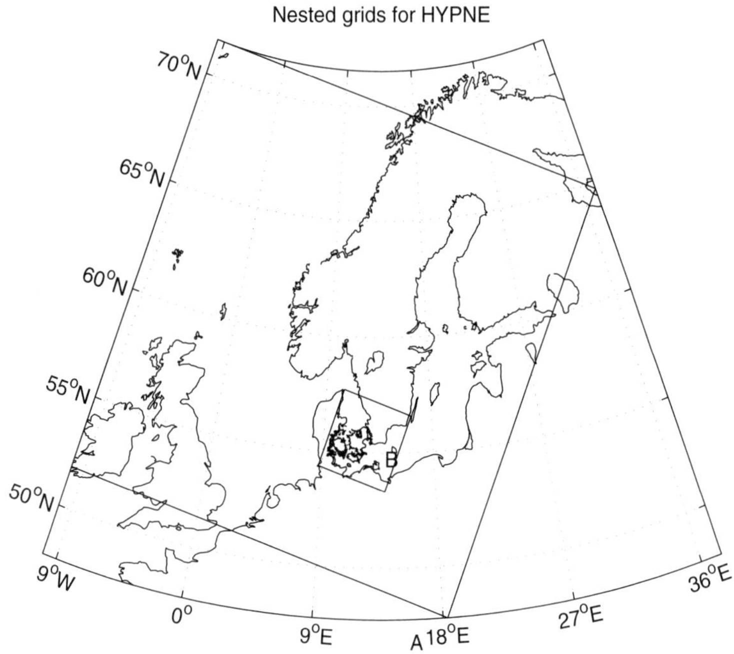

Fig. 12.4 HYPNE is run on 4 grids, 22, 11, 5 and 0.5 km. Here the 22 and 11 km grids are merged together (A). The 5 km grid is marked B and the 500 m grid is not visible on this resolution..

12.7 Post-processing

The output from the oceanographic models is normally accessible after 5 to 6 hours from the start of the forecast run and then immediately distributed to the marine forecast department, where it is transformed into various products, like ship routing and flood warnings. The Coast Guard receives currents and surface winds for their oil-drift models and for assistance in emergency situations. To easily access all model results, a WEB-oriented system (OceanView) has been developed. This also includes an easy downloading system, which is of particular value for the HIROMB partners. An example of a real-time validation of the model results through OceanView is shown in Fig. 12.5.

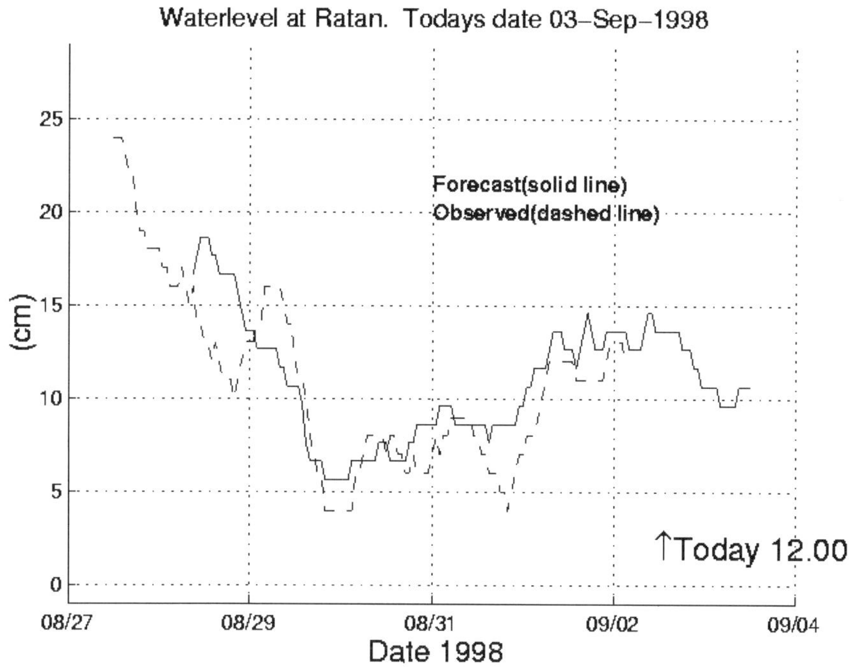

Fig. 12.5 Example from OceanView: real-time validation of water level.

12.8 Conclusions

There are existing and potential customers waiting for forecasts or better ocean forecasts in many areas. The development has to continue and at least the present authors believe that an increased co-operation between researcher as well as operational units is necessary for the development of the complex tools for the production and quality assurance of these forecasts.

The development of oceanographic models is certainly an important part of ocean forecasting, but equally important is the ability to produce the forecasts when they actually are needed. A drift forecast to support a rescue operation is needed immediately, not tomorrow. The forecasting system has to be truly operational and has to resolve the time and space requirements from different users. This means that ocean forecasting is not only a task for researchers, it is a combination of different kinds of skills including also routine observations, data transmission, computer management and product dissemination. In institutes running operational forecasts, the finalised and tested version of a model is transferred from the research department to an operational department which is running the model according to a fixed schedule, monitoring the performance and quality, and rectifying errors. Simultaneously, the further development of the model continues on a research version.

The effort needed to transfer a research model to an ocean forecasting system is in general very underestimated.

Acronyms

BOBA (Bohai and Baltic Sea ice model)
HIROMB (High Resolution Operational Model for the Baltic Sea)
HYPAS (Hybrid Parametric Shallow Water Wave Model)
HYPNE (Hybrid Parametric Shallow Water Wave Model in Nested Version)
PROBE (Program for Boundary Layers in the Environment)
MATCH (Multiple scale Atmospheric Transport and Chemistry Model)
MESAN (Mesoscale Analysis System)

13 Predictions in the North Sea

DAVID PRANDLE
Proudman Oceanographic Laboratory, Bidston Observatory,Birkenhead, UK, CH43 7RA

13.1 Introduction

13.1.1 Approach

This chapter aims to address the following

a)What do we want to predict in shelf seas and why?

b)What can we predict now?

c)What can't we predict?

d)An identification of the underlying enabling factors and constraints in (b) and (c) and thence ways forward.

For convenience, applications described here tend to focus on simulations of the North West European Shelf Seas undertaken at the Proudman Oceanographic Laboratory.

13.1.2 What do we want to predict in shelf seas?

Table 13.1 summarises requirements described at a EuroGOOS User Group meeting held in June 1999. It is necessary to protect people from flooding and the environment from accidental, unplanned or ill-advised contamination. Industry requires marine environmental information in both design and operational phases e.g. defence, navigation, offshore oil, gas and aggregate extraction. These usages must be accommodated within a shared environment in which overlapping and occasionally conflicting interests are viewed in the broader perspective outlined below.

13.1.3 Shelf Seas - a global perspective

Increased concentration of atmospheric CO_2 along with other indicators, provide a warning that relatively rapid changes in the global environment may already be in progress. The likelihood that industrialisation has contributed to such changes has focused scientific research into the causal mechanisms, possible remedial actions and the future impacts. Since the world's oceans are the dominant feature in moderating global environmental mechanisms, related international programmes have been established (e.g. WOCE, World Ocean Circulation Experiment, http://www.soc.soton.ac.uk/OTHERS/woceipo/ipo.html), involving observations, global circulation models and studies of specific processes (e.g. BOFS, Biogeochemical Ocean Flux Study, www.bodc.ac.uk/project/bofs.html).

Forecast For (Time Ahead)	Geographical Region	Parameters	Application (Examples)	User
1 Day 5 Day	Localised Coast Regional Synopsis	Sea-State Water Quality (O_2 Fecal Counts) Bacteria Count	Recreation + Waste Management Discharge	Shared Community Interest
30 Day	Shelf Sea	Temperature Nutrients Plankton	River Management	
1 Year	(European) Shelf Seas	Fish Stocks	Conservation Management	
10 Years	Global	Ecological	Climate & Environment Energy Forecasts	
1 - 6 Hours 1 - 5 Days	Harbour Approach Coastline	Sea State, Water Levels And Currents Contaminant Spill Flood Levels	Navigation Contingency Planning	Harbour Authority Civic Agencies
1 - 5 Days	Sea Loch, Fjord	Temperature, Salinity, O_2	Feeding Rate/ Tank Location	Fish Farming
1 - 10 Days	Ship Route	Sea State, Ice	Ship Routeing	Shipping Co.S
1 - 10 Days	Site Specific	Sea State, Currents Temperatures	Operation Of Off-Shore Rigs	Off-Shore Exploration And Production

Table 13.1 Range of user requirements

While such programmes are clearly essential in establishing causal mechanisms, additional efforts are necessary in the coastal zone. Shelf seas are important both globally and nationally since, cumulatively, they form part of the 'causal mechanisms' and, locally, they suffer most directly the 'impacts'. Thus maintaining the balance between exploitation and conservation is especially difficult. Effective management requires an improved scientific understanding of the sea, expressed in computer-based models able to differentiate (and predict) the impact of man's activities from natural variability.

While the atmosphere is the major route by which most pollutants reach the open oceans, the situation in coastal areas like the North Sea with major rivers is less clear, but riverine, direct dumping, and atmospheric routes are all likely to be important.

13.1.4 Chapter Contents

Section 13.2 contains a description of the existing forecasting systems relating to dynamics and mixing in the seas of N.W. Europe. Details of the dynamical processes are relegated to appendices: Appendix A provides definitions of the various modes and applications of model simulations, Appendix B contais Dynamics (tides and surges) and Contaminant Mixing, Appendix C Residual Horizontal Circulation; Appendix D Vertical Exchange and Appendix E the Seasonal Cycle. In section 13.3 we offer the discussion of mixing of tracers includes: temperature, salinity, caesium 137, sediments, metals in both dissolved and particulate phases and nutrients.

Section 13.3 discusses future developments.

13.2 Forecasting dynamics and contaminant mixing in the Seas of North Western Europe

Table 13.2 is a summary of the characteristic features and stage of development of forecast systems in the N.W. European Shelf Seas ranging from dynamics to the mixing of both conservative and non-conservative tracers.

Indirectly, this table answers the questions (b) and (c) raised in section 13.1, i.e. what we can and can't predict now. A system shown to be 'Operational' clearly indicates what can be done, a system shown to be 'Experimental' is clearly under development. (See Appendix A for an explanation of the terms: Operational, Pre-operational, Climate Mean and Experimental). For some parameters, such as storm surges and waves, their energy (or mass) dissipates rapidly. This enables simulations to be made, in the same manner, for time scales from hours to decades. Other parameters, such as temperature or salinity, have a longer memory and simulations are restricted by the difficulties of initialisation (or re-initialisation/chronology). The temporal and spatial resolutions may be limited by the corresponding resolution of initial boundary and forcing conditions.

As an illustration of the above, a major constraint in forecasting mixing of contaminants is the limited knowledge of the amount, timing and location of inputs. For non-conservative substances, these limitations in defining sources, sinks and open boundary and exchange rates are exacerbated by the difficulties in quantifying the complex interactions between dissolved and particulate phases and between chemical and biological parameters. Oceanic and atmospheric exchanges may constitute a source or a sink (or both) for mass, momentum or energy. Thus as a background to appreciating such constraints, Table 13.2 aims to indicate the dependency of the various simulations on exchanges within the Atlantic, the atmosphere, coasts and rivers.

Since the N.W. European Shelf Seas are semi-enclosed, the longer term distributions of many tracers entering from the ocean or the coasts are governed, overwhelmingly, by the persistent pattern of residual horizontal circulation. Conversely the distributions of tracers entering via the atmosphere or seabed may be primarily

	ATLANTIC	ATMOSPHERE	RIVERS & COASTS	DYNAMICS	MONITORING	MODEL DEVELOPMENT	ASSOCIATED APPLICATIONS
PARAMETER							
A WAVES							
SEA LEVEL & CURRENTS		(Wind)					
i Waves	BC2	FC1	-		Sa/In-S	OP	
ii Tides	BC1	-	-		In-S	OP	
iii Storm Surges	BC2	FC1	-		In-S	OP	
iv Seasonal	BC2	FC1	-		Sa	POP	
v Mean	BC2	FC1	-		Sa	POP	
B TURBULENCE	-	FC2	-	VE	RS	POP	
C TEMPERATURE		thermal exchange					
i Event		FC1	-	VE	Sa	POP	tranquil stratification
ii Seasonal	BC2	FC1	-	VE	Sa	CM	autumnal overturning
iii Mean	BC2	FC1	-				
D SALINITY		precipitation/ evaporation					
i Event	-	-	So	VE/HC	RS	EX	Baltic, Rhine outflow C(i)
ii Seasonal	-	SiSo	only coastal	VE/HC	RS/In-S	CM	confined to the coast
iii Decadal	So	-	So	HC	RS	EX	
iv Mean	So	-	So	HC	RS	CM	
E CAESIUM 137							
i Event	-	So		HC	RS	EX	Chernobyl
ii Seasonal	-	-	-	HC	RS	POP	
iii Mean	Si	-	-	HC	RS	POP	
F METALS							
i Events	-	So	So	VE/HC	RS	EX	wet deposition
ii Mean	Si	So	So	HC	RS	CM	
G SEDIMENTS							
i Event	-	-	So	VE/HC	Sa/Ai	EX	storm/run off
ii Mean	So (fines)	-	So/Si	VE/HC	Sa/RS	CM	
H NUTRIENTS							
i Event	-	So	So	VE/HC	Sa/Ai	EX	thunderstorms(?)
ii Seasonal	So	-	So	VE/HC	Sa/RS	CM	blooms
iii Mean	So	-	So	HC	RS	CM	

FC (Forcing Condition), BC (Boundary Condition 1(st order), 2(nd order)), FC1, BC2 indicates a systems where the forcing condition is predominant (1st order) with a boundary condition of secondary influence.
Si(nk), So(urce), SiSo (mixed); Stage of Model Simulations: OP(erational), PreOP(erational), C(limate), M(ean), EX(perimental)
Primary Monitoring: Sa(tellite), Ai(craft), In-S(itu mooring), R(esearch),S(hip)
Predominant Dynamical Process: H(orizontal), C(irculation), V(ertical), E(xchange)

Table 13.2 Characteristics and Operational Modes of Shelf Sea Models (see Appendix A for explanation).

determined by the localised rate of vertical exchange. Table 13.2 indicates the relative contributions of these two processes. Note their roles may be different depending on the time scale of interest - vertical exchange often being of interest in the short term (over the tidal cycle for sediments on the seasonal cycle for temperature) but with horizontal circulation tending to predominate in the longer term. Appendix C describes general features of Residual Horizontal Circulation and Appendix D of Vertical Mixing.

Finally the table also indicates the primary technique used for monitoring. The range of capabilities to monitor, whether in real-time or via subsequent processing, is a major constraint in initialising (including re-initialising via assimilation techniques), providing boundary conditions and in verifying forecasts.

13.2.1 Dynamics

Hydrodynamic models are required as the basis for transporting and mixing contaminants both horizontally and vertically. The dynamical processes involved occur over time scales of seconds (turbulent motions), to hours (tidal oscillations), to months and years (seasonal cycle) to inter-annual variability; with corresponding space scales from millimetres to thousands of kilometres - thus a range of models is required. Appendix B presents the governing equations for dynamics and mixing.

13.3 Tides, surges, waves and turbulence intensity

While the Proudman-Doodson (1924) co-tidal map provided the basis for predicting tidal levels for the major M_2 constituent, present-day models extend this capability to upwards of 100 constituents and can include both surge-tide and tide-wave interactions (Fig. 13.1). Gravitational attraction within the N.W. shelf seas can be neglected with these models depending solely on tidal constituents prescribed along the shelf edge. The associated surge-tide models depend primarily on the prescription of surface wind data together with some smaller reliance on (dynamical) atmospheric pressure variations and 'external' oceanic surges (Heaps, 1967). These systems use conventional shore-based tide gauges both for subsequent verification and, using real-time interrogation, for assimilation purposes.

The tide-surge simulations can be run continuously for decades (subject to relevant data on wind forcing) to indicate seasonal and mean sea level variations (Fig. 13.2) and to calculate long-term statistics. Alternatively tidally-averaged forms of equations (B1) to (B3) can be used for more extensive sensitivity tests (Prandle 1984). The seasonal variation in mean sea level is primarily associated with the seasonality of the associated winds. Perhaps surprisingly, persistent sea level gradients along the Atlantic boundaries do not propagate into the shelf seas, the steepness of this shelf results in a balancing residual circulation confined to the shelf edge region, Fig.13.3 (Prandle, 1987).

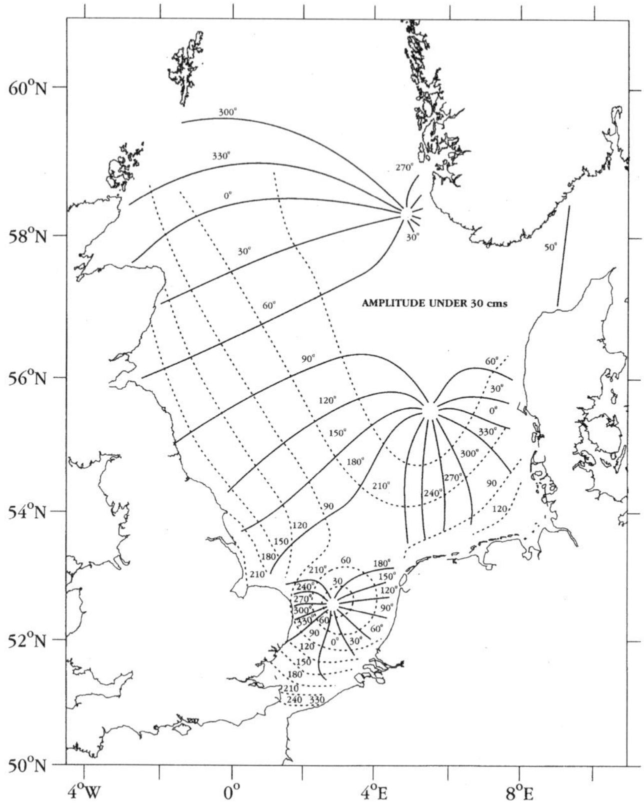

Fig. 13.1 The M2 tide in the N.W. European Shelf Seas, (Proudman and Doodson, 1924).

Wave modelling is well-developed with several operational forecasting systems linked to sea surface wind forecasts derived from meteorological models (Fig. 13.4). Intercomparisons of a number of these systems have been reported by Monbaliu et al. (1997). Some account of swell waves propagating from the Atlantic is a necessary boundary condition. Both satellite altimeters and SAR systems can be used to evaluate the forecasts together with conventional wave buoys at selected locations. High Frequency (H.F.) Radar (Wyatt, 1966) offers a convenient way of measuring, in real time, the diverse adjustment of the spectra as the waves propagate into shallower water. These monitoring developments allied to the shallow water development of the existing deep water wave models provide exciting opportunities for studying rapidly evolving near-shore bathymetry i.e. simulations which include sediment movement and resultant adjustments to both bathymetry and thence to the prevailing wave dynamics.

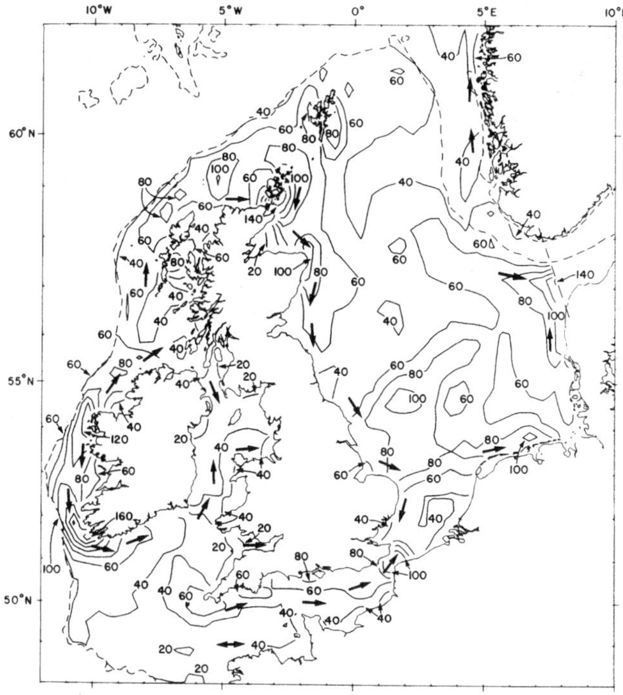

Fig. 13.2 Statistics of surge currents (cm/s) in the North Sea (Flather, 1987).

In these strongly tidal seas, turbulence levels are generally determined locally by tidal currents at the sea-bed, the rapid dissipation rate minimising the effect of horizontal advection. A range of turbulence closure models are used, often based on the work of Mellor and Yamada (1974). Generic models (Burchard and Baumert 1995) are well validated against free-fall sensors (Simpson et al. 1996). Some difficulties remain in the reproduction of the background levels which determine layer exchange rates in stratified conditions.

13.3.1 Temperature and Salinity

The distributions of temperature and salinity jointly determine the density distributions and any associated stratification. Temperature is primarily determined by a localised balance i.e. by the exchange of heat at the sea surface modulated by any vertical stratification. This exchange is a function of the difference in temperature between the sea surface, T_S, and the ambient air, T_a. On a seasonal basis, T_a is modulated by T_s (Prandle, 1998). Whilst, this seasonality can be predicted, shorter term forecasts for T_S are limited by the ability to forecast T_a. Temperatures along the western limits of the N.W. Shelf Seas show some influence of Atlantic bound-

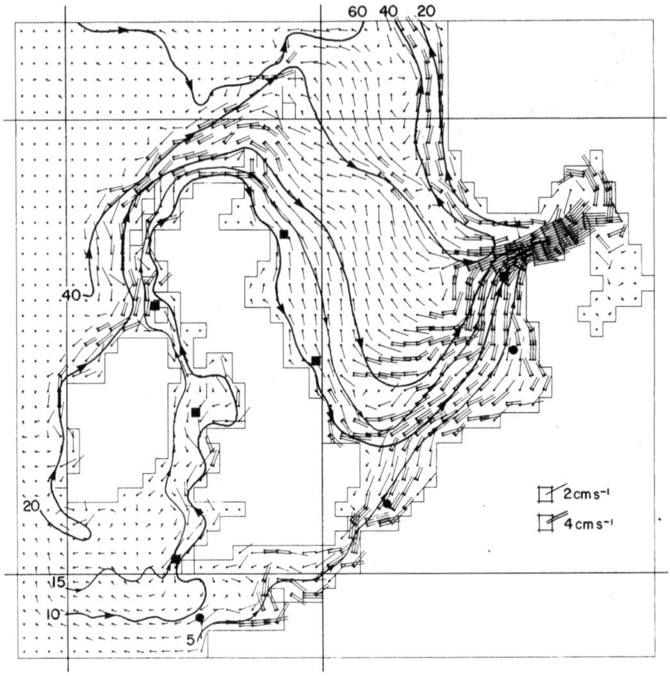

Fig. 13.3 Residual circulation in the North Sea (Prandle, 1984).

ary conditions, but within the North Sea, this Atlantic influence is effected via its impact on the prevailing westerly air streams, Fig.13.5 . Long term historical observations enable 'climate mean', CM, seasonal cycle data sets to be assembled. Thence CM simulations can be made together with associated sensitivity analyses. However CM spatial and temporal distributions can mask the importance of transient or fluctuating intensifications such as thermal fronts.

In clear sky conditions, satellite AVHRR images provide good verification data that potentially can be used for model assimilation. However, the surface distribution of temperature masks more interesting vertical structure (with associated longer term memory) and operational modelling will require in-situ monitoring of these vertical profiles.

By contrast to the large seasonal cycle in temperature, away from the coastal fringes there is little seasonality in salinity. Likewise, away from the Rhine plume and the Baltic outflow region, stratification is almost solely determined by temperature. Hence, the primary interest in salinity is in the annual mean distribution and its longer term variability. These are determined by the residual horizontal circulation, the rates of coastal mixing and the variability in Atlantic and sea surface (net evaporation/precipitation) exchanges. Remote sensing of salinity is still under

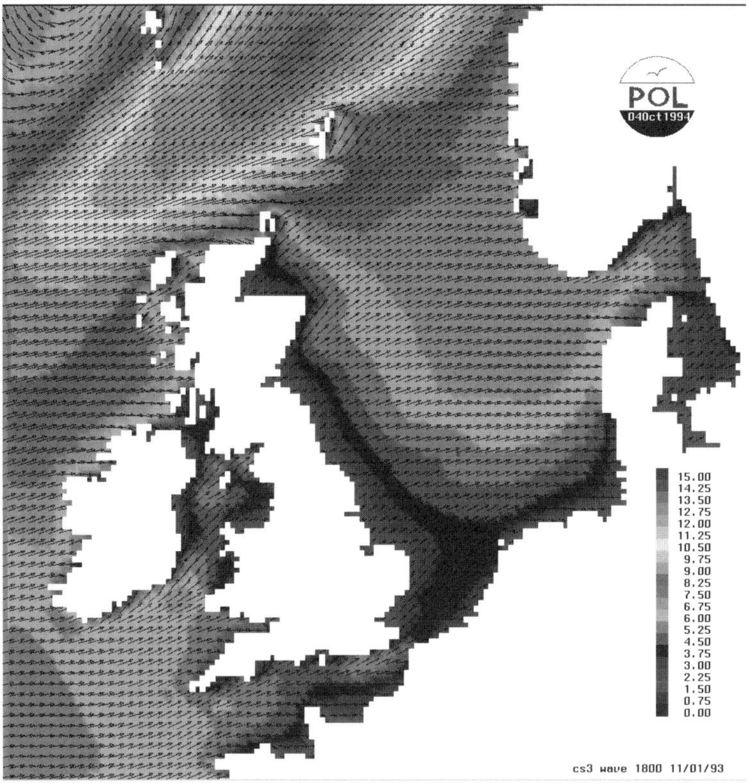

Fig. 13.4 Forecast of wave height (HS) and direction in the North Sea (done at U.K. Meterological Office), (Flather, R. (personal communication)).

development. In open seas, the more stationary and broader scale variability of salinity enables useful spatial distributions to be assembled from research vessels.

13.3.2 Conservative tracers, Caesium 137 and dissolved metals

Simulation of the long term mean distribution of dissolved conservative tracers such as Caesium 137 or, non particle reactive, metals such as copper is analogous to that for salinity described above. For ^{137}Cs, simulation (Prandle and Beechey 1991) is straightforward since the location and input rate of the source was well known. Fig. 13.6 shows observed versus computed distributions of ^{137}Cs over the period 1978-1988. The large scale, long term patterns are in good agreement. By contrast, calculation of metal distributions is complicated by uncertainties about the various sources from: the Atlantic, the atmosphere and rivers/coasts. With adequate 'synoptic' monitoring programmes, inverse methods can be used to circumvent this problem, Fig. 13.7 (McManus and Prandle, 1994).

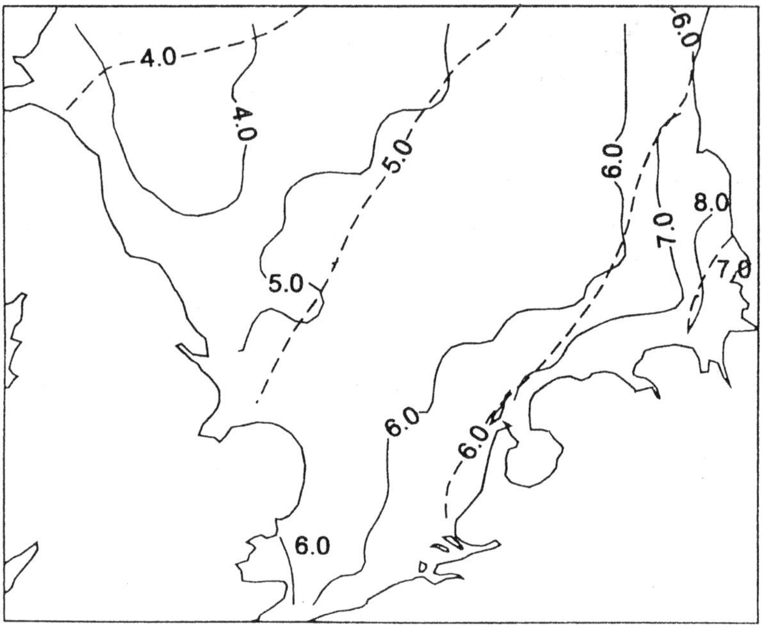

Fig. 13.5 Observed vs forecast (dashed) seasonal temperature amplitudes in N.W. European Shelf Sea (Prandle and Lane, 1995).

Whilst salinity and temperature can be measured directly by autonomous in-situ sensors, no such direct in-situ measurement of radionuclide or metal concentrations is possible. This need for subsequent laboratory analyses, is likely to limit simulation of such tracers to the pre-operational mode.

The Atlantic boundary provides a convenient 'infinite' sink in many such simulations.

13.3.3 Non-Conservative Tracers

Many tracers are effectively transported and 'mixed' by their adsorption onto fine sediments, plutonium and lead are examples (in direct contrast to the essentially dissolved tracers discussed above). Thence, two problems arise in simulating such adsorbed tracers, first the relative amount adsorbed - conveniently, if crudely, represented by the appropriate partition coefficient Kd (Balls, 1989). Secondly, a simulation is then required of the transport of suspended sediment.

Because of the enormous increase in the ratio of surface area:mass for fine clay particles compared to coarse sand, only the transport of 'fines' need to be considered. Moreover, whilst coastal sediment movement can involve a full spectrum of particle sizes, in water depths greater than about 20m, the transport is largely by

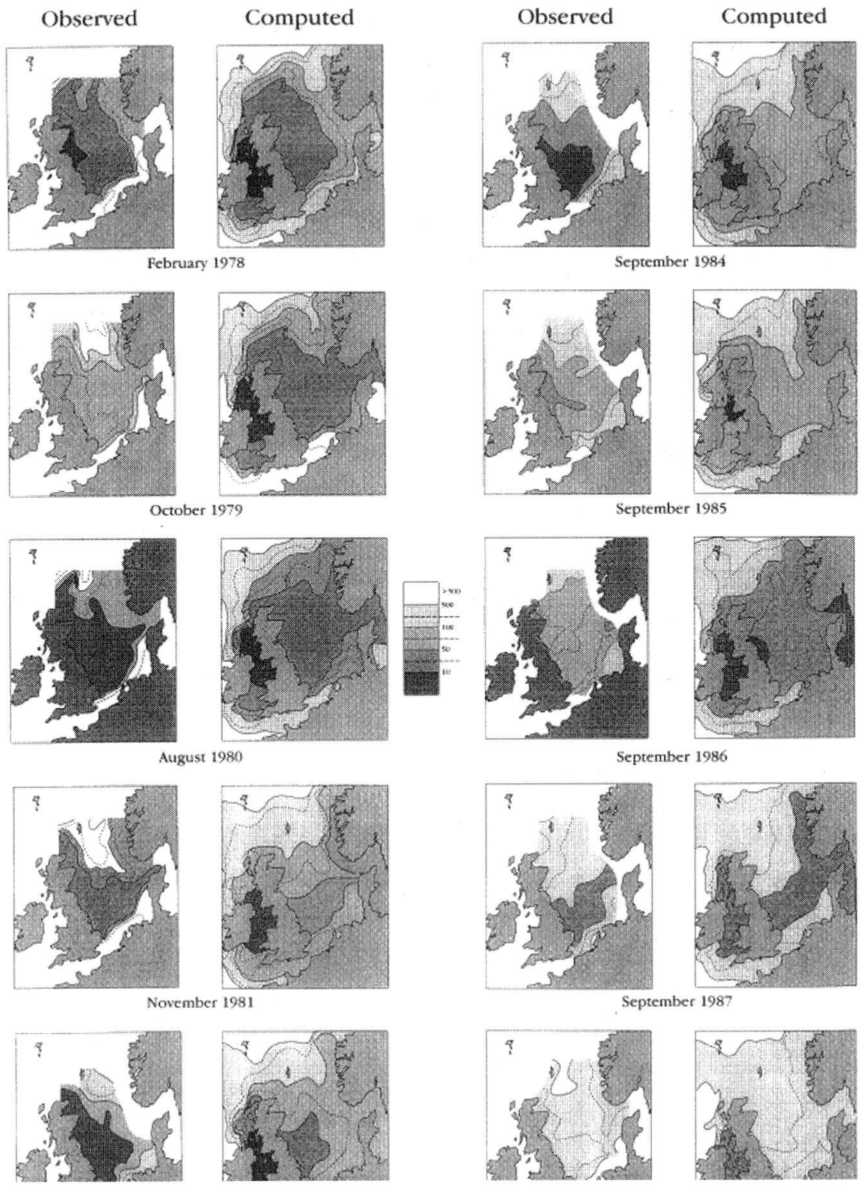

Fig. 13.6 Observed and forecast distributions of ^{137}Cs in N.W. European Shelf Seas (Prandle and Beechey, 1991).

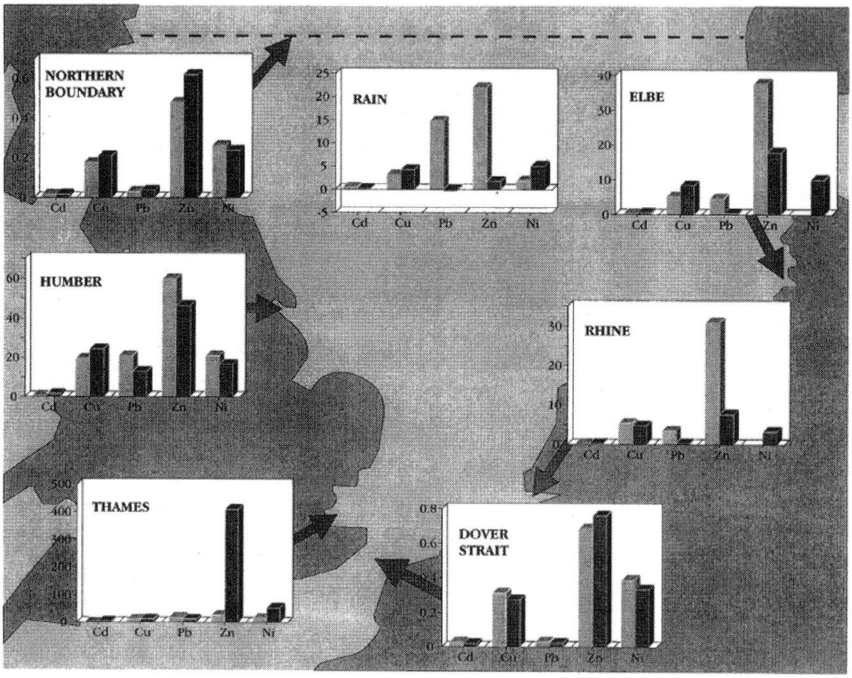

Fig. 13.7 Derived (right) vs observed (left) sources of metals in the North Sea (McManus and Prandle, 1994).

tidal and surge currents. In the shallower areas, wave stirring increasingly predominates and combined tidal-surge-wave dynamics must be considered. A further simplifying factor is that in such deeper water, fine material generally stays in suspension for periods in excess of the tidal cycle and hence the net sediment movement can be simulated using horizontal circulation models similar to those described earlier (McManus and Prandle, 1996). However specification of the sources and sinks in Suspended Particulate Matter (SPM) simulations is severely complicated with seabed sources complicated by the chronology of previous deposition and associated consolidation rates.

Aircraft surveillance using multi-wavelength imagery promises to differentiate between the reflectance associated with chlorophyll and various sediment fractions and allow model verifications. However the need for atmosphere corrections is likely to involve continued reliance on in-situ calibrations.

13.3.4 Biological tracers

This review is essentially limited to physical and chemical tracers, another chapter examines ecological models in which nutrients constitute the essential fuel. However, a recent study (Prandle et al., 1997) shows how the techniques described above can be used to indicate salient features of the annual cycle of nutrients. The approach involves combining model simulations of the seasonal cycle with related observations and thereby inferring grazing and re-cycling rates. Such simulations emphasise the importance of the oceanic inputs and the limited effect, away from the coastal fringe, of riverine and coastal inputs.

The development of nutrient modelling promises to be accelerated by the rapid development of autonomous sensors combined with the aircraft surveillance noted earlier.

13.4 Ways forward - future developments

This paper has generally described discrete simulations (i.e. one parameter in isolation) or with 'information flow' (i.e. interaction) in one direction only, e.g. transport paths defining tracer mixing. Ultimately fully coupled models run operationally will be required. As an example of this interactive requirement, consider fine sediments which are important as the medium by which many adsorbed pollutants are carried from river to estuary, estuary to coastal zone and thence to shelf sea/ocean. Suspended sediments also influence the biology of the coastal zone by their extinction of light (limiting photosynthesis in deeper turbid waters).

One exciting 'Grand Challenge' is to understand and thence predict coastal bathymetric evolution. New approaches are necessary to bridge the gulf between geomorphologists and process modellers. The 'Soft' Defence approach to coastal protection (i.e. avoidance of fixed installations), now widely adopted, accentuates this Challenge. Excluding engineering influences, our present-day coastlines generally reflect the balance between geology, fluvial discharge, supply of mobile sediments, exposure to tides, storm surges and waves and the rate of change of mean sea level. The transport of mobile coastal sediment is the result of complex interactions between each of these dynamical processes within the governing framework of the instantaneous bathymetry - which is itself evolving via their integrated impacts. Over longer time scales some (evolving) balance will be achieved between the evolving bathymetry due to erosion and settlement of sediment and the subsequent modification of the forcing mechanisms. To reproduce this evolution requires fine resolution models run in operational or pre-operational mode. Tidal analyses of suspended sediment time-series (Fig. 13.8), is a related challenge. Relict features such as submerged sand banks can provide indications of long-term bathymetry evolution providing their governing characteristics can be accurately determined (Huthnance 1982). Long term predictions in Shelf Seas must consider the contribution to and effect from global pollution, specifically possible greenhouse effects such as rise in mean sea level and meteorological changes. The widespread impact of such phenomena emphasises the necessity for international co-operation in all aspects of these studies. Long-term data sets are vital for such programmes, systematic marine monitoring programmes must be initiated, involving combinations of remote sensing, sea-truth moorings and coastal stations.

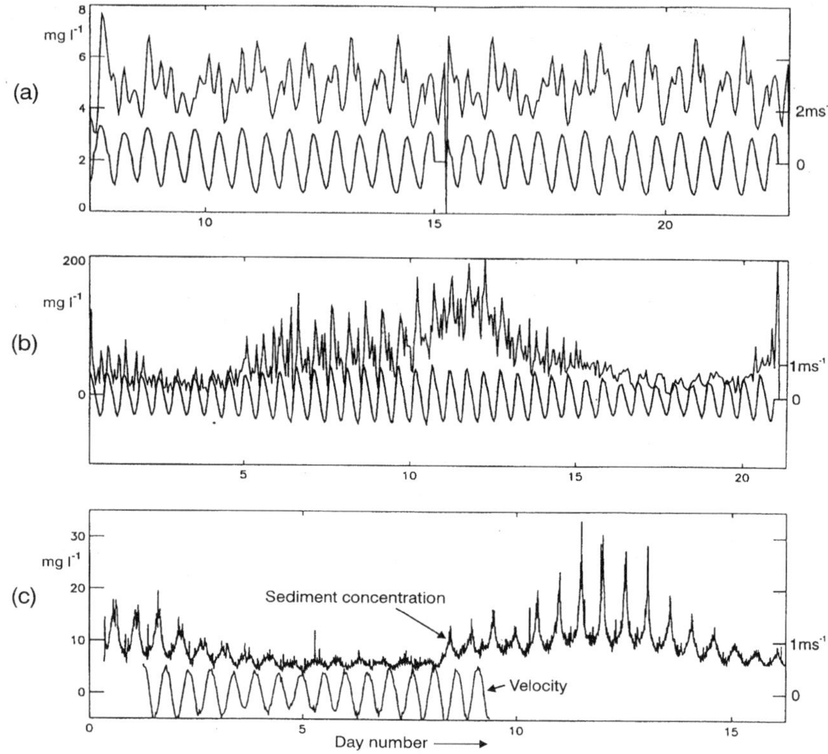

Fig. 13.8 Near-bed suspended sediment observations in tidal seas. Scales: left - SPM concentrations, right - velocities, (Prandle, 1997b)

Incorporation of interactive coupling requires a minimum level of accuracy in each of the dynamical processes. Evaluation of such dynamically linked models is becoming possible via a suite of remote sensing, satellite and aircraft sensors, coastal radar, in-situ profiling instruments etc. Over the next decade, operational oceanography will extend rapidly (Woods et al. 1996). This will provide the infrastructure to establish the requisite monitoring networks and maintain linkages to the associated meteorological, marine (ocean) and terrestrial (hydrological) models. Thus new challenges and opportunities will be presented building and extending on the developments described here.

Acknowledgement

This paper relies on background studies supported by the European Commission, Marine Science and Technology (MAST) programmes, namely SCAWVEX (MAS2 CT94 0103) and PROMISE (MAS3 CT950025).

Appendix A Supplement to Table 13.2

A brief note on the parlance of Marine Forecasting is required in relation to Table 13.2 .

'Operational Modelling' indicates real-time simulation with future extrapolation in time to produce forecasts - analogous to daily weather forecasts.

'Pre-operational modelling' infers model simulations similar to that above (i.e. with synoptic specifications of initial, boundary and forcing conditions) but run for past events. Since some synoptic data, e.g. remote sensing images or chemical concentrations, can only be obtained subsequently, some simulations are restricted to this mode. The complex interactions between physical, chemical and biological parameters increasingly requires this mode of simulation even when the interest is to understand isolated specific aspects of a marine system.

'Climate Mean', many simulations are restricted by the availability of the associated data (referred to above), by simulating 'climatic mean' conditions, data gathered over many years can be interpolated to provide representative descriptions. Moreover, in annual cycle simulations, climatic mean conditions can be used for repeated yearly cycles to achieve 'cyclic convergence' - often independent of initial conditions. Unfortunately these mean conditions may differ in character from actual conditions found in any one year, for example sharp density fronts that migrate annually in location might disappear.

'Experimental' is used here to indicate specific simulations applicable to a limited set of observations as collected in intensive experimental campaigns.

In contrast to the above definition for Operational Modelling, 'Operational Oceanography' extends beyond such **'forecasting'** to include **'nowcasting'** and **'hindcasting'**. **'Nowcasting'** might typically involve Pre-operational Modelling whereby models are run subsequently (when non real-time data processing has been completed) to provide an up-to-date description of, for instance, the net thermal or nutrient budget of a sea region. **'Hindcasting'** reflects the capability and enhanced value of using the most advanced models to simulate historical sequences (with perhaps climate-mean prescribed data where gaps exist) to both re-interpret specific events or construct relevant statistical data sets.

Appendix B Dynamics And Mixing

This appendix on 'dynamics' presents the basic equations and provides references which show how these can be simplified for use in specific applications.

The dynamics of tides and surges are briefly described as a pre-cursor to Appendix C and D dealing with residual circulation and vertical exchange respectively. The dynamics of surface waves and baroclinic processes are described elsewhere in this volume.

Neglecting vertical acceleration, the equations of motion at any height z (measured vertically upwards above the bed) along orthogonal horizontal x and y axes may be written in Cartesian accelerations in x-direction:

$$\frac{\partial u}{\partial t} + u\frac{\partial u}{\partial x} + v\frac{\partial u}{\partial y} + g\frac{\partial \zeta}{\partial x} - \Omega v = \frac{\partial}{\partial z}\left(E\frac{\partial u}{\partial z}\right) \tag{B1}$$

accelerations in y-direction:

$$\frac{\partial v}{\partial t} + u\frac{\partial v}{\partial x} + v\frac{\partial v}{\partial y} + g\frac{\partial \zeta}{\partial x} + \Omega u = \frac{\partial}{\partial z}\left(E\frac{\partial v}{\partial z}\right) \tag{B2}$$

continuity:

$$\frac{\partial u}{\partial x} + \frac{\partial v}{\partial y} + \frac{\partial w}{\partial z} = 0 \tag{B3}$$

Horizontal eddy viscosity terms are omitted in (B1) and (B2), these are often included in numerical simulations to enhance stability. Forcing due to wind or variations in density or atmospheric pressure are omitted in (B1) and (B2). u, v and w are velocities along x, y and z, ζ is surface elevation, $\Omega = 2\omega\sin\phi$ is the Coriolis parameter representing the influence of the earth's rotation ($\omega = 2\pi/24$h), ϕ is latitude, and E is a vertical eddy viscosity coefficient.

Tides and Surges

Perhaps the simplest and earliest solutions to the above equations are for the tidal propagation of a single constituent in a 'cross-sectionally averaged' estuary. Prandle (1991a) reviews solutions for the above equations for a range of estuarine shapes. A theoretical framework can be constructed to explain the nature of the tidal response in all funnel-shaped estuaries. The theory extends to all major tidal constituents and indicates the likely sensitivity of these responses to system modifications.

G.I. Taylor (1921) provided analytical solutions for tidal propagation for a depth-averaged version of (1) - (3) in a rectangular basin of constant depth. This solution reproduces almost all of the salient features of amphidromic systems and associated current ellipse distributions.

While accurate maps of tidal elevations have existed for many shelf seas for over 70 years, the complexities of the associated tidal currents are still unfolding. Tidal currents exhibit much greater small scale variability, both spatially and temporally, and hence accurate descriptions require more sophisticated measurement techniques and finer resolution 3D numerical models. Recent technology developments in both H.F. Radar and ADCP (Acoustic Doppler Current Profiler) illustrate these intricacies.

The dynamics of storm surges are dependent on the magnitude (depth of depression), spatial scale and speed of travel of the associated meteorological system in relation to the water depths and extent of the tidal basin. For the North Sea, these dynamics are similar to those of the semi-diurnal tides (Heaps, 1967).

Mixing

The conservation of mass equation is:

$$\frac{\partial C}{\partial t} + u\frac{\partial C}{\partial x} + v\frac{\partial C}{\partial y} + w\frac{\partial C}{\partial z} = \frac{\partial}{\partial x}\left(K_x\frac{\partial C}{\partial x}\right) + \frac{\partial}{\partial y}\left(K_y\frac{\partial C}{\partial y}\right) + \frac{\partial}{\partial z}\left(K_z\frac{\partial C}{\partial z}\right) \qquad \text{(B4)}$$

Short term turbulent fluctuations are not necessarily the primary mechanisms for horizontal dispersion in shelf seas. Dispersion often arises from the interacting 3-D variations in phase and amplitude of both the currents and contaminants (though these variations are themselves highly dependent on both topography and the vertical turbulence levels). Simulation of mixing processes is much more difficult than for tidal dynamics since a much wider spectrum of length scales is involved. Simulation of mixing is often de-coupled from simulation of tidal dynamics with velocities and elevations from the latter stored and used both to specify appropriate advective transports and to determine appropriate dispersion coefficients (Prandle, 1984). Alternatively the tidal dynamics may be specified from observations. Whatever approach is used it is generally convenient to define separately the various tidal constituents and the residual. Unfortunately the definition of a residual current in a shallow region with pronounced spatial variability in tidal currents is complicated.

Several (depth-averaged) definitions are possible, e.g. Eulerian, Lagrangian, Stokes Drift (Longuet-Higgins 1969). Lane et al. (1997) discuss these definitions and suggest a framework for relating estimates from models and observations.

Appendix C Residual (Horizontal) Circulation

Tidal Component

By convention, discrepancies between observed (or simulated) sea levels (or currents) and associated tidal predictions are referred to as 'residuals'. However, the residual current components described here relate to longer term components associated with tidally-averaged values. In strongly tidal shallow seas, a substantial residual originates from non-linear terms associated with tidal propagation (Nihoul and Ronday, 1975). Three residual tidal components exist associated with: shallow water, bed friction and advection. The latter is particularly pronounced near sharp changes in bathymetry leading to pronounced gyres (Pingree and Maddock, 1980). In shallow water, the other terms generate net circulations which are linked together across the deeper areas by topographically steered streamlines (via the pertaining geostrophic balance) (Fig. 13.3).

For barotropic tidal propagation in water depths greater than 50 m, 2D models are generally adequate. In shallower water, the increasing importance of both bottom friction and effective vertical eddy viscosity requires fully 3D models (Prandle 1997a).

The net transport, Q_T due to an oscillatory current $\hat{U}$ accompanying an oscillatory elevation $\hat{\zeta}$ is

$$Q_T = \frac{1}{2}\hat{U}\hat{\zeta}\cos\theta \tag{C1}$$

where θ is the phase lag between the two. Since the related net energy propagated, Q_E is

$$Q_E = \rho g Q_T H \tag{C2}$$

it follows that the propagation of tidal energy from ocean to shelf seas is accompanied by a related residual net transport.

Wind and Density Driven Components

In open seas, the observed steady state current, response R for a constant eddy viscosity coefficient, E, follows from Ekman theory, (Prandle 1991b).

$$R_{(z)} = \frac{T_w}{\rho E b e^{bH}}\left\{e^{bz} + \frac{Eb}{k} - 1\right\} \tag{C3}$$

where

$$b^2 = \frac{if}{E} \tag{C4}$$

T_W is wind stress, ρ water density, f Coriolis parameter, H water depth, k bed friction coefficient and z height above the bed. In deep water (bH >>1) the first term predominates and the current veers at 45° to the wind. In shallow water the second term predominates, the amplitude of the current response reduces and the direction is aligned with the wind.

The wind stress T_W may be approximated by

$$T_W = 1.3\ 10^{-5} \rho W^2 \left(\frac{N}{m^2}\right) \qquad \text{(C5)}$$

where W is the wind speed.

In enclosed seas, T_W generates both residual currents and gradients in sea level. In strongly tidal seas, the tidally-averaged nature of these responses is approximately linearly proportional to T_W - allowing unit response streamline functions to be calculated for orthogonal components of T_W.

Density

Features such as frontal gyres and internal waves associated with baroclinic dynamics are not considered here. However the persistent residuals associated with horizontal density gradients can be readily described (Prandle, 1991b).

For (large) $E/H^2 f > 10$ the (small) surface velocities are orthogonal to the density gradients, i.e. in geostrophic balance. However for (small) $E/H^2 f < 0.02$ the much larger surface velocities are more closely aligned with the density gradient.

In less well-mixed regions, density effects can have a more spectacular influence on current ellipses (Souza and Simpson, 1996). In the extreme case of two uncoupled layers, currents in the surface layer show a (deep) frictionless response to the surface gradients while those in the lower layer respond as if the surface layer did not exist. Nearshore density gradients are primarily associated with salinity variations, these may produce pronounced influences on circulation (Prandle, 1991b). Further offshore, density gradients tend to be the result of temperature differences.

Oceanic sea level gradients

Somewhat conveniently, the large scale circulation in the N.W. European Shelf Seas is largely isolated from sub-tidal forcing originating in the adjacent Atlantic. Thence long term patterns of residual circulation can be calculated by superposition of the cited tidal and wind-factored components. The validity of the resulting general pattern (Fig. 13.3) is supported by long-term drogue tracks (Nichols and Brander 1989). For tracers that are reasonably well-mixed vertically, flushing times associated with these combined circulation patterns can be used to estimate concentrations for specified inputs.

Appendix D: Vertical Exchange

In general vertical velocities are small and the effect of vertical advection can be accounted for within the vertical dispersion coefficient K_Z. The time, T_V, for complete vertical mixing of a tracer introduced at the sea surface or seabed is H^2/K_Z. For the North Sea, T_V varies from a few hours in shallow strongly tidal seas to the order of a year in the deep sluggish waters of the Northern North Sea. Vertical stratification induced by solar heating exists over much of the deeper regions from May until October. In the region of the thermocline, such stratification can reduce K_Z to close to the kinematic viscosity level.

In practice K_Z varies continuously with depth and time over both the tidal and seasonal cycle with significant input from surface waves and surge currents. As indicated in Table 13.2 , the various models for simulating these variations in K_Z are now well developed (Burchard and Baumert 1995).

For sediment modelling the fall velocity can be regarded as an effective advective term - allowing semi-analytical solutions (Prandle 1997b).

Prandle and Lane (1995) show how temperatures in the North Sea are determined by the dynamics of vertical exchange assuming a localised balance. Prandle (1998) extends this analysis to incorporate a coupled ocean-sea thermal model to provide global expressions for the seasonal cycle of both sea surface and ambient air temperature. This study shows that, in general away from the immediate influence of pronounced ocean current systems, the temperatures of the ambient air is also determined by these same localised vertical exchange rates.

Appendix E - Seasonal Cycles

Influence of horizontal circulation on the seasonal cycle

For a cyclical input of C equal to $I\cos wt$ and an external exchange rate $-VC/F$ then (Prandle et al. 1993)

$$\frac{dC}{dt} = \frac{I}{V}\cos wt - \frac{C}{F} \tag{E1}$$

(V volume, C concentration and F flushing time).

For $C = 0$ at $t = 0$, Laplace transforms provide the solution

$$C(t) = \frac{FI}{V(1 + F^2 w^2)}\{\cos wt + Fw\sin wt - e^{-t/F}\} \tag{E2}$$

For the case of a steady inflow $w = 0$ and thus (E2) gives

$$C(t) = \frac{FI}{V}\{1 - e^{-t/F}\} \tag{E3}$$

and 90% of maximum concentration is reached after $t = 2.3$F.

From E2, seasonal concentrations associated with the annual cycle in river or rain inflows are reduced by the factor $1 / (1 + F^2 w^2)^{1/2}$ compared to a continuous inflow. For the southern North Sea (south of 56°N) a flushing time of 240 days has been calculated. From (E2), the seasonal cycles is reduced to a factor of 0.23 hence the observation of little seasonality in salinity away from the coastal zone. Conversely, for decadal variations this factor is 0.92, i.e. little diminished.

Influence of vertical mixing on the seasonal cycle

For equation (B4) without advective terms and applied to the vertical dimension z the general solution for a cyclical surface exchange Ie^{iwt} is

$$C(z, t) = \frac{bIe^{bz} + e^{-bz}}{iw(e^{bH} - e^{-bH})}e^{iwt} \tag{E4}$$

(with K_z constant and $\partial C/\partial z = 0$ at the bed $z = 0$), where H is the water depth and $b = (iw/K_z)^{1/2}$.

Figs. 13.9a and 13.9b show, respectively, the amplitude and phase of $C\,(z,t)$ at the surface normalized against the depth averaged value, for a range of values of both K_z and D with w taken as an annual cycle.

In many areas of the southern North Sea, where strong tidal action ensures $K_z > 10^{-2}\mathrm{m}^2\mathrm{s}^{-1}$, since $H < 100$m, vertical homogeneity for the annual cycle is sensibly maintained. However, in areas with weaker tides, thermal stratification can greatly reduce the effective depth-averaged value of K_z (Simpson and Hunter 1974). For a shorter period event of P days duration, the response is equivalent to a dispersion coefficient K_z reduced by P/365 in Fig. 13.9 . Hence, vertical variability is likely to be of primary concern where inputs are concentrated within a period of a month or

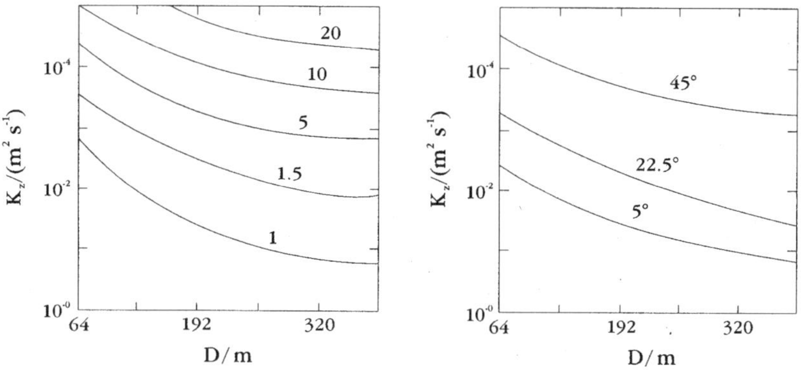

Fig. 13.9 Amplitude (left) and phase (right) of seasonal signal, K_z vertical diffusion coefficient, D depth. Contours indicate ratio (left) and difference (right) between sea surface and depth-average.

less. As an example, for the tranquil clear-sky conditions conducive to plankton blooms, residual circulation is likely to be a secondary influence (although tidal advection and vertical structure may be important).

14 Forecasting Wind-driven Ocean Waves

GERBRAND J. KOMEN
KNMI, Koninklijk Nederlands Meteorologisch Instituut,
De Bilt, The Netherlands

14.1 Introduction

The study of wind-generated surface waves on the ocean is important from a physical, a geophysical and a practical point of view. In these lectures I will concentrate on their dynamics and prediction. Over the last 50 years enormous progress has been made in this field (see, for example: Phillips, 1977; WMO, 1988; and Komen et al., 1994), which has resulted in the availability of advanced measurement tools and advanced numerical models, such as the WAM, WAve Model (WAMDI, 1988).

In the not so distant past, wave predictions were made manually. This required considerable insight in processes responsible for wave growth, propagation and decay. Now there is a real danger that this experience is lost, as one can use a wave prediction model as a black box: winds in - waves out. To avoid this unhealthy situation, I will begin with a discussion of some basic phenomenology. I will start by considering certain simplified cases. However, the main message will be that wave spectra are usually quite complex in realistic situations.

In the second part I will show how our understanding of basic wave dynamics and air/sea interaction has been used to derive the WAM model, and I will discuss some aspects of the operational implementation of this model. I will also discuss its performance and its usefulness for wave forecasting.

14.2 Phenomenology

Our knowledge of waves is based on many observations. *Visual observations* of wind speed and direction, significant wave height, wave period, and wave direction (wind sea and swell) have always been of importance, and will continue to be so. In addition, there are valuable *instrumental observations* from buoys and platforms of the (two-dimensional) wave spectrum. Certain areas, such as the North Sea, are covered reasonably well with instruments. Instrumental observations in other areas, notably in the Southern Hemisphere are extremely scarce. Fortunately, a wealth of ocean wave observations has become available from *satellites* such as Geosat, Topex-Poseidon and ERS-1 and ERS-2. In particular, the radar altimeter has proven to be quite reliable for measurement of the significant wave height. Synthetic aperture radars (SAR) are able to provide two-dimensional information about the wave spectrum.

Conventional and satellite observations are complementary. Satellites have a global coverage, which is a great advantage, but their repeat cycle (the time lapse between two observations at the same position) is rather low. Conventional observations, on the other hand, have excellent temporal resolution, but they are point measurements, so their spatial coverage is restricted by the number of instruments one has.

Fig. 14.1 shows a scatter plot in which for a large number of cases the wave height is given as a function of the wind speed. (The cases are model simulations rather than observations, but they are sufficiently realistic to be used as an illustration of the point.) It is important to note that there is no clear relationship. The solid line is an attempt to draw a relation[1] . But as you can see, it fails. At low wind speed waves are higher than expected. These are cases in which advection of energy (swell) is important. At high wind speed most wave heights lie below the curve. This is because they are not yet fully grown.

To illustrate this we will discuss wave generation in a number of simple situations.

14.2.1 Duration- and fetch-limited wave growth

Imagine an infinite flat windless ocean. At a given time, say $t = 0$, a uniform wind starts blowing. Due to instability (Kawai, 1979) waves begin to form. They grow with time. The significant wave height H_S, everywhere equal, will be a simple function of two variables, the wind speed u and time t:

$$H_S = H_S(u, t) \tag{1}$$

Wind-generated surface waves are gravity waves, i.e. waves in which gravity is the restoring force. Therefore, one may expect that their behaviour scales with the strength of the gravitational acceleration g. This allows one to rewrite (1) as

$$\frac{H_s}{u^2}g = f\left(\frac{tg}{u}\right), \tag{2}$$

where f is a function that has been determined from observations. Wave modellers can not agree on the exact shape of f but typically $H_S\, g/u^2$ grows as a function of tg/u to reach a maximum level of $H_s\, g/u^2 = 0.22$ at $tg/u = 2 \times 10^5$. (For longer durations the growth becomes negligible, i.e. at $tg/u = 2 \times 10^5$ when the wave height reaches its asymptotic value the waves are fully grown, and can't grow higher for that given wind speed.) For a typical (strong) wind of 20 m/s, it takes

1. This particular curve is a parametrisation proposed by Sanders in the 1970s for the maximum significant wave height that would be reached under steady wind conditions ($H_{S,\,max} = \beta\, U^2{}_{10} / g$, with a mean winter value for β of 0.22, as opposed to the summer value which was only 0.14 (Janssen, Komen and de Voogt, 1984)).

about two days until the waves reach their maximum height of about 9 m. Younger waves will be less high. This gives a qualitative explanation for the scatter at the high-wind end of Fig. 14.1.

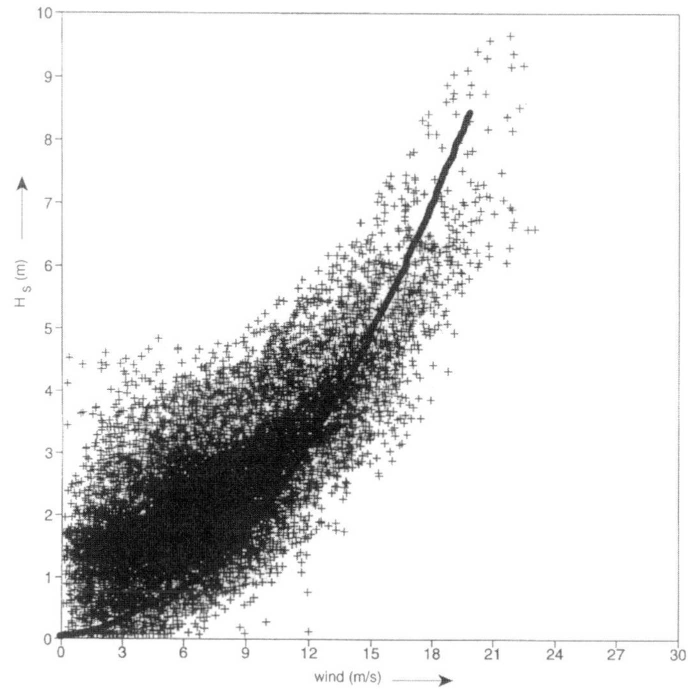

Fig. 14.1 The significant wave height as a function of wind speed for a large number of (simulated) situations. There is no relation. The curve is an estimate of the maximum wave height that would be obtained under steady homogeneous wind conditions.

As an alternative to duration-limited growth one often considers fetch limited growth. In this case one considers a half plane ocean, with a constant off-shore wind of infinite duration. The wave height then is a function of wind speed and distance to shore, the so-called fetch. This situation can be treated in a similar way, i.e. the relevant quantities can be made non-dimensional, and one can consider the nondimensional wave height as a function of nondimensional fetch. This fetch-limited growth curve has the same asymptotic value as the duration-limited growth curve. At a wind speed of 20 m/s one needs a fetch well over 1000 km to reach asymptotics.

The concepts of fetch- and duration-limited growth have only limited value, because in reality they are never encountered: winds are not homogeneous and constant, and coasts are not straight.

Another problem arises from the choice of wind speed parameter. Scaling with u is disputable, because there is another quantity with the dimension of velocity, which is probably more relevant for wave growth, namely the friction velocity u_*. This parameter is defined as the square root of the (kinematic) surface stress. It can be measured, although not easily. One has found that the drag coefficient C_D, defined by

$$u_*^2 = C_D u^2 \tag{3}$$

varies roughly between 1×10^{-3} (low wind speeds) and 3×10^{-3} (high speeds). This implies that u-scaling and u_*-scaling can not both be correct. When u scaling is correct - as assumed above - one obtains an asymptotic wave height which is proportional to u^2. In the case of u_* scaling it will be proportional to u_*^2. Suppose one has two scaling relation, one based on u-scaling, the other on u_*-scaling, giving the same asymptotic wave height at low wind speed. Then they will give asymptotic wave heights that differ by a factor of three at high wind speed. For practical applications this difference is unacceptably large. There are arguments and observations that favour u_*-scaling (Janssen, Komen, de Voogt, 1987), but even until today the superiority of u_*-scaling has not been proven convincingly from wind and wave observations.

14.2.2 Spectral evolution

Of course, the wave height is not the only quantity that matters. Individual, monochromatic, waves are characterised by their height (which equals 2 × the amplitude), period, length and direction. Period and length are related through the dispersion relation (see next section), so one may choose height, period and direction as the basic characteristics. In general, the sea surface consists of a superposition of many waves. To characterise these one has to specify how the energy is distributed over different directions. This is done with the help of the two-dimensional wave spectrum. The total energy is proportional to the square of the significant wave height. This allows one to write:

$$H_S^2 \sim \int F_{(2)}(f, \theta) df d\theta. \tag{4}$$

Here $F_{(2)}$ is the two-dimensional wave spectrum, $f = 1/T$ is the frequency (the inverse period) and θ is the wave direction. There exist different ways of representing the two-dimensional wave spectrum. In one representation the spectral level is contoured in the $f - \theta$ plane. The other representation is a polar one, in which the distance to the origin is proportional to the frequency and the angle with the horizontal axis corresponds with θ.

The one-dimensional spectrum is also often used. This is defined by integration over angles:

$$F_{(1)}(f) = \int F_{(2)}(f, \theta) d\theta. \tag{5}$$

The shape of the spectrum provides the basic mathematical and physical characterisation of the sea state. One may now wonder what determines this spectral shape. A first answer to this question came from the Jonswap experiment (Hasselmann et al., 1973). In this experiment the one-dimensional wave spectrum was measured under conditions that were close to the ideal situation of fetch-limited growth. Three important features were observed: 1. all spectra were sharply peaked; 2. the peak-frequency decreased with increasing fetch; 3. the high frequency tail approached an asymptotic universal level, but an overshoot occurs before this level is reached (Fig. 14.2). Other measurements provided evidence that the spectrum reaches a stationary shape for very large fetches. It was also found that, while, in general, individual waves propagate at an angle to the wind, they do travel in the wind direction in the mean. Duration-limited growth is thought to have a similar spectral characteristic. This relatively simple picture has stimulated the development of so-called parametric wave models in which the spectral shape was parametrised in terms of a small number of parameters.

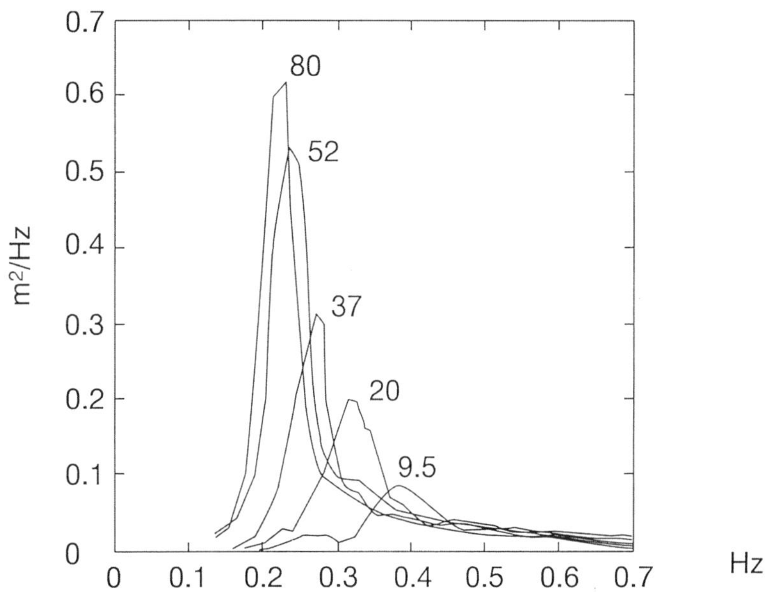

Fig. 14.2 Spectral evolution as observed during Jonswap (Hasselmann et al., 1973), under conditions which were close to the idealisation of fetch-limited growth.

As already noted, the concepts of fetch- and duration limited growth have only limited value, because in reality they are never encountered. When the conditions are not ideal, spectra are no longer simple. This became first apparent in the SWAMP (1985) study, when the behaviour of many wave models was compared in situations that were still idealised but which differed from fetch- and duration-limited wave growth. The simplest such case is the slanting fetch situation in which the coastline is still straight and the wind is constant but blowing under an angle. This was found already to introduce a significant asymmetry in the spectrum.

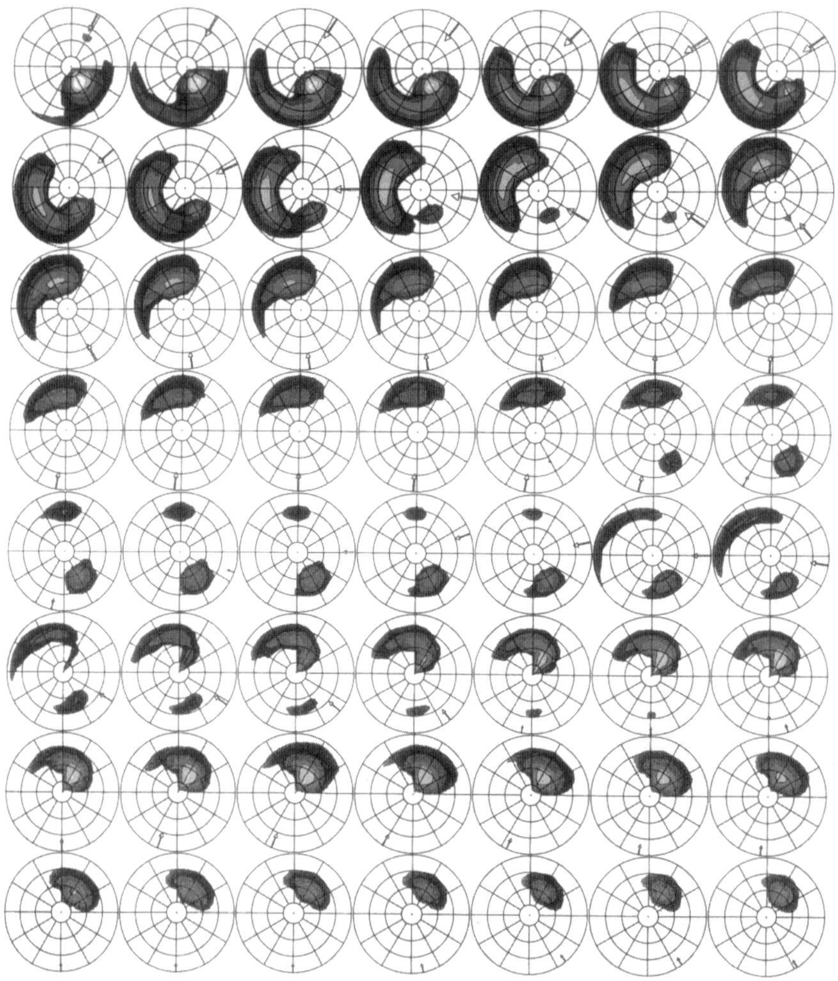

Fig. 14.3 Time sequence of two-dimensional spectra as simulated by Tom Gerling (1991). The arrows indicate the wind direction. This sequence illustrates the complexity of the spectral shapes under realistic conditions.

Another simple situation is one in which the wind is turning. This gives rise to a complex spectral response. Things become even more complicated when the wind field has a spatial and time-varying structure. In such realistic situations one has to add advected wave energy (usually called swell) to the locally generated wind sea.

In extreme swell situations one may locally have large waves without (or even against) the wind. It is these waves, which can often be observed as breakers near a beach, that account for the high wave heights at low wind speeds in Fig. 14.1.

In general, wave spectra can have very complex shapes indeed. This is nicely illustrated in Fig. 14.3 (Gerling, 1991), which shows a time sequence of spectra for realistic conditions. The central question then is whether we can understand and predict this type of spectral evolution on the basis of the fundamental laws governing ocean wave dynamics. The answer is yes, as we will see in the next section.

14.3 The WAM model

The first operational wave predictions were based on the work of Sverdrup and Munk (1947), who introduced a parametrical description of the sea state and who used empirical relations to describe wind sea and swell. An important advance was the introduction of the concept of a wave spectrum (Pierson et al., 1955). This was not yet accompanied, however, by a corresponding dynamical equation describing the evolution of the spectrum. This step was made by Gelci et al. (1956) who introduced the concept of the spectral transport equation. This equation is used today as the basis for wave prediction models. The derivation of this spectral transport equation is rather lengthy. Details can be found, for example, in Komen et al. (1994). Here we summarise the main steps that are made in its derivation.

1) The linearised equations of fluid dynamics allow for free harmonic wave type solutions. The dynamics determines the dispersion relation, relating wavenumber k and angular frequency ω:

$$\omega^2 = gk \tanh kD. \tag{6}$$

This relation depends explicitly on the depth D. A general solution takes the form of a superposition of monochromatic waves:

$$\eta(\mathbf{x}, t) = \sum_{\boldsymbol{\kappa}} a_{\boldsymbol{\kappa}} e^{i(\boldsymbol{\kappa} \cdot \mathbf{x} - \omega t)} + c.c., \tag{7}$$

Here η is the height of the sea surface. Of course, realistic waves are not free and linear. But one may assume that nonlinearities and interaction with the environment are weak – at least in the mean – and can be accommodated by allowing for slow variations of amplitudes and corresponding wavenumbers. This is known as the geometrical optics approximation.

2) One does not go for a deterministic description, but merely seeks to make statements about average quantities. So one does not want to predict the height of the sea surface at a particular position and a particular time, but one wants to predict the probability that the waves have a particular height at a particular place and time. Statistics is introduced by assuming that the different components are - and remain - independent:

$$\langle a_k a_{k'}^* \rangle = |a_k|^2 \delta_{k,k'} = \frac{1}{2} F_k(x,t) \delta_{k,k'} . \tag{8}$$

3) It can now be shown that the spectrum $F(\boldsymbol{k})$ satisfies the following equation:

$$\left\{ \frac{\partial}{\partial t} + \mathbf{c}_g \cdot \frac{\partial}{\partial \mathbf{x}} \right\} F(\mathbf{k}, \mathbf{x}, t) = \sum_l S_l(F;\mathbf{u}) = S_{in} + S_{nl} + S_{ds} \tag{9}$$

The form given here applies to deep water. A more general version with shallow water effects and wave-current interactions is given in Komen et al. (1994). The left-hand side describes propagation; the source terms on the right describe the interaction of wave components with the atmosphere (S_{in}; the input), among themselves (S_{nl}) and with the ocean (S_{ds}; dissipation). Thanks to major research efforts made over the last 30 years we now have suitable parametrisations for the different source terms. In general, these are based on a mixture of theoretical arguments and empirical observations. All source terms are functionals of the wave spectrum, F. Dissipation and wave-wave interactions are nonlinear in F. The wind input is linear in F, but it depends also on the local wind.

It seems appropriate to repeat the main assumptions that went into the derivation of equation 9 because they presumably specify its validity range. These assumptions were: 1) the geometrical optics approximations; 2) weak nonlinearity (in the mean); and 3) (local) statistical homogeneity. These conditions seem to be well met at sea, but they surely break down in situations in which the waves change on a scale which is short with respect to their wavelength or period, such as may happen close to the shore.

Equation 9 can be solved numerically by forward integration in time. To do this, one needs to specify forcing wind fields at each grid point and for each time, as well as initial and boundary conditions. Wind fields are usually taken from NWP (numerical weather prediction) models. Coasts are normally treated as fully absorbing. Open boundaries may be avoided by a suitable choice of the integration area. If this is not possible one may nest a limited area model within a larger scale model.

The initial conditions need some discussion. Physically, the instability of air flow over a flat sea occurs at centimetre wave lengths. At these short length scales surface tension still plays a role. Also the physics of the interaction between the boundary layer and these short waves is rather different from the physics governing the growth of longer waves. Therefore, the very short wave scales are not resolved in the WAM model. It is then consistent to assume that there is always a minimum spectral energy level, which can be parametrised in terms of the grid size, the time step and the wind speed. When a cold start is made the model is initialised at this level.

14.3.1 The energy balance

The first applications of the WAM model concerned the idealised cases discussed in the previous section. It is instructive here to consider the case of duration-limited growth. The phenomenology of this type of growth was discussed above: a peaked spectrum; a peak that moves to the left when duration increases; overshoot at high frequencies; and an approach to a fully grown stage in which the spectrum does not change anymore. All of these features have been reproduced (Komen, Hasselmann and Hasselmann, 1984). An analysis of these results has greatly improved our understanding of what is going on. An illustration of this is given in Fig. 14.4 which shows the different source terms during the stage of active growth. The total source term is characterised by a positive lobe at frequencies below the peak frequency of the spectrum.

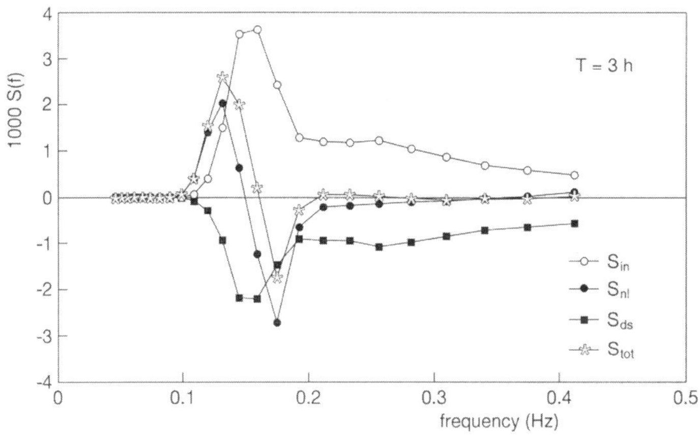

Fig. 14.4 The energy balance (equation 9) in growing waves (from Komen et al., 1994). The source terms were computed for a spectrum that peaked at about 0.17 Hz.

This is a combined effect of the different source terms. Wind input is at relatively high frequencies. The non-linear interactions transfer this energy to lower frequencies. A similar analysis can be made of the fully grown stage, when the three source terms, wind input, nonlinear transfer and dissipation balance each other to within several orders of magnitude, i.e. $(S_{in}+S_{nl}+S_{ds}) / S_{in} \ll 0.01$.

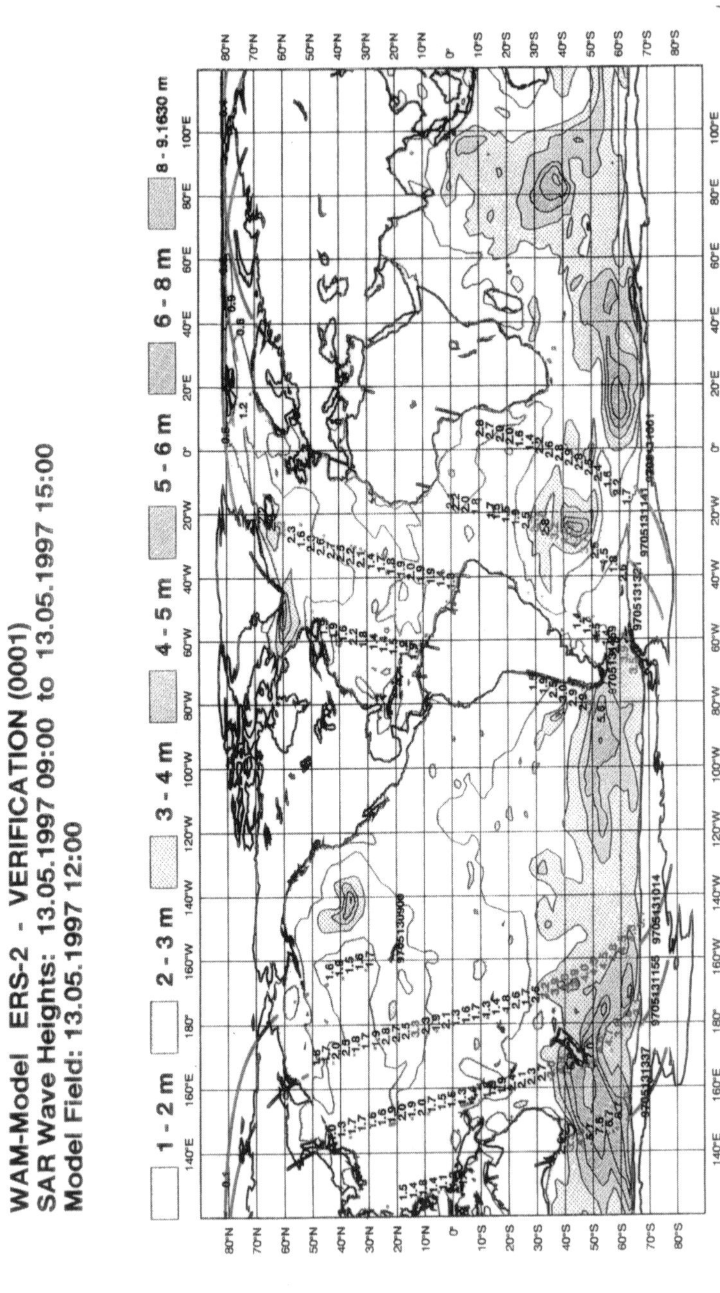

Fig. 14.5 An example of a global wave forecast for one particular time. Isolines give the significant wave height. Wave height observations by ERS-2 are also plotted.

14.3.2 Operational applications

The WAM model has been implemented operationally in many forecasting centres. The European Centre for Medium-range Weather Forecasts (ECMWF) runs a global version of the model with a half-degree resolution to make 10-day forecasts. KNMI runs a North Sea version of the model. An example of the output of the global model is given in Fig. 14.5, which shows isolines of significant wave height, as computed for one particular time. Quasi-synchronous wave height observations by ERS-2 are also plotted. The performance of the model has been tested against buoy and satellite observations. As an example, Fig. 14.6 shows the bias as computed for every month from early 1995 until March 1997. The comparison is made both for all observations and for separate regions. Fig. 14.7 shows, for the northern hemisphere, the anomaly correlations between the forecast and the verifying analysis for the 3-, 5-, 7- and 10-day forecast (January 1996 though March 1997). These correlations are a measure of the forecast skill. More details on model validation can be found in Komen et al. (1994) and in Janssen, Hanssen and Bidlot (1996).

Wave *forecasting* is only one application of a wave model. Another important application is the use of a wave model to study the wave climate (Guenther et al., 1998). Interest in wave climate studies has recently increased when it was realised that the wave climate may exhibit a considerable amount of decadal variability (Carter and Draper, 1988; Bouws et al., 1996; WASA, 1998). This variability is associated with the North Atlantic Oscillation (NAO), which is characterized by the mean pressure difference between Iceland and the Acores. The WAM model has also been used to estimate the effect of CO2 doubling on the wave climate of the North Atlantic (Rider et al., 1996) using surface winds from global change simulations with coupled atmosphere/ocean general circulation models.

14.4 Outlook

To make useful wave fore- or hindcasts one needs good quality input winds. Fortunately, atmospheric models have considerable skill, which allows us to make useful wave forecasts. A further improvement of wind analysis and forecast over the ocean will be crucial to increase the quality of wave computations.

Wave models can be further improved. There are three broad areas for improvement: numerical resolution (both spatial and spectral), numerics (propagation scheme and the integration of the non-linear transfer integral) and physics (i.e. the representation of wave growth and dissipation in the source terms). One should not forget to test "improvements". After all, what really counts is the quality of the wave predictions, when compared with observations. The WAM model has been tested in numerous hindcasts and in many years of operational application. Repeating such tests is a formidable task. Maybe inverse modelling, using an adjoint (Hersbach, 1998) can be useful here.

Wave forecasts can also improve by the assimilation of wave observations. Present operational methods are still based on optimum interpolation (see, for

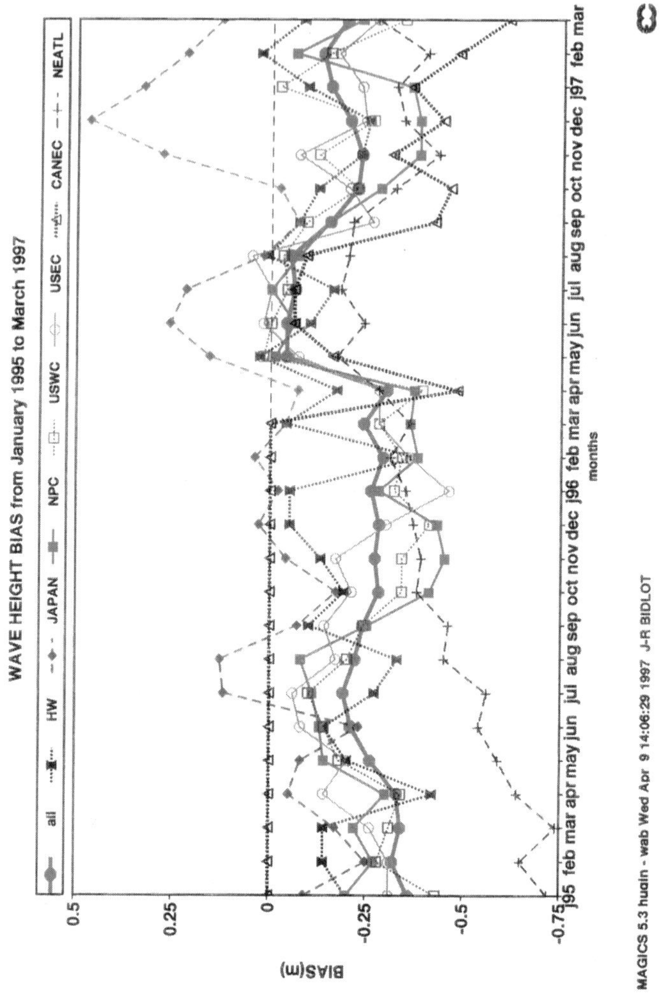

Fig. 14.6 Validation statistics of the global WAM model as run by ECMWF. HW = Hawaii; NPC = North Pacific; USWC = US West Coast; USES = US East Coast; CANEC = Canadian East Coast; NEATL = Northeast Atlantic.

example, Lionello et al., 1992; or Voorrips et al., 1997). But the practical applicability of more advanced methods (adjoint, Kalman filter, etc) should be explored.

Finally, there remain some very basic questions. What exactly is the mechanism of wave generation? How important is turbulence, and the interaction between wave motion and turbulence in the air? How can one better understand the role of breaking waves? These questions have been addressed in many studies (see, for some recent studies: Janssen, 1994; Belcher and Hunt, 1993; Makin et al., 1995),

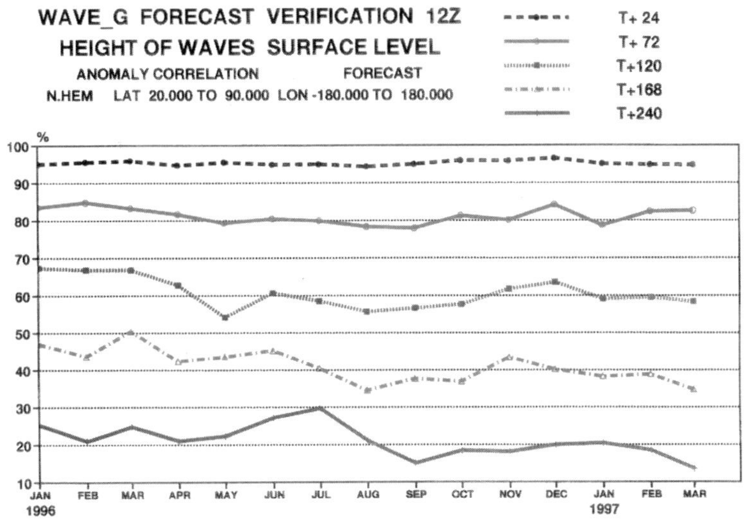

Fig. 14.7 Forecast skill of the global WAM model as run by ECMWF. Shown are northern hemisphere anomaly correlations between the forecast and the verifying analysis for the 3-, 5-, 7- and 10-day forecast (January 1996 through March 1997).

and much progress has been made, but many questions remain unanswered: theoretical ideas wait experimental confirmation and experimental results wait for an explanation. There can be no doubt that waves play an essential role in air/sea interaction. But a question that has not been settled is whether their role can be parametrised in terms of wind speed alone. From a theoretical point of view this is hard to believe, but experimental evidence is conflicting (Smith et al., 1992; Janssen, 1992; Yelland and Taylor, 1996; see also review papers by Donelan et al., 1993; and Komen et al., 1997). Answers to these questions will not only help improve the quality of wave prediction, they will also explain C_D and provide better forcing for ocean circulation models of the ocean (Mastenbroek et al., 1993; Burgers et al., 1996) and better boundary conditions for atmospheric GCM's (Doyle 1994, 1995, Janssen and Viterbo, 1995), and improved coupling for coupled atmosphere/ocean models.

Acknowledgements

I learned much in discussions with Klaus Hasselmann and Peter Janssen. I am grateful to Peter for providing Figs. 14.1, 14.5, 14.6 and 14.7. This chapter was written during the MAST school in Torregrande and during a visit at ECMWF. I would like to thank Nadia Pinardi and David Burridge for their hospitality. Some of the material of this lecture was also presented at the 1997 ECMWF Seminar.

15 A Multivariate Reduced-order Optimal Interpolation Method and its Application to the Mediterranean Basin-scale Circulation

PIERRE DE MEY AND MOUNIR BENKIRAN
LEGOS, 18 avenue Edouard Belin, 31401 Toulouse Cedex, France

15.1 Introduction

For more than a decade, the Ocean Circulation and Prediction Team at LEGOS, Toulouse, has been developing data assimilation methods and conducting data assimilation experiments in various basins of the World Ocean, and in particular in the Mediterranean. Our aims are to study the feasibility of multivariate control of a model trajectory, and to characterize the predictability of the general circulation, seasonal and interannual variability mesoscale eddies, meanders, sub-basin-scale gyres, and response to wind. This chapter deals with the design and *modus operandi* of a practical algorithm for data assimilation and application to the Mediterranean.

Data assimilation (or "state estimation") consists in calculating the "best" estimate of the state of a physical system (usually the atmosphere or the ocean) and its evolution in time, given observations and a prognostic numerical model. An increasing number of ocean modelling projects now involve data assimilation. Simultaneously, numerical models become more sophisticated and more expensive, and the choice of an assimilation approach becomes critical. Specific reanalysis efforts left aside, the cost of rigorous, physically consistent state estimation algorithms such as the Kalman Filter, Kalman Smoother and adjoint variational methods is usually still too high, even if reduced-order variants of these methods have been proposed for meteorological applications (e.g. Courtier et al., 1994). Suboptimal algorithms such as nudging and optimal interpolation (OI) have been used in the ocean (e.g. Carton and Hackert, 1989; Derber and Rosati, 1989) and are much more economical for large-scale problems, but they do not usually ensure that the solution is consistent with physics and with our understanding of how the ocean works.

The method discussed and applied here is based on four-dimensional Optimal Interpolation on a base of pre-calculated EOFs (Empirical Orthogonal Functions). Since this is an OI-based algorithm, time-dependent dynamical constraints are not explicitly enforced. However, the dominant coherent physical and statistical relationships between variables can be introduced via the EOFs. Reduced-order estimation has been reviewed e.g. by De Mey (1997) in the ocean and by Bernstein and Hyland (1985) presenting the algorithmic details. For instance, the use of EOFs in isopycnal coordinates for the basis functions seems promising (e.g. Gavart and De Mey, 1997) and consistent with the large-scale conservation of properties (Cooper and Haines, 1996). One of the advantages of the method presented here over previous statistical methods based on regression in the ocean (e.g. Mellor and Ezer, 1991; Oschlies and Willebrand, 1996) lies in the fact that it directly derives from the Reduced-Order Extended Kalman Filter (ROEKF) equations. Therefore the assumptions and limitations can be clearly stated, and it is relatively straightforward to add a forecast error estimation scheme.

Section 15.2 describes the assimilation algorithm, rationale and limitations. Section 15.3 deals with some results of applying the algorithm to basin-scale circulation estimation in the Mediterranean in twin experiments and with real observations. A brief perspective is given in section 15.4.

15.2 Optimal Interpolation on a base of EOFs

15.2.1 Optimal interpolation

We will use the Ide et al. (1997) classic notations. In the Extended Kalman Filter (EKF, e.g. Gelb, 1974), *observations* $\mathbf{y}^{o}$ are formally linked to the *true state* $\mathbf{x}^{t}$ by the stochastic equation

$$\mathbf{y}^{o} = H(\mathbf{x}^{t}) + \varepsilon \tag{1}$$

where $H(\ )$ is the nonlinear *observation operator*, and the observational noise process ε is assumed to be with zero mean and covariance matrix $\mathbf{R}$. The complete nonlinear numerical model $M(\ )$ is used to produce a *state forecast* $\mathbf{x}^{f}$ at successive filter time steps:

$$\mathbf{x}^{f}(t + \delta t) = M(\mathbf{x}^{a}(t)) \tag{2}$$

where $\mathbf{x}^{a}$ denotes the *analyzed* estimate and δt is the filter time step. The state update (*analysis*) step at time t+δt is given by:

$$\mathbf{x}^{a} = \mathbf{x}^{f} + \mathbf{K}\mathbf{d} \tag{3}$$

where $\mathbf{K}$ is the *Kalman gain* and

$$\mathbf{d} \equiv \mathbf{y}^{o} - \mathbf{y}^{f} = \mathbf{y}^{o} - H(\mathbf{x}^{f}) \tag{4}$$

is the *innovation vector.*

Following e.g. Daley (1991), Optimal Interpolation (OI) is a particular suboptimal filter in which the EKF forecast error covariance matrix is replaced by an approximate *background* error covariance matrix $\mathbf{B}^{f}$, such as:

$$\mathbf{B}^{f} = (\mathbf{D}^{f})^{1/2}\mathbf{C}(\mathbf{D}^{f})^{1/2} \tag{5}$$

where only the *background error variances* in the diagonal matrix $\mathbf{D}^f$ are predicted, using an external scheme, from the previous analysis error variances. The *correlations* in matrix $\mathbf{C}$ are assumed to be constant in time. Since OI does not evolve dynamically the errors, the modelling and parameterization of $\mathbf{B}^f$ contains most of the physics of the estimation problem and should be treated very carefully.

Equations (1-4) hold for OI as well as for the EKF. The OI gain has the same structure as the EKF gain:

$$\mathbf{K}^{OI} = \mathbf{B}^f\mathbf{H}^T(\mathbf{H}\mathbf{B}^f\mathbf{H}^T + \mathbf{R})^{-1} \tag{6}$$

where $\mathbf{R}$ is the observational error covariance matrix, and an approximation is used instead of the predicted EKF forecast error. As in the EKF gain, the full observation operator $H(\,)$ is replaced in (6) by the tangent linear observation operator $\mathbf{H}$.

Finally, let us turn to the *analysis errors*. In most OI schemes, the forecast error variances are derived from the analysis error variances at the previous filter step (e.g. Ide et al., 1997) using an *ad hoc* scheme. Considering any linear (optimal or suboptimal) filter, the analysis error covariance matrix can be written:

$$\mathbf{B}^a = (\mathbf{I} - \mathbf{K}\mathbf{H})\mathbf{B}^f(\mathbf{I} - \mathbf{K}\mathbf{H})^T + \mathbf{K}\mathbf{R}\mathbf{K}^T \tag{7}$$

This is valid for all forms of the gain $\mathbf{K}$ and therefore for the form (6), with the best available estimates of the forecast (background) and observational error covariance matrices. This assumes that both types of errors are uncorrelated to each other, which is a reasonable good assumption for error of a purely observational nature, but might fail if e.g. observations are mapped prior to assimilation.

15.2.2 3-D EOFs

Numerical weather forecasting, for the purposes of which OI schemes were brought in and continuously improved over the years, has had to face the question of how to model realistic, three-dimensional, multivariate error correlations, and how to predict error variances, using expressions such as (5). Historically, many "recipes" chiefly aimed at splitting the problem into simpler ones have been used by the various meteorological agencies and research groups (see e.g. Gustavsson, 1981; Lorenc, 1986). In the atmosphere, many attempts have revolved around the idea of using (5) for a basic set of increments (mass field, wind field), and implementing adjustments or balance relationships such as geostrophy for the other increments.

In the ocean, and in particular in the Mediterranean, such a "splitting" approach is more complicated to set up, because of the lack of experience on what works best, spatial inhomogeneities such as coasts, islands and straits, and relatively poor, irregular data coverage. Here we will review methods to introduce empirical, spatially coherent multivariate structure functions in the modelling of errors.

Let us write the spectral factorization of the background error covariance matrix $\mathbf{B}^f$:

$$\mathbf{B}^f = \mathbf{V}\Delta\mathbf{V}^T \tag{8}$$

where **V** is an orthogonal matrix whose columns are the eigenvectors (EOFs) of background errors and Δ is a diagonal matrix whose diagonal elements are the eigenvalues.

In an analogy with the assumptions in (5), let us first assume that the eigenvectors (not necessarily the eigenvalues) are known and stationary. As a way to model $\mathbf{B}^f$, expression (8) has no fewer degrees of freedom than (5). However it has two advantages over (5): EOFs define spatially coherent, multivariate modes of variability which are physically more meaningful than correlations, and the spectrum of eigenvalues is usually very red if the norm is well chosen, henceforth permitting to truncate the problem to some order.

Unfortunately the "true" background error covariances and therefore the "true" background error EOFs are usually unknown and probably not stationary. One promising but costly way around this difficulty is given by the SEEK filter (e.g. Pham et al., 1998), which is a reduced-order Kalman filter evolving part of the base of EOFs with the dynamics.

Remaining within the context of optimal interpolation for the practical purposes of the present chapter, our suggestion for the modelling of $\mathbf{B}^f$ will use two practical approximations:

(1) At any time, we write the decomposition of $\mathbf{B}^f$ over a set of pre-calculated orthonormal vectors $\tilde{\mathbf{V}}$ (which we can physically validate before use), for instance approximating the "true" eigenvectors in **V**.

(2) We truncate the problem to the dominant modes, using an external criterion (we will come back to this) as well as physical insight. A subset of the vectors in $\tilde{\mathbf{V}}$ is assumed to generate the *reduced state space*. The remaining vectors generate the corresponding unresolved space or *null space*. In other words, we write:

$$\tilde{\mathbf{V}} = \mathbf{S}^+ | \bar{\mathbf{S}}^+ \tag{9}$$

where the vertical bar denoting the column-wise juxtaposition of both matrices. The matrices **S** and $\bar{\mathbf{S}}$ are respectively the reduced-space and null-space simplification operator. From the orthonormal character of the vectors in $\tilde{\mathbf{V}}$, the following properties are true:

$$\begin{aligned} \mathbf{S}^+ &\equiv \mathbf{S}^T(\mathbf{S}\mathbf{S}^T)^{-1} = \mathbf{S}^T \\ \bar{\mathbf{S}}^+ &\equiv \bar{\mathbf{S}}^T(\bar{\mathbf{S}}\bar{\mathbf{S}}^T)^{-1} = \bar{\mathbf{S}}^T \\ \mathbf{S}^{+T}\bar{\mathbf{S}}^+ &\equiv \mathbf{S}\bar{\mathbf{S}}^+ = \mathbf{0} \\ \bar{\mathbf{S}}^{+T}\mathbf{S}^+ &\equiv \bar{\mathbf{S}}\mathbf{S}^+ = \mathbf{0} \end{aligned} \tag{10}$$

Appendix A gives the form of state increments if such order reduction is applied (equation A1). It also addresses in some detail the desired properties of the simplification operator **S**. In brief, the reduced-space processes must be observable, and the errors in the reduced space and null-space must be weakly correlated to each other through the dynamics. The latter condition can be achieved e.g. if the dominant physical processes in both subspaces are weakly coupled over the time scale considered. Considering these assumptions to be valid, we explicitly discard any attempt to control the null space (we can bring this back later if needed by means of an external algorithm). The background error covariance matrix is therefore modelled as:

$$\mathbf{B}^f = \mathbf{S}^T\mathbf{Br}^f\mathbf{S}. \tag{11}$$

The $\mathbf{Br}^f$ matrix contains the background error covariances in the reduced space. It is modelled as a diagonal matrix in (11), which comes back to assuming that the columns of $\mathbf{S}^T$ are eigenvectors of $\mathbf{B}^f$. In other words, in the 3-D EOF case, $\mathbf{Br}^f$ contains the error variances. Expression (11) also implies that the null-space background errors are assumed to be zero.

Given the above assumptions, the OI problem can now be entirely expressed in the reduced space. The Reduced-Order Optimal Interpolation (ROOI) gain in the full-state space can be written:

$$\begin{aligned} \mathbf{K}^{ROOI} &= \mathbf{S}^T\mathbf{Kr} \\ \mathbf{Kr} &= \mathbf{Br}^f\mathbf{Hr}^T(\mathbf{HrBr}^f\mathbf{Hr}^T + \mathbf{Rr})^{-1}. \end{aligned} \tag{12}$$

Observations are related to the reduced state space by means of the reduced-order observation operator $\mathbf{Hr}=\mathbf{HS}^T$. The reduced-order observational error covariance matrix $\mathbf{Rr}$ includes the representativity errors in the reduced space due to the observability of the null space through the $\mathbf{H}\nu$ term in (A3), in addition to the measurement errors. At analysis time, the model restarts from its previous fields plus a correction in reduced space which is converted back to full-state space by $\mathbf{S}^T$. Whatever properties were present in the null space are preserved. The method does not provide a built-in evolution of $\mathbf{S}$ such as in more sophisticated algorithms, but one can impose for instance a dependency on the season.

The analysis errors now become:

$$\mathbf{B}^a = \mathbf{S}^T\mathbf{Br}^a\mathbf{S} \tag{13}$$

with

$$\mathbf{Br}^a = (\mathbf{I} - \mathbf{KrHr})\mathbf{Br}^f(\mathbf{I} - \mathbf{KrHr})^T + \mathbf{KrRrKr}^T \tag{14}$$

The null-space errors can be added in (13) but are not accounted for by the ROOI scheme.

As noted above, $\mathbf{B}^f$ and its EOFs are generally unknown. One practical possibility to generate useful orthonormal vectors is to calculate EOFs of departures from a climatology, which we will hereafter call *perturbation EOFs*. Either data or model outputs can be used for this purpose. One must of course make the assumption that

$\mathbf{B}^f$ can be efficiently expressed on these perturbation EOFs. Behind that assumption lies the very widespread, but not often expressed idea (see e.g. Rienecker and Adamec, 1995) that if the perturbation EOFs are able to separate distinct physical processes, then the forecast errors corresponding to these processes in the model will be only weakly correlated, which is one of the things that we want (see the above discussion). This perturbation approach also provides an external criterion to truncate the assimilation problem, since one can think of using the eigenvalues attached to the perturbation EOFs as an indication of the dominant modes of errors. In the application sections of this chapter, we used perturbation EOFs to set up the assimilation in the Mediterranean (albeit in a 1-D case – see next section).

The methodology is of course applicable to multivariate problems (this is one of its advantages when compared to classic OI error modelling such as (5)). When calculating multivariate EOFs, the state vector is usually scaled, in order to avoid making the numerical estimate depend on the respective units in which variables are expressed. In the calculation of EOFs, scaling ensures that the eigendecomposition is well behaved. However there is no single way to scale the variables. For instance, with appropiate scaling, the numerically small but dynamically significant variations at depth in the ocean can be accounted for in the dominant modes, instead of being rejected to the tail of the spectrum. Alternatively, the variables can be scaled according to type, but regardless of the vertical level (this is the approach we will follow later in this chapter).

Let us derive the scaled filter equations for reference. We define the diagonal scaling matrix Σ containing the "scales" of the variables. The optimality criterion for the estimate and the whole method above can be expressed in terms of the scaled state vector $\mathbf{x}' = \mathbf{\Sigma}^{-1}\mathbf{x}$. Once back in dimensional space, this writes:

$$\begin{aligned} \mathbf{x}^a &= \mathbf{x}^f + (\mathbf{\Sigma S}'^{+})\mathbf{Kr}'\mathbf{d} \\ \mathbf{Kr}' &= \mathbf{Br}'^f\mathbf{Hr}'^T(\mathbf{Hr}'\mathbf{Br}'^f\mathbf{Hr}'^T + \mathbf{Rr})^{-1} \\ \mathbf{Hr}' &= \mathbf{H}'\mathbf{S}'^{+} = \mathbf{H}(\mathbf{\Sigma S}'^{+}) \end{aligned} \tag{15}$$

i.e. everything now expressed in terms of scaled variables, except for the observation operator and the gain which involve the dimensional modes $\mathbf{\Sigma S}'^{+}$.

15.2.3 Vertical EOFs

We will now consider the case of order reduction performed on the vertical. This approach was first explored in the ocean by De Mey and Robinson (1987), and later by many others. It is close to the method of separating the 3-D correlations into horizontal and vertical structure functions, as is (was) currently done in numerical weather forecasting. It has some advantages when compared to the 3-D case explored in the previous section. When calculating the modes empirically, the reduction to one multivariate dimension in the vertical leads to a better convergence than three-dimensional modes given the scarcity of ocean data. Besides, the vertical is a special dimension in oceanic problems, and vertical modes in the appropriate vertical coordinate system seems to have quite extended horizontal domains of validity. Examples of vertical simplification approaches in oceanic estimation problems, and discussions on those issues, are given in Gavart and De Mey (1997), Faucher et al. (2000), and Cooper and Haines (1996). We will also see later that vertical simplification helps make the assimilation "local" in space, which leads to a better fit to data and straightforward efficient solution on massively parallel computers.

The question of whether purely vertical order reduction would work in areas such as the Tropics, where the coherent variability and response of the ocean is non-local, still has to be explored. Other yet unclear domains of application are the coastal areas, shelf seas and upper ocean. In those contexts, a hybrid 3-D/1-D approach could be envisaged. However the vertical order reduction has been found useful in the Mediterranean. It is applied in the next sections of this chapter.

At any horizontal grid point $j \in [1 \ldots h]$ we define the local simplification operator $\tilde{\mathbf{S}}(j)$ acting on the vertical only (but across all variables defined at that location). The global simplification operator $\mathbf{S}$ defined in the previous section can be written:

$$\mathbf{S} = \begin{bmatrix} \tilde{\mathbf{S}}(1) & 0 & \ldots & \ldots & 0 \\ 0 & \ldots & \ldots & \ldots & \ldots \\ \ldots & \ldots & \tilde{\mathbf{S}}(j) & \ldots & \ldots \\ \ldots & \ldots & \ldots & \ldots & 0 \\ 0 & \ldots & \ldots & 0 & \tilde{\mathbf{S}}(h) \end{bmatrix} \tag{16}$$

i.e. a block-diagonal matrix made of the $\tilde{\mathbf{S}}(j)$. Locally, the columns of $\tilde{\mathbf{S}}(j)^T$ are orthonormal vectors. They can for instance be calculated as EOFs of departures from climatology as in the previous section. It is clear that the columns of $\mathbf{S}^T$ are also orthonormal. However the main difference with the previous section lies in the fact that the columns of $\mathbf{S}^T$ do <u>not</u> approximate the 3-D EOFs of $\mathbf{B}^f$ any more. Therefore the modelling of the $\mathbf{Br}^f$ matrix as a diagonal matrix is not valid any more.

Reorganizing the lines and columns of $\mathbf{B}^f$ and $\mathbf{Br}^f$ to form local blocks, we get:

$$\mathbf{B}^f = \begin{bmatrix} \ldots & \ldots & \ldots \\ \ldots & \tilde{\mathbf{B}}^f(j,j') & \ldots \\ \ldots & \ldots & \ldots \end{bmatrix}$$
$$\mathbf{Br}^f = \begin{bmatrix} \ldots & \ldots & \ldots \\ \ldots & \tilde{\mathbf{Br}}^f(j,j') & \ldots \\ \ldots & \ldots & \ldots \end{bmatrix} \tag{17}$$

with

$$\tilde{\mathbf{B}}^f(j,j') = \tilde{\mathbf{S}}(j)^T\tilde{\mathbf{B}}\mathbf{r}^f(j,j')\tilde{\mathbf{S}}(j') \tag{18}$$

The condition for all sub-blocks $\tilde{\mathbf{B}}\mathbf{r}^f(j,j')$ to be diagonal is for the vertical modes to be the singular vectors of the local error covariance matrices. In the following, we will only consider the case of spatially invariant vertical modes, in which case (18) expresses a local equivalent of (11), with the local modes being approximations of EOFs of the local error covariance matrices. As we will see, we do not lose much generality by doing so.

It is straightforward to make $\mathbf{Br}^f$ block-diagonal by reorganizing its lines and columns in mode-based blocks, each one modelled using the OI parameterization (5):

$$\mathbf{Br}^{f,i} = \mathbf{D}_i^{1/2}\mathbf{C}_i\mathbf{D}_i^{1/2} \tag{19}$$

where i is the rank of the mode, $\mathbf{D}_i$ is a diagonal matrix composed of the background error variances for mode i, and $\mathbf{C}_i$ is a symmetric correlation matrix generated for instance from one space-time correlation function per mode. It can be also seen that the $\mathbf{Br}^i$ and $\mathbf{C}_i$ matrices now only contain horizontal-time covariances and corrections on the model grid.

This statistical model of errors has of course its limits. We can certainly restore some of the spatial variations of the order reduction by splitting the geographical domain into regions, and allowing the background error variances to be zero everywhere except in the region where we want the corresponding EOF to apply. However we cannot allow the background error features to be tilted vertically because of the separation into horizontal and vertical structure functions. Echevin et al. (2000) show the limits of such a statistical model in the coastal areas, but in the open ocean far from frontal regions the limitation should not be too severe.

The ROOI forms of the gain (12) and (15) still apply here. The only algorithmic difference with the previous section is the background error model, now expressed by (17-19).

15.2.4 Local statistical inverse

One last practical decision we will make will be to solve a local problem at each grid point using the neighboring data only. This is again common in numerical weather forecasting (see e.g. Gustavsson, 1981; Lorenc, 1986). The procedure involves data selection based on space-time correlation between the analyzed variable and observations, but also on the relative multivariate correlations between observations. One reason for doing so is that in practice the quality of the analysis depends to a large extent on which observations are selected to influence the analysis at each gridpoint. However, by doing so, one better "adapts" to the local observations, and the splitting of the workload on massively parallel computers becomes straightforward, permitting the solution of very high-dimensional problems.

The influence functions of the observations in OI are the columns of the $\mathbf{B}^f\mathbf{H}^T$ matrix in (6), which are called the *representers* of the observations. From (11), the ROOI representers are the columns of $(\mathbf{S}^T\mathbf{Br}^f\mathbf{S})\mathbf{H}^T$. Assuming that we use vertical order reduction as in the previous section and that the spectral error correlations in

$\mathbf{C}_i$ vanish after some distance, it is clear that the ROOI representers can only be made local if the observation operators are local. In practice, there are three kinds of observation operators in the ocean: *vertical* observation operators (e.g. temperature, salinity, density profiles), operators based on *local gradients* (e.g. geostrophic velocities), and *global* observation operators (e.g. surface pressure and sea-level anomaly in most rigid-lid models). Therefore if we wish to solve local inverse problems we have to exclude the last category. However, the sea-level anomaly (SLA) observation operator can be considered to be local in some cases such as the following:

(1) If the model is a rigid-lid model, if only the geostrophic contribution to SLA is subject to errors in the model, and if errors can be considered small below some level h_0, the so-called geostrophic, flat bottom limit of the surface pressure formulation (e.g. Pinardi et al., 1995) can be used locally relate changes in SLA to changes in barotropic streamfunction and density:

$$\delta SLA = \frac{f}{gh(x,y)}\delta\psi + \frac{1}{\rho_0 h_0}\int_{-h_0}^{0}(z' + h_0)\delta\rho dz' \tag{20}$$

with the usual notations. By differentiating the equation of state, changes in density can be formally related to changes in temperature and salinity . Expression (20) stays mathematically valid close to the equator with $f \to 0$ but alternate approximations can be introduced in the low latitudes.

(2) If SLA is part of the full state (in that case the vertical EOFs must include the SLA).

(3) If the model is a free-surface model.

In the case of relation (20) at non-equatorial latitudes, the full state and hence the vertical modes must contain the baroclinic variables and the barotropic transport. However the vertical modes calculated from hydrographic databases will only contain baroclinic information. In that case, one can either add a zero barotropic component to all baroclinic EOFs and adding a barotropic EOF to the set (but the multivariate character will be lost), impose a level of no motion in (20), or consider calculating the EOFs from an alternate data set such as model outputs (this is what is done below).

15.3 Assimilation of satellite altimetry in 1993-97 in the Mediterranean

15.3.1 How is the assimilation configured in the Mediterranean model

The principles in section 15.2 have been put to work in the form of a multivariate, reduced-order optimal interpolation Fortran code named SOFA (for "System for Ocean Forecasting and Analysis"). It is a flexible and model-independent code. We will now review a few examples using SOFA with an eddy-permitting numerical model of the Mediterranean large-scale circulation, in years 1993-1998. The

SOFA data assimilation code has is also used for projects outside the Mediterranean (chiefly North Atlantic and Global).

The Mediterranean General Circulation Model (GCM) is an eddy-permitting 1/8°, 31-level version of MOM-1 implemented by both the Toulouse and Bologna teams (see e.g. Pinardi and Masetti, 2000). It is forced with ECMWF 6-hour operational analyses: 10-meter winds, air temperature to calculate interactive heat fluxes. Baroclinic exchanges are allowed at Gibraltar and Sicily Straits. Constant vertical diffusion is used for efficiency. In all experiments reviewed below, a 5-year spin-up from climatology was carried out, followed by one year forced with ECMWF analyses before the assimilation experiments start. Reanalyzed ERA forcing fields were avaible up to 1993, the operational analyses starting in 1994. The salinity relaxation at the surface typically uses the MED5 MODB database (www.modb.oce.ulg.ac.be).

The sequence of operations is the following: (1) the MOM model runs for a week while collecting misfits between model and observations (the "innovation"); (2) the forecast model fields (the "background solution") are corrected by SOFA from the innovation vector along the lines of the algorithm described in the previous section, i.e. OI on a base of pre-calculated vertical EOFs. Estimates are calculated at each model grid point using the local inverse approach. The time differences between the observations and the analysis time are taken into account in the estimates. Then the sequence restarts as MOM is re-initialized from the corrected fields and a Euler foward time step. Only past data are used for each correction step. Therefore the statistics of the misfits for any cycle are truly a measure of the capacity of the assimilating model to predict the data which have *not yet* been assimilated. We will use second-order misfit statistics, spatially averaged over regions or temporally averaged, to illustrate the benefit of the assimilation in the examples below.

In SOFA, the model grid must be nonsingular (no poles) and can be periodic in one dimension; it is not restricted to being rectangular, although it is in the cases studied here. In the calculation of the innovation vector in step (1), the full observation operator $H(\ )$ is used. This can be a global, non-linear operator. On the contrary, in the analysis step (2), local, tangent linear observation operators as discussed in section 15.2.4 are used to form the ROOI gain. When needed to calculate misfits, the model variables are interpolated at the data points by means of hyperbolic paraboloid functions (which come back to bilinear interpolation in the case of a rectangular grid). Data search on the grid is performed by a Kd-Tree algorithm (e.g. Skiena, 1997).

We schematically present the interface between SOFA and the Mediterranean GCM. The state vector in MOM, i.e. the list of independent and dependent variables, is defined as

$$\chi = \{\psi, \mathbf{u}_c, w, T, S, \rho\} \tag{21}$$

i.e. the transport streamfunction, baroclinic horizontal velocities, vertical velocity, temperature, salinity, and density. The *estimation* problem is entirely written in terms of transport streamfunction, temperature and salinity defined on depth levels. The *estimation* state vector (our state vector for all practical means) is

$$\mathbf{x} = \{\psi, T, S\} \tag{22}$$

The EOFs we use below are therefore trivariate in these variables. What this means is that for instance temperature measurements will influence model temperature, but also salinity and streamfunction, in an empirically coherent manner. Once the trivariate correction has been calculated, we correct the density, apply geostrophy on the increments (hence assuming no error on the ageostrophic part – a very common, robust assumption), and perform ancillary restart tasks such as recalculating the value of streamfunction around islands and applying convective adjustment. A simple scheme was used to update the model error variances in $\mathbf{D}_i$ from one analysis to the next (adding a constant to the previous analysis error variance as in De Mey, 1994), whereas the space-time error correlations and observational error variances were kept fixed, in accordance with standard OI practice. The guess error correlation (in EOF modal space) derived from a space-time analytical model with isotropic and homogeneous correlation radii of 100 km and 30 days for all the modes. The observational errors were assumed to be uncorrelated in space and time.

We present simulations assimilating separately two kinds of observations: Sea-Level Anomaly (SLA), using form (20) of the local observation operator, and temperature profiles. SLA data were real altimeter data processed by CLS (www.cls.fr), while XBT data were simulated in a twin experiment setup. The difference in surface steric heights between the model and the "truth" observed by the altimeter was accounted for by removing the average of the innovation vector at each time step of the filter. The reference or "Mean Sea-Surface Height" (MSSH) for altimetry was a model annual or multi-year average (depending on the case). In addition, a special processing was used by CLS to correct the altimeter SLA for the long-wavelength, high-frequency departures from the inverted barometer effect.

15.3.2 Assimilation of simulated XBT profiles

We first present a multivariate assimilation experiment of XBT profiles as well as the effect of using regional, regime-dependent EOFs. The twin experiment is made of three simulations:

- The *control* simulation is initialized on Jan 1, 1998 and is forced by ECMWF 1998 winds and interactive fluxes. Fifty randomly-distributed XBT profiles are simulated from this run every week at all model levels between surface layer (120m) and 1000m.
- The *free* run is initialized on Jan 1, 1993 (wrong initial conditions) but uses the correct 1998 forcings. The initial conditions are obtained from a simulation using ERA (ECMWF Re-analysis, www.ecmwf.int/research/era/index.html)

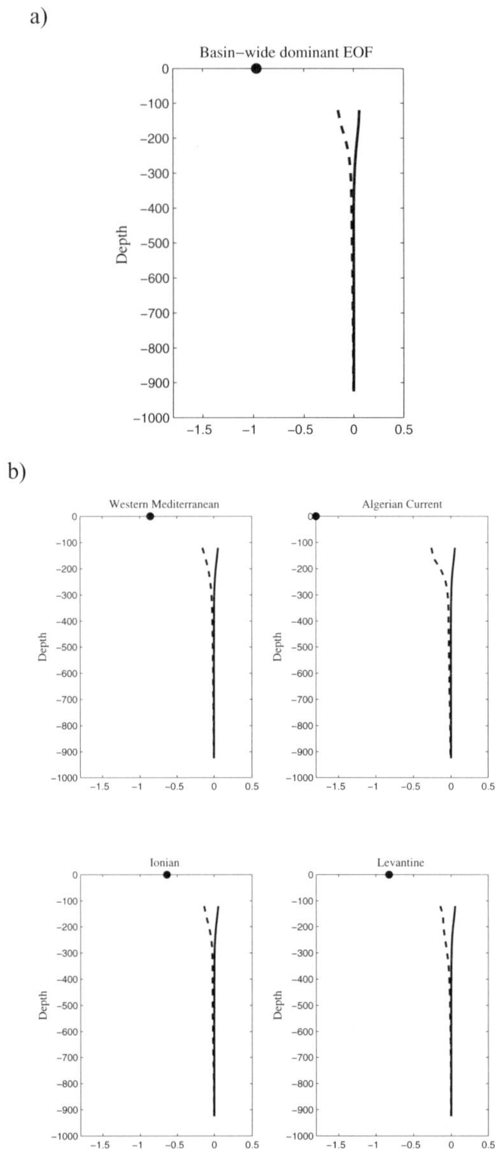

Fig. 15.1 Trivariate EOFs calculated from a model simulation in 1998: temperature (dotted, °C), salinity (solid, psu), barotropic streamfunction (dot at z=0). (a) Basin-wide dominant EOF. (b) Regional dominant EOFs. Regions are: Algerian Current (1), Western Mediterranean (2), Ionian (3), Levantine (4). X-axis: °C, *psu, Sv.* The graduation are the same for all three units.

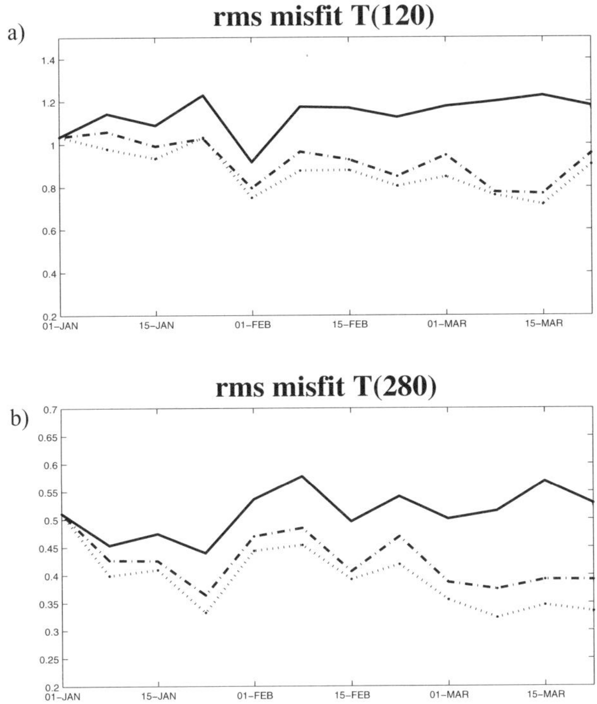

Fig. 15.2 Mean-square 7-day-range prediction error for temperature at two different depths over the first 3 months of 1998 for several simulations initialized with the Jan 1, 1993 conditions: free run (solid line), assimilation of simulated XBT profiles using one basin-wide EOF (dash-dot line), assimilation of profiles using regional EOFs (dotted line). (a) 120m; (b) 280m. Unit: °C.

winds up to the end of December 1992. The free run is meant to show the ability of the model to "fall back on track" by itself, and therefore to serve as a reference for the assimilation diagnostics.

- The *assimilation* run is the same as the free run, except that it assimilates the simulated XBT profiles with a weekly cycle (7-day "forecast", then analysis, then repeat). Temperature was assumed to be available at all model levels in the profiles.

As in (22), our EOFs are trivariate in temperature, salinity and transport streamfunction. They are calculated from the control run. We will consider two sets of EOFs: one dominant basin-wide EOF (Fig. 15.1a), and four regionally dominant EOFs (Fig. 15.1b). The latter case is by no means limiting in terms of the number of regions; it is meant to show on a simple example the benefit of tuning the statistics to the regional regimes and physics. As noted in section 15.2.3, we define regional masks corresponding to each one of the four regions: Algerian Current

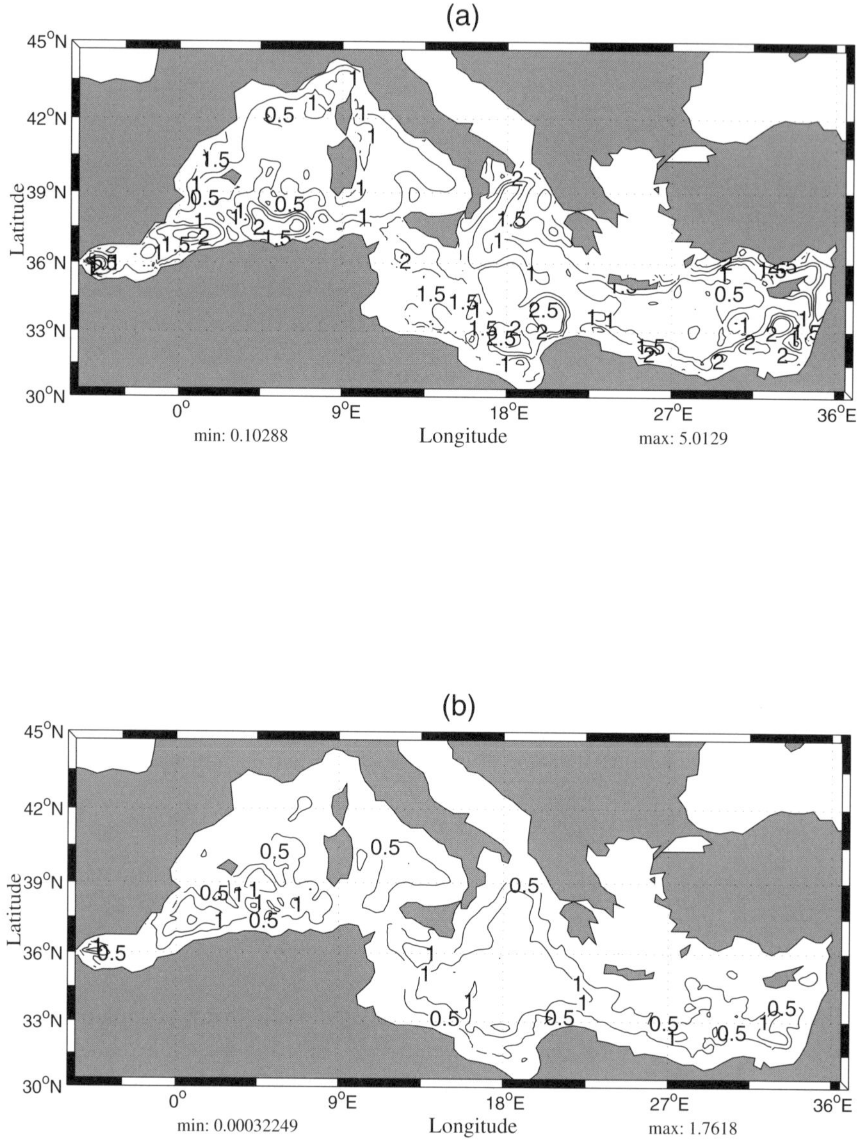

Fig. 15.3 Maps of scalar property departures from the control run (3-month rms, the closer to zero the better). The assimilation run uses regional EOFs. (a) T_{free}-$T_{control}$ at 120m, to be compared to (b) T_{assim}-$T_{control}$ at the same depth; (c) S_{free}-$S_{control}$ at 120m, to be compared to (d) S_{assim}-$S_{control}$ at the same depth.

(AC), Western Mediterranean excluding AC, Ionian basin, Levantine basin. In effect, at each model grid point, we solve a 4-EOF problem ($\mathbf{S}^T$ has 4 columns), but the background error variances in $\mathbf{D}_i$ are zero except on the corresponding mask. An hyperbolic tangent ramp in the background error variances provides a smooth connection between regions.

In the present case where only one EOF is used at one given location, the EOFs were not defined in the surface layer, since the Ekman layer variability is largely decorrelated from the deeper variability and would require more EOFs. Therefore the surface layer variables are not directly changed by the assimilation. However they are influenced by the assimilation, e.g. through barotropic velocities.

Table 15.1 shows the percentages of variance explained by the various first EOFs in model outputs, as well as the coefficients used to scale the variables (matrix Σ in (15)). For one given variable, all levels were scaled with the same coefficient. The modes shown in Fig. 15.1 are the columns of $\Sigma \mathbf{S}'^{+}$. It can be seen from Fig. 15.1 that the dominant modes are surface-intensified and that the signs are pretty homogeneous across all regions. Therefore our test case with only one basin-wide EOF is certainly not widely erroneous. The fact that T and S are of opposite sign is mostly the signature of the Modified Atlantic Water's important contribution to the basin-wide variability. From Table 15.1, it can also be seen that in the model simulation the barotropic mode plays a specially important role in the Western basin. The Algerian Current area shows the largest variabilities in all variables. In the Eastern basin, the spectrum of eigenvalues is not as "red" as in the Western basin (more significant EOFs, more complex system), and the salinity variability increases as one moves eastward.

Figs. 15.2 and 15.3 show assimilation results with reference to the free simulation. From Fig. 15.2, the assimilation does improve the capability of the model to predict the temperature on the short term. In addition the regional EOFs do an even better job at reducing the residual error, which is of the order of 0.35°C at 280m (down from 0.55°C). It can be seen from Fig. 15.3 that the assimilation improves the temperature (which is the assimilated variable) in most regions, but also the salinity (not assimilated) in some regions such as the Levantine basin and the axis of the Algerian Current. These results could very likely be improved by fine- tuning the regions and parameters.

15.3.3 Assimilation of altimeter data in 1993-97

We now illustrate the influence of altimeter data assimilation in the same GCM, in a 5-year period simulation in 1993-97. For this purpose, we use TOPEX/POSEIDON and ERS-1/2 along-track sea-level anomaly from CLS (G. Larnicol, pers. comm, 1999). The processing includes long-wavelength error removal. The data are not gridded prior to assimilating. The assimilation is carried out with one model-derived trivariate basin-wide EOF as in Fig. 15.1a. As before, no assimilation goes on in surface layers (0-100m) and at depth (below 850m) except for barotropic velocities. The average sea-level is not changed in the model. The assimilation starts on Jan 1, 1993 and proceeds with a 7-day cycle to Dec 31, 1997.

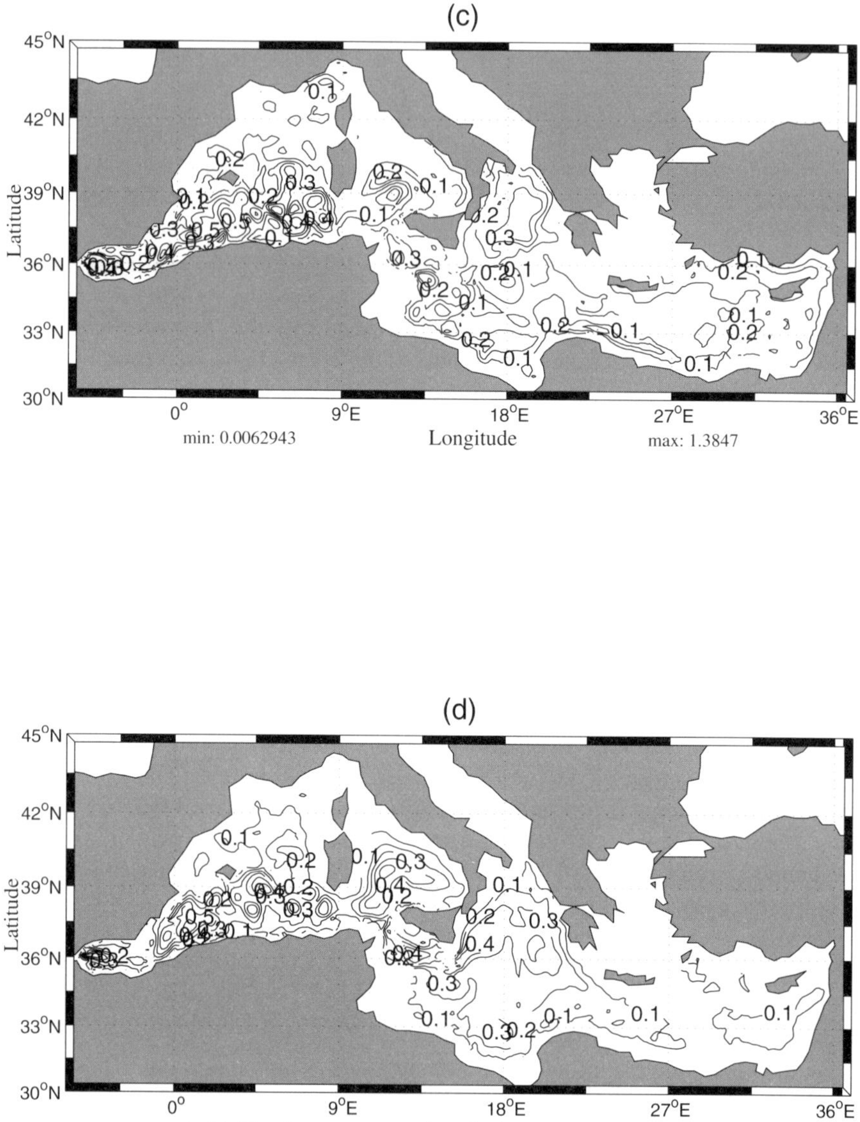

Fig. 15.3 (continued).

	Explained variance (%)	Matrix Σ coefficients σ_ψ (Sv)	σ_T (°C)	σ_S (psu)
Basin-wide	**58.5**	**1.885**	**0.611**	**0.122**
Western Med.	62.4	1.966	0.451	0.099
Algerian Current	70.5	3.612	0.856	0.250
Ionian	44.2	1.300	0.746	0.067
Levantine	43.5	1.049	0.516	0.081

Table 15.1 First EOF ensemble statistic.

No correction is calculated in the Aegean sea and on the Tunisian shelf because of the shallow depth and complex coastlines.

Fig. 15.4 shows a salinity field at 50 meters with and without assimilation for November 4, 1993. The assimilation is able to correct for a few deficiencies of the model and forcings, such as the too high Atlantic water flux between the Balearic Islands and the Spanish coast, and the unrealistic large scales present everywhere in the basin (see e.g. the anomalous structures around 29°E, 33°N). The Ierapetra gyre (26°E, 35°N) can now be distinguished. The Atlantic water flow is closer to the African coast and more realistic in the assimilation case. One important effect of assimilation is to bring in the small scales missing from the wind forcing and imperfect instability physics.

Fig. 15.5(a,b) show that the assimilation does improve the capability of the model to predict the sea-level anomaly on the short term. The spin-up of the assimilation for SLA is O(1 month). The skill is weaker during winter where the wind plays a specially important role in the ocean dynamics. This confirms the fact that the ECMWF forcing fields and therefore the wind stress and fluxes in the model (Fig. 15.5c-e) lack short-scale information. The along-track altimetry assimilation seems to be able to (at least partly) introduce those scales, but the largest effect of the assimilation, in particular for mesoscale features, is mostly in summer (Fig. 15.5f). A long-term trend of sea-level anomaly appears in the Eastern basin (Fig. 15.5b); it is partly corrected by the assimilation. The heat storage time series (Fig. 15.5g,h) shows a significant interannual drift, mostly in the Eastern basin, not corrected by assimilation, but clearly not caused by it. This may be caused by the constant vertical diffusion scheme used in our simulations.

Figure 15.6 shows longitude-time plot of sea-level anomaly in the model with assimilation. The longitude-time plot at 38°N (Fig.15.6a) shows the space and time scales of the meanders and eddies in the Algerian Current. Some of the structures persist for several months and Westward propagation is clear on some occurences

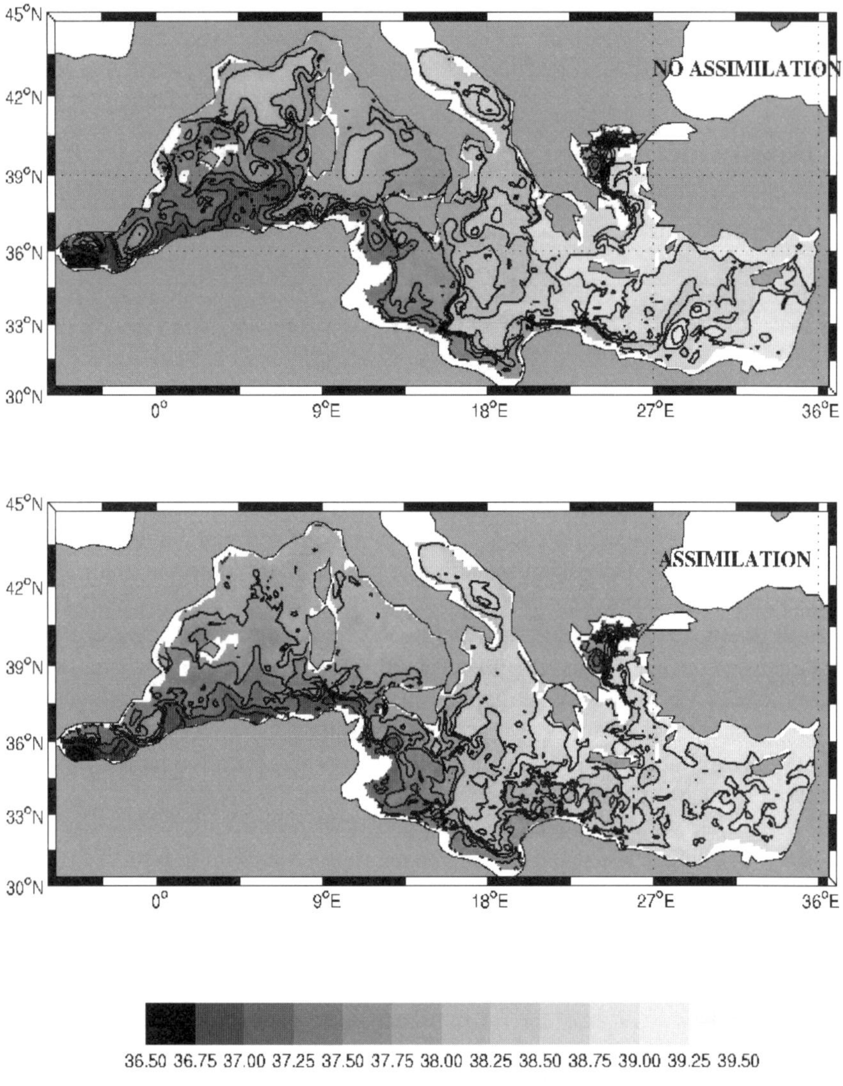

Fig. 15.4 Salinity field at 50 meters, November 4, 1993. Top: run without assimilation, bottom: run with assimilation of TOPEX/POSEIDON and ERS-1 altimetry. Unit: *psu*.

(e.g. 8°E/Jan.97). The longitude-time plot at 33°N (Fig. 15.6b) shows the annual and interannual variabilities of the Mersa-Matruh and Shikmona systems as well as a recurrent winter meander off Ras'Amir (22°E) propagating into the Ionian basin.

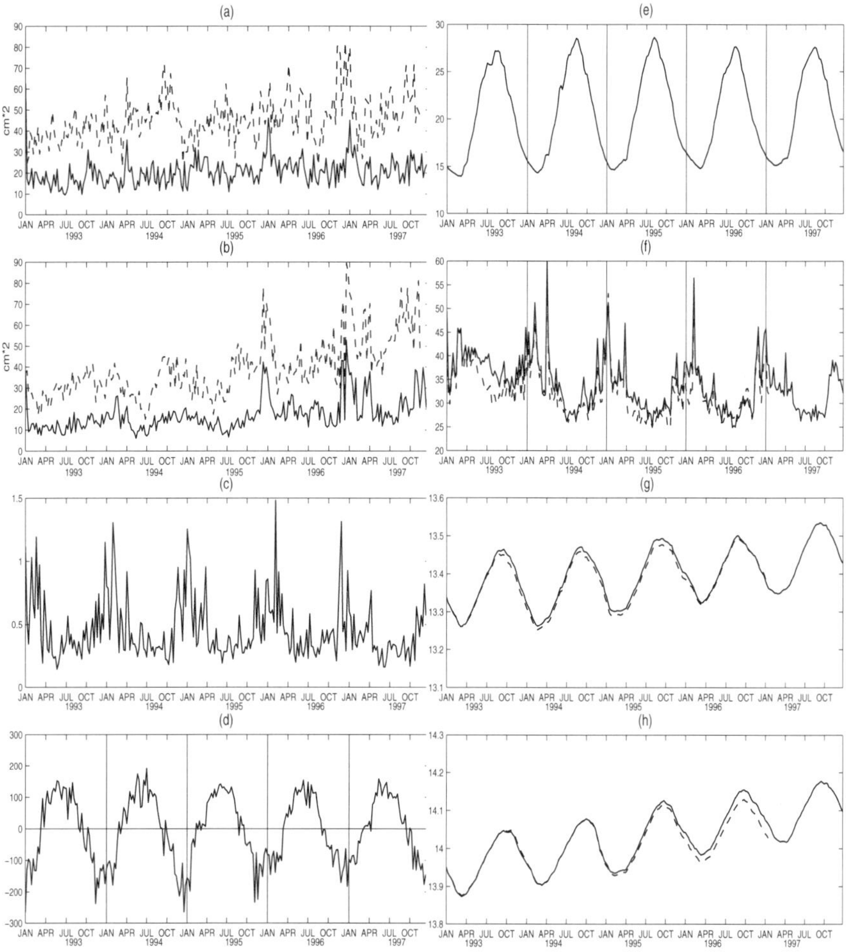

Fig. 15.5 Time series in 1993-97 for the altimetry assimilation run. Norm of innovation vector in the (a) Western, (b) Eastern basin (unit: cm^2); (c) mean wind stress (unit: $dyne/cm^2$); (d) mean heat flux (unit: W/m^2); (e) mean sea-surface temperature (unit: °C); (f) mean kinetic energy (unit: cm^2/s^2); mean temperature in the (g) Western basin), (h) Eastern basin (unit: °C). All plots: solid line: assimilation run; dotted line: no assimilation (reference).

The salinity section at 18°W in the Ionian basin in winter 1993-94 with assimilation (Fig.15.7a) shows a core of Levantine Intermediate Water (LIW) near 400-500m depth and the Modified Atlantic Water closer to the surface to the South. In the 5-year 1993-97 average of winters (Fig. 15.7b), the vertical gradient and LIW characteristics have been diffused with respect to Fig. 15.7a. This is probably due to the combined effects of constant vertical diffusion, grid resolution, horizontal diffusion, vertical resolution and horizontal eddy-induced mixing in the assimilation run.

15.3.4 Influence of horizontal and vertical resolution in the assimilation

Prior to using the 1/8°, 31-level model, the team had started configuring and

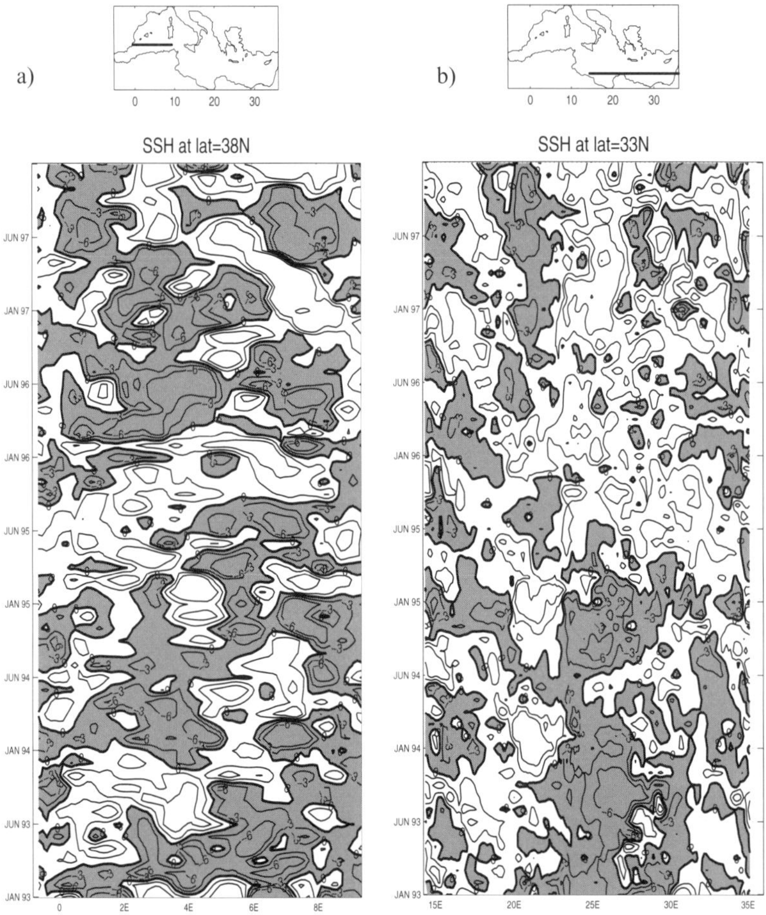

Fig. 15.6 Longitude-time plots of sea-level anomaly in the model: (a) at 38°N in the Algerian Current; (b) at 33°N across the Mersa-Matruh and Shikmona systems. Unit: *cm*. Contour interval: 3 *cm*. Negative anomalies are shaded.

testing the assimilation on a lower-resolution 1/4°, 19-level version of the same model (Benkiran, 1997; Benkiran et al., 1997). The beginnings were rather frustrating because the low-resolution model did not respond in a significant manner to configuration changes in the assimilation. It was found very difficult to test the ideas put forward in section 15.2 in such a configuration.

Moving to the higher-resolution model, even if still only eddy-permitting given the small scales of the Rossby radii in the Mediterranean, changed everything as far as the "response" of the assimilation system was concerned. In practice, the 1/8°, 31-level model was found to "keep" the information inserted by the assimilation. An example is given in Fig. 15.8, showing the Crete-Rhodes area with both resolutions. The small-scale but energetic Ierapetra gyre shows up at the at the southeastern tip of Crete (26°E, 34.5°N) at the end of summer almost every year. It is not present in the simulations without assimilation at any of the resolutions tested, because its origin is thought to be related to unresolved small-scale wind curl effects. The Ierapetra gyre is seen by satellite altimetry and corrected by assimilation. However the 1/4° model can barely resolve it (Fig. 15.8b) and is unable to evolve it dynamically. Assimilation in the 1/8° model (Fig. 15.8a) does a much better job; in particular, it is able to produce a much more realist guess of the gyre one week after correction.

15.4 A brief summary and perspective

The assimilated results show the general circulation variability of the Mediterranean in a way which can be improved, but which is coherent between variables and fairly realistic in many respects. In particular, recurrent seasonal patterns of variability as well as interannual variations of those patterns not present in the free simulation are evidenced in the '90s.

Tests of sensitivity to the choice of the "simplification operator" **S** as well as to more realist guess error correlation are underway, in particular as part of the Mediterranean Forecasting System Pilot Project (MFSPP). In MFSPP, altimetry as well as in situ measurements are being assimilated in real-time for demonstration purposes and to get more scientific insight into the relative benefits of the possible *modus operandi* and of the various observations. A higher-resolution (1/20°), truly eddy-resolving model of the Mediterranean is also being implemented in the framework of the French MERCATOR project (www.mercator.com.fr/html/information/frame_produit_fr.html). It is fully coupled with a North Atlantic model of a similar resolution, and uses the same assimilation approach.

The SOFA Fortran code is available free from the author (Pierre.De-Mey@cnes.fr). An alternate version of the assimilation code with isopycnal vertical representation as in Gavart and De Mey (1997) has also been implemented in the 1/8° model, where isopycnal vertical EOFs are used as "lifting-lowering" modes. The SOFA algorithm has very simplified error propagation characteristics. We have performed ensemble integration tests to show the potential of the method for retrieving the yet unknown forecast error variances. The method has been applied in the Western Mediterranean to various situations in winter and spring of 1993, and also in a limited-area coastal model in the Gulf of Lyons and Liguro-

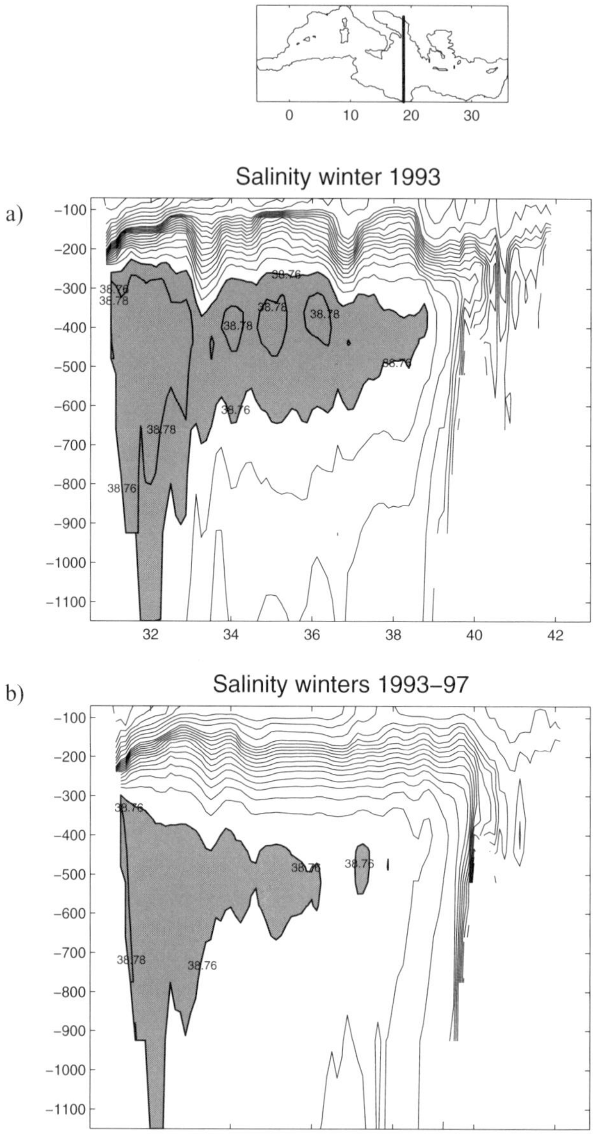

Fig. 15.7 Salinity sections at 18°W (Ionian basin) in the model: (a) winter 1993 average; (b) 5-year average of the 1993-97 winters. Unit: *psu*. Contour interval: .02 *psu*. Salinities higher than 38.76 *psu* are shaded.

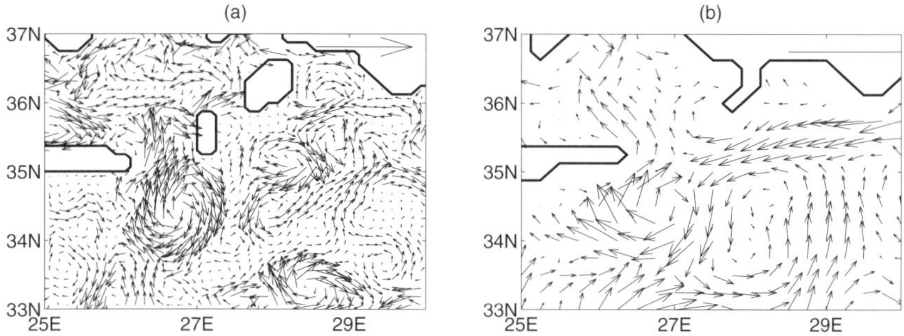

Fig. 15.8 50-m current velocities in (a) 1/8°, 31-level and (b) 1/4°, 19-level configurations in the Crete-Rhodes area on Oct. 22, 1993. In both cases, the model was last corrected on Oct. 15 using the same altimeter data.

Provençal-Catalan current (Echevin et al., 2000). In the GCM ensemble runs, initial conditions have been randomly perturbed along the dominant vertical EOF and an ensemble of 21 model runs have been conducted to forecast time. Not too surprisingly, the largest forecast error variances in the Western basin were found in the high variability areas such as the Algerian current. The ensemble method shows barotropic error growth in the Algerian Current and in the Alboran gyres (a result compatible with the results shown in this chapter), and baroclinic/tracer error growth in the Lyons gyre where deep convection occurs in late winter 1993. The method is very costly, but is very general in exploring the validity of the simple hypotheses usually assumed in OI.

Acknowledgments

Support of the authors under European Union contracts MAS3-CT96-0051, ENV4-CT98-0744, and MAS3-CT98-0171 is acknowledged.

Appendix A. Properties of the order-reduction operator S

Let us call an "increment vector" any variation of that vector, i.e. $\delta\mathbf{x}$ a variation of the state $\mathbf{x}$. The problem at hand is to find the state increment that minimizes some optimality criterion. In the order-reduction approach followed in this chapter, the optimal solution is sought in the subspace defined by the simplification operator. The full-space state increments are modelled as:

$$\delta\mathbf{x} = \mathbf{S}^{+}\delta\mathbf{w} + \nu \tag{A1}$$

where $\mathbf{x}$ is the full-space state vector, $\mathbf{w}$ is the reduced-space state vector, and ν is a null-space vector. Let us examine what this modelling translates into as far as the influence of observations and forecast are concerned in an Extended Kalman Filter (EKF) and in Optimal Interpolation (OI).

The classical observation equation becomes:

$$\mathbf{y}^{o} = H(\mathbf{x}^{t}) + \varepsilon \tag{A2}$$

where $\mathbf{y}^{o}$ are the observations, $\mathbf{x}^{t}$ is the "true" state, H is the observation operator, and the observational noise process ε is assumed to have zero mean and covariance matrix $\mathbf{R}$. Using (A1) in (A2) for the data space increments:

$$\delta\mathbf{y} = \mathbf{H}\mathbf{S}^{+}\delta\mathbf{w}^{t} + \mathbf{H}\nu + \varepsilon \tag{A3}$$

where $\mathbf{H}$ is the tangent linear observation operator. The reduced state is therefore seen through the reduced-order observation operator $\mathbf{Hr} = \mathbf{HS}^{+}$, and a new observational noise process $\mathbf{H}\nu$ supplements the original observational noise term.

In the EKF and OI, the complete nonlinear model M is used to produce a state forecast $\mathbf{x}^{f}$:

$$\mathbf{x}^{f}(t+\delta t) = M(\mathbf{x}^{a}(t)) \tag{A4}$$

where $\mathbf{x}^{a}$ denotes the analyzed estimate and δt is the filter time step. Again using (A1) in (A4) for the reduced state space increments:

$$\begin{aligned} \delta\mathbf{w}^{f}(t+\delta t) &\equiv \mathbf{S}\delta\mathbf{x}^{f}(t+\delta t) \\ &= \mathbf{SMS}^{+}\delta\mathbf{w}^{a}(t) + \mathbf{SM}\nu^{a} + \mathbf{S}\zeta \\ &= \mathbf{SMS}^{+}\delta\mathbf{w}^{a}(t) + \mathbf{SM}\bar{\mathbf{S}}^{+}\delta\bar{\mathbf{w}}^{a} + \mathbf{S}\zeta \end{aligned} \tag{A5}$$

where $\mathbf{M}$ is the dynamical resolvent (or tangent linear model) and ζ models the internal sources of errors (system errors in the EKF formalism). The coupling of the null space errors with the reduced space errors via the dynamics, illustrated by the conjugacy operator $\mathbf{SM}\bar{\mathbf{S}}^{+}$, is found in (A5) to provide a new process which can for instance be included in the system or background errors.

From the above results we can list some desirable properties for choosing the simplification operator $\mathbf{S}$:

(1) The $\mathbf{HS}^+$ product must be well conditioned, i.e. the columns or *modes* of $\mathbf{S}^+$ must be chosen in the observable subspace, which comes back to the requirement that $\mathbf{Hr}$ be of full rank (all eigenvalues of $\mathbf{Hr}^T\mathbf{Hr}$ must be nonzero). One example of the exploration of a reduced-order observability problem is the work of Gavart and De Mey (1997) on isopycnal EOFs and their signature in sea-level anomaly.

(2) The $\mathbf{SMS}^+$ product must be well-conditioned, i.e. the columns or *modes* of $\mathbf{S}^+$ must be chosen in the control space of the full-state problem. For a nonlinear model, however, the properties of the eigenvectors of $\mathbf{SMS}^+$ are not sufficient, and one must turn to singular vector analysis (e.g. Ehrendorfer and Tribbia, 1997). In the case of optimal interpolation, which does not have a built-in error propagation scheme, this discussion may seem out of scope. However it is still important to make sure that the order reduction remains efficient and meaningful even after adjustment and time evolution by the dynamics.

(3) The observability of the null space must be marginal (the signature of null space $\mathbf{H}v$ must be a "small" contribution to $\delta\mathbf{y}$; the eigenvalues of $\bar{\mathbf{H}}\mathbf{r}^T\bar{\mathbf{H}}\mathbf{r}$ must be "small", with $\bar{\mathbf{H}}\mathbf{r} = \mathbf{H}\bar{\mathbf{S}}^+$) and close to normality so as to easily include the null-space signature in the observational noise. Otherwise the observations will not properly be taken into account by the reduced-order scheme. This condition and the condition (1) together could for instance be automatically achieved by an adaptive method (such as found in Blanchet et al., 1997), applied to the adaptive estimation of $\mathbf{S}$ from the sequence of innovation residuals.

(4) The dynamical coupling between reduced space and null space errors (illustrated e.g. in Fukumori et al., 1993) must be small and close to normality so as to easily include the corresponding error term in the system errors (for the EKF) or background errors (for OI). Again for a nonlinear model singular vector analysis or an ensemble approach may be useful in exhibiting the most rapidly growing errors and their correlations in both subspaces. There are at least two contexts in which the dynamical coupling is weak: (a) if the null space only contains stable or neutral modes or slowly growing errors; (b) if the *physical* processes in the null space and reduced space are only weakly coupled for physical reasons (as in Gavart and De Mey, 1997). Both contexts can actually be the same.

(5) Other properties can be desirable, e.g. the full-state stability of the filter. OI methods usually have no useful properties in that respect, but the nature of $\mathbf{S}$ and in particular the null space are expected to play an important role on full-state stability.

16 ENSO Predictions with Coupled Ocean Atmosphere Models

MARTIN FISCHER
Consiglio Nazionale delle Ricerche, Istituto per lo studio delle Metodologie Geofisiche Ambientali

16.1 Introduction

The strongest signal on the short range climatic time scale, ranging from a few months to several years, is the El Niño/Southern Oscillation Phenomenon (ENSO). It is characterized by a weakening of the trade winds along the equator and a huge redistribution of heat from the western to the eastern tropical Pacific. The impacts of ENSO are felt worldwide through a disruption of the atmospheric general circulation pattern (Ropelewski and Halpert 1987), which leads, for instance, to severe droughts in Northwest Australia and South East Asia during an El Niño event. The Northeast region of Brazil and Zimbabwe are other regions where rainfall variations are highly correlated with ENSO indices and there are many other examples. Off the Peruvian coast, the marine ecosystem is directly affected by the oceanic variations in upwelling, which has a large impact on the fishery industry. For all these regions, reliable ENSO forecasts would offer decision-makers an opportunity to take account of anticipated climate variations, in order to reduce impacts of ENSO on the economy. Thus, it is a logical consequence that the study of ENSO predictability has become a field of major research.

ENSO can be regarded as a slow oscillation of the coupled ocean atmosphere system. According to the theory of the *delayed action oscillator* (Schopf and Suarez 1988) the evolution of warm and cold events is determined by the propagation of equatorial waves and their reflection at the eastern and western boundaries, which implies that ENSO is inherently predictable on time scales up to a few years.

Since ENSO is a phenomenon of the coupled system, coupled ocean-atmosphere models are required for its prediction. Such models have been developed over the last 10 years by different groups and applied to ENSO forecasts (e.g. Cane et al. 1986, Barnett et al. 1988, Leetmaa and Ji 1989, Latif et al. 1994). ENSO forecasting is an initial value problem, i.e. the further evolution of the system depends highly on the initial state from which it started. Although ENSO is a phenomenon of the coupled system, its evolution is determined mainly by the ocean. Therefore the initialization can be considered as an initial value problem for the ocean. Thus, a very important task in carrying out an ENSO forecast is to determine the oceanic initial state as accurately as possible.

The usual way to initialize an ENSO forecast with a coupled ocean-atmosphere model is as follows. The oceanic model is integrated in an uncoupled mode and forced by observed wind stresses, heat fluxes, and fresh water fluxes which yields the initial state from which the coupled model prediction is started. The quality of the wind stress field over the whole initialization period of about one to two years is crucial for the quality of the initial state, and thus for the forecast.

Assimilating oceanic observations during the initialization phase yields an additional source of information, and may thus compensate for errors and/or uncertainties in the forcing data or even for model errors. Over the last few years several groups started to develop integrated data assimilation/coupled general circulation model forecast systems and investigated the impact of different types of ocean measurements (Ji et al. 1997, Rosati et al. 1996, Fischer et al. 1997).

The paper is organized as follows. In section 16.2 a description of ENSO and the impact of ENSO on climate and economy is given, and the possible predictability of ENSO is discussed. In section 16.3 different types of ENSO prediction models are presented and in section 16.4 we discuss several initialization methods and present results from different groups. Especially the assimilation of oceanic surface observations like sea surface height and sea surface temperature is discussed in more detail. This is an important issue, since these data can be measured with satellite based instruments which provide a complete and regular coverage in space and time. In section 16.5 the interpretation of forecast results is addressed and examples for skill measurements and corrections of systematic errors are given. The paper concludes with a summary in section 16.6.

16.2 The El Niño / Southern Oscillation Phenomenon

16.2.1 Description

The El Niño/Southern Oscillation phenomenon is the strongest climate signal on time scales of a few months up to several years. It is characterized by a weakening of the trade winds in the equatorial Pacific and a huge redistribution of heat from the western to the eastern equatorial Pacific.

In Fig. 16.1 the Pacific ocean is presented schematically in two different states. During normal conditions, relatively warm surface water is driven from the east Pacific to the west Pacific by the trade winds. Thus an east-west pressure gradient is built up, which is in the depth balanced by a west-east tilt of the thermocline. The depth of the thermocline is about 150 to 200 meters in the west Pacific and only about 30 to 50 meters in the east Pacific. Due to the westerly flow of surface water cold subsurface water upwells at the equator, and thus leads to the so called cold tongue region in the eastern equatorial Pacific. The temperature difference between the warm pool and the cold tongue region can reach values up to about six degrees centigrade. Under these conditions the coupled air-sea system is in balance (Fig. 16.1a .

If, for some reason, the trade winds get weaker, or the temperature maximum of the warm pool is shifted to the east, the whole system may get out of balance. Due to the shift of warm water further to the east, the center of the convective cell is also shifted to the east and thus decreases the strength of the trade winds. This leads

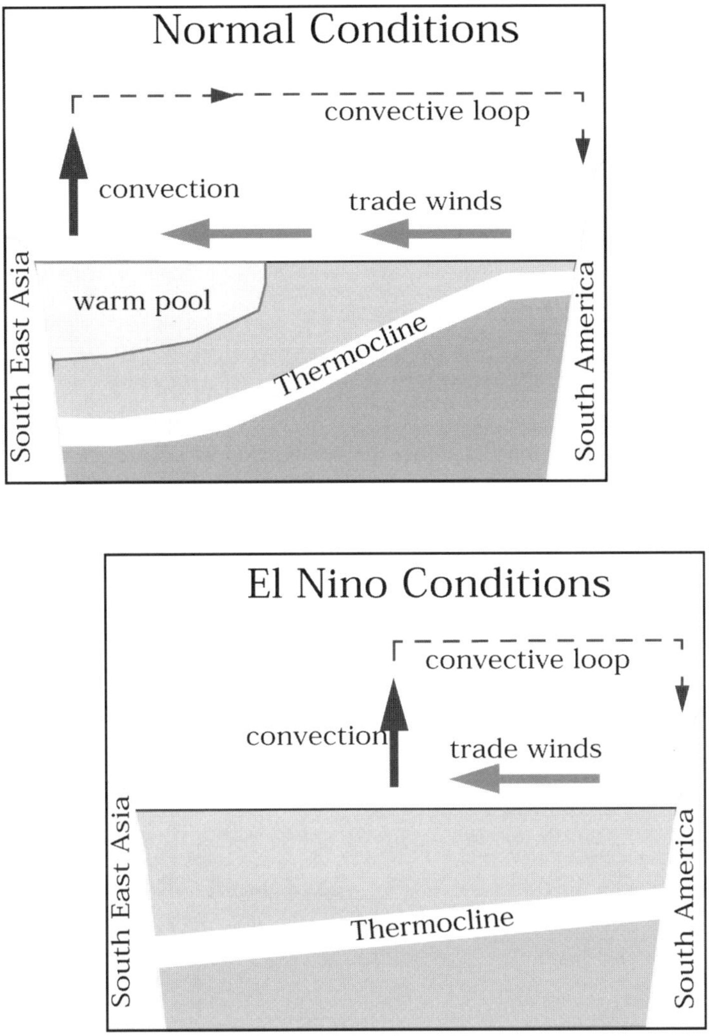

Fig. 16.1 Schematic description of the Pacific ocean during normal conditions and El Niño conditions. The coupled atmosphere ocean system at the equator is shown.

again to a temperature increase further east, and thus decreases the trade winds even more. This positive feedback between upper ocean heat content anomalies and wind stress anomalies finally may lead to a complete collapse of the east-west pressure gradient. In the end the slope of the thermocline almost disappears, the trade winds get very weak or even reverse their direction from west to east, as observed during the very strong El Niño 1982/83. A huge amount of heat is shifted from the west Pacific to the east Pacific, and the sea surface temperature increases

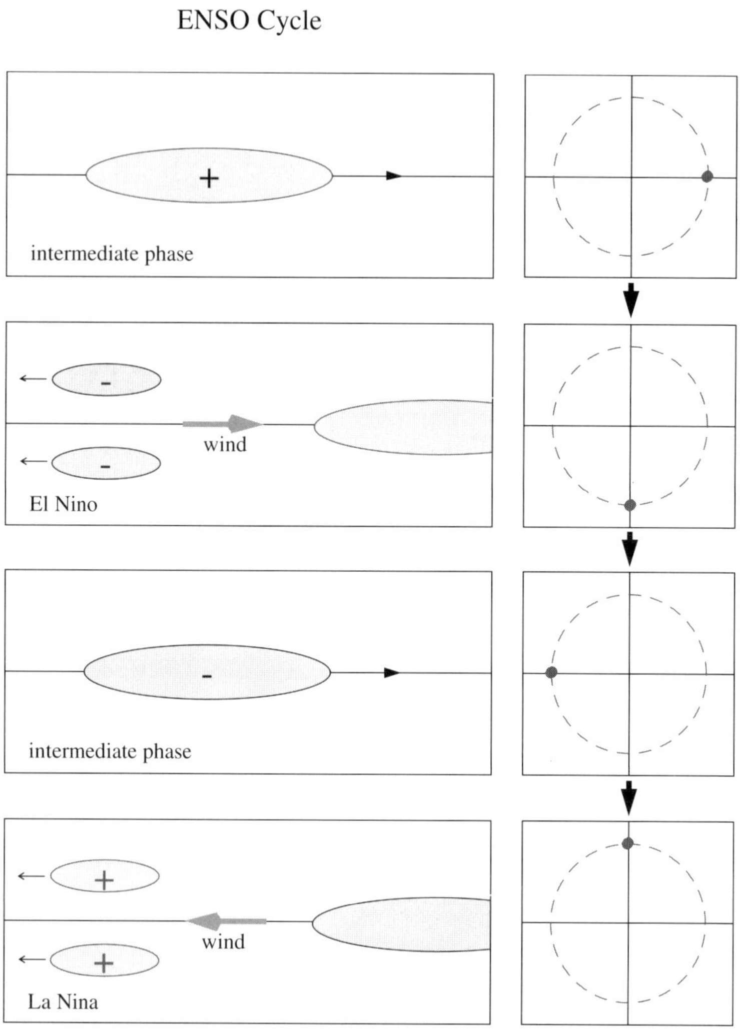

Fig. 16.2 Schematic description of heat content anomalies, following the theory of thr *delayed action oscillator*. Due to easterly wind anomalies an eastward propagating Kelvin wave packet and westward propagating Rossby waves are generated. The Rossby waves are reflected at the western boundary and cause a westward propagating equatorial Kelvin wave packet of the opposite sign. The wind arrows represent the dominant wind anomalies.

by about one to three degrees centigrade in the equatorial east Pacific. It is important to note that this phenomenon occurs due to unstable air-sea interactions. It is a feature of the coupled ocean-atmosphere system in the tropical Pacific. However, the inertia of the ocean governs the evolution and time scale of ENSO. The impor-

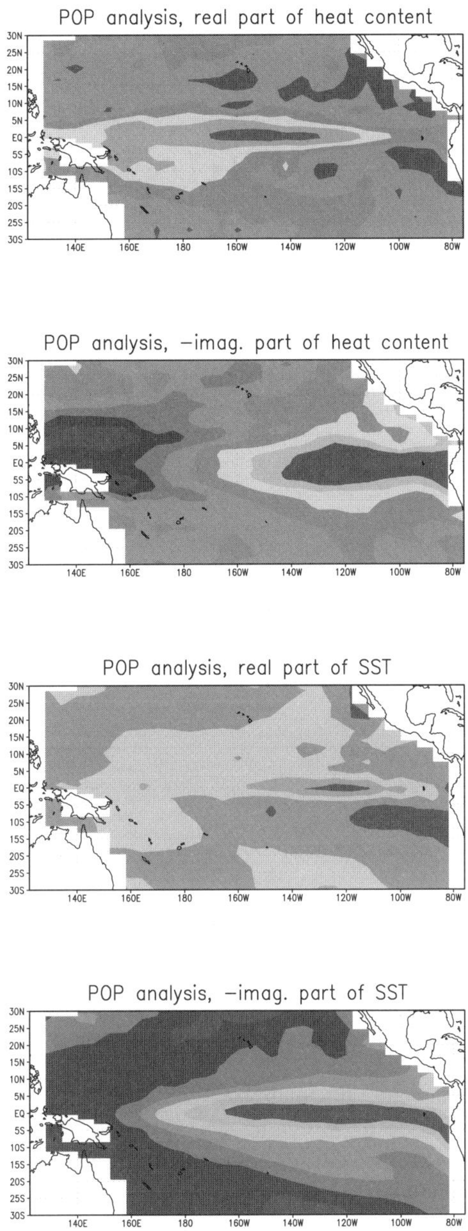

Fig. 16.3 Leading POP of a combined POP analysis of heat content anomalies, SST anomalies, and wind stress anomalies. The NCEP (National Center for Environmental Predictions, Washington, D.C.) ocean reanalysis and the FSU pseudo wind stress data were used. Heat content anomaly and SST anomaly patterns are presented.

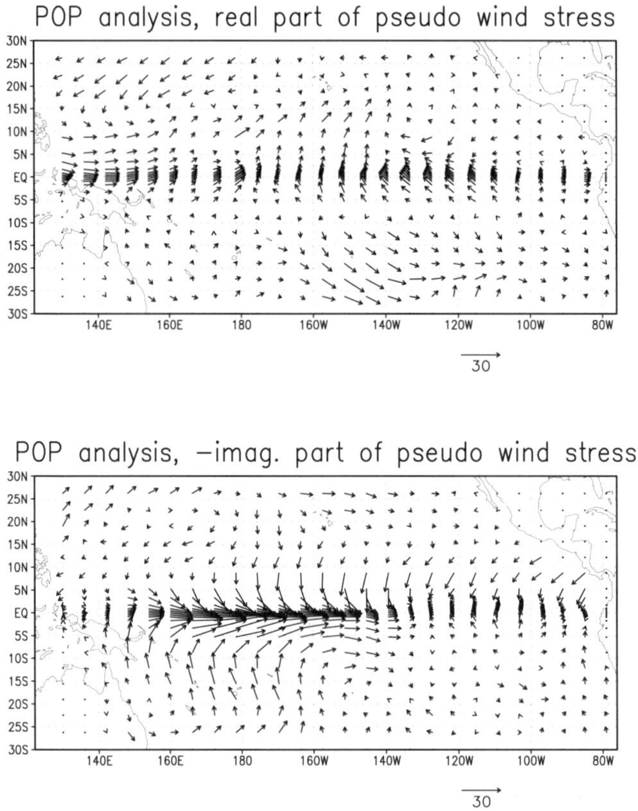

Fig. 16.4 Leading POP of a combined POP analysis of heat content anomalies, SST anomalies, and wind stress anomalies. The NCEP ocean reanalysis and the FSU pseudo wind stress data were used. Pseudo wind stress anomalies patterns are presented.

tance of the subsurface memory, as expressed by slow variations in the equatorial Pacific upper ocean heat content, for the generation of ENSO has been shown in many observational and modeling studies (e.g. Wyrtki 1985, Schopf and Suarez 1988, Latif et al. 1992).

One of the main characteristics of ENSO is its quasi-periodicity. ENSO can be regarded as an eigenmode of the coupled ocean-atmosphere system, where the memory of the system resides entirely in the ocean. To explain the quasi-periodic nature of ENSO a conceptual model, the *delayed action oscillator* was proposed by Schopf and Suarez (Schopf and Suarez 1988) which summarizes the fundamental zeroth-order physics involved in ENSO (Fig. 16.2). According to this model, the easterly wind anomalies over the western Pacific prevailing during the cold phase (La Niña) of the ENSO cycle generate an upwelling Kelvin wave packet which propagates eastward along the equator. In the eastern Pacific, where the ther-

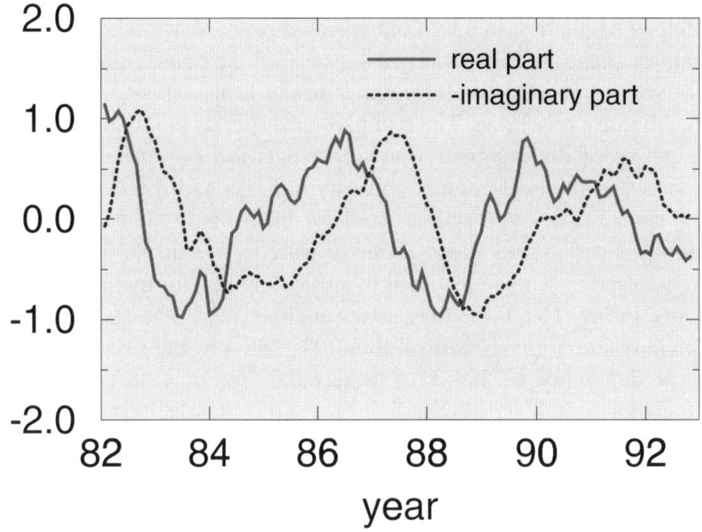

Fig. 16.5 Time series of the leading POP of a combined POP analysis of heat content anomalies, SST anomalies, and wind stress anomalies. The NCEP ocean reanalysis and the FSU pseudo wind stress data were used.

mocline is shallow this causes cooling at the sea surface. The ocean response to the easterly wind anomalies, however, also consists of downwelling Rossby waves off the equator which propagate westward. They do not affect the sea surface temperature, since the thermocline is too deep in this region. The Rossby waves then reflect at the western boundary into a downwelling eastward propagating Kelvin wave packet. In the western Pacific this does again not effect the sea surface temperature, since the thermocline in the western pacific is too deep. Once the wave packet has propagated far enough into the east, where the thermocline is shallow, SST is affected, and positive SST anomalies develop which may grow by unstable air sea interactions (as described above) into an El Niño event. Thereafter, the sequence of events repeats itself, but with opposite sign. In this context ENSO can be regarded as a cyclic trajectory in a two dimensional phase space which is spanned by the two basis vectors entitled *intermediate phase* and *extreme phase* or El Niño. The four phases of such a cycle and the corresponding positions in the phase space are presented in Fig. 16.2. Following this model, ENSO is a perfectly predictable regular oscillation. This cycle, however, is disturbed by factors like random noise, non linear interactions with the annual cycle, or decadal variations of the mean state which thus limit the predictability of ENSO.

The features of the delayed action oscillator scenario may also be extracted from observations. A powerful method to obtain the dominant oscillating patterns from complex data sets is the *Principal Oscillation Pattern* (POP) analysis (Hasselmann

1988, Storch et al. 1995). A very brief description of the method is given in Appendix B. The POP analysis yields a set of oscillation patterns. Each of them represents a damped oscillation in a two dimensional phase space which is represented by the complex plane. In Figs. 16.3, 16.4, and 16.5 the results of a combined POP analysis of sea surface temperature, upper ocean heat content, and pseudo wind stress anomalies are presented. The real part corresponds to the intermediate phase and the negative of the imaginary part represents the extreme warm phase. The similarity between the heat content anomaly patterns and the theoretical patterns that follow from the delayed action oscillator theory is obvious. In Fig. 16.4 the corresponding pseudo wind stress anomaly patterns are displayed. To each POP mode two time series can be attributed by projecting the original data sets onto the POP patterns. In Fig. 16.5 these time series are presented. The quasi periodic character is obvious and a phase shift of about 90° between the two time series indicated the predictability of ENSO. Furthermore, we may infer from the POP patterns, together with the corresponding time series that the heat content and wind stress anomalies propagate, whereas the SST anomaly is a standing pattern with varying amplitude. This also agrees well with the delayed action oscillator theory. These characteristics become even more evident in a Hovmöller diagram (Fig. 16.6). Instead of the original data, the time evolution as described by the POP model is presented. This filters out patterns that are not related to ENSO. Especially in the heat content it is obvious how anomalies in the west Pacific propagate within one to two years to the east Pacific and finally result in an extreme warm or cold event. Thus, heat content anomalies in the west Pacific may be used as a precursor for El Niño/La Niña predictions.

16.2.2 ENSO predictability

From observational studies as well as from modeling studies it is evident that ENSO is predictable on lead times up to a year. The easiest way to prove this is to look at the time series of upper ocean heat content in the west and in the east Pacific. In the following, averages over the two index regions NINO-3 (5°N - 5°S, 150°E - 150°W) and NINO-4 (5°N - 5°S, 150°W - 90°W) are frequently used (Fig. 16.7). Most of the scientific groups working on ENSO and ENSO predictions use averages over these regions to quantify the quality of simulations and predictions. From Fig. 16.8 one may infer that extreme ENSO events are predictable on lead times up to about one year. The averaged temperature anomaly down to a depth of 275 meters is used as a measure for upper ocean heat content. The anomalous heat content in the west Pacific leads the corresponding signal in the east Pacific by about one year. The maximum correlation of about 0.82 between the two time series (not shown here) is observed at a time lag of 11 months which is a clear indication for the predictability of ENSO. Even more evidence for the predictability of ENSO comes from the POP analysis from the previous section. The leading POP mode explains about 32% of the total variance in the combined data set which demonstrates that this is the dominant pattern. The rotation period is about four years and the decay time about 3 years. The two POP coefficient time series vary

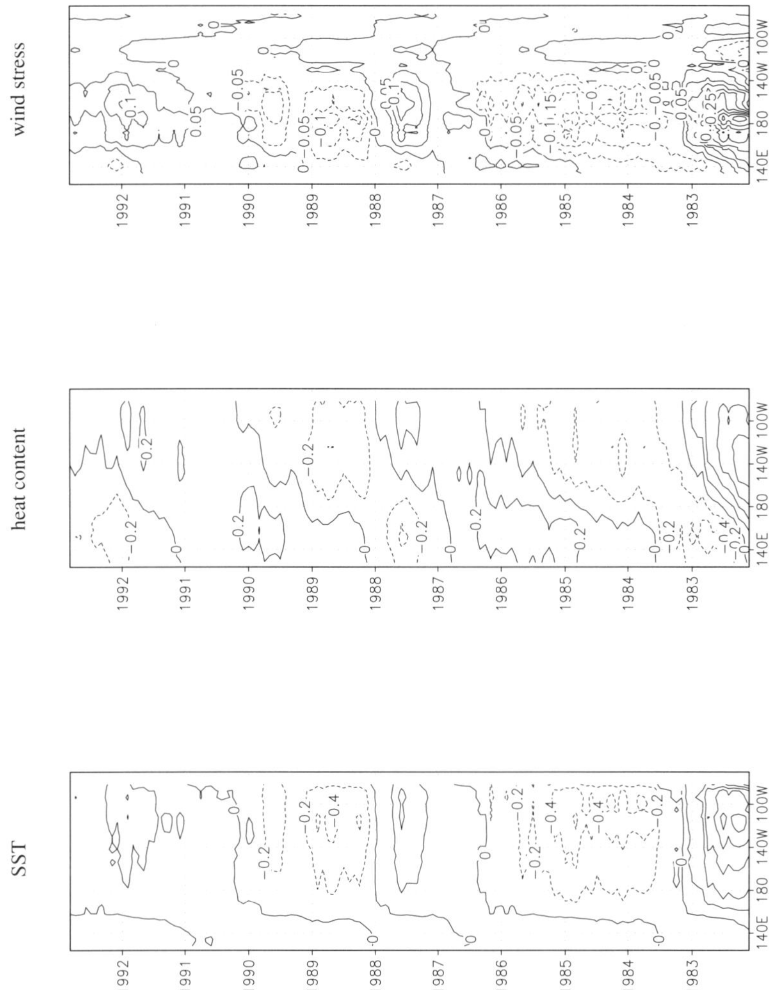

Fig. 16.6 Hovmöller diagram of anomalous heat content, sea surface temperature, and zonal pseudo wind stress as described by the POP model. The data are averaged between 5°N and 5°S, the vertical axis represents the time, the horizontal axis represents the longitude

coherently with a phase difference of about 90° (as theoretically expected) over most of the time which in itself is an expression of predictability.

Model studies also indicate the predictability of ENSO. Such numerical experiments were conducted in the mid-eighties at the Geophysical Fluid Dynamics Laboratory (GFDL) to simulate interannual variability. Observed SST's for the period 1962-1976 were specified as lower boundary condition for an atmospheric general

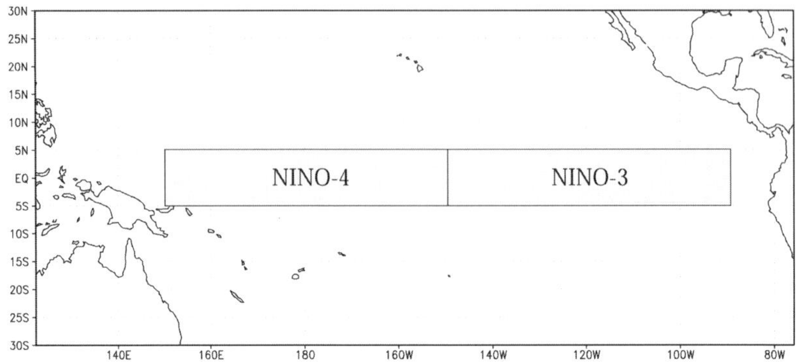

Fig. 16.7 The NINO-3 and NINO-4 index regions.

circulation model integration. This model integration successfully reproduced the observed interannual ENSO related variability. In a second experiment with the same boundary conditions but different initial conditions the low frequency interannual variability was equally well reproduced. The principal result from these experiments is that the ENSO related variability is not governed by internal atmospheric instabilities. The low frequency variations in tropical Pacific SST, on the other hand, can be reproduced by ocean models, provided the surface wind stress variations are specified from observations (e.g. Latif 1987). Thus, low frequency variations in both the atmosphere and the ocean are strongly controlled by changes in the boundary conditions which demonstrates the coupled nature of ENSO (see also Latif at al. 1997 and references therein).

Coupled ocean atmosphere models of different complexities also have been used for ENSO predictions and predictability studies (e.g. Zebiak and Cane 1987, Eckert and Latif 1997). These models generally simulate fairly robust ENSO cycles which share many similarities with the observations. Predictability studies with such models also indicate an ENSO predictability of about one to two years, and thus confirm the results obtained from observational studies, and investigations with uncoupled models.

16.2.3 The impact of ENSO on climate and economy

The impacts of ENSO are felt worldwide through a disruption of the atmosphere general circulation patterns (Ropelewski and Halpert 1987), which leads to climate changes in many regions of the world (Fig. 16.9) ENSO related climate variations and the corresponding economical impacts are correlated worldwide so that amplification effects play an important role. Severe droughts are observed during El Niño events in Northwest Australia, South East Asia and Northeast Brazil. Zimbabwe and the south of the United States are other regions where rainfall variations are highly correlated with ENSO indices. The agriculture, and thus the crop yield is directly and strongly affected in these regions. The variability of crop yield in Zim-

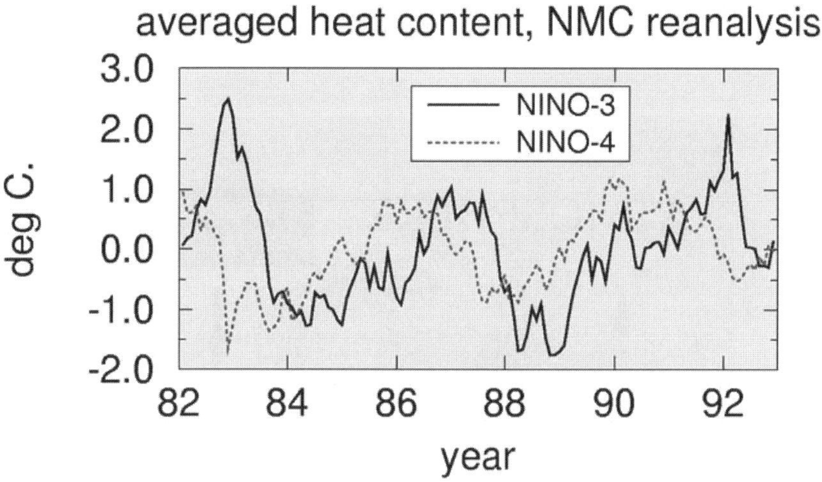

Fig. 16.8 Monthly means of temperature anomalies averaged over the upper 275 meters, separately for the NINO-3 and the NINO-4 region. The data were taken from the NCEP ocean reanalysis.

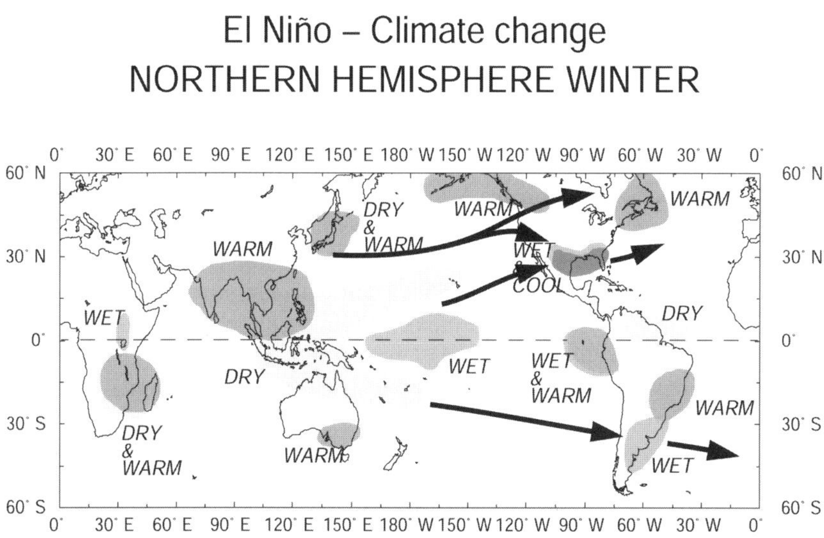

Fig. 16.9 The global impact on climate during an El Niño event. Only regions which show significant correlations with ENSO are displaied (after Ropelewski and Halpert 1987).

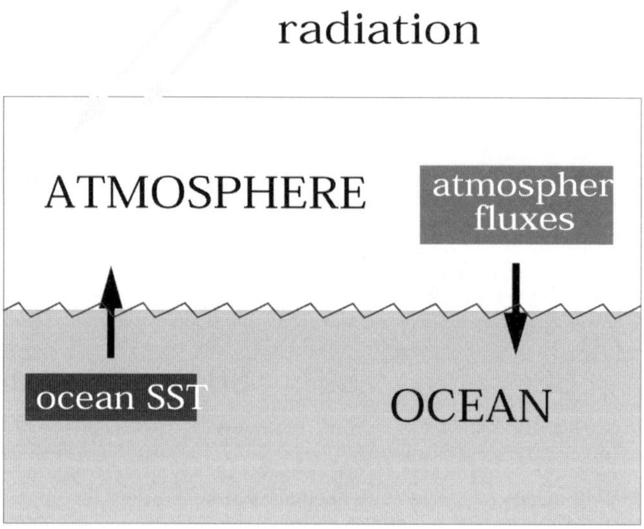

Fig. 16.10 Schematic view of a coupled ocean-atmosphere general circulation model. Each component is constrained with boundary conditions computed by the other model. Thus, solar radiation is the only external forcing that is needed.

babwe, for instances, shows a correlation of about 0.8 with the SST anomalies averaged over NINO-3. In the south of the United States which is an important region for the supply of wheat for the world population, farmers already use ENSO predictions routinely to optimize their results. Also in Australia ENSO predictions are routinely used in agriculture. In South East Asia reliable ENSO predictions are needed for the water reservoir management. For instance, in Thailand each year before the monsoon season the government has to make the decision, whether to empty the water reservoirs or not. A failure may either lead to catastrophic flooding or to a severe lack of water which both have dramatic consequences for the population. Off the Peruvian coast the marine ecosystem is directly affected by the oceanic variations in upwelling which has large impact on the fishery industry. And also decision makers in Europe are very much interested in potential long range forecasts over Europe. For the energy supplying companies it is, for instance, very important to know whether the winter will be cold or not. A difference of one degree centigrade less in temperature leads to an additional energy demand of about 1000 MW only in France which corresponds approximately to one big nuclear power plant.

For all these regions, reliable ENSO forecasts would offer decision makers an opportunity to take account of anticipated climate variations, in order to reduce impacts of ENSO on the economy. The successful prediction of ENSO is therefore not only of scientific but also of large public interest.

16.3 ENSO prediction models

In the past a large variety of successful ENSO prediction models has been developed and applied. The models can be divided into purely statistical models and physical models. In the following I describe very briefly statistical models, how they are initialized and what potential forecast skill they have. A more detailed and complete explanation of statistical models and methods would be beyond the scope of this paper. I therefore refer to the literature (e.g. (Latif et al. 2001 and references therein), (Barnett and Hasselmann 1979), (Navarra 2001)). Physical models are described in more detail in later sections.

16.3.1 Statistical ENSO prediction models

Statistical methods of different complexity have been used over the last decades to identify the principal modes of climate variability. The same statistical models may be also applied for climate predictions. Common to all statistical methods is that they need a so called training data set to fit prediction parameters. Most of the data records used in ENSO predictions do not go back earlier than 1950, and thus capture only a few realizations of ENSO extremes. Therefore the prediction parameters might be difficult to estimate.

The most simple statistical model that can be applied to ENSO predictions is persistence. From the heat content time series in Fig. 16.8 it may be inferred that useful prediction skill is obtainable up to lead times of about 3 months. This is far below the 12 to 18 months of predictability that were obtained from predictability studies.

The next step in a hierarchy of statistical schemes is a correlation model. In section 16.2.2 it was shown that the heat content anomalies in NINO-4 lead NINIO-3 temperature anomalies by about 11 months, and are thus useful as a precursor for El Niño events. However, the extremely high lag correlation of 0.82 was only found in the eighties which was a period of very regular and strong ENSO events. During the nineties the lag correlations are much lower and useful prediction skill can be expected for only a few months.

A more sophisticated scheme would be the POP analysis presented in section 16.2.1. By projecting a combined data set of observed heat content, SST and pseudo wind stress anomalies onto the leading POP patterns the actual position in the two dimensional phase space can be identified, and thus the further evolution of the system can be predicted. Alternatively other variables may be used as predictors. Xu and Storch (Xu and Storch 1990) used in their study sea level pressure anomalies in the latitude band 15°S to 40°N, and achieved a prediction skill (anomaly correlation between prediction and observation) of about 0.5 after a lead time of about 10 months.

A large variety of other sophisticated statistical methods with different predictors have been applied to forecast the state of the ENSO cycle. A good and complete overview is given in the review paper of (Latif et al. 1997 and references therein).

16.3.2 Physical ENSO prediction models

Since ENSO is a phenomenon of the coupled ocean-atmosphere system, coupled ocean-atmosphere models are required for its prediction. Such models have been developed over the last 10 to 15 years by different groups and applied to ENSO forecasts (e.g. Cane et al. 1986, Barnett et al. 1988, Leetmaa and Ji 1989, Latif et al. 1994). The coupled models can be categorized into three classes of models: Intermediate models, hybrid coupled models (HCM) and coupled general circulation models (CGCM).

Intermediate models

The development of the *Lamont-model* (Zebiak and Cane 1987) was a milestone in ENSO forecasting. It is a non linear anomaly model of intermediate complexity, and has been used operationally for ENSO forecasts since 1986. It is a limited domain model covering the tropical Pacific. Thus there is an inherent assumption that the cause of ENSO lies entirely in this region. The model simulates realistically the interannual variability in tropical Pacific SST. The mechanism for the interannual variability is closely related to the subsurface memory of the system and is consistent with the *delayed action oscillator* scenario. This model is one of the most successful ENSO forecast models. The correlation skill of this model for a large number of predictions, initialized monthly during the period 1972 to 1995 is very encouraging (Fig. 16.16). For most of the period a useful correlation skill of 0.5 is still found after one year. The prediction skill of this model will be discussed in more detail in section 16.4.

Hybrid coupled models

In section 16.2.1 it was pointed out that ENSO is a coupled ocean-atmosphere phenomenon but that the time evolution of the system is mainly determined by the ocean. At low frequencies the atmosphere over the tropical Pacific can be regarded as a strongly forced quasi equilibrium system governed by the state of the tropical Pacific SST. Theoretical investigations, observations, and numerical studies with coupled ocean atmosphere models support this view. Thus it is possible to express the atmosphere's response to variations in the oceanic boundary conditions by a statistical (empirical) model inferred from data. In the following the statistical atmosphere model used at the Max-Planck-Institut in Hamburg is described as an example (Latif and Flügel 1991, Flügel 1994).

We assume that the memory of the coupled ocean-atmosphere system resides entirely in the tropical Pacific. It is therefore possible to use a *minimum* atmospheric model with no internal dynamics which passively responds to the anomalous boundary conditions provided by the ocean. Further, we assume a linear relation between the dominant anomaly patterns of sea surface temperature and wind stress. In the following, we explain how the atmosphere model was constructed (see also Barnett et al. 1993). We decompose observed sea surface temperature $T'(x,t)$ and wind stress anomalies $\tau'(x,t)$ into empirical orthogonal functions (EOF):

$$T'(x, t) = \sum_{n=1}^{N} \alpha_n(t) E_n(x) \tag{1}$$

$$\tau'(x, t) = \sum_{n=1}^{N} \beta_n(t) F_n(x) \tag{2}$$

Here α_n and β_n are the principal components and E_n and F_n are the spatial patterns, respectively. A matrix of regression coefficients $\underline{\underline{C}}$ relating the two vectors of principal components is obtained by minimizing the function :

$$\langle(\underline{\underline{C}}\underline{\alpha} - \underline{\beta})^2\rangle \tag{3}$$

where $\langle\ldots\rangle$ denotes the time expectation operator. During a coupled model run the actual wind stress anomaly field is computed as follows. From the SST anomalies, the corresponding principal components are computed by projection onto the corresponding EOFs :

$$\alpha_n(t) = \sum_{x} T'(x, t) E_n(x) \tag{4}$$

from which the principal components for the EOF expansion of the wind stress anomaly fields are determined as:

$$\beta_n = \sum_{m=1}^{N} C_{nm} \alpha_m \tag{5}$$

which are finally inserted into equation (2). The leading five EOFs were retained for the prediction experiments. In order to obtain self sustained oscillations in a coupled control run, the wind stress anomalies computed with the statistical model were scaled by a factor of 1.4. The performance of the ocean model as measured by its ability to reproduce the observed SST in response to prescribed observed wind stresses is good but not perfect. This means that the ocean model SSTs passed to the statistical atmosphere differ in some respect from those that were used to develop the atmosphere. This error was corrected by constructing an interface between the ocean and the atmosphere model (Barnett et al. 1993, Flügel 1994). Again, a regression approach was used, by which the relation between the simulated and observed SST patterns was approximated. The leading five EOFs of the observed and simulated SST anomalies (obtained from a run forced with observed wind stress fields) were used to derive the *correction matrix*. The interface or *correction matrix* was constructed similarly to the regression matrix C, described above. The simulated SST anomalies are corrected by this interface before they are passed to the statistical atmosphere model.

The coupled model simulates a regular ENSO cycle with a period of approximately five years, which was inferred from a multi decadal control run (Latif and Flügel 1991, Flügel 1994). This is only one possible approach for a statistical atmosphere. Other statistical methods can be used and also non linear approaches may be useful. This simple atmospheric model is then coupled to a limited domain primitive equation model of the tropical Pacific. Useful skill is achieved for lead times up to one year. A problem with this kind of models is that the atmospheric feedback has to be determined from data, that means from the past. If the relation between SST and wind stress anomalies changes for some reason, the prediction skill of these models drops dramatically. This happens for instance during the early nineties. It seems that during that time the mean state of the Pacific ocean is different from the state in the eighties, and thus also the feedback between ocean and atmosphere changed. The parameters of the statistical atmosphere described above were determined from data from the seventies and eighties. For this period the coupled model was successful in forecasting the evolution of the ENSO cycle. In the nineties, however, the forecast skill at a lead time of one year dropped from about 0.7 for the eighties to about 0.2 to 0.3.

Changes of the mean state or of feedback mechanisms are a general problem for every statistical method. To catch them sophisticated statistical methods would be necessary and very long observational time series would be needed to obtain reliable estimates of the parameters. However, most ENSO related observations do not go further back than 1950, and thus are too short. The only way to overcome this problem is to use fully physical models which are presented in the following section.

Coupled general circulation models

Coupled general circulation models are the most complex models applied to ENSO forecasting. Such models have been developed over the last 10 to 15 years by scientific groups all over the world. Examples are the ECHO model (ECHAM (ECMWF Hamburg Model) + HOPE (Hamburg Ocean Primitive Equation model)) (Frey et al. 1997), the GFDL model (Rosati et al. 1996), the COLA model (Kirtman et al. 1997), or the NCEP model (Ji and Leetmaa 1996) to name just a few coupled general circulation models. Usually they consist of a global atmosphere general circulation model and a global ocean general circulation model. The coupling is done by forcing the ocean model with momentum fluxes (wind stress), heat fluxes, and fresh water fluxes as computed by the atmospheric model, and the SST computed by the ocean model is used as lower boundary condition for the atmosphere. The time intervals to exchange informations between the two models typically range from a few hours up to one day. Thus, the only external forcing that is needed to run a coupled general circulation model is the solar radiation which may be easily computed even for the next millennium (Fig. 16.16). The advantage of coupled general circulation models relative to statistical approaches, either in purely statistical models or in hybrid coupled models is that most of the tunable parameters have strong physical constraints. Parameters in these models are not tuned for best prediction skill but to give the best representation of the physical laws. Thus, a change of the mean state of different feedbacks over a certain period are in principle not a problem for these models.

However, a lot of other problems are encountered with coupled general circulation models. First of all they are extremely expensive in terms of computer time which makes it difficult to tune these models and to obtain time series long enough for reliable statistics. Second, most of the coupled models show a more or less severe *climate drift*. This means that the model drifts into a new and unrealistic climatology after coupling. Typical climate drift values in terms of SST are between 0.1 and 1 degree centigrade per year. The drift is due to an imbalance of the energy fluxes that are exchanged between the models. Therefore flux corrections are applied to many coupled models. Another approach to avoid climate drift is to couple only anomalies relative to a mean annual cycle which is determined separately for the ocean and atmosphere model. However, both methods are questionable if the model is used for ENSO studies. Flux correction methods generally reduce or even suppress interannual variability, and anomaly coupling ignores non-linear interactions between the ENSO cycle and the annual cycle and cannot handle a change of the mean state. Therefore models without any flux correction are usually used for ENSO studies and forecasts. Since the integration time for ENSO studies is typically of the order of one to two decades and for forecasts about one year, a climate drift of about 1 degree centigrade in 10 year, as it is, for instance, found in the ECHO model can be accepted. To achieve such a relatively low climate drift, most ENSO models do not exchange informations at high latitudes, like poleward of 50°N/S. In these regions climatological boundary conditions are prescribed for the atmosphere and ocean. The reason is that during the formation and melting of sea ice strongly nonlinear processes play an important role, and errors in the description of these processes may lead to a very severe climate drift.

Nowadays ENSO prediction skills are achieved with coupled general circulation models that are comparable with the results from intermediate or hybrid coupled models.

16.4 ENSO predictions

16.4.1 Initialization methods

A very important issue in carrying out an ENSO forecast is the initialization of the forecast system. ENSO prediction is like numerical weather prediction an initial value problem. The further evolution of the system highly depends on the initial state, and thus an initial state as realistic as possible is crucial for a successful forecast. In section 16.3.2 it was already pointed out that the atmosphere in the tropics is strongly forced by sea surface temperature anomalies and that the memory of the system resides entirely in the upper ocean. Thus, the initialization of the coupled system can be reduced to an initial value problem of the upper ocean only. The usual way to initialize an ENSO forecast model for a prediction is as follows (Fig. 16.11a). The oceanic model is integrated in an uncoupled mode and forced over a period of typically one to two years with observed wind stresses, heat fluxes, and fresh water fluxes. Latif and Graham (Latif and Graham 1992) showed that the ocean acts during this phase as a complicated time-space filter integrating basically the information contained in the wind stress fields. That means that the initial oceanic state does not only depend on the actual boundary conditions, but also depends on the history of the wind field prior to the start of a forecast. Therefore it follows that the quality of the wind stress fields, especially the low frequency component, is crucial for a successful initialization of a forecast. Errors in the initial state are mainly due to model errors or to errors in the wind forcing, since it is not an easy task to obtain high quality wind observations over the open ocean regularly in space in time.

Ocean data assimilation can be an additional source of information during the initialization process in order to compensate for model errors and uncertainties or errors in the wind forcing fields (Fig. 16.11b). Since the upper ocean heat content, down to a depth of about 300 meters is crucial, useful observations must contain informations about the vertical structure of the upper ocean. Such data are for example subsurface temperature measurements or velocity measurements. But the vertical density structure can also be reconstructed from sea surface height data (Mellor and Ezer 1991, Cooper and Haines 1996, Fischer et al. 1997). In section 16.4.2 forecast experiments from different groups with and without ocean data assimilation are presented and compared to each other. In the following subsection special emphasis is put on the assimilation of sea level measurements, since it is not as straight forward as the assimilation of subsurface temperatures for instance.

Common to both methods is that the forecast system is initialized in uncoupled mode. One problem of this approach is the so called *coupling shock* that occurs immediately after coupling an degrades the correlation skill over the first 6 to 8 weeks significantly. At the beginning of the coupling the two models are still not in balance relative to each other, and the models need some time to adjust to each other. A way to avoid this coupling shock is to initialize the coupled system directly. In this case atmospheric and/or oceanic observations are assimilated during the initialization period of one to two years into the coupled model, in order to constrain it (Fig. 16.11c). This method is relatively new but first results with the intermediate Cane/Zebiak model (Chen et al 1997) are very encouraging.

Assimilation of sea level data into an ocean primitive equation model

To assimilate observations that have a prognostic equivalent in the model formulation, like for instance temperatures, is straight forward. From the actual model solution and the observations a correction field is computed, and this correction is added to the model solution. Then the integration of the model is continued from the corrected model state. With sea surface height (SSH) measurements this is not possible. Most of the ocean primitive equation models used today make the so

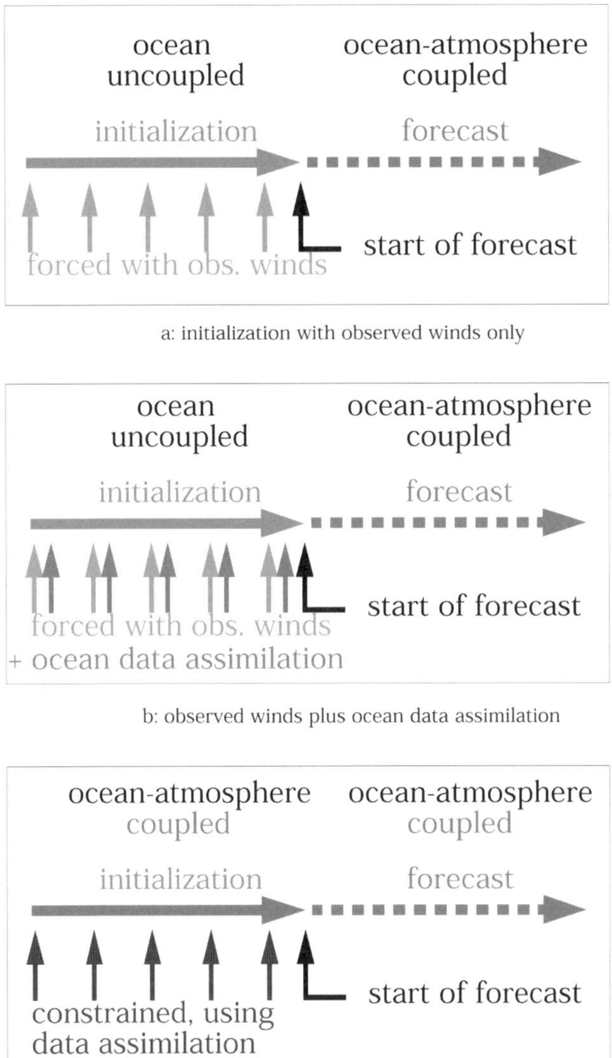

Fig. 16.11 Three different initialization methods for coupled general circulation models. In the upper panel the conventional method is shown. Only observed wind fields are used to initialize the model. In panel b additional information is obtained from oceanic observations. In the lower panel the coupled system is initialized directly. Assimilation of oceanic or atmospheric observations constrains the model.

called rigid lid approximation. That means that the sea surface height is a diagnostic variable. But even if no rigid lid assumption is made it is not sensible to insert sea surface height data directly into the model. This would affect the barotropic mode only, and the information would be lost within a few time steps. Sea surface elevation is an integral quantity that contains information about the vertical density structure. To assimilate SSH observations, the information contained in the data must be transformed into a correction of the mass field. Different methods have been developed to solve this problem (Mellor and Ezer 1991, Cooper and Haines 1996). In the following the method of (Fischer et al. 1997) is presented.

For the tropics, a linear relation between small changes in sea level and the corresponding changes in the vertical mass structure is a reasonable assumption. Further, salinity variations can be neglected, which yields a linear relation between small changes in the vertical temperature stratification and sea level. This relationship can be estimated from a long ocean model integration (here 10 years) using a linear regression method. For each horizontal grid point, an EOF analysis (compare appendix 1) of the anomalous vertical temperature profile was performed. Five EOFs were retained, explaining about 95% of the local variance in most of the model domain. Regression coefficients A^n between SSH anomalies and the principal components γ of the EOFs $G_n(z)$ were computed and stored separately for every horizontal grid point. To insert the sea level observations into the model, first a correction field for the sea surface height anomalies $\Delta\zeta'$ was computed. The principal components of the EOF expansion of the vertical temperature corrections are obtained by multiplying the sea level corrections by the A^n (eq. 6). To obtain the desired temperature correction field, the principal components must be inserted into the EOF expansion of the anomalous temperature profile (eq. 7):

$$(\gamma^n)_i = (A^n)_i \cdot (\Delta\zeta')_i \tag{6}$$

$$(\Delta T'(z))_i = \sum_{n=1}^{5} (\gamma)_i \cdot (G(z)^n)_i \tag{7}$$

The index i denotes the actual horizontal model grid point. Finally, the temperature corrections were added to the original model solution, and a convective adjustment routine was applied to avoid unstable stratifications. Thus, the surface height correction was transformed into a correction of the vertical temperature structure.

The extension of this method to include salinity is straight forward. Then for each grid point a combined EOF analysis of (T,S) has to be performed.

16.4.2 Results of ENSO prediction experiments

In this section prediction results obtained with different models and with different initialization methods are presented and compared to each other. In Fig.16.12 The forecast skill obtained from two statistical models is compared with the persistence of a slightly smoothed Southern Oscillation Index (SOI). Both statistical models use the time history of one year of sea level pressure anomalies as predictors. The POP model (thick line) is based on monthly predictions of the southern oscillation index (SOI), whereas the canonical correlation model (CCA) is based on seasonal predictions of SST anomalies averaged over NINO-3. To avoid artificial skill the forecast period was excluded from the data set to fit the statistical parameters. Therefore the skill presented here can be regarded as real forecast skill. The POP and the CCA model give significantly better skill than the persistence model on lead times larger than 4 months. A useful skill (correlation larger than 0.5) is obtained up to about 3 seasons, and the correlation drops significantly slower with increasing lag relative to persistence.

Fig. 16.13 shows forecast results obtained with different versions of the NCEP Coupled Mdel Project (CMP) forecast model (Ji et al. 1996 and Ji and Leetmaa 1997). The correlations and rout mean square (RMS) errors between predictions and observations, presented in the first panel (Fig. 16.13a), were obtained by initializing the model in a conventional way. The ocean model was initialized in uncoupled mode and only wind observations were used to force the model. The initial correlation at short lead times is remarkably high. However, this is due to the fact that a relaxation towards observed SSTs is applied as a parameterization of heat fluxes. In terms of anomaly correlation and error growth the model results are significantly better than persistence, and a clear improvement going to more recent versions of their model is seen. With the latest version CMP10 a correlation skill above 0.5 is achieved at a lead time of 11 months. This is remarkable since the results are based on forecasts for the period June 1982 to March 1995. That means the extremely difficult period 1991 to 1994, in terms of ENSO predictions, is included. The situation becomes even better when surface and subsurface temperatures are assimilated into the ocean model during the initialization period as seen in Fig. 16.13b. The results were obtained with the CMP6 version, and forecasts were carried out for the period May 1989 to October 1993. In one experiment (F-RA2) all available temperature measurements were assimilated to initialize the ocean

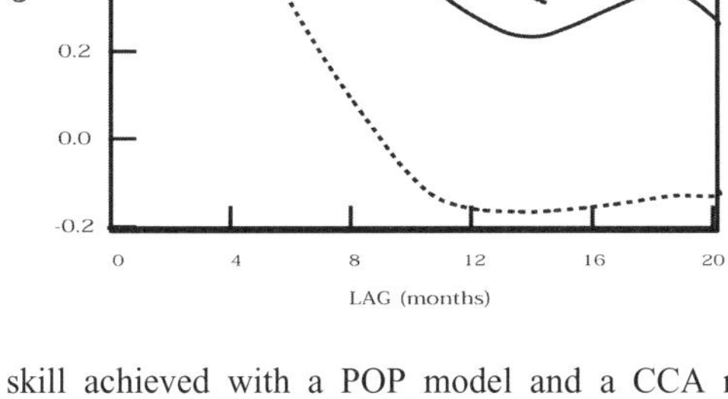

Fig. 16.12 Forecast skill achieved with a POP model and a CCA model, compared to persistence. Useful skill is achieved up to lead times of 3 seasons. From Latif et al. 1994.

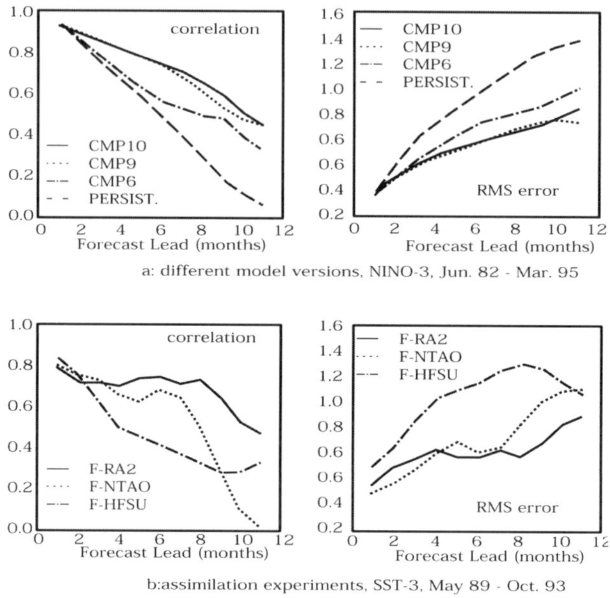

Fig. 16.13 Forecast results obtained with different versions of the NCEP forecast system. In a) correlations and RMS errors in predicting NINO-3 SST anomalies are presented. Only wind observations were used to initialize the model. The results in panel b) were obtained without ocean data assimilation (F-HFSU), assimilation of XBT data (F-NTAO), and assimilation of XBT and TOGA-TAO temperature measurements (F-RA2). From Ji et al. 1996 and Ji and Leetmaa 1997.

model. That included XBT data and measurements obtained from the TOGA-TAO array (Mc Phaden 1993). The prediction skill remains at an extremely high level of above 0.8 for lead times up to 9 months, and even at 11 months forecast lead it is still at about 0.7. If the TOGA-TAO data are excluded (i.e. only XBT measurements) the correlation is still very high (F-NTAO) but significantly lower than in the previous experiment. This shows clearly that the higher accuracy of the TOGA-TAO data, relative to XBT measurements, is an important issue. As a reference (F-HFSU) the skill curve obtained without ocean data assimilation is shown which is clearly lower than in the other two experiments. The positive impact of subsurface temperature data assimilation is not only seen in the correlation curves but also in the error growth. These results prove that subsurface temperature measurements may compensate for errors in the forcing fields and the model itself, and it shows that the vertical structure of the upper ocean is crucial for the further evolution of the coupled air-sea system in the tropical Pacific.

Since the coverage in space and time of subsurface measurements is quite irregular, and in the central Pacific, off the ship tracks between the United States and Asia, very sparse, it is an interesting question whether the assimilation of surface measurements may also improve ENSO predictions. SST data are not sufficient to initialize an ENSO forecast system (Fischer and Latif 1995), since they contain no information about the subsurface structure of the ocean. But the upper ocean heat content anomalies impose a signature on the sea surface height anomaly, and these data are nowadays available from satellite measurements almost continuously in space and time. A method to extract the subsurface information from sea level measurements has been presented in section 16.4.1. Below, three forecast ensembles are presented (Fig. 16.14): predictions initialized only with observed winds (a), with observed winds and subsurface temperature assimilation (b), and initialized with observed winds and sea level assimilation (c) (Fischer et al. 1997). Each forecast was initialized for at least one year, to ensure that the ocean model was properly preconditioned. Although the SST variability in the control ensemble (Fig. 16.14a) is underestimated by a factor of about two, the correlation between predictions and observations is quite high. Even at lead times of one year it remains above 0.7 (not shown here). However, the 1986/87 El Niño was not predicted at all. The coupled model simulated anomalous warm conditions only after the warm event is fully developed in early 1987. If subsurface temperatures are assimilated during the initialization, the situation improves considerably (Fig.16.14b) and the amplitudes of the predicted SST anomalies are quite realistic representations. The positive impact of subsurface temperature assimilation on ENSO prediction skill, found also by other groups (e.g. Ji et al. 1996, Ji and Leetmaa 1997, Kleeman et al. 1995) is confirmed by these results. If sea level measurements are assimilated to initialize the model comparable results as in the temperature case are obtained (Fig. 16.14c). Especially the transition from anomalous cold conditions in the mid eighties to the warm conditions which finally led to the strong El Niño 1986/87 are well represented. This shows that the subsurface information contained in sea level measurements can be successfully inserted into an ocean primitive equation model, and thus may improve the performance of ENSO prediction models.

At the GFDL experiments with three different assimilation schemes were performed (Rosati et al. 1996) with a coupled general circulation model. In two nudging experiments either SST (SDA2) or SST and surface winds (SDA1) were nudged directly into the coupled model. In a third experiment a three dimensional variational assimilation method was applied to insert temperature data into the ocean model integrated in uncoupled mode (FDA). Forecasts in each experiment started either in January or July of the years 1982 to 1989, which yields ensembles of 7 members (Fig. 16.15). The correlation skill from these experiments was quite different. In the case of mere SST nudging the skill is rather poor and only marginally better than persistence. Useful skill is only obtained up to 5 months. From this experiment one may infer that SSTs alone are not sufficient to precondition an ENSO forecast system. The subsurface structure is not properly initialized in this case. In the SDA1 experiment (nudging of SSTs and surface winds) the results are slightly better but not satisfying. This disappointing result, however, probably must be attributed to relatively poor wind fields. In the FDA experiment the ocean model was forced with the same wind fields that were nudged in the SDA1 case but addi-

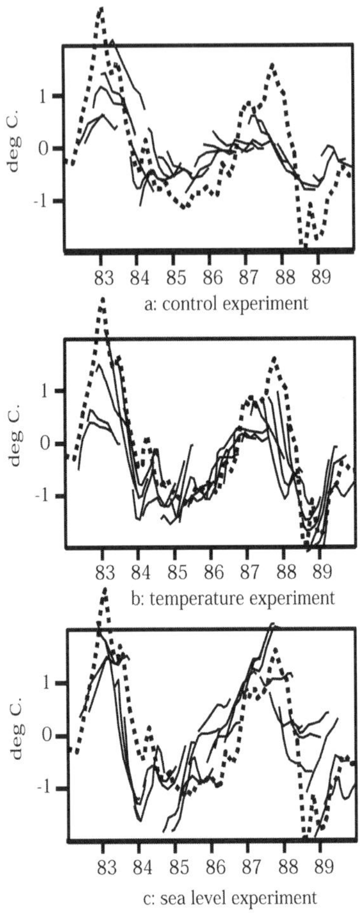

Fig. 16.14 Observed and predicted SST anomalies averaged over the NINO-3 region for the period 1982 to 1989. Each individual forecast was carried out over one year. From Fischer et al. 1997.

tionally subsurface temperature measurements were assimilated. The skill is clearly superior and does only slightly degrade with increasing lead time in the January ensemble.

A point to mention is the local correlation minimum at a lead time of 2 to 3 months in both ensembles of the FDA experiment. Directly after the start of the forecasts, the correlation skill drops even quicker than persistence, and recovers after it reached a minimum after 2 months. This behavior was not found in the other two experiments. The reason is the coupling shock that was already mentioned in section 16.3.1. Since the components of the forecast system are initialized separately the ocean and atmosphere are not balanced relative to each other. They

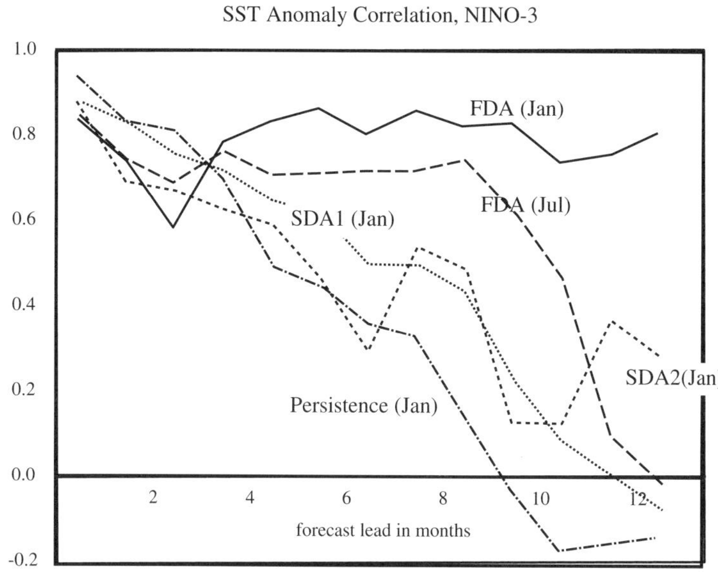

Fig. 16.15 Correlation skill obtained with the GFDL model applying three different initialization methods. In the experiments indicated with FDA subsurface temperature data were assimilated into the ocean model. In this case the initialization was carried out in uncoupled mode whereas in the other two experiments either SSTs only (SDA2) or SSTs and surface wind stress fields (SDA1) were nudged into the coupled ocean atmosphere model. From Rosati et al. 1996.

need about 2 to 3 months to achieve a new equilibrium which is expressed by the local skill minimum at the beginning of the forecast. In SDA1 and SDA2 such a minimum is not found since the coupled model was initialized directly. This feature is not only found in this model but in most of the ENSO prediction systems initialized in uncoupled mode. As already indicated by the skill curves of experiment SDA1 and SDA2 a possibility to overcome this problem is to do a coupled initialization of the forecast system.

This was demonstrated with the Lamont model (Chen et al. 1997). In a control ensemble of forecasts the initialization was done in a conventional way. The ocean model was forced with observed winds to obtain initial conditions for the ocean and to perform a forecast the coupled model was started from this state. In a second experiment the same winds as before were used but they were nudged into the coupled ocean-atmosphere model. The results are presented in Fig. 16.16. Skill correlation and RMS as a function of lead time are shown. As a reference persistence is also presented. The coupled initialization method outperforms the conventional method in terms of correlation skill and also in terms of RMS error and error growths. (Chen et al. 1997) attribute the improvement mainly to a reduction of the

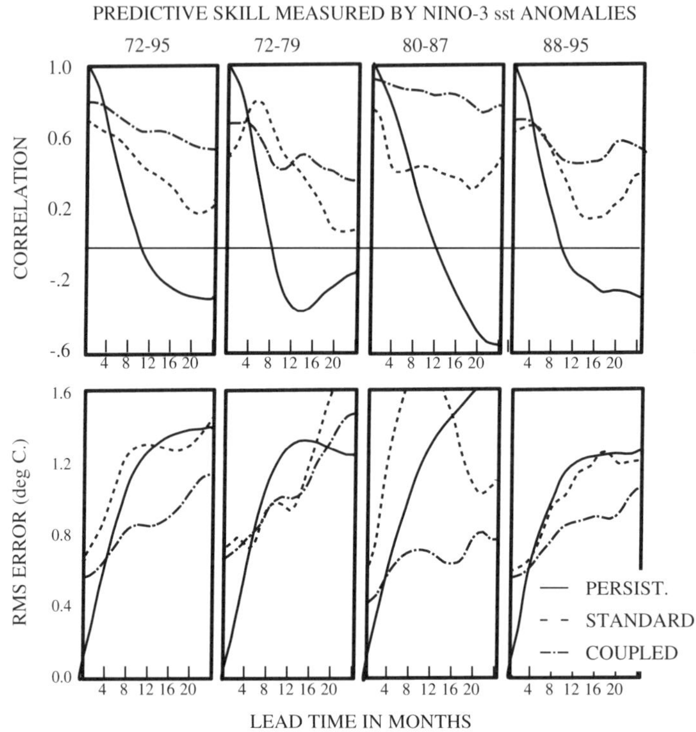

Fig. 16.16 Forecast skill obtained with the Lamont model using two different initialization methods. The results of the standard model (uncoupled initialization) are shown as dashed lines and those for the new model (coupled initialization) as thin solid lines. As a reference persistence is shown as a thick solid line. From Chen et al. 1997.

coupling shock and random noise. There is potential to improve the forecasts further, since the initial errors are still relatively large. This is a very encouraging study and further investigations also incorporating more complex models should be done. It should be mentioned, however, that the results with the coupled initialization might suffer from some artificial skill, since the nudging parameters were optimized with respect to the forecast skill.

16.5 Interpretation of ENSO forecast results

16.5.1 Definition of an appropriate skill measure

An objective measure that is commonly used to evaluate the performance of a forecast system is the correlation skill. In the case of ENSO forecasts most groups have agreed on the correlation between the observed and predicted sea surface temperature anomalies averaged over the NINO-3 region:

$$r(\tau) = \frac{\langle T'_{pred}(\tau) T'_{obs}(\tau) \rangle}{\sqrt{\langle T'_{pred}(\tau) T'_{pred}(\tau) \rangle \langle T'_{obs}(\tau) T'_{obs}(\tau) \rangle}} \tag{8}$$

In this equation, $T'_{pred}(\tau)$ and $T'_{obs}(\tau)$ denote the predicted and observed SST anomalies, respectively, averaged over the NINO-3 region. The lead time is denoted by τ , and the angle brackets denote the forecast ensemble average. If we apply this skill measure to the three forecast ensembles obtained with the hybrid coupled model (Fig. 16.14) the result in the case of sea level assimilation is rather disappointing (Fischer et al. 1997). The correlation is lower than in the control ensemble (no assimilation) for lead times between 2 and 7 months (not shown). So it seems that there is a contradiction between the objective and subjective evaluation of the forecast quality. Comparing the individual forecasts, it is noticeable that in the sea level ensemble the predicted temperature trends are quite realistic. They agree well with observations, even if the forecast itself is poor. An example is the prediction started in July 1985. An objective measure that evaluates the capability to predict trends is the correlation between differences. In our case, we define an alternative skill measure by computing the correlation between the observed and predicted temperature differences as a function of lead time:

$$r(\tau) = \frac{\langle \Delta T'_{pred}(\tau) \Delta T'_{obs}(\tau) \rangle}{\sqrt{\langle \Delta T'_{pred}(\tau) \Delta T'_{pred}(\tau) \rangle \langle \Delta T'_{obs}(\tau) \Delta T'_{obs}(\tau) \rangle}} \tag{9}$$

where $\Delta T'_{pred}(\tau)$ and $\Delta T'_{obs}(\tau)$ are the differences between the temperature anomaly at a certain lead time τ minus the temperature anomaly at the initialization time τ_o as predicted or observed, respectively:

$$\Delta T' = T'(\tau) - T'(\tau_o) \tag{10}$$

At medium and long lead times this alternative skill measure captures the capability to predict low frequency trends of the temperature in NINO-3. Using this definition of a skill measure supports the subjective impression of positive impact of sea level assimilation.

This shows at least that the question of forecast verification is more complicated than it seems to be, and that it should be addressed in more detail. It is unsatisfactory that the interpretation of the forecast result critically depends on the definition of the skill measure.

16.5.2 Correction of systematic errors

Another important issue is the correction of systematic, and therefore known forecast errors. Many coupled ocean atmosphere models tend to simulate the maximum of interannual SST variability too far in the west. Thus, the variability in NINO-3 which covers the eastern equatorial Pacific is often highly underestimated relative to observations. A simple correction could be, to compare the observed SST variability in NINO-3 with predicted SST anomalies averaged over the central equatorial Pacific.

The coupled forecast system at the ECMWF encounters a very strong drift of up to 1 °C in terms of NINO-3 SST within the first three months. However, this drift seems to be systematic, since skillful forecasts are obtained with this system if the mean forecast drift, as determined from a large number of forecast experiments, is subtracted.

A more sophisticated correction method is the canonical correlation analysis. This method looks for patterns in different data sets, here the forecasts and the observations, that show maximum correlation to each other in time. Then the forecast skill can be computed as the correlation between the time series of the patterns. The singular value decomposition is another possible method. This maximizes the covariance instead of the correlation. With a multidimensional regression analysis, similar to the approach used for the statistical atmosphere model in the hybrid coupled model, predicted SST patterns can be projected onto observed SST patterns. Thus, more than just one pair of corresponding patterns can be taken into account.

All these methods do not give further scientific insight into the problem of ENSO prediction but they may improve the quality of the forecasts substantially, and make it much easier for customers who are not familiar with the scientific issues of ENSO, to understand and use these forecasts.

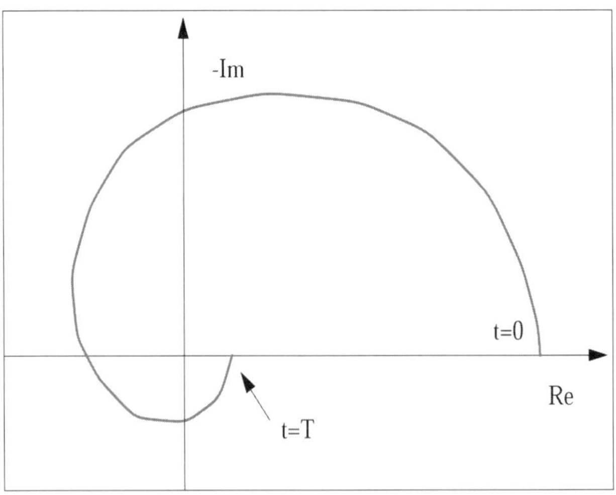

Fig. 16.17 Damped oscillation as described by a POP model in the complex plane.

16.6 Summary

In the first part of this paper the El Niño/Southern Oscillation phenomenon was described, and it was shown that the ENSO cycle is predictable on lead times up to one year. This is an important issue, since the climatic and economic impacts are tremendous.

Different types of ENSO prediction models were presented in the second part of the paper, and their specific strong and weak points were discussed. Nowadays the forecast skill obtained with coupled general circulation models is comparable to the skill achievable with much simpler and much more economical models like statistical methods or hybrid coupled models. However, coupled general circulation models show the highest potential for further improvements, and therefore further research should be devoted to these complicated and expensive models.

It is evident from the results found by different groups that data assimilation techniques may help to compensate for model errors or for wind forcing errors during the initialization phase. However, the main goal must be to improve the performance of the models in order to achieve higher forecast skills. An interesting and very promising approach is the coupled initialization of an ENSO forecast system. In this case the coupling shock at the beginning of a forecast can be avoided. Preliminary results further indicate that the low frequency component of the wind fields, essential for ENSO initializations, was better extracted from the wind data than in the case of uncoupled initialization.

This paper also indicates that the question of forecast verification needs to be addressed in more detail. In one example it was shown that the interpretation of the results critically depends on the definition of the skill measure.

Correcting known systematic errors of a forecast system is necessary to make the final product more understandable for people who are not familiar with the scientific issues of ENSO predictions.

Appendix A - EOF Expansion

Here we shall give a very brief description of the idea of Empirical Orthogonal Function (EOF) expansion. For a more detailed description we refer to the literature, e.g. (Lorenz 1956, Gallagher et al. 1991). The EOF expansion is a method to extract patterns of maximum variability from an arbitrary data set. Let us assume we have a system that can be described by a state vector $X(r,t)$ with n degrees of freedom (e.g. n is the number of points). Here r and t are arbitrary indices. In our context they stand in general for a spatial index and for a time index, respectively . Then this system has n normal modes. For a large system, these normal modes represent a huge number of data from which the important characteristics, i.e. the dominant normal modes, may be hard to extract. To overcome this problem, a coordinate transformation can be carried out. The state of the system can be described in a new orthonormal basis, the so called Empirical Orthogonal Functions (EOFs). The EOFs are defined as the eigenvectors of the covariance matrix of $X(r,t)$ i.e. the eigenvectors of :

$$\underline{\underline{C}}(r, r') = \langle X((r, t), X(r', t))\rangle \tag{A1}$$

where $\langle \ldots \rangle$ denotes an average over the index t, e.g. a time average. The first EOF is defined as the eigenvector associated with the largest eigenvalue the second EOF with the second largest and so on. In the new orthonormal basis an arbitrary state of the system can be expressed as :

$$X(r, t) = \sum_{r=1}^{n} y_j(t) F^i(r) \tag{A2}$$

in this equations the time invariant EOFs are denoted by $F^i(r)$ and the time dependent coefficients $y_j(t)$ are called *Principal Components* given by the projection of $X(r,t)$ onto the EOFs:

$$y_j(t) = \sum_{r=1}^{n} X(r, t) F^i(r) \tag{A3}$$

This new basis also spans the whole phase space but they have the advantage that they are ordered such that the first EOF accounts for the most variance, the second for the most of the residual variance of the data which is not accounted for by the first, and so on. Thus, most of the variance of a system can be expressed in a subspace spanned by the first m EOFs with m being considerably smaller than n.

Appendix B - POP Analysis

The *Principal Oscillation Pattern* (POP) analysis is a powerful method to extract oscillating patterns from an arbitrary data set. It is the linear approximation of the more general *Principal Interaction Pattern* (PIP) analysis (Hasselmann 1988, Gallagher et al. 1991, Storch et al. 1995). With the POP method the normal modes of a system with unknown dynamics, or very many degrees of freedom can be estimated. We assume to have a dynamical system with n degrees of freedom, which can be described by a state vector $X(t)$ of dimension n. Just for the sake of simplicity we further assume that t is a time index. The time evolution of such a system can be modeled as a linear vector process which can be written as :

$$X(t+1) = \underline{\underline{B}} \cdot X(t) + \zeta(t) \tag{B1}$$

where B is a constant (n x n) matrix, the so called process matrix, and ζ is some noise vector driving the system. The process matrix B can is estimated by fitting a data set to the linear vector process. It can be shown that B is given as :

$$\underline{\underline{B}} = \underline{\underline{C}}_1 \underline{\underline{C}}_o^{-1} \tag{B2}$$

where C_o and C_1 are the lag-0 and lag-1 covariance matrices :

$$\underline{\underline{C}}_o = \langle X(t), X(t) \rangle \tag{B3}$$

and

$$\underline{\underline{C}}_1 = \langle X((t+1), X(t)) \rangle \tag{B4}$$

The eigenvectors P^k of B are called *Principal Oscillation Patterns* (POPs). The matrix B is in general not symmetric so that in general the eigenvalues λ^k and eigenvectors are complex. Normally this set of eigenvectors is complete but not necessarily orthogonal. Since X and B are real, any complex eigenvalues and the corresponding eigenvectors occur in conjugate pairs. We may reconstruct the original time series in terms of the eigenvectors :

$$X(t) = \sum_{k=1}^{n} a_k(t) P^k \tag{B5}$$

The time evolution coefficients obey :

$$a_k(t+1) = \lambda_k a_k(t) + \zeta(t) \tag{B6}$$

In the absence of noise this becomes :

$$a_k(t) = \lambda_k^t a_k(0) \tag{B7}$$

If we express the complex eigenvalues λ^k as

$$\lambda^k = |\lambda^k| \exp(i\omega_k) = \exp(-t/\tau)\exp((i2\pi)/T_k) \tag{B8}$$

we obtain for each POP a characteristic e-folding (decay) time

$$\tau_k = -1/\ln|\lambda_k| \tag{B9}$$

and an oscillation period T_k

$$T_k = 2\pi/\omega_k \tag{B10}$$

Thus, the POP model describes a damped oscillation in the complex plane.

17 Toward Marine Environmental Predictions in the Mediterranean Sea Coastal Areas: A Monitoring Approach

NADIA PINARDI[1], FRANCIS AUCLAIR[2], CLAUDIA CESARINI[3], ENCHO DEMIROV[4], SERENA FONDA UMANI[5], MICHELE GIANI[6], GIUSEPPE MONTANARI[3], PAOLO ODDO[1], MARINA TONANI[4] AND MARCO ZAVATARELLI[1]

[1]*Corso di Scienze Ambientali, University of Bologna, Ravenna, Italy*

[2]*Istituto di Geologia Marina, CNR, Bologna, Italy*

[3]*ARPA, Bologna, Italy*

[4]*Istituto Nazionale di Geofisica, Rome, Italy*

[5]*Istituto di Biologia Marina, Trieste, Italy*

[6]*ICRAM, Venice, Italy*

17.1 Introduction

Marine environmental predictions involve the observation and modeling of physical, biogeochemical processes and parameters, fused by advanced data assimilation schemes that optimally merge the observational and numerical modeling information in order to produce forecasts. The result of such system should be interfaced with socio-economic models of sustainable development and management of marine resources. Fusing the deterministic dynamical information about the marine ecosystem with the socio-economic and political knowledge of the marine environment has not been tried yet but it is one of the outstanding challenges in integrated coastal management studies. This paper tries to show a scientific strategy to predict the physical and biochemical components toward marine environmental predictions, being aware that some considerations could be changed in view of the integration with socio-economic models and issues. Here we use prediction in total analogy with forecasting and thus as synonymous of deterministic prognosis of dynamical variables from a pre-defined initial state of the system.

The far reaching goal of marine environmental predictions is to advance the understanding of marine ecosystem dynamical processes, manage resources, handle emergencies in the coastal areas and mitigate their impact on the marine ecosystem. Users of such environmental prediction system should be the coastal guard services involved in the management and protection of the coastal zone waters, the environmental protection agencies responsible for the control of marine water quality and health status, fishery agencies and tourist industries. Last but not least, such systems should be part of the information database that could support the handling of marine disasters and emergencies.

The scientific strategy pursued in this paper is the one that considers monitoring networks producing the bulk of the observations capable of sustaining the modeling and prediction system. The modeling and prediction system are also permanent, meaning that a careful calibration of model parameters and validation of model simulation results is done before the observing and the model system is changed, based upon the experience with the previous permanent system components. A coastal environmental prediction system is, in our mind, based upon permanent observing systems, efficiently and scientifically designed to fit the modeling needs and continuously re-evaluated by producing forecasts which will enable to check the quality of both the observing and modeling components. This permanent system will support, in the future, the climate observing system in the coastal marine areas, producing long time series of data for trend analysis and global change signals detection.

In addition to the choice of a sustained monitoring and modeling system as a base to understand and predict the marine ecosystem, we concentrate our attention in the marine coastal areas, where hydrodynamic and biogeochemical cycles have short and small time and space scales. The observing system should resolve a relevant part of the dynamical field variability so that optimal estimates of the ecosystem state variables can be done in order to initialize forecasts. Both the measurement technologies and the modeling tools are barely adequate for coastal marine predictions, since there is a need to measure simultaneously physical and biogeochemical parameters at the synoptic time scales and the hydrodynamic models need high resolution.

Our study area is the Mediterranean Sea where a Mediterranean Forecasting System approach has started to be developed, which aims at the 'understanding and prediction of coastal primary producers biomass variability from the time scales of days to months' (Pinardi and Flemming, 1998). Brink and Robinson (1998) have recently discussed the coastal areas of the World Ocean in a global overview. In our view, the different coastal area ecosystems could be classified on the basis of the different values taken by a ratio between river and large scale flow field forcing in the shelf/coastal areas. This ratio in fact sets the basic hydrodynamic response, the nutrient and light levels. In the Mediterranean Sea very different coastal ecosystems coexist and in general this ratio varies an order of magnitude. We concentrate our test case in a river dominated shelf area such as the Northern Adriatic Sea where a monitoring approach toward environmental predictions has started.

Following Robinson and Leslie (1985), the forecasting capability can be achieved if three basic phases of scientific research and technology development are attained. The first phase consists of historical data collection, specific process oriented data collection exercises in dynamically active areas, numerical and theoretical dynamical studies of relevant processes. This phase is called 'descriptive' and in fisheries system analysis can be called 'conceptual model definition' (Grant, 1986). The second phase considers the calibration/validation of the numerical mod-

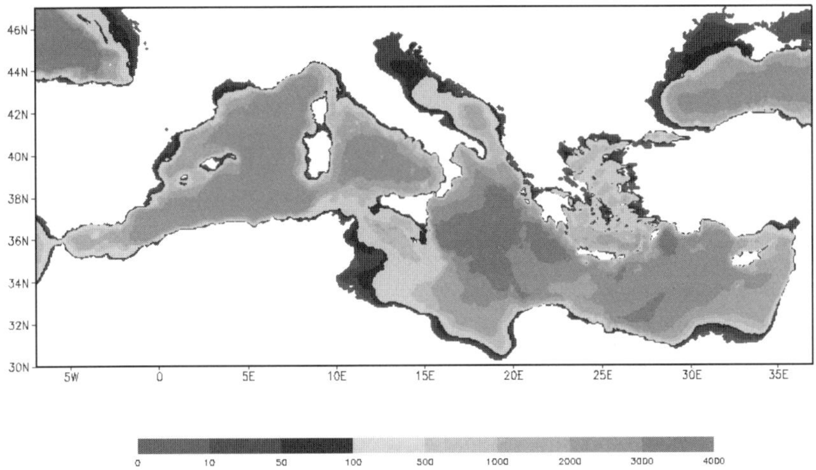

Fig. 17.1 Map of the Mediterranean Sea bathymetry with dark colors indicating areas shallower than 200 meters.

els based upon the capability to reproduce both the historical and newly collected observations. The third phase consists of active predictions where forecast skill is assessed on the basis of observations collected at a relevant time frequency and products are developed for users. During the second phase, the observing system is streamlined in order to become sustained and permanent in support of the third phase of forecasting. The optimal way to design the permanent observing system remains through Observational System Simulation Experiments (OSSE, Lozano et al., 1996) where it is possible to study of the impact of synthetic data assimilated into a numerical model in order to reduce the forecast error.

In the Mediterranean Sea the three methodological phases toward environmental forecasting have been partially achieved and results will be described in this chapter. The first section contains the description of the physical circulation and primary producer variability in the Mediterranean and the Adriatic Sea coastal areas with a particular emphasis to known and understood dynamical mechanisms. The second describes the overall system components for coastal environmental predictions. The third and fourth sections describe the observational and modeling system components implemented in the Mediterranean and Northern Adriatic Sea coastal areas. The fifth section illustrates the application of the methodology to the shelf areas of the Northern Adriatic Sea.

17.2 The Mediterranean Sea shelf and coastal ecosystems

The shelf areas of the Mediterranean Sea are depicted in Fig. 17.1, limited by the 200-m. isobath contour. They involve narrow coastal strips in the majority of the cases but some extended shelves are also present. In the Northern part, the extended shelf areas are adjacent to the river deltas such as the Ebro, the Rhone, the Po and the Neretva to mention only the major rivers. In addition to that, the Northern Aegean Sea has an extended shelf area partially forced by the low salinity water inflow from the Dardanelles and other local rivers. On the Southern part, the extended shelf areas consist of the Nile delta region and the Tunisian shelf on the Sicilian plateau.

The remaining basin regions are deep ocean areas with steep continental slopes. The abyssal basins can reach depths of 3500 m over extended areas. The Mediterranean Sea is considered by enlarge a deep ocean basin containing analogues of deep ocean circulation processes such as deep open ocean formation and meridional thermohaline overturning. Such deep, open ocean circulation is influencing the shelf/coastal circulation and its primary productivity especially when the slope is steep as in general is for the Mediterranean case (Csanady, 1982).

The winter average surface circulation produced by the general circulation model which will be described few sections later, is shown in Fig. 17.2a. In the same picture the winter average surface chlorophyll map from a satellite sensor in the visible range is also shown (Sea-viewing Wide Field-of-view Sensor, SeaWiFS, http://seawifs.gsfc.nasa.gov/SEAWIFS.html). First of all, the circulation is composed of intensified free and boundary currents and sub-basin scale gyres, both cyclonic and anticyclonic (Pinardi and Masetti, 2000). These circulation structures impinge on the narrow shelf areas partially driving the shelf flow field. Thus the shelf area circulation is connected to remote forcing such as the Gibraltar Strait inflow, the center of the basin wind vorticity input (Molcard et al., 2001) and deep/intermediate water formation processes (Nittis and Lascaratos, 1998, Haines and Wu, 1998).

The ecosystem response at the level of primary producers is large on the extended shelf areas, in correspondence with the major river estuaries and deltas, along the Northern and Southern coasts. However, the large scale flow field forces the ecosystem response inducing biomass horizontal transport, vertical upwelling of nutrients and determining the depth of the nutricline, associated with the seasonal thermocline formation. The influence of the large scale geostrophic flow field can be seen in the Gulf of Lyon Gyre area of Fig. 17.2b, where a relative maximum in Chlorophyll is reached around the border of the gyre. The negative zonal gradient between Western and Eastern Mediterranean basins is connected to the Atlantic water influence. The positive meridional gradient across the basin is sustained by isopycnal slopes in the first 100-200 meters favoring northern upwelling of nutrient rich waters and downwelling in the Southern regions and the presence of rivers in the Northern shelves (Crise et al., 1998).

Thus the overall basin primary productivity is dominated by the contribution of river inflow in the shelf areas and by the large-scale upwelling/downwelling circulation systems. Naturally we expect the primary production associated to mesoscale eddies to be very important, as already shown for the world ocean (McGillicuddy et al., 1998) and for the case of deep water formation processes (Levy et al., 2000). The evidence of important interactions between physical variability and biochemical cycles is evident from the satellite chlorophyll picture shown in Fig. 17.2c where the surface chlorophyll pigments concentrate in patches geographically correlated with vortices and gyres. However, more in situ work is still required to quantitatively estimate the importance of mesoscale variability on the primary productivity levels of the Mediterranean.

The absolute maximum in Chlorophyll concentration is achieved in the Adriatic Sea. The winter average SeaWIFS Chlorophyll distribution is shown again in Fig. 17.3a together with the winter mean circulation resulting from the regional circulation model of the Adriatic Sea described later. In correspondence with the intense Western Adriatic Coastal Current (WACC) the largest chlorophyll values are achieved along the Italian coastlines due to the advection of nutrients and chlorophyll (Zavatarelli et al., 2000, 2001). Here the interplay between general circulation transport and river sources is very clear showing how important are the concepts of downscaling properly from the large to the shelf regions and then to the coastal areas. The Italian side shelf slope is in fact relatively gentle but the high chlorophyll band extends up to the 100 meters isobath in the Middle and Southern Adriatic showing the effect of the coastal area in the deeper basin parts.

In Fig. 17.3b we show a single day picture of the surface chlorophyll which shows another important mesoscale to sub-mesoscale phenomena due to the Po river run-off plume (Kourafalou, 1999) in contrast with the average distribution of chlorophyll in the basin. These high frequency events may have the same influence on the primary productivity of the shelf areas as the long-term trends due to general circulation physical forcing.

The chlorophyll distributions shown above should be interpreted also as the manifestation of complex food web interactions which have been described in the past for the Mediterranean Sea by Legendre and Rassoulzadegan (1995). The schematic shown in Fig. 17.4 represents two different pathways that may be taken either in the open ocean or coastal areas of the Mediterranean Sea depending on the physical water column conditions. In the Mediterranean, at the contrary of what happens in other deep basin areas, these conditions can be present simultaneously in the shelf areas and the deep ocean. Both the shelf and deep ocean areas loose almost completely their stratification during winter thus producing a traditional herbivorous food web. In summer both the coastal and open ocean water column is very well stratified, thus favoring the set up of a microbial food web in competition with the herbivorous one. This microbial food web is common in shelf/coastal areas such as the Northern Adriatic (Vichi et al., 1998) and in the deep ocean areas where a subsurface Chlorophyll maximum is formed during summer. The major difference in the development of the food web between the open ocean and the shelf areas is given by the light conditions that in the coastal areas are strongly affected by suspended inorganic sediments from rivers. Light limitation is a very efficient process

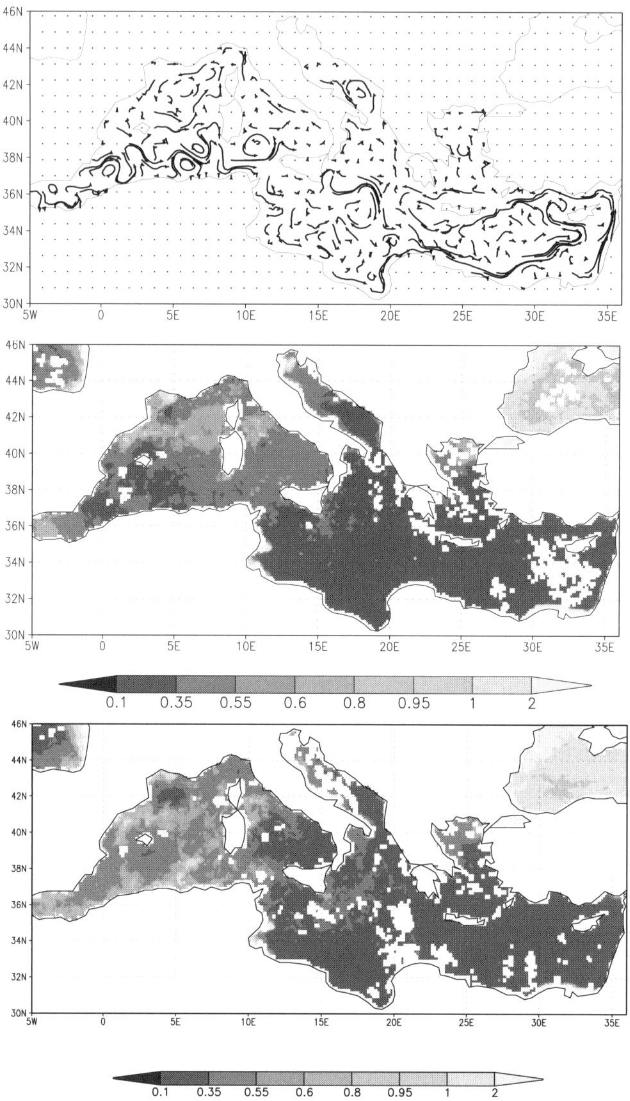

Fig. 17.2 Upper panel: winter mean circulation in the Mediterranean Sea as depicted from a model simulation that uses climatological atmospheric forcing. The model used is described in section 17.5. Middle panel: winter average surface chlorophyll from the SeaWiFS sensor. The average is made for the months of December 1999, January and February 2000 from 10 days composites of data. The completely white rectangles indicate absence of values due to clouds. The 10 days composites were produced during the Mediterranean Forecasting System Pilot Project by Dr. D. Antoine of LPCM, Villefranche, France. Lower panel: ten-day composite chlorophyll map from SeaWiFS sensor for the period 4-14 of February 2000.

in the coastal areas, capable of reducing the potential productivity of the local nutrient rich waters, while in the Mediterranean open ocean areas the light availability shifts the Chlorophyll maximum at the subsurface.

In synthesis, the shelf/coastal areas of the Mediterranean Sea are coupled to the open ocean flow field characteristics and forcing by many processes, such as:

1) The shelf net transport can be connected in several areas to basin scale features (boundary currents);
2) The open ocean currents are modified by their interaction with the shelf areas due to bottom stress compensation effects (Mellor and Wang, 1996) and/or JEBAR processes (Joint Effects of Baroclinicity And Relief, Huthnance, 1984) that can modify the proportion of barotropic/baroclinic energy in the flow field;
3) The shelf areas can produce rectification or damping of wave energy coming from the open ocean through tidal internal or external waves, waves generated by storms far off-shore, etc., enhancing also wave-tidal components energy interactions;
4) The temperature extreme in the shelf areas can drive amplified air-sea flux exchanges as shown by Roussenov et al. (1995) and Castellari et al. (1998). Several upwelling areas are known to exist on the southern and eastern boundaries of the Mediterranean (Eastern Aegean Sea, Eastern Adriatic Sea and southern Sicilian coasts) and this creates large air-sea temperature differences that enhance the turbulent surface flux exchanges. The Northern African coasts have a relative maximum of temperatures produced by downwelling favorable winds and large positive heat net fluxes. Due to this, the Northern African shelf areas show the basin largest heat gains during summer and largest heat losses during winter;
5) Lateral heat and salt fluxes at the slope/open ocean interface can drive the shelf areas. An example of this process is realized in the Adriatic Sea which looses heat during winter on its Northern shelf areas and regains it from the Otranto Strait at the subsurface (Artegiani et al., 1997). This lateral flux compensation mechanism can be delayed in time thus giving rise to a possible nonlinear feedback between the open ocean and the shelf areas. The lateral salt flux feedback can also be active due to the river runoff effects that are in the long term compensated by saltier waters arriving on the shelf;
6) The shelf areas serve as a source of dissolved nutrients, particulate and dissolved organic matter and inorganic matter for the open ocean areas. This source controls the amount of nutrients exported in the deep ocean areas by simple transport mechanisms and/or regeneration of organic material by the pelagic food web. On the other hand, in upwelling coastal areas, the shelf areas work as a sink of nutrients for the deep ocean. This complex interaction between shelf and open ocean areas is mediated by the pelagic and benthic food web dynamics;

a)

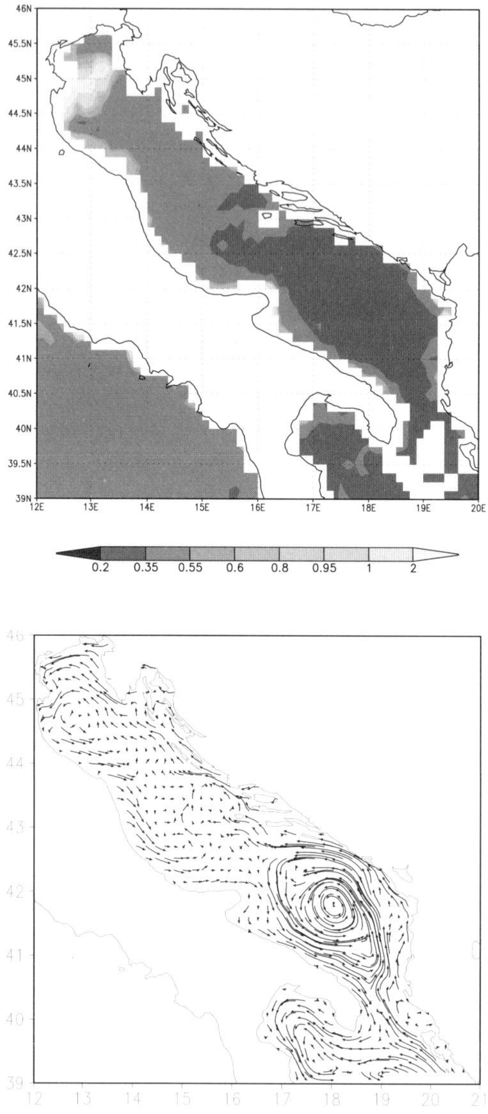

Fig. 17.3 a) Upper panel: winter average surface chlorophyll from the SeaWiFS sensor. The average is made for the months of December 1999, January and February 2000 from 10 days composites of data. The white rectangles indicate absence of data due to clouds. The 10 days composites were produced during the Mediterranean Forecasting System Pilot Project by Dr. D. Antoine of LPCM, Villefranche, France. Lower panel: winter mean circulation in the Adriatic Sea as depicted by a model simulation with climatological atmospheric forcing. The model is described in section 4. b) Instantaneous Chlorophyll distribution for june 16, 2000 produced by Dr. R.Santoleri of IFA-CNR, Rome, Italy.

b)

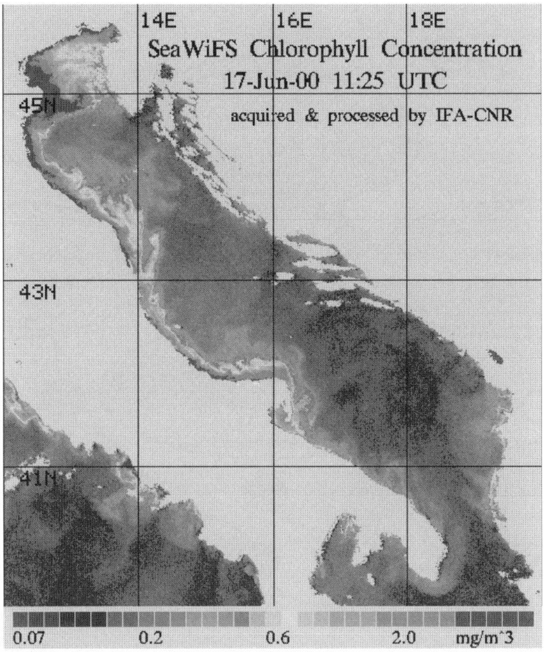

Fig. 17.3 (continued).

7) All sedimentary processes and material fluxes from the coastal areas to the deep ocean are a complex interaction of local effects and remote forcing from the deep basin currents. In the Mediterranean Sea canyons are believed to be the site of these important carbon exchange mechanisms (DeMadron, 1994).

The importance of the interaction between the shelf and open ocean areas forces us to take a precise observational and modeling strategy that tries to develop both the large scale and shelf/coastal scales systems together. The system components will be described in the section that follows.

17.3 The coastal environmental predictions system

Any geophysical prediction problem has inherent scales of unpredictability connected to the nonlinearity of the dynamical equations and the errors in the knowledge of the initial conditions over the domain of interest. The classical problem of predictability for the atmosphere has been studied first by Lorenz (1963) and re-examined later by Farrell (1988) and Farrell and Moore (1992) for the ocean. The intrinsic nonlinearity of the hydrodynamic equations and physical parameterizations amplifies the errors in the initial conditions so that approximately after few days the difference between the forecasted fields and the observations is not differ-

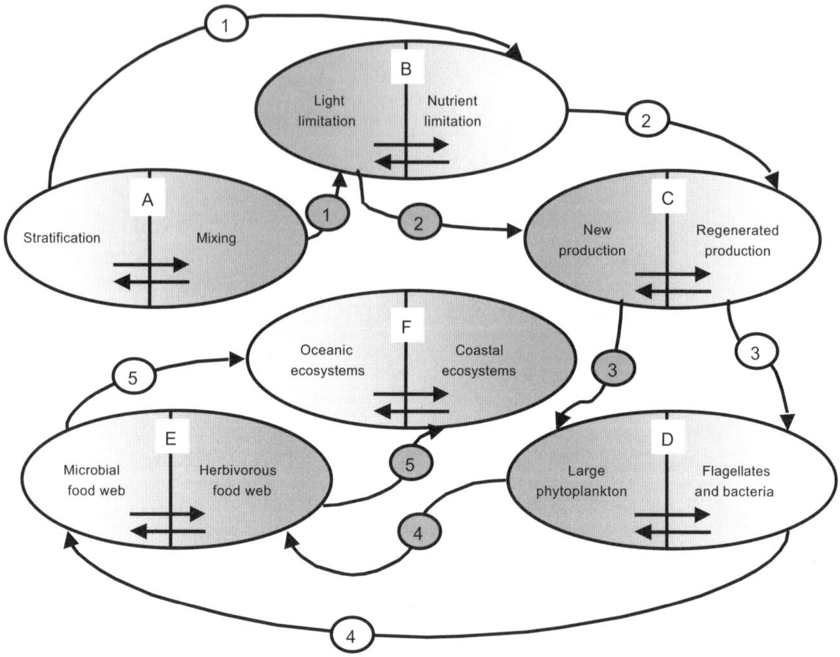

Fig. 17.4 Redrawn from Legendre and Rassoulzadegan (1995) schematic of trophic pathways for open ocean and coastal regions.

ent from what it could have been generated by a random initial condition. The measure of the difference between forecasts and observations is calculated in terms of root mean square and correlation errors.

The predictability time scale depends on many factors such as the observing system that provides the observations, the model and the assimilation scheme used. Scientific and operational advancements in each of these components have extended the limit of predictability for the atmosphere to several days instead of few. Lately coupled ocean-atmosphere models have extended useful atmospheric predictions to several months for some key dynamical variables such as atmospheric precipitation and sea temperature (Fischer, 2001).

The ocean currents predictability studies have started in the eighties with the forecasting of mesoscale eddy fields (Robinson, 1999). The time scales of predictability was found to be of few weeks and it was related to the knowledge of the initial condition. The longer oceanic predictability time scale with respect to the atmosphere can be heuristically related to the different stability of the two flow fields and the higher nonlinearity of the atmospheric physical parameterizations, such as the cloud convection processes. The ocean mesoscale prediction problem considers also the loss of predictability due to open boundary condition effects. This factor can be as large as the internal nonlinearity of the flow field in limiting the predictability of the flow field (Walstad and Robinson, 1990).

The first assessment of the predictability of the flow field was done without considering the effects of atmospheric forcing on the oceanic mesoscale flow field. For the coastal areas, the uncertainties in the definition of the atmospheric forcing, with respect to those in the definition of the initial condition, could be the most severe limit to the predictability of the flow field. In addition to that, the memory mechanism of the coastal circulation is most likely coming from the lateral boundary conditions especially if boundaries are put on the shelf slope or even in deeper parts of the basin. The imprecise knowledge of these lateral boundary conditions could be another serious limitation to the predictability time scales of the flow field. Furthermore, nesting techniques could induce adjustment processes in the interior flow field that could severely limit the predictability of the shelf areas. Lastly, if the coastal area is dominated by river discharges, the specification of the characteristics of the river runoff can be seen as a serious limitation to the predictability time of the flow field.

In conclusion, the coastal environmental prediction problem has a multiplicity of system components that could limit the predictability time. They are: 1) the limited predictability of the atmospheric forcing directly influencing the coastal dynamics; 2) the lateral boundary condition uncertainties, meaning with that both the open boundary conditions and river runoff uncertainties, affecting the long term memory of the system and the major driving mechanisms; 3) the initial adjustment processes to the nesting techniques; 4) the initial condition specification for all the dynamical variables of interest; 5) the flow field nonlinearities.

In order to reduce some or all the possible sources of unpredictability for the coastal flow field, the coastal environmental problem has to consider the design of system components for both the large and coastal scales. In Fig. 17.5 we illustrate the two major system components, e.g., observing and modeling, for the large and shelf scales, with data assimilation being part of modeling.

Starting from the observing system components, the large-scale part has been developed in the past twenty years and it consists of interlinked networks of different measurements providing high quality observations of the surface and ocean interior dynamical variables. As said in the introduction, our attention is focused on the permanent and sustained observing systems and thus Fig. 17.5 lists only these kind of measuring networks. The major global systems are satellite based with sensors indirectly measuring sea surface temperature, sea surface height, surface winds, and surface color. For ice covered areas these satellites retrieve also ice extent and thickness. Today, the satellite data sets provide the highest quality synoptic information at global scale and they are the essential component of the future permanent observing system for the oceans. However, these global satellite systems do not provide high quality data in near coastal areas due to problems of signal interference with the nearby land areas. This is a technological problem that should be solved in the near future.

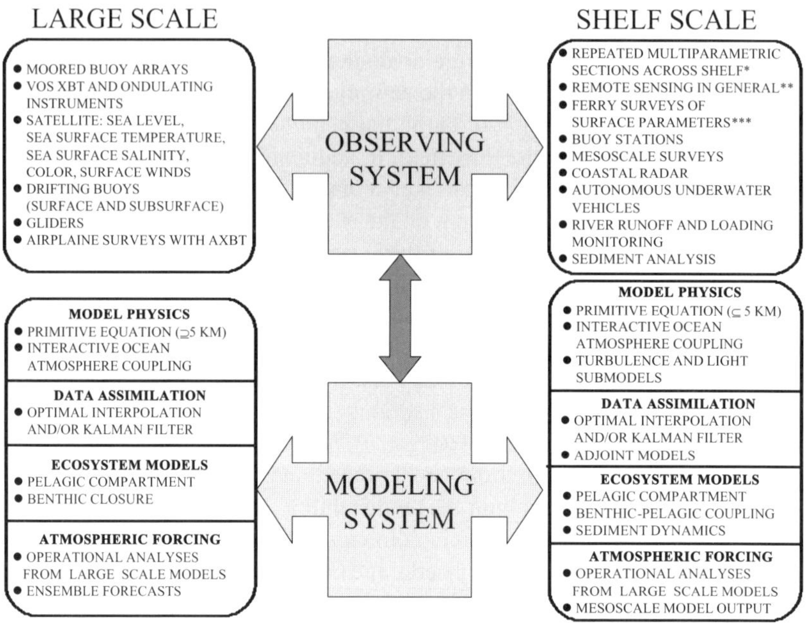

Fig. 17.5 System components and methods schematic for environmental predictions. Note: (*) multiparametric measurements could include: CTD profiles, dissolved oxygen profiles, florescence profiles, dissolved nutrient profiles, current profiles with ship bottom mounted ADCP (Acoustic Doppler Current Profiler). (**) Remote sensing in general includes both satellite imagery and airplane surveys. (***) Ferry surveys are the analog of the VOS system for the shelf areas, allowing for very frequent and automatic recording of surface waters parameters, including the measurement of several physical and biochemical variables.

The other global permanent and sustained systems are connected to the Ship of Opportunity Program network that has both surface and subsurface measurements of temperature and air-sea parameters (Molinari et al., 1999). The subsurface measurements are made with expandable sensors, which sample approximately the first 1000 meters of the water column. The second global in situ observing platform is associated with moored buoys where different subsurface instruments sample the temperature, salinity and velocity field. The most successful example of such system is working in the tropical Pacific with TAO/TRITON buoys (McPhaden, 1995). The other systems listed in Fig. 17.5 are largely under research development and they have never been used yet operationally. In the next five years a major world wide experiment will release thousands of subsurface free-drifting floats that measure temperature-salinity profiles in the first two kilometers of the water column (ARGO, http:\\www.argo.ucsd.edu). Gliders, that are autonomous underwater vehicles changing their buoyancy while they travel, can offer an open ocean monitoring strategy in the future, complementing the VOS, ARGO e satellite systems.

The global ecosystem monitoring is out of the scope of this paper and it is still at its infancy. The satellite based surface chlorophyll estimates from radiative measurements (ocean color imagers) are the first truly global observations of the ocean primary producers biomass at high space and time resolution (1x1 km^2 and every day). The routine multidisciplinary measurements from Ship of Opportunities are done in few regions with towed or undulating instruments, which however are far from being proven reliable in a world wide operational sense, with near real time data transmission. The most advanced technologies to measure multiple ecosystem dynamical variables, such as Photosinthetically Available Radiation (PAR), florescence, turbidity, nutrients (in particular nitrates), small pelagic fish, etc. are placed on moored platforms, an example of which is the Bermuda Testbed Mooring station (Dickey et al., 1998).

In the coastal areas, the monitoring systems are mainly based upon dedicated ships CTD surveys across the coastal gradients, on the continental shelf. Multidisciplinary observations are more easily done again by traditional water sampling methods on board of the same dedicated ship which measures the physical parameters. To this traditional measuring system, there is associated a moored buoy system, which records the surface air-sea interaction parameters, the waves and subsurface current, temperature and salinity structure. Recently, lands fixed HF-Radar systems have shown to produce high quality data of surface currents in coastal areas. The coastal area has been the site of recent technological advancements such as Autonomous Underwater Vehicles (AUV) and gliders that record different quantities and are designed to accomplish several complete surveys of the coastal areas (Glenn et al. 2000a). The most important and unique monitoring aspect of the coastal areas is the measurement of river runoff and water quality parameters (meaning nutrient loads) since it is a major driving element of the coastal ecosystem. This may need measuring stations at upstream river checkpoints and buoy stations at sea to monitor suspended inorganic sediment loads, nutrients and organic material outflow. Different systems are at work now and they are reviewed by Glenn et al. (2000b).

Bearing in mind the complexity of the observing system, the modeling and data assimilation components should be designed to consider all the dynamical and diagnostic variables measured by the observing system components outlined above (Fig. 17.5). This means that both the large and coastal/shelf scale hydrodynamics should be primitive equation, with sophisticated air-sea interaction physics that parameterize adequately the coupling with the atmosphere. In addition, several physical submodels should be interfaced with the resolved hydrodynamics, such as the turbulence, the bottom boundary layer, light and heat radiative penetration and the sediment resuspension, transport and deposition. Above all, the important question of nesting of finer grid models into coarser ones is crucial to progress in the direction of quality forecast in the coastal areas. At the same time the atmospheric forcing fields should be at relatively high resolution (10 km or less) and used at relatively high frequency (only 1-2 hours instead of six or more) which adds the need for a systematic increase of the atmospheric operational models resolution.

The ecosystem models for the coastal areas should consider quite complex food web interactions and several species of primary producers functional groups to cope with the relatively high biodiversity of the coastal ecosystem at the pelagic and benthic levels. The relative availability of multidisciplinary data sets in coastal areas with respect to the open ocean makes it possible to check and calibrate several parameters of complex ecosystem models. Such complex models could be able in the future to predict harmful algae blooms and help to assess health and productivity of the ocean issues associated with changes in community structure and species.

The data assimilation component is a relatively young branch of research in oceanography and a very new one for ecosystem models. However, in the past ten years, the large-scale hydrodynamic community has developed robust assimilation methods for all the observational data sets discussed above and pre-operational ocean forecasting systems are now capable to assimilate a combination of satellite and in situ measurements collected at the global scale. Data assimilation for coastal hydrodynamic models is very novel and only few examples exist of successful initialization with in situ data sets (Lermusiaux, 2001). In addition to the traditional problem of assimilating the available data, the nesting of the coastal model within the coarser model introduces the coarser model fields as an additional type of data to be considered for assimilation. The coarser grid information is very important and should be melded efficiently with the observational data in order to specify an optimal initial condition for the high-resolution coastal model. A first solution to this problem has been found by using a tangent linear model to downscale the initial and boundary conditions from a coarse to an high-resolution model (Auclair, 1999). This technique is very promising but in the future it should be used in conjunction with the assimilation of observed data.

In synthesis, the coastal environmental prediction problem is dependent upon the existence of: 1) large scale optimal estimates for initial and boundary conditions; 2) high resolution atmospheric models; 3) complex ecosystem models which require multidisciplinary data sets and river run-off and nutrient loading data for initialization and forcing; 4) data assimilation techniques which could consider coarse resolution fields and field observations in the right sequence.

In the following section we show the application of some of the concepts of Fig. 17.5 in the Mediterranean Sea.

17.4 The Mediterranean large scale and Adriatic coastal scale observing systems

The Mediterranean Sea has been explored in the past twenty years by many international research programs that have defined the structure of the general circulation, collected and analyzed historical data sets and constructed numerical models of the large and regional scale circulation. This has put the basis for the initiation of data assimilation experiments, the start of calibration/validation of the modeling components and the set up of the permanent observing system elements.

A plan on the lines explained in Fig. 17.5 has been lied down (Pinardi and Flemming, 1998) which, at the end, aims at the forecast of primary production in the Mediterranean Sea coastal areas from days to months. Such a prediction could have consequences for the tourist industry and the fisheries but the way the prediction will affect such activities is not yet clear. This is because policy and socio-economic elements should be studied together with the deterministic predictions in order to be able to see the impact of forecasting in the management of the coastal areas. Forecasting the Mediterranean marine areas state variables makes possible to initialize coupled regional ocean-atmosphere models with assimilated ocean initial conditions. Coupled ocean-atmosphere models at Mediterranean scale may be important to increase the predictability time scale for the hydrological cycle over these areas.

Following the plan, a large scale observing system for the Mediterranean has been set up and it is schematized in Fig. 17.6a. Its single components work in near real time meaning that observed variables are exchanged with a short delay (1-2 days maximum) between data collection and the modeling centers. The forecasting/modeling center receives both the satellite and in situ data sets with a one-two days delay with respect the last day of data collection every week. Two are the novelties of the system implemented in the Mediterranean: first the Voluntary Observing System (VOS) was designed at high resolution and secondly a new type of moored buoy was designed for multidisciplinary measurements.

The VOS tracks for the Mediterranean are described in Fig. 17.6b; they are occupied twice a month with 12 n.m. resolution XBT down to 700 m (for a complete description, see chapter in this book). This spatial frequency requires technical personnel on board of the ships allowing for a much greater control of the quality of the XBT profile. The north-south tracks take less than one day to be done while the one across the longitudinal extension of the Mediterranean is two weeks long. The implementation of the observing system at the Mediterranean basin scale has been shown to be feasible but major data transmission problems have been encountered during the winter months where the temperature profiles are practically uniform and the standard software does not easily compute decimated data. The XBT data is sent to the Global Teleconnection System (GTS, Daley, 1991) but at the same time it is stored at a local Mediterranean data collection center.

The buoy system designed for the Mediterranean is similar to the TAO/TRITON buoys (McPhaden, 1995) but with the addition of biochemical measurements. The

a)

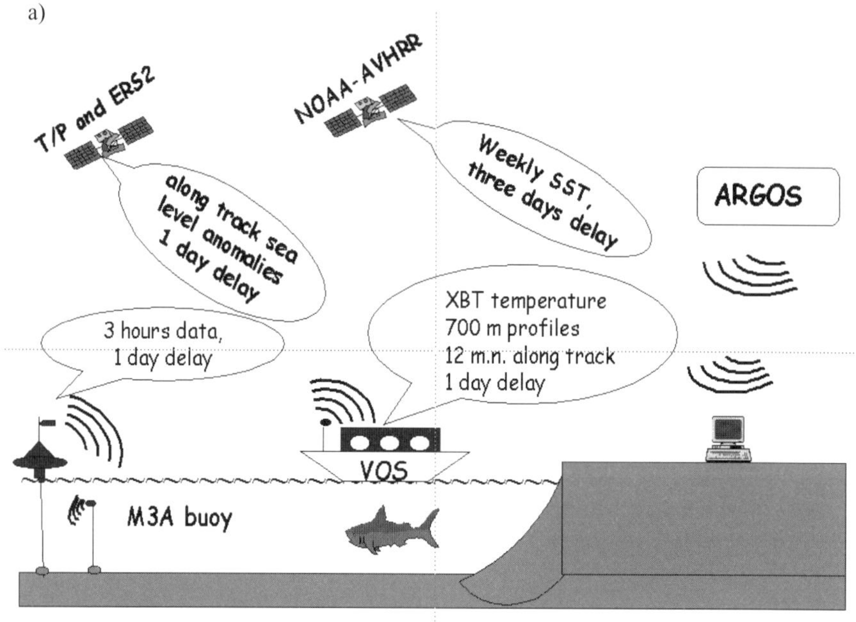

b)

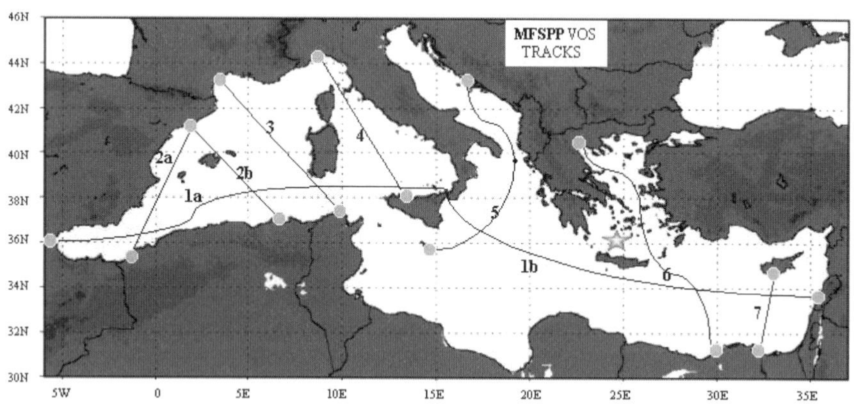

Fig. 17.6 The large scale monitoring system of the Mediterranean Sea working in near real time. a) the schematic of all the components; b) the implemented tracks for the Voluntary Observing Ship subsystem (see Manzella et al., 2001).

Multidisciplinary Mediterranean Moored Array (M3A) is a buoy system with three moorings, each of them specialized for a selected task, at a distance of less than a kilometer from each other, communicating with underwater acoustic telemetry. The "mother" buoy is the air-sea interaction buoy and it is made of a large surface buoy with air-sea interaction parameter sensors and subsurface fixed depth sensors measuring temperature, conductivity, dissolved oxygen down to 500 meters. The second buoy is constituted of a upward looking ADCP located at about 500 meters which measures the current profile with a resolution of about 20 meters. The third mooring has sensors placed at high resolution in the first 100 meters corresponding to the upper thermocline region and the euphotic zone for the Mediterranean open ocean waters. The fixed depth sensors measure temperature, conductivity, and dissolved oxygen in the second mooring and in addition to those turbidity, fluorescence, PAR and nitrates on the third mooring line. This new design allows for easy maintenance of the third mooring line every two months and the continuation of the observations even in case of failure of parts of the system. The first and third line are connected with subsurface acoustic telemetry, which allows the real time transmission of data from the mother buoy to the land stations via cellular phone technologies or satellite links.

The most important part of the real time observing system in the Mediterranean is given by the satellite sea level anomalies (SLA) that are computed along the tracks of the two different satellites having altimeters on board. As an example, the track coverage given by Topex/Poseidon and ERS-2 for a two week period is shown in Fig. 17.7. This data set gives information on the mesoscale eddy field but for the Mediterranean the signal is small (order of several cm). In the past, the signal was analyzed in order to show that the orbit corrections could be done with the required accuracy (Larnicol et al., 1995) and finally the sea level anomalies were shown to be consistent with in situ observations. In addition to this important data set, the remotely sensed sea surface temperature is also analyzed in near real time with multichannel algorithms and released for assimilation every week.

The satellite data sets together with the VOS system for the Mediterranean form the basis of the large scale observing system. However, there are unsolved problems related to this system that it is interesting to discuss. First of all the system does not consider any salinity measurements yet that is a serious lack of information for the Mediterranean. Recent developments in satellite sensors, the ARGO floats and gliders could solve the problem but the sampling strategy should be evaluated. Second of all, the satellite altimeter still gives only sea surface topography anomalies and there is a problem in the definition of an optimal "synthetic geoid" to retrieve the absolute dynamic topography which is the observable for the dynamical model. Thirdly, the VOS system is quite expensive because it is man-managed and environmentally polluting due to the loss of instruments on the ocean floor. Concentrating technological development into these areas should solve these problems in the near future.

Coupled to this large-scale system we conceive a coastal system that monitors the same quantities of the basin scale system but at higher space-time resolution to fit the needs of the high-resolution coastal modeling. This coastal observing system

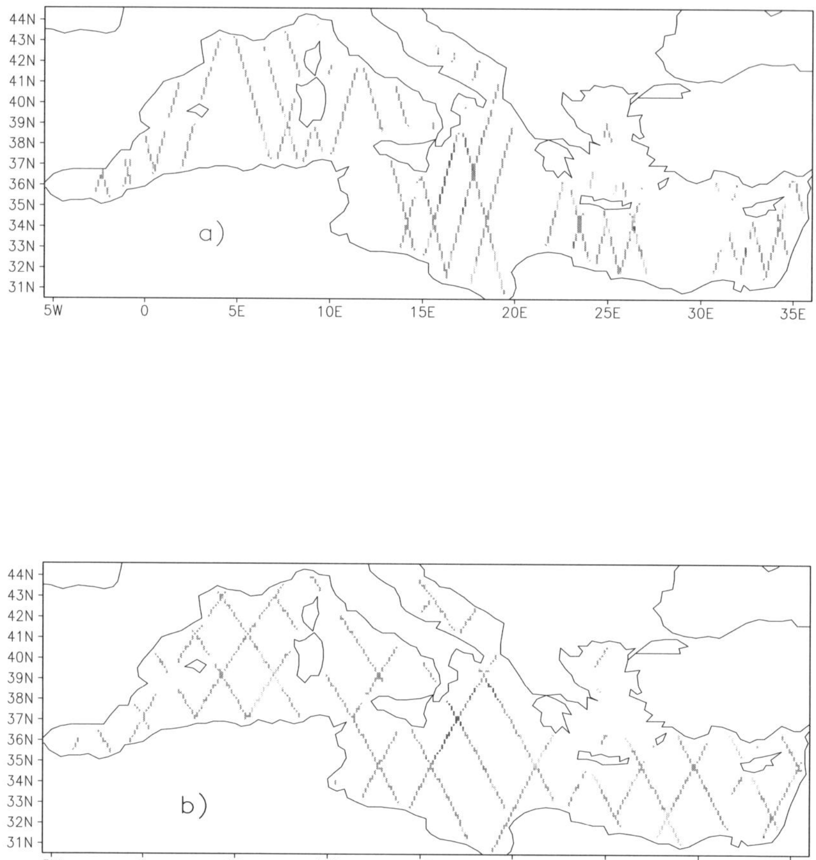

Fig. 17.7 Number of radar altimeter tracks crossing the Mediterranean every two weeks from ERS-2 (a) and Topex/Poseidon (b) satellites. The data along track are nominally spaced 7 km a part.

already exists in the Northern Adriatic coastal areas due to the societal pressure to monitor and preserve the marine environment for the tourist industry and the fisheries. Thus the Northern Adriatic will be a perfect test site to try to implement the concepts of environmental forecasting explained in the section above. The basic coastal observing system components are shown in Fig. 17.8 for the Emilia-Romagna coastal areas and the Trieste Gulf.

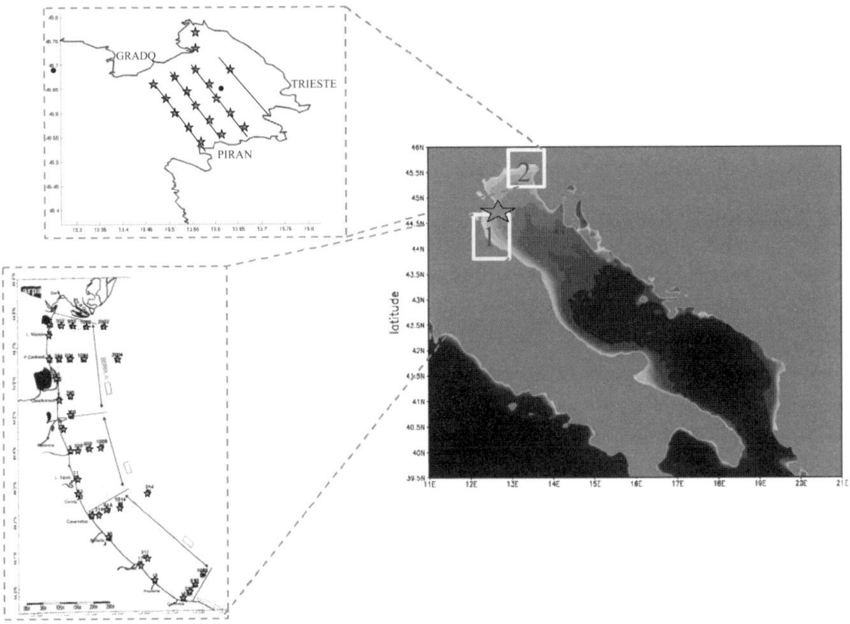

Fig. 17.8 The coastal observing systems in the shallow coastal areas of the Northern Adriatic. The Emilia Romagna panel illustrates the CTD stations covered in half a day every week by the Daphne motorboat of the Agenzia Regionale Protezione Ambiente (ARPA) of Cesenatico (Italy). The Trieste Gulf panel illustrates the monitoring stations covered at least once a month by the group of the Laboratorio di Biologia Marina of the Trieste University. The star shows the location of the S1 monitoring station.

The coastal monitoring consists of across shelf repeated sections at specified time and space frequency. For the Northern Adriatic, eutrophication in the eighties and mucilage (Sellner and Fonda Umani, 1999) in the nineties was important enough that a monitoring system was established in the near coastal areas. The monitoring stations acquire both physical, chemical and biological data, such as temperature, salinity, dissolved oxygen and nutrients, phytoplankton biomass and species, PAR profiles, etc. On the basis of such data sets, empirical indices of the system health status and change have started to emerge. One example is the TRIX, Trophic Index (Vollenweider et al., 1998), which can describe the status of the coastal system in cases of abnormal algae blooms and anoxia in the water column.

The combination of this coastal scale monitoring with the basin scale system is at the base of the forecasting activities that will be illustrated below. It is worth to point out that the coastal and basin scale observing systems are complementary in space and time, with the coastal system focusing on the narrow coastal area regions. This fits the perception, demonstrated later by the modeling activities, that observations are most needed in the very shallow coastal strips, where the dynamical regimes could be local and the satellite data fails to give accurate information.

17.5 The modeling system and the data assimilation components

The ocean models for predictions started to be developed at the beginning of the eigthies and they consisted of simplified physics equations (quasigeostrophy, Robinson and Walstad, 1987, multilayer reduced gravity, Hurlburt, 1986, shallow water, Prandle, 2001). They were used to forecast the mesoscale variability and the coastal sea level. At the contrary of the atmosphere, ocean forecasting started from the mesoscale and the coastal areas, in view of the high observing system requirements and following the users needs.

Data assimilation techniques were developed first for mesoscale forecasting experiments to control the problems of initialization that can severely limit the predictability time (Walstad and Robinson, 1990). The assimilation techniques concentrated on the usage of satellite altimetry (De Mey and Robinson, 1987), sea surface temperature and temperature profiles (Derber and Rosati, 1989). They all set up the scene for the advancements in ocean data assimilation of the second half of the eighties and the nineties (Fukumori et al., 1993, Carton et al., 2000, Masina et al., 2001, De Mey and Benkiran, 2001).

In the following sections we will try to overview the state of the art theoretical framework for the hydrodynamic and ecosystem modeling. The data assimilation component will be only mentioned and the interested reader should consult the data assimilation chapters of this book.

17.5.1 The large and coastal scale hydrodynamic modeling

Hydrodynamic modeling in the ocean is based upon numerical models developed in the seventies. Recently new models have been developed (Haidvogel and Beckman, 1998) but they have not been extensively used yet for forecasting. Here we illustrate two of the most commonly used models, one for the large and the other for the shelf/coastal scale. We distinguish between the two scales because historically models were developed in a different way for the two different environments, even if they are quite similar at the level of governing equations. The large scale modeling community concentrated mainly in the along isopycnal and non-viscous dynamics, thus building models with different subgrid scale and physical turbulence parameterizations from the coastal models. The latter gave attention to the bottom boundary layer and the mixing parameterizations, the sea level variability connected to ageostrophic effects, such as storm surge and tidal dynamics. The two modeling communities started to merge in the nineties but several differences remain in model formulation even today.

The large scale model described here is a version of the Modular Ocean Model (MOM, Bryan, 1969, Pacanowsky et al., 1993) that is a rigid lid, primitive equation model with centered finite differences in horizontal and vertical. The coastal scale model is the Princeton Ocean Model (POM, Blumberg and Mellor, 1983) which is free surface and uses finite difference both in vertical and horizontal but with a vertical coordinate transformation, which will be discussed later. Outstanding differences between the two models are: the consideration of metric terms in the basic equations to account for earth's spherical geometry, the assumption of free and rigid lid air-sea interface, the usage of tidal forcing. We will describe in detail both modeling systems outlining the similarities and differences.

The basic model formulation use the hydrostatic, incompressible and Boussinesq approximation to the Navier-Stokes equations written in a rotating frame. Thermodynamics state variables are temperature, T, and salinity, S, and the first law of thermodynamics is written neglecting compressible effects. The equation of state for seawater density, ρ, is normally a polynomial approximation in T, S and pressure, p, to the complete equation of state (Fofonoff and Millero, 1993).

The basic large-scale equations are written in spherical coordinates (λ,θ,z), that is, the longitude, latitude and the local normal to the earth surface, and they are:

$$\frac{\partial u}{\partial t} + \vec{u} \cdot \vec{\nabla} u - tg\theta\frac{uv}{r} - 2\Omega\sin\theta v = -\frac{1}{\rho_0}\frac{1}{r\cos\theta}\frac{\partial p}{\partial\lambda} - A_h\nabla^4 u + A_v\frac{\partial^2 u}{\partial z^2} \tag{1}$$

$$\frac{\partial v}{\partial t} + \vec{v} \cdot \vec{\nabla} v + tg\theta\frac{u^2}{r} + 2\Omega\sin\theta v = -\frac{1}{\rho_0}\frac{1}{r}\frac{\partial p}{\partial\theta} - A_h\nabla^4 v + A_v\frac{\partial^2 v}{\partial z^2} \tag{2}$$

$$\frac{\partial p}{\partial z} = -\rho(x, y, z, t)g \tag{3}$$

$$\frac{1}{r\cos\theta}\left\{\frac{\partial u}{\partial\lambda} + \frac{\partial}{\partial\theta}(v\cos\theta)\right\} + \frac{\partial w}{\partial z} = 0 \tag{4}$$

$$\rho = f(T, S, p) \tag{5}$$

$$\frac{\partial T}{\partial t} + \vec{u} \cdot \vec{\nabla} T = -K_h\nabla^4 T + (\delta + K_v)\frac{\partial^2 T}{\partial z^2} \tag{6}$$

$$\frac{\partial S}{\partial t} + \vec{u} \cdot \vec{\nabla} S = -K_h\nabla^4 S + (\delta + K_v)\frac{\partial^2 S}{\partial z^2} \tag{7}$$

where: $\vec{u} \cdot \vec{\nabla} \equiv \frac{u}{r\cos\theta}\frac{\partial}{\partial\lambda} + \frac{v}{r}\frac{\partial}{\partial\theta} + w\frac{\partial}{\partial z}$, r is the earth's radius, Ω the earth's rotation rate, ρ_0 is a constant density value, A_h, A_v the constant eddy viscosity coefficients, K_h, K_v the constant eddy diffusivity coefficients, δ the parameterization of convective adjustment and $\vec{u} = (u, v, w)$ the three dimensional velocity vector.

To write these equations three major assumptions are considered:

1) The expect ratio, $\delta = \frac{L}{r}$ where L is the horizontal length scale, is small so that terms with w drop out from the velocity equations (Pedlosky, 1979).

2) The horizontal eddy viscosity and diffusivity are scale selective, e.g., biharmonic, and coefficients should be kept small depending on the numerical resolution to allow for along isopycnal transport of water masses.

3) Vertical eddy viscosity and diffusion coefficients are constant but a parameterization of convective adjustment is provided. This assumption is empirical, however for intermediate and deep water convection events it is found (H. Buchard, personal communication) that this parameterization gives similar results to turbulence submodels such as Mellor-Yamada schemes (Mellor and Yamada, 1982).

Another large-scale modeling assumption is that the model is rigid lid, e.g. w=0, thus inhibiting external gravity waves effects on the general circulation. This implies that the free surface is a diagnostic quantity (Pinardi et al., 1995) and that sea surface slope is determined indirectly by the general circulation forcing.

The coastal scale models normally do not use spherical coordinates since they apply to limited areas of the ocean basins, these regions being normally nested within large-scale models. Thus coastal coordinates are rectangular with $\delta x = r\cos\theta\delta\lambda$, $\delta y = r\delta\theta$, and the equations are written as:

$$\frac{\partial u}{\partial t} + \vec{u} \cdot \vec{\nabla} u - 2\Omega\sin\theta v = -\frac{1}{\rho_0}\frac{\partial p}{\partial x} + \nabla \cdot (A_h \nabla u) + \frac{\partial}{\partial z}\left(A_v \frac{\partial u}{\partial z}\right) \quad (8)$$

$$\frac{\partial v}{\partial t} + \vec{u} \cdot \vec{\nabla} v + 2\Omega\sin\theta u = -\frac{1}{\rho_0}\frac{\partial p}{\partial y} + \nabla \cdot (A_h \nabla v) + \frac{\partial}{\partial z}\left(A_v \frac{\partial v}{\partial z}\right) \quad (9)$$

$$\frac{\partial p}{\partial z} = -\rho(x, y, z, t)g \quad (10)$$

$$\frac{\partial u}{\partial x} + \frac{\partial v}{\partial y} + \frac{\partial w}{\partial z} = 0 \quad (11)$$

$$\rho = f(T, S, p) \quad (12)$$

$$\frac{\partial T}{\partial t} + \vec{u} \cdot \vec{\nabla} T = \nabla \cdot (K_h \nabla T) + \frac{\partial}{\partial z}\left(K_v \frac{\partial T}{\partial z}\right) + \frac{1}{\rho_0 C_p}\frac{\partial I}{\partial z} \quad (13)$$

$$\frac{\partial S}{\partial t} + \vec{u} \cdot \vec{\nabla} S = \nabla \cdot (K_h \nabla S) + \frac{\partial}{\partial z}\left(K_v \frac{\partial S}{\partial z}\right) \quad (14)$$

where now simply $\vec{u} \cdot \vec{\nabla} \equiv u\frac{\partial}{\partial x} + v\frac{\partial}{\partial y} + w\frac{\partial}{\partial z}$ and C_p is the specificnheat of the sea water. The heat equation considers radiative heat penetration assuming that

$$I(z) = Q_s\left\{\sum_{i=1}^{N} Tr_i e^{\gamma_i z}\right\} \tag{15}$$

where Q_s is the incident solar radiation in W/m^2 and the solar spectrum is subdivided into N wavelength bands, each containing a fraction Tr of the incident irradiance. Furthermore, each band has an associated γ attenuation coefficient classified by Jerlov (1976) for different open ocean and coastal water types. Normally, only two bands are considered. If we suppose Tr to be identically zero, then all radiation is absorbed at the surface and heat penetrates the water column only by turbulent fluxes.

In shallow water, shelf scale models the heat equation has been traditionally used in a more complete form with respect to the large scale equations in view also of the more sophisticated turbulent physical parameterizations for vertical mixing. In equations (8)-(14) in fact the horizontal eddy viscosity and diffusion coefficients are spatially dependent, following for example the Smagorinsky (1993) formulation of nonlinear viscosity, and the vertical eddy viscosity and diffusion coefficients are calculated by a turbulence submodel. The latter normally expresses the vertical diffusion and viscosity terms as:

$$K_v = lqS_H$$

$$A_v = lqS_M$$

where S_H and S_M are stability functions and lq is the product of the turbulence macroscale and turbulent velocity field. Mellor and Yamada (1982) defined a two equations turbulence scheme for lq^2 and q^2 by appealing to a second order closure hypothesis. This parameterization has been shown in the past to reproduce well the upper mixed layer dynamics and the shallow water mixing typical of coastal/shelf areas. More modern approaches consider other boundary layer parameterizations that are reviewed in Large et al. (1994).

In coastal models, the air-sea interface or sea level, h, is free, e.g.,

$$w|_{z=h} = \frac{Dh}{Dt} = \frac{\partial h}{\partial t} + u\frac{\partial h}{\partial x} + v\frac{\partial h}{\partial y} \tag{16}$$

This allows for Ekman barotropic dynamics to develop and external gravity wave to propagate on the continental shelf.

The important modeling consideration at the level of both large and coastal scale dynamics is the air-sea physics parameterization used for the surface vertical boundary conditions. They are:

$$\rho_0 A_v \frac{\partial \vec{u}}{\partial z}\bigg|_{z=h} = \vec{\tau}_w \tag{17}$$

$$\rho_0 K_v \frac{\partial T}{\partial z}\bigg|_{z=h} = -\frac{1}{C_p}\{Q_B(T_a, T_s, C, rh) + LE(T_a, T_s, rh, |\vec{v}_w|) + H(T_a, T_s, |\vec{v}_w|)\} \tag{18a}$$

$$\rho_0 K_v \frac{\partial T}{\partial z}\bigg|_{z=h} = \frac{1}{C_p}\{Q_s(C, \alpha, t) - Q_B(T_a, T_s, C, rh) - LE(T_a, T_s, rh, |\vec{v}_w|) - H(T_a, T_s, |\vec{v}_w|)\} \tag{18b}$$

$$K_v \frac{\partial S}{\partial z}\bigg|_{z=h} = S_o(E - P - R) \tag{19}$$

$$q^2 l\big|_{z=h} = 0 \tag{20}$$

$$q^2\big|_{z=h} = B_1^{2/3} \frac{|\vec{\tau}_w|}{\rho_0} \tag{21}$$

where (18a) is valid for shelf scale models with radiative heat penetration while (18b) is used in large-scale models without that process. Furthermore, $\vec{\tau}_w$ is the wind stress, Q_B is the net longwave emitted by the surface, *LE* is the latent and *H* the sensible heat flux, *E,P,R* are the evaporation, precipitation and river runoff rates in terms of velocities, S_o the surface salinity computed by the model, B_1 is an empirical constant from the turbulence closure relations and C_p is the specific heat at constant pressure. We note that R is different from zero only at the river mouth grid points and this contribution is normally considered only for shelf models. The turbulent kinetic energy and turbulent mixing length boundary conditions expressed by (20) and (21) are used in the turbulence closure submodel not described here. Furthermore, we note that for large-scale rigid lid models the boundary condition is imposed at $z=0$ instead of *h*.

The terms on the right hand side of (18a) and (18b) depend upon the air temperature at 2 meters, T_a, the sea surface temperature computed by the model, T_s, the total cloudiness, *C*, the relative humidity computed from dew point temperature at 2 meters, *rh*, the 10 meters wind velocity modulus, $|\vec{v}_w|$, the albedo, α, and the time, *t*, of the day, month and year. The incident solar radiation term, Q_S, is normally computed with astronomical formulas and considers a constant transmission coefficient. The different heat and momentum bulk expressions for the terms in 18) were empirically determined and explicit formulas are given in Rosati and Miyakoda (1988) and Castellari et al. (1998) for the large scale and Zavatarelli et al (2001) for the coastal scale. The important concept is that T_s and S_o are coming from the model integration itself while all the other meteorological parameters are assigned from independent data. This surface flux formulation is called interactive since the heat and salt fluxes depend upon the state of the ocean directly. Naturally also the atmospheric parameters should in principle depend upon the surface oceanic state and this is in fact how the coupled atmosphere-ocean models work.

We remind that the correct water flux boundary condition should consider the general imposition of $w|_{z=h} - \left(\frac{\partial h}{\partial t} + \vec{v} \cdot \nabla h\right)\Big|_{z=h} = (E-P-R)$ (Beron-Vera et al., 1999) in free surface models but this is not normally considered.

A relevant difference between boundary conditions for large and coastal scale models is the imposition of a realistic river runoff conditions. In coarse resolution models, the rivers appear only from he surface salinity flux, making R in (19) different from zero there. However, if the resolution is high, than the river mouth is equivalent to an open boundary condition and we shall specify:

$$S|_{(xo,\,yo,\,z\,=\,h)} = S_R \tag{22}$$

$$T|_{(xo,\,yo,\,z\,=\,h)} = T_R \tag{23}$$

where *(xo,yo)* are the latitude and longitude points of the river mouth or estuary and S_R , T_R are the salinity and temperature profiles at these same points. In addiction, at *(xo,yo)* the velocity profile should be also impose as for any inflow point.

The bottom vertical boundary conditions, located at $z = -H(x, y)$, where H is the bathymetry, are:

$$\rho_0 A_v \frac{\partial \vec{u}}{\partial z}\bigg|_{z=-H} = \vec{\tau}_b \tag{24}$$

$$w|_{z=-H} = -\vec{u}_b \cdot \nabla H \tag{25}$$

$$K_v \frac{\partial(T, S)}{\partial z}\bigg|_{z=-H} = 0 \tag{26}$$

where $\vec{\tau}_b$ is the bottom stress and $\vec{u}_b$ is the bottom velocity field. Normally the bottom stress considers the 'law of the wall' logarithmic velocity profile which may be not explicitly resolved by the vertical discretization of the model.

At the lateral boundaries, the conditions for large-scale models are normally rigid walls or parameterized heat and salt climatological fluxes. This is a principal difference with the regional/coastal models where the interaction with the flow field external to the domain needs to be at high temporal and spatial frequency. Thus large-scale models make use of no-slip boundary conditions, e.g. $\vec{u} = 0$, along the domain rigid walls and/or have a series of sponge layers where temperature and salinity are fixed to climatological values. These sponge layers may parameterize unresolved physics (such as water-ice interactions at high latitudes), parts of oceans connected through narrow straits to the large-scale basin of interest and parts of ocean basins dynamically uncoupled on the time scales of interest. This is the case for example of the Atlantic ocean and the Mediterranean Sea, connected by the narrow Gibraltar Strait, the Atlantic with the Gulf of Mexico Strait, the Pacific and Atlantic basin themselves. The sponge layers are represented by additional terms in equations 6 and 7 written as:

$$\lambda(\phi - \phi^*) \tag{27}$$

where λ is a relaxation function, corresponding to an inverse of time, and ϕ can be temperature or salinity and ϕ^* the reference or relaxation field. The time scales for relaxation can range from the time step value to several days or months and λ can be different from zero for several grid points defining the sponge layer extension.

Lateral boundary conditions for the coastal models differ from large-scale models in a relevant way. Due to the high numerical resolution required, coastal/shelf models normally concentrate on a small part of the flow field that is changing at the lateral boundaries almost at the same space and time scales of the flow field internal to the coastal model domain. This requires a sophisticated technique of 'nesting' of the coastal model into the large-scale model. In the past decade several studies have established different nesting boundary conditions but every different model implementation requires a careful tuning of such conditions to the specific kind of flow conditions and open boundaries. As a rule of thumb, it was found that the difference between the large and coastal scale grids should not be higher than a factor of three-four. An alternative to nesting is the usage of finite element models or non constant orthogonal grids. However, wave propagation in strongly irregular grids could be affected by the changes in grid spacing (Waisowicz, 1986).

Numerically, both the large and regional scale model equations have been solved with a relevant number of numerical methods, ranging from finite differences to Galerkin methods (Haidvogel and Beckman, 1998). The fundamental difference with respect to numerical fluid dynamics problems is that generally two different time steps are used to solve for the internal and external modes contained in the equations. By definition, the external mode is the velocity field solution of the vertically integrated equations (1)-(7) or (8) through (14). The internal mode is the remaining part of the velocity field that includes explicitly baroclinic effects. The external mode time steps are forced to match every once in a while the internal mode time step, which is normally one-two orders of magnitude larger than the external mode time step. In the large-scale models with rigid lid assumption the external and internal time steps coincide since there are no fast external gravity modes.

Another important modeling consideration is the choice of the vertical coordinate system that may be a rationale function of position and depth, otherwise called a bottom following coordinate system. The most commonly used transformed coordinate system is called "sigma" and is defined by the transformation:

$$\sigma = \frac{z-h}{H+h} \tag{28}$$

The equations (8)-(14) are rewritten using this transformation and then discretized with the chosen numerical scheme. This coordinate system clearly works well when bathymetric changes are small, e.g., the continental shelf up to the slope for example, but it has been used also in steep bathymetry problems. For an overview of the vertical coordinate systems see Mellor et al. (2001) in this book.

Specifying initial and boundary conditions solves the equations of motion (1)-(7) or (8)-(14). Thus a part from the validity of the hydrodynamic and physical approximations contained in these equations, the specification of the initial and boundary fields itself is a major consideration to recover a realistic solution. The initial value problem that starts from a full state of rest and null available potential energy (constant initial density field) is normally not done. Instead, it is common to prepare initial conditions for the temperature and salinity fields from observations. Here two different lines of modeling begin: the first, which uses climatological fields from historical data sets, and the second, which tries to have the best guess initial and boundary fields at a specified instant. The first choice is normally taken when a long-term integration is done, usually at the large-scale. This choice involves the fact that several of the initial model integration years are disregarded, as part of the spin up of the model integration, because they contain 'spurious numerical information'.

The second choice conducts very rapidly to the need to define the optimal observing system and model-data melding procedure in order to determine the initial and boundary fields. This is a very important part of the coastal environmental prediction problem, the optimal estimation scheme that could be used to define the best guess initial and boundary fields. For large scale models the assimilation schemes are already very sophisticated, see the other chapters of this book. However, for the coastal nested modeling approach only few examples exist (Auclair, 1999).

The coastal assimilation problem need to consider two distinct sets of information for the estimation of initial and boundary conditions: the first is the large scale flow field resulting from a coarser model integration and the second is the in situ and satellite observed data. Up to now, in the coastal areas, the two problems have been solved separately, e.g., either the initial condition is defined only by coarser model simulations or enough synoptic observations are available for the definition of the initial condition for the dynamical variables. However, the general solution of the coastal environmental prediction problem clearly requires the optimal combination of coarser model first guess fields and observations.

17.5.2 The ecosystem modeling

Large and coastal scale ecosystem models have been developed during the past two decades. Experimental evidence in the last decade has shown how important is the coupling between physical conditions and pelagic food web processes (McGillicuddy et al., 1998). Thus a major effort has been put in the three-dimensional coupling of hydrodynamics and biogeochemical cycles and its numerical representation. At the same time, laboratory and field observations have empirically determined trophic interaction rates within functional group parameterizations of the marine food web. For an overview of ecosystem modeling see Hoffman et al.(1998). Here we will try to outline some basic formalism used in complex ecosystem models such as the European Regional Seas Ecosystem Model (ERSEM) described in Baretta et al. (1995) and Vichi (1998).

The basic equations describe the rate of change of an ecosystem state variable $C(x,y,z,t)$ as composed of two parts:

$$\frac{\partial C}{\partial t} = \left.\frac{\partial C}{\partial t}\right|_P + \left.\frac{\partial C}{\partial t}\right|_B \tag{29}$$

where the P subscript defines the time rate of change due to physical processes and the B subscript indicates the rate of change due to biochemical processes. In the model of Baretta et al. (1995) C represents concentrations of chemical elements either dissolved in the water or in the functional organism groups. Chemical constituents and their rate of change describe the organisms and its evolution in each functional group. The chemical currencies considered up to now are carbon, C, nitrogen, N, phosphorus, P, and silica (Si). Naturally this list should be enlarged to consider various calcium (Ca), iron (Fe) and other nutrients which are already known to limit phytoplankton growth in certain open ocean areas.

The physical rate of change is written in general terms as:

$$\left.\frac{\partial C}{\partial t}\right|_P = -\vec{u} \cdot \vec{\nabla} C + \nabla \cdot (K_h \nabla C) + \frac{\partial}{\partial z}\left(K_v \frac{\partial C}{\partial z}\right) \tag{30}$$

The particle sedimentation velocity could augment the vertical part of the advection term if this is relevantly different from fluid velocity. Advective effects in ecosystem equations have been studied theoretically by Robinson (1997) and differentiated from biochemical rate terms.

The time rate of change due to biological processes is given by a complex food web representation for both the pelagic and benthic compartments and it is here reproduced in Fig. 17.9a. The model contains dissolved nutrients described above and ammonium, a representation of the pelagic phytoplankton functional group subdivided into different size classes, a pelagic bacteria group, dissolved and particulate organic matters, mesozooplankton and dissolved gases compartment. The pelagic environment interacts with the physical environment not only through the advection and diffusion processes but also responding to the temperature and salinity field values, as indicated in Fig. 17.9b. Another important coupling with the physical environment is given by the light radiative propagation submodel which normally follows a simple exponential law which itself is dependent upon inorganic matter and phytoplankton pigments concentration.

The pelagic submodel communicates with the benthic system through pore water diffusive processes. The benthic system is described as a multiple layer system with pore water nutrients, dissolved gases, benthic bacteria, zoobenthic organisms, detritus, dissolved organic matter and particulate organic matter. Important developments in the future should consider benthic-pelagic coupling through resuspension processes of cohesive sediments. Coastal ecosystems almost certainly require the complexity of interactions illustrated above while open ocean ecosystems may be more simply schematized, as done in the past (Fasham et al., 1990). Another level of complexity in coastal ecosystem modeling is given by the imposition of land derived lateral fluxes of nutrients, particulate and dissolved organic matter and inorganic sediments, as outlined in Fig. 17.9b.

In order to solve the initial and boundary condition problem given by equation (29) we need to give initial estimates of each state variable, the lateral fluxes from land sources and the vertical fluxes determined by atmospheric transport. The latter can be relevant for open ocean areas and/or shelf areas in absence of importance river sources. The second dominates over atmospheric transport in river runoff dominated shelf areas. The land derived lateral flux is only considered for dissolved nutrients, dissolved and particulate organic matter and inorganic material. The flux is parameterized in the form of prescribed values, e.g.:

$$C|_{(x_0,\, y_0,\, z\, =\, 0)} = C^* \tag{31}$$

where x_0, y_0 are the nutrients discharge land points and C^* is the measured concentration. For atmospheric input, the problem should be solved with the imposition of observed nutrient fluxes at the surface,

$$K_v \frac{\partial C}{\partial z}\bigg|_{z\,=\,0} = Q_C \tag{32}$$

where Q_C is the flux of that particular nutrient at the surface.

The initial condition problem can be again posed by specifying climatological estimates of state variables or by assimilation techniques. If the assimilation methods are being developed for the shelf areas hydrodynamic modeling, data assimilation for field estimation in ecosystem models is at its infancy. Initially the observations have been used to solve an inverse problem where best estimates of parameters for the biochemical rates were sought. Only recently, multidisciplinary observations have been used to estimate three-dimensional initial fields for some of the state variables. The large dimensionality and the multivariate character of the field estimation problem does not allow simple approaches such as nudging or univariate optimal interpolation schemes. However, experiments are starting which consider the assimilation of satellite color data to estimate several initial ecosystem dynamical model variables.

a)

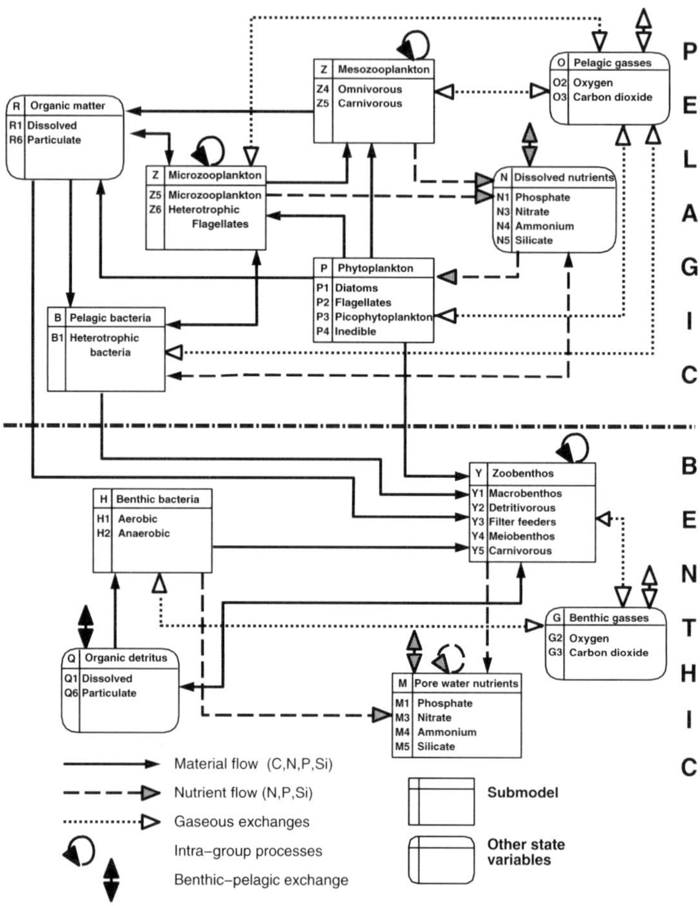

Fig. 17.9 a) The schematic of ERSEM food webs from Baretta et al. (1995). b) The schematic of interactions between the different inputs to the coastal ecosystem and the three constitutive compartments of the ecosystem, the hydrodynamics, the pelagic and benthic food webs. The inputs are both vertical (radiation and air-sea interaction fluxes) and lateral (land inputs to the sea, advection and horizontal diffusion).

b)

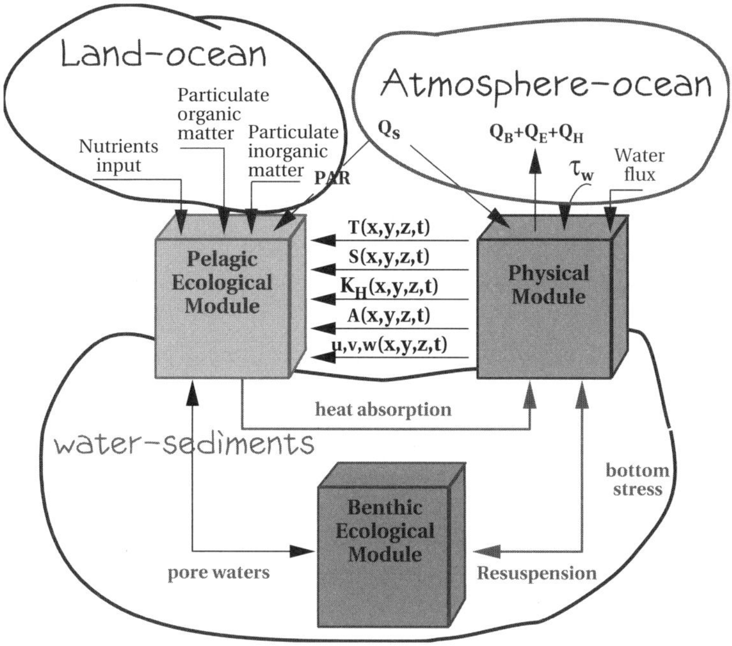

Fig. 17.9 (continued).

17.6 The calibration/validation exercise in the Northern Adriatic coastal areas

The Adriatic Sea shelf ecosystem variability has been and is now investigated at the level of coordinated observing programs and monitoring activities. The scientific observing programs such as PRISMA (Progetto di RIcerca per la Salvaguardia del Mare Adriatico) and MAT (Mucillaggine Adriatico e Tirreno) have collected an impressive set of multidisciplinary data that are now being exploited for validation/calibration of future environmental prediction systems of different complexity. In this chapter we review the current effort to design and calibrate such a system

The Adriatic Sea has a relatively wide northern shelf area, associated with the large river runoff, the largest contribution coming from the Po with an average yearly runoff rate of 1.7 $10^3 m^3/s$. This shelf area has frequent events of algae blooms and sea-bottom anoxia that could severely affect the tourist and fish industry. Thus the implementation of coastal forecasting is needed to manage resources in a sustainable way and protect the marine ecosystem.

In the Northern Adriatic Sea the first phase of system analysis toward environmental predictions has been activated. The system is composed of five elements:

1) The large scale observing and forecasting system, hereafter called large scale MFS, based upon the monitoring system of Fig. 17.6 and 17.7, a large scale MOM implementation and the SOFA data assimilation system (De Mey and Benkiran, 2001);
2) The shelf scale observing system;
3) The regional Adriatic Sea model (AREG) nested within the large-scale MFS model and coupled asynchronously with atmospheric analyses that produce surface forcing fluxes. It is based upon the POM code;
4) The Northern Adriatic Sea shelf model (ASHELF) nested within AREG and coupled asynchronously with atmospheric analyses. It is based upon the POM code;
5) An ecosystem model implemented in a selected coastal area station with realistic atmospheric forcing.

The five components are at a different stage of development. The first element is operational at the weekly time scale, i.e., once a week a 10 days forecast is disseminated in near real time. The horizontal resolution of the analyses and forecasts is 12.5x12.5 km and all the elements of the observing system are assimilated in the initial condition. Some of the results are presented in Pinardi et al. (2001) and are available on the Web (http://www.cineca.it/mfspp).

The second element is composed of the CTD stations in the near coastal areas shown in Fig. 17.8, giving temperature and salinity measurements every two-three weeks. The system is not automated yet but information is easily exchanged due to the relatively few data collected each time (20 meters maximum depth and 20 stations). In addition to the monitoring stations on the narrow coastal and shelf break areas, a quasi-permanent S1 station is also present. Different multidisciplinary data sets have been collected at this station during the past ten years and we used such data to calibrate/validate the ecosystem models. The CTD stations are used for the hydrodynamic model validation but are not yet assimilated.

The third element of the system, the AREG model, is at 5-km resolution and is coupled at an open boundary to the large-scale MFS model. In Fig. 17.10 the AREG open boundary is shown with respect to the MFS grid nesting. The AREG model is coupled asynchronously to atmospheric surface analyses that are downscaled to 5 km by simple interpolation. The lateral nesting boundary conditions are the imposition from MFS fields of the total velocity, temperature and salinity at inflow points and advection of temperature and salinity at outflow points. Several interpolation and consistency constraints have been constructed to allow the models to communicate the appropriate values of volume transport and temperature, salinity across the open boundary.

The ASHELF model is nested within the AREG (Fig. 17.10) and is at the resolution of 1.5 km. The ASHELF model has the same air-sea interaction physics of the AREG model and it is coupled asynchronously to atmospheric forcing analyses. The minimum depth of ASHELF is 3 m while for AREG the minimum depth is 10 meters. Thus the two model bathymetries differ in the narrow coastal strip area so

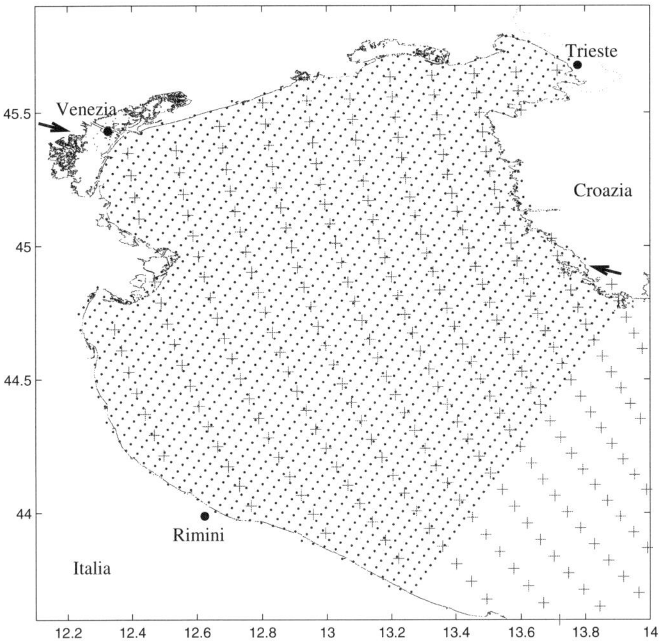

Fig. 17.10 The computational grid points of the nested 1.5 km shelf model (ASHELF) of the Northern Adriatic Sea within the 5 km regional model (AREG) of the Adriatic Sea. The latter is nested at the Northern Ionian boundary within the large scale-forecasting model of the Mediterranean Sea. Both grids are plotted every other points. Crosses: AREG model. Dots: ASHELF model. The arrows near Venice and on the Croatian coasts indicate the start of the section illustrated in Fig. 17.11.

that the initialization of ASHELF from AREG fields has to consider extrapolation. In order to overcome this difficulty Auclair et al (2000a) proposed a 3D variational method that limits the amplitude of the transient processes following the initialization. The tangent linear equations to the ASHELF are used to decrease the initial time step tendency of the model physical parameters down to realistic values. The amplitude of some of the model biases, like the horizontal pressure gradient truncation error due to the use of sigma coordinates in regions with large topographic slopes (Auclair et al, 2000b), is also drastically reduced. A comparison of the time evolution of the surface elevation anomaly along a section in the Northern Adriatic basin with and without the tangent linear model initialization method is given on Fig. 17.11. The results obtained with the initialization technique (Fig. 17.11b) show a drastic decrease of the amplitude of the external gravity waves with respect to a crude initialization (Fig. 17.11a). The consequences of these transients on a 15-days forecast are studied in details in Auclair (1999). This initialization method is the first step toward an initialization/assimilation scheme that should in a near future lead to optimal set of initial and open boundary conditions for the ASHELF forecasting.

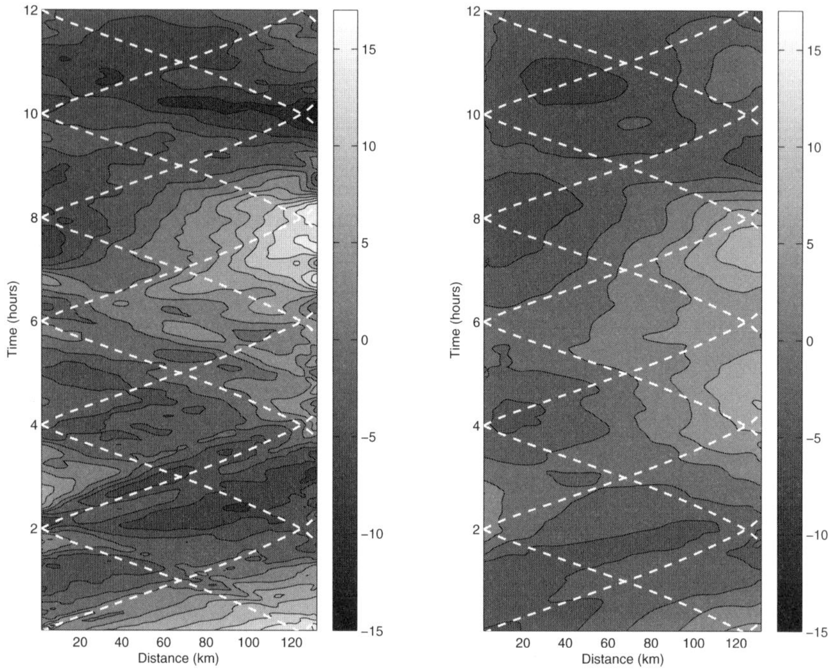

Fig. 17.11 Hindcast results for the ASHELF domain (1.5-km resolution) with and without tangent linear model initialization. The section location is indicated in Fig. 17.10. Initial and lateral boundary conditions are taken from the AREG model (5-km resolution) starting from May 1, 1995. The sea level time evolution is shown for 12 hours along a section in the Northern Adriatic basin in the ASHELF model. Contour interval is 2 cm. The initial and lateral boundary conditions are obtained from the coarse resolution fields of AREG with standard interpolation (left panel), 3D variational initialization (right panel) based on the tangent linear model. The white dashed lines are the characteristic lines for the external gravity waves, they are tangent at each point to $\sqrt{gH}$, H being the local topography.

The fifth element of the Northern Adriatic environmental prediction system is the ecosystem model that could simulate the dynamics of the primary producers for the region. The ecosystem model described in the previous section has been implemented in the coastal area of the Northern Adriatic, in particular at the S1 station of Fig. 17.8. At the moment, the model is one-dimensional and only the phase of calibration and validation has started. In the past five years, a very comprehensive set of observations has been collected at this site, from dissolved nutrient profiles to microzooplankton, dissolved and particulate matter, bacterial heteronanoflagellates (HNF) biomass and phytoplankton species. This is a unique data set to calibrate model parameters and understand which initial field are important to achieve predictability in different state variables, in particular the phytoplankton biomass and its different groups.

The data set consisted of high temporal resolution (4 hours) profiles taken at the station S1 during a two days period. The model was initialized with temperature, salinity, dissolved nutrient profiles from the observations, dissolved and particulate organic matter, bacterial and etheroflagellates biomass and predicted the ecosystem state variables for two days, under realistic external atmospheric forcing conditions, the same used for the hydrodynamics models. The results are shown in Fig. 17.12 and 17.13 for a two days periods during October 1995. It is evident that the model is capable of capturing the bulk of the biomass values for both bacteria, diatoms and flagellates but it is missing some small temporal scale dynamics and the large lateral advection of phytoplankton in the region. We see in fact that observations indicate large subsurface patches of diatoms entering the region between midnight and 6 a.m., thus forcing us to conclude that they were laterally advected into the region represented by our water column model. Errors, defined as the root mean square values of the difference between observations and model predictions, range overall between 10% and 30%, normalized with the mean observed values.

This experiment shows that there is a potential predictability in the ecosystem state variables at short time scales, in particular for the primary producers biomass. The overall structure of the phytoplankton, bacterial and microzooplankton functional groups is captured and simulated correctly, hinting to the fact that this coastal ecosystem is mainly driven by nutrient supply by the river sources. During summer however, internal food web dynamics could prevail, such as microbial carbon dynamics competing with phytoplankton uptake and benthic-pelagic coupling which could become a relevant source of dissolved nutrients from benthic remineralization of particulate and dissolved organic matter.

The results of this modeling experiment are difficult to transpose to the future environmental forecasting system for this area since the observing system requirements are high (many parameters need to be monitored) and measurement techniques not yet developed for automatic observations of such quantities. If the model would have not been initialized with, let's say etheronanoflagellates biomass, the bacterial biomass would not have been maintained within the range of observed values (see Fig. 17.13). The coastal observing system components for model initialization should be carefully planned in the future and major technological development will be required to realize measurements in operational mode.

17.7 Conclusions

In this chapter we have described a potential scheme to develop and implement environmental predictions in the coastal areas of the world ocean. The coastal envi-

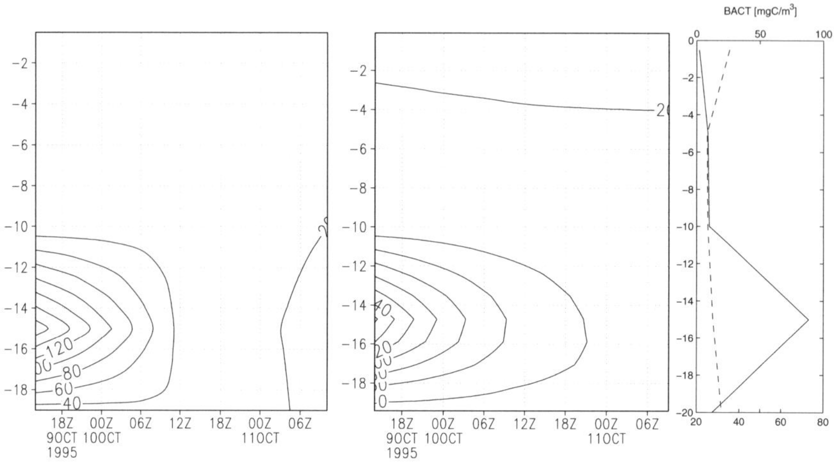

Fig. 17.12 Ecosystem model simulation/prediction and observations for the period 9-11 October, 1995 at station S1 of Fig. 17.8. Observations were collected at intervals of four hours in time and every 5 meters in depth. The model prediction is displayed at the same time/space intervals of the observations. The results are displayed for bacterial biomass in mgC/m^3 as a function of time (abscissa) and depth (ordinate). Model root mean square errors are also displayed as a function of depth (dashed line) together with average values of the ecosystem variable (continuos line) for the two days period. The root mean square values are computed for the difference between observations and model predictions every four hours and are normalized with the mean observed values. The upper abscissa reports the percentage of error and the lower the values of the ecosystem variable.

ronmental predictions aim at forecasting the primary producers biomass fluctuations from days to weeks. The design of such system starts from the basic understanding of the dynamics of the coastal areas and capitalizes on the developments that have occurred in the past twenty years both in the observing and modeling communities. The basic methodological approach focuses on the monitoring practices and the downscaling of the information from large-scale to the shelf areas. Such methodology privileges the networking of existing technologies for monitoring the large and shelf scale dynamical field variables and considers the satellite observations as a fundamental source of synoptic information for the correct initialization of models. Furthermore it develops nesting procedures for numerical models which could efficiently downscale the information to the high-resolution models required in every world ocean coastal areas. Finally, it develops a strategy for the usage of deterministic ecosystem models to evaluate the potential predictability of ecosystem state variables such as phytoplankton and bacterial biomass.

The Mediterranean Sea is shown as a basin test case for the implementation of this strategy. First of all, 10 days large-scale forecasts of hydrodynamic state vari-

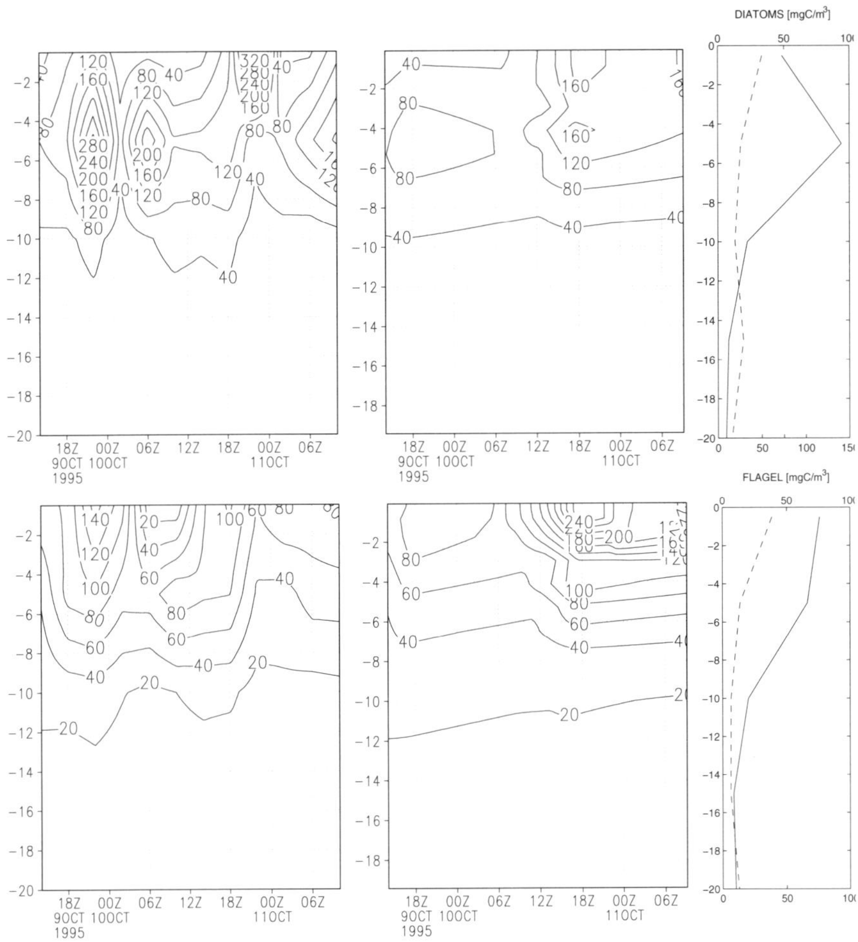

Fig. 17.13 Ecosystem model simulation/prediction and observations for the period 9-11 October, 1995 at station S1 of Fig. 17.8. Observations were collected at intervals of four hours in time and every 5 meters in depth. The model prediction is displayed at the same time/space intervals of the observations. The results are displayed for diatoms (a) and flagellates (b) biomass in mgC/m^3 as a function of time (abscissa) and depth (ordinate). Model root mean square errors are also displayed as a function of depth (dashed line) together with average values of the ecosystem variable (continuos line) for the two days period. The root mean square values are computed for the difference between observations and model predictions every four hours and are normalized with the mean observed values. The upper abscissa reports the percentage of error and the lower the values of the ecosystem variable.

ables have started and are now weekly available. High resolution models of the Northern Adriatic are then nested within this forecasting system to arrive at 1.5-km resolution, that is, the minimum resolution required to capture hydrodynamic processes in this area. The initialization of coastal high resolution models from coarser information is done with a tangent linear modeling approach that is capable of dynamically extrapolate information in data void regions. In the future this methodology should be integrated with more traditional data assimilation schemes to merge the in situ and satellite data with the coarser model fields. The one dimensional ecosystem model implemented in the coastal area shows a potential predictability of few days after initialization from observations of temperature, salinity, nutrients, dissolved and particulate organic matter concentrations and some phytoplankton groups biomass.

Major technological and scientific developments are still required before we can advance toward operational environmental forecasting in the coastal areas even if the major tools are ready to be assembled. The future challenges are: 1) the development of sensors which could sample in near real time several dynamical ecosystem state variables, 2) the extension of the data assimilation techniques to the coastal areas for both physical and biochemical variables, 3) the improvement of the atmospheric forcing resolution at the scale of the coastal areas air-sea interaction processes.

Acknowledgment

The first author has been partially supported by the EU-MFSPP (Mediterranean Forecasting System Pilot Project) contract MAS3-CT98-0171 and MAS3-CT98-0189 (MEDNET). The MURST-ENEA Progetto 'Ambiente Mediterraneo' sponsored the second author during his visit in Italy. This work would not have been possible without the continuos support and work of all the MFSPP partners who made operational the MFS large-scale system in the Mediterranean. Particular thanks are for Dr. Manzella, responsible for the VOS network, Dr. P.Yves Le Traon, responsible for the satellite data network, Dr. D.Antoine responsible for the SeaWiFS data, Dr. R. Santoleri for the SeaWiFS data in the Adriatic, Dr. P. De Mey responsible for the data assimilation system, Dr. C Tziavos responsible for the M3A buoy system, Prof. A. Lascaratos for the regional modeling, Dr. I. Allen for the ecosystem modeling and Dr. C.Maillard for the data management in MFSPP.

18 Primitive Equation Modelling of Plankton Ecosystems

JOHN WOODS
Imperial College, London

18.1 Introduction

We start with a technical problem that hindered progress in biological oceanography throughout the 20th century, and show how it will be circumvented in the 21st century. Because of inadequate sampling, very little of our knowledge of population dynamics in the plankton ecosystem is derived from hypothesis-free analysis of observations. On the contrary, most of our knowledge comes from mathematical simulation. But here comes Catch 22: the same sampling problems make it very difficult to collect data that can effectively test the models used to make those simulations. So their credibility depends mainly on the equations used in the models. Equations used in 20th century models did not have secure scientific foundations of the kind that made modern physics work. They were not based on reproducible laboratory experiments.

18.1.1 Biological Primitive Equations

This lecture will show how using primitive equations in ecosystem models can bring biological oceanography up to the standards by which we address complex physical phenomena like the weather. It starts by considering whether there are any primitive equations for biology, of the kind that have been so effective in physics. This question was first considered by Descartes (1664). Today we answer that they do not exist for higher animals, because we cannot perform reproducible experiments to describe how animals change their minds when responding to events in their ambient environment. However, it is possible to establish biological primitive equations for micro-organisms, which do not have minds to change.

Most plankton are mindless organisms, whose physiology and behaviour are governed by rules fixed throughout their lives. These genetically-controlled rules can be discovered by reproducible laboratory experiments. They are described by Biological Primitive Equations (BPEs), which are as trustworthy as the primitive equations of physics and chemistry. An ecosystem model, which is based on physical, chemical and biological primitive equations, has sound scientific foundations. Simulations of the plankton ecosystem based on BPE models have credibility, even when formal verification is skimpy. That is the fundamental message of my lecture.

18.1.2 Individual-Based Modelling

My lecture will introduce the modelling technology that is needed to exploit the inherent strength of biological primitive equations; technology to create trustworthy simulations of the plankton ecosystem. Because BPEs describe the interaction between a single micro-organism and its ambient environment, the modelling technology must be individually-based. The advantages of individual-based modelling are well known. It has been widely used to simulate particular processes, especially in zooplankton. This involves Biological Lagrangian Integration (BLI), in which the life history of an individual organism is simulated by integrating the BPEs along its trajectory through the water.

Here we are concerned with the more ambitious goal of using BLI to create comprehensive simulations of the plankton ecosystem (Fig. 18.1). To achieve that, we must use the information contained in the life histories of individual plankters to derive accurate simulation of (1) the demographic history of each population, and (2) the biofeedback to the physical and chemical environments. The modelling technology developed to achieve that goal is called the Lagrangian Ensemble (LE) method (Woods & Onken 1982). It links a dynamical sub-population of identical plankters to each of a large number of particles behaving like individual plankters. Techniques have been developed to monitor and control the errors in the demography and environment simulated by this method.

18.1.3 Virtual Plankton Ecosystems

Simulations of the plankton ecosystem based on the LE method are called Virtual Plankton Ecosystems (VPE) to distinguish them from simpler (Eulerian) simulations that contain no information about individual plankters. This distinction is illustrated by the Azores VPE, derived from a simple food chain (NPZD) model. The Azores VPE has a stable demographic attractor, which does not exhibit chaotic fluctuations of the kind found in simulations by the Eulerian method. VPEs can be used to parametrize Eulerian models (18.15). The computational cost of creating VPEs was daunting in the last century, but the arrival of cheap parallel computers is making it affordable in the 21st century. You can explore the Azores VPE for yourself with a graphical analysis tool, called VEWdata, provided in a CD-ROM (18.19).

The LE method was originally developed to simulate the plankton ecosystem one-dimensionally in a vertical water column in the open ocean. It can be extended to simulate the plankton ecosystem in shallow water (18.13) and in three-dimensions (18.14). The LE method is suitable for modelling a number of phenomena of practical interest in which plankton dynamics play an important role. These include fisheries recruitment, pollution, epidemics of diseases that are transmitted to larger animals, including humans, climate change by the greenhouse effect, and the impact of enhanced UV radiation on life in the ocean. We now have mature technology that allows us to use biological primitive equation models to address practical problems.

18.2 Primitive equation modelling

The concept of primitive equations originated with Galileo (1638), who showed how the basic Laws of Nature could be established by reproducible laboratory experiments. This new approach to science swept away the intuitive, but false, laws propagated by Aristotle. It provided a sound scientific basis for modern physics. Applications of primitive equations for simulating natural phenomena began with Newton's study of gravity. In the last century they were used to simulate increasingly complex systems in nature and engineering. The primitive equations of real systems are often non-linear, which prevents analytical solutions. Richardson (1910) showed how to integrate non-linear equations numerically, and he went on to publish a scheme for numerical weather forecasting, which is now in operational use. Richardson's (1922) method of modelling complex physical systems by integrating a primitive equation model has become standard practice in physics and engineering. It is the method used to simulate ocean circulation, and it has an important role to play in biological oceanography.

18.2.1 Credibility and Verification

Primitive equation (PE) models have scientific credibility because they are based on reproducible laboratory experiments. Whenever possible, credibility is bolstered by verification. This is done by the Ecological Turing Test (see 18.17), which compares features of the ecosystem, as they are described by two data sets generated respectively by mathematical simulation and observation. In practice this kind of formal verification is very difficult. It sets high standards for both simulation and the observations. In biological oceanography it is rare to obtain adequate observations for proper verification of models.

Until recently, models have not been sufficiently comprehensive to create simulations suitable for true verification. Consider, for example, Sverdrup's (1953) model of light-limited growth during the spring bloom, which has had an enormous impact on biological oceanographers (Platt et al 1994). Although we believe the scientific idea behind Sverdrup's model, it cannot be verified because its equations describe the seasonal variation of only two processes, light and turbulence: it omits other processes which influence the increase of phytoplankton biomass during the spring bloom: nutrient limitation, grazing and diurnal variation. Verification demands comprehensive simulation of the ecosystem in winter (Woods & Barkmann 1994).

It is not enough to claim credibility by noting similarities between a pair of time series from observation and simulation: verification is about differences, not similarities. And no comparison is valid without explicit treatment of the uncertainties in each description of the chosen feature of the ecosystem (18.17 & 18.18). The near impossibility of formally verifying plankton ecology models puts the onus for credibility on model design. Confidence acquired by using primitive equations is particularly important.

18.2.2 Software sensing

Many important features of the plankton ecosystem cannot be observed with sufficient clarity to permit an accurate empirical description in the sense of Bacon (1605). So our knowledge of biological oceanography depends largely on mathematical simulation. The reliable scientific base provided by primitive equations encourages us to believe simulated features that cannot be observed. Modellers give the name "software sensing" to the procedure by which mathematical simulation is used to measure unobservable features of complex systems. Even when we do have observations, the information obtained from software sensing is usually more comprehensive. Such completeness and seeming realism are seductive. So we must take care to control its quality by identifying and quantifying the uncertainties in our modelling technique (see below 18.17-18). With that note of caution, we proceed on the basis that PE modelling will provide a scientifically sound basis for software sensing. Models that are not based on primitive equations will tend to be less satisfactory.

18.3 Biological primitive equations

Biologists often suppose that modelling with primitive equations applies only to physical systems, like planetary motion, tides, ocean circulation and weather. My aim is to show how it can also be applied to systems, which involve not only physics, but also chemistry and biology.

I define Biological Primitive Equations (BPE) in the sense of Galileo (1638). They are mathematical equations established by analysing data from reproducible laboratory experiments designed to describe fundamental laws of nature for living organisms. Descartes (1664) was the first scientist to contemplate PE modelling of biological systems. Rather ambitiously he thought of doing so for the physiology of Man. Later, Turing (1950) and others seriously contemplated an even more ambitious challenge: PE modelling of the mind. We now know that the sheer complexity of human physiology and mind, and of the brain which controls them, places such goals beyond our reach for the foreseeable future. Nevertheless, there is growing research activity in modelling complex systems involving people (Casti 1997).

18.3.1 Complex Adaptive Systems

It has became standard practice to divide complex systems into two categories: (1) those that are based on models comprising the immutable equations of physics, and (2) those that are based on models with equations for the model agents (usually people) that can adopt new rules when circumstances change. Weather forecasting falls into the first category, financial markets fall into the second. The latter are called Complex Adaptive Systems, because they involve decisions by people who can change their rules of behaviour. The equations remain conjectures because it is not yet possible to perform reproducible laboratory experiments to describe mathematically how people change their minds. Casti (1997) states that the subject is still in a "pre-Galileo" era.

That may be true of complex systems that depend on the behaviour of higher animals. However, most species on Earth are micro-organisms. The behaviour of a simple micro-organism is not mediated by a brain that can change its mind. It follows rules of behaviour that are hard-coded into its genes. Their behaviour may adapt to circumstances, but they do so in a predictable way, which is also encoded in their genes. All these rules can be established by reproducible laboratory experiments. So ecosystems that involve only micro-organisms may be classified as Complex Non-adaptive Systems. They offer the promise of predictability, even though they involve biology. The plankton ecosystem falls into that category.

18.3.2 Evolution

Here I make a short digression to deal with concern that biology is not conservative like physics, which obeys laws frozen since the Big Bang. Evolution is the quintessence of biology, so biological primitive equations change with time. That being the case, how can we describe the plankton ecosystem as being non-adaptive? The answer is that the statistical rules governing evolutionary changes in the biological primitive equations are known, and can be incorporated into our models. In fact it is natural to do so with individual-based models, in which integration of the biological equations follows individual organisms and their offspring through lineages.

The genetically-controlled rules governing the physiology and behaviour of a micro-organism remain unchanged during its life. In the case of non-sexual reproduction, the rules are normally passed on unchanged to the individual's offspring. Occasionally the genes may suffer mutation, leading to changes in the rules. Accumulation of mutations can lead to competitive advantage within a limited range of environmental conditions (see 18.9); that is how bio-zonation evolves (Longhurst 1998). How does mutation affect our plan to use primitive equations to model an evolving plankton ecosystem? Biological Lagrangian Integration has the essential property of accurately simulating inheritance between the generations in a lineage. We can include statistical rules governing occasional abrupt changes in the biological primitive equations of individuals (or rather in their parameter values; see 18.16). Sometimes these changes are fatal, often they are neutral, occasionally they are beneficial. We can also modify the model to include sexual reproduction, which creates new sets of parameters in the primitive equations drawn haphazardly from those encoded in the genes of the parents.

Both mutation and sexual reproduction introduce a stochastic element into inheritance of primitive equations. The statistical rules governing these changes are well known and can be found in standard textbooks, such as Roughgarden (1996). It is technically straightforward to extend PE models to include them. It is possible to design a modelling technique to accurately calculate the demographic consequences of competition between varieties with primitive equations modified by sex or mutation. The requirement is that the simulation has a significant demographic signal: noise ratio. I shall describe a Virtual Plankton Ecosystem with mutating plankton in 18.9.3. For now, we conclude that evolution does not prevent primitive- equation modelling of the plankton ecosystem.

18.4 Plankton

Plankton were discovered at Scheveningen on 27th July 1676 by Antony van Leeuwenhoeck (1677). They are micro-organisms that drift in the sea. Or more precisely (my definition), plankton are pelagic organisms, which cannot swim fast enough to change their ambient environment usefully by swimming horizontally. Of course, many species of plankton can swim fast enough to benefit from migrating vertically through the water. This anisotropy arises from the fact that the correlation scale of environmental variables is much shorter in the vertical than in the horizontal, in the ratio of the Coriolis frequency to the Brunt-Vaisala frequency.

18.4.1 One-dimensional modelling

Thus the plankton ecosystem is essentially one-dimensional. The environment is controlled by vertical fluxes from the sea surface, as in a classical boundary layer, and the plankton move up and down through the water. This suggests that to first order we can treat vertical and horizontal variation separately in our models. We can do this by simulating the ecological processes with strictly one-dimensional equations. If horizontal fluxes exist, their divergence is assumed to be negligible. This is like simulating the plankton ecosystem in a vertical water column, which outcrops at the sea surface and extends down into the ocean. The water column has an arbitrary horizontal area: one square kilometre is convenient. We assume that the fluxes through the sea surface are uniform across that area. The plankton populations in the water column are expressed as either the number of organisms or biomass per unit area (e.g. copepods/m^2 or mgC/m^2). Vertical variation is expressed per unit volume (e.g. copepods/m^3). The water column has open sidewalls: water flows freely through them. That is unlike experimental ecosystem enclosures, the sidewalls of which cause unwanted disturbance (Lalli & Parsons 1997). In one-dimensional modelling, the neglect of horizontal flux divergence is assumed to cause insignificant error in the simulation of plankton demography.

However, observations do show horizontal variation. We can divide it into two categories: smaller and larger than the Rhines scale, which is about 100 km. That is the maximum size of transient eddies in the ocean; any larger and they lose energy to Rossby waves (Rhines 1979). Satellite images of ocean colour show that there is substantial horizontal variation of phytoplankton concentration on scales smaller than the Rhines limit. This is known as "patchiness" (Fasham 1978; Steele 1978). It is caused by the transient eddies and jets of mesoscale turbulence, the oceanic equivalent of weather (Woods 1988). Simulating patchiness demands three-dimensional models. Martin et al. (2001) show that horizontally-averaged properties of a patchiness-resolving three-dimensional model differ significantly from that simulated by a one-dimensional model. I shall describe the first attempt to simulate patchiness with a three-dimensional primitive-equation model in 18.14.

At scales greater than the Rhines limit we have geographical variation, which is the traditional concern of biological oceanography (Longhurst 1998). One-dimensional modelling can be used to simulate this large-scale horizontal variation, provided we use the appropriate surface fluxes to drive the model at each geographical location. This raises technical problems, which can be solved by a method called geographically-Lagrangian integration (18.11). It also raises scientific questions of population dynamics in a non-stationary environment, which I shall discuss in 18.12 and 18.16.

18.4.2 Zooplankton

Most plankton fall into the category of mindless micro-organisms for which we can establish biological primitive equations. Many common zooplankton, such as calanoid copepods, do have brains (Mauchline 1998), which have evolved to control their bodily functions in fixed ways, hard coded in their genes. The brains of these zooplankton are like the computers in motorcars, which manage the smooth functioning of the engine, gears and brakes, but which do not have the capability to change their prescribed behaviour arbitrarily (a rather daunting prospect for motorists!). Some "neural network" computer programs are designed to learn from experience. However, such programmed learning follows formal rules, which can be included in PE models. There is no reason to exclude plankton species that have evolved to include pre-programmed adaptation in their portfolio of primitive equations.

18.4.3 Fish larvae

So where do we draw the line between mindless plankton, which obey biological primitive equations, and higher marine organisms that do not? The line must be drawn between plankton and fish, but where exactly? I suspect that newly-hatched fish larvae fall into the category of mindless plankton, but that as they grow they soon pass the line to become young fish with minds of their own. Some fisheries biologists draw the line much higher, and assume that it is possible to make useful models of the migration of adult fish (Steele 1977, Cushing 1995). Perhaps. The test is not whether such models describe observed patterns of migration, but whether they can predict how the migration will respond to a changing climate (Laevastu 1993). Even if the line is drawn while the fish larvae are quite small, they may still behave like mindless plankton during the early weeks of life when most of their mortality occurs. If so, we can design primitive equation models to compute the relationship between such losses and ambient weather. That would make a significant contribution to introducing more science into fisheries recruitment (Cushing 1996).

There are many other problems of practical importance that depend on the ecology of plankton, and which can be addressed with greater scientific confidence through the application of biological primitive equations. To give just a few examples, plankton species that fall into the mindless category are responsible for toxic algal blooms, eutrophication, cholera epidemics, and the biological pump in the global carbon cycle (and therefore global warming in the 21st century and at the end of ice ages). Primitive equation modelling is the most promising technique for predicting these phenomena.

18.4.4 Reproducible laboratory experiments

It will first be necessary to devise reproducible laboratory experiments to establish biological primitive equations for each of the basic functions of the physiology and behaviour of plankton. Because these organisms are tiny, the experiments will most likely be performed on populations rather than individuals. However, that is not a problem if the individuals in the populations are identical, being clones from a single egg, and therefore identical genetically. Dr.Carlotti has drawn my attention to the existence of significant intra-population variation in experiments with plankton from the same clutch of eggs, rather than from the same egg.

The BPEs extracted from such experiments describe aspects of the physiology and behaviour of a single organism acting in response to its ambient environment (which will be carefully controlled in the experiment). Note that the ambient environment includes physical, chemical and biological variables. The ambient biological environment experienced by a single plankter comprises the local concentration and biological state of its prey, its predators, its own species and competitors. Experiments designed to produce biological primitive equations require considerable ingenuity. They present a serious technical challenge to marine biologists because plankton have complex physiology and behaviour (Geider et al 1997). But the experimental challenges are no greater than those confronting particle physicists in search of the Higgs boson.

The task of establishing a portfolio of biological primitive equations to be used in modelling plankton is also daunting because of the remarkable diversity of species (Ormond, Gage and Angel 1997), and the complexity of their physiology and behaviour (for an example see Geider et al. 1997). A substantial research programme will be needed to establish BPEs for the thousand most common plankton species. Marine biologists have already taken useful steps towards that goal. For recent reviews of the biology of phytoplankton see Falkowski & Raven (1997), for zooplankton see Harris et al. (2000) and Mauchline (1998). But, given the scale of the task, the present rate of progress is too slow, and needs to be accelerated if BPE modelling of the plankton ecosystem is to become a reality in the next few decades.

18.5 Modelling with biological primitive equations

Biological primitive equations describe the basic physiological and behavioural functions of individual plankters. They do not describe the populations of those organisms. Changes in populations are caused by complex interactions between many biological primitive equations. It is the task of BPE modelling to compute those demographic changes. Modelling also describes biofeedback: the impact that populations have on the physical and chemical environment, through physical and biological biofeedback. This link between the three elements of BPE modelling is illustrated in Fig. 18.1.

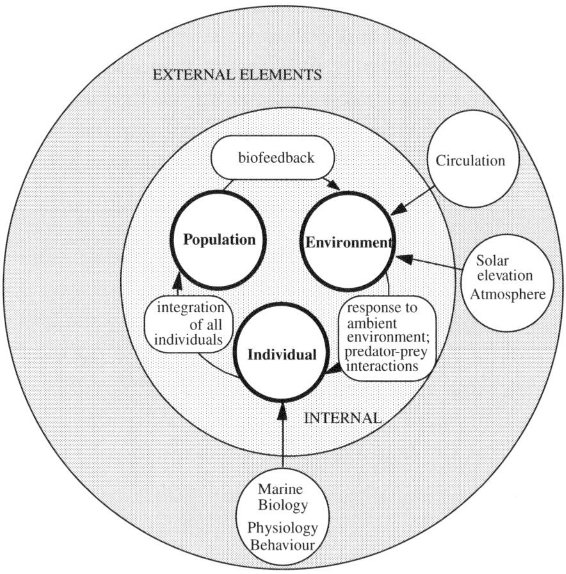

Fig. 18.1 Modelling the plankton ecosystem with biological primitive equations derived from reproducible experiments in marine biology. In this Lagrangian Ensemble method (LE), the biological equations are applied to individual organisms. Leeuwenhoek's three elements of the plankton ecosystem are internalized in simulations by Lagrangian Ensemble method (LE). The changing properties of plankton populations are emergent properties of the simulation.

18.5.1 Individual-based modelling

Modelling with biological primitive equations is individual-based modelling. The biological equations are integrated separately along the trajectories of individual organisms. This is called Biological Lagrangian Integration. There is a considerable literature on individual-based modelling (e.g. DeAngelis & Gross 1992; Lomicki 1988, 1999, Sutherland 1997, Fryxwell & Lundberg 1998). Carlotti et al. (2000) review its use in modelling zooplankton. Biological Lagrangian Integration has three main advantages compared with the classical Eulerian method of modelling:

1) The main benefit is the scientific credibility that results from using biological primitive equations. That is not possible in Eulerian models, which use population-based equations that are not controlled solely by the genes of the target species.

2) Biological Lagrangian Integration retains the intrinsic dynamical stability of the plankton ecosystem, because the biological equations are integrated along the trajectories of individual organisms. Stability is a pre-requisite for successful prediction, and therefore for effective use of plankton ecosystem models in practical applications.
3) One of the emergent properties of simulations derived from Biological Lagrangian Integration is intra-population variability. This arises from turbulent dispersion of particle trajectories. It is absent from simulations based on Eulerian integration. Lomnicki (1988) was among the first to draw attention to the statistical bias that arises from neglecting intra-population variability. Barkmann & Woods (1996) give an example in which the bias approaches a factor of two.

18.5.2 The WB food chain model

Twenty years ago, I needed a simple model to use while developing a practical method of individual-based simulation in plankton ecology. On the advice of colleagues in Kiel (Bernd Zeitzschell and Juergen Lenz), I based it on the classical food chain in which copepods graze on diatoms (Cushing 1995). This WB model Woods & Barkmann (1994) (Fig. 18.2) contains four explicit (i.e. particle-based) populations: phytoplankton, herbivorous zooplankton, dead plankton (detritus) and faecal pellets. The model also contains two implicit populations: carnivorous zooplankton, and bacteria attached to the detritus and faecal pellets. The WB model has a modular structure; each module describes one basic process. It is designed to accept biological primitive equations, when they are available. In practice, the WB model incorporated pseudo-BPEs similar to those used in Eulerian modelling (Evans & Fasham 1993). It is sufficiently realistic to justify analysis of simulations based on it, such as the Azores VPE (18.7.2). Trials showed that substituting true BPEs (i.e. those based on reproducible laboratory experiments) can lead to significantly different results. That was the case in (unpublished) models by Wolfgang Barkmann and Lucas Partridge who used BPEs for photosynthesis derived from laboratory experiment by Geider et al. (1997). However, the precise form of the equations does not matter for my lecture, which is concerned with the technology of primitive equation modelling, and not with the scientific nuances of particular models.

18.6 The Lagrangian Ensemble method

The Lagrangian-Ensemble (LE) method (Woods & Barkmann 1994) provides a well-engineered implementation of individual-based, biological-Lagrangian modelling for applications in plankton ecology. It is designed to use biological primitive equations when they are available. Where they are not available, it uses "consensus" equations from the marine biology literature, as in the WB model. Integrating the WB model by the LE method generates the demographic history of

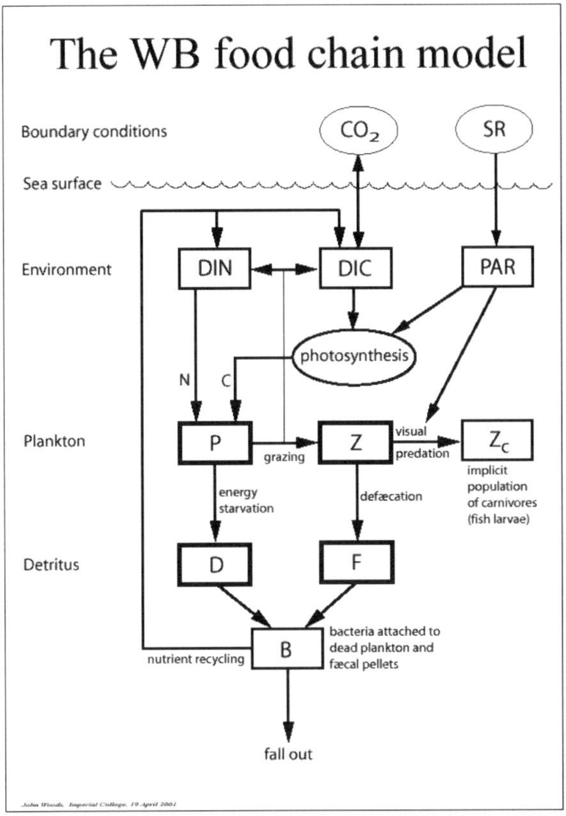

Fig. 18.2 The WB food chain model.

each population from its individuals, and biofeedback to the environment. It provides tools for monitoring and controlling errors. Extensions of the WB model exist for simulating the influence of plankton in a variety of practical problems, including: global climate change, local pollution, response to UV radiation, epidemiology of infectious diseases, and fisheries recruitment. For more information about the Lagrangian Ensemble method and for examples of its products see the website http://www.huxley.ic.ac.uk/Oceans/.

In its basic form the LE method is used in one dimension to simulate the plankton ecosystem in a vertical water column which outcrops at the sea surface and extends down to an open boundary half a kilometre below. This version is designed for the open ocean where the seabed is several kilometres deep. A shallow water version is discussed in 18.13.

18.6.1 Particles and sub-populations

The LE method is an example of agent-based computer modelling. The plankton populations are represented by a set of particles, each of which behaves like an individual plankter, and carries information about a dynamic sub-population of identical plankton. The trajectory of each particle is computed in (typically) 30 minute time steps. During each step, the particle is displaced by three processes: (1) advection by the ocean current or upwelling, (2) random displacement by turbulence in the mixing layer, and (3) vertical motion relative to the water by sinking or swimming. Turbulence in the mixing layer causes trajectories to diverge.

The LE method includes particle management rules designed to control errors in the simulated environment and demography due to local unevenness in the number of particles and in the size of the sub-populations. As the integration proceeds, particles are split (and their sub-populations shared out), or two or more particles (and their sub-populations) are combined. Ultimately the accuracy of the simulation depends on the number of independent trajectories.

18.6.2 Physiology and behaviour

Phytoplankton physiology is based on energy and nitrogen pools. When the energy pool becomes negative, all the plankters in a particle's sub-population die of energy starvation (natural mortality). When the pools both exceed their respective thresholds for reproduction, the number of plankters in a particle's sub-population is doubled. Grazing reduces the number of plankters in a sub-population. Zooplankton physiology is based on body weight, which is increased by grazing. On reaching the threshold weight for reproduction, the zooplankton produces offspring, which comprise a new sub-population carried by a new particle. The number of zooplankton in a particle's sub-population decreases as the result of carnivorous predation, which occurs at a rate determined by the prey visibility. The zooplankton have simple control rules for diel migration (to reduce losses to predation) and for foraging (to maximize ingestion).

18.6.3 Ambient environment

The biological equations use a particle's ambient environment to update the biological attributes of the plankton in its sub-population. The ambient environment is computed by interpolating the fields of environmental variables (light, temperature, nutrient concentration, prey concentration, etc.) to the particle's location. These fields are described by an Eulerian grid with a one-metre resolution. The profiles of physical and chemical variables change in response to fluxes through the sea surface, and biofeedback in the water column. The biological environment is computed from the plankton populations in each layer of the Eulerian grid.

18.6.4 Initial and boundary conditions

In order to integrate the WB model by the LE method, we need initial and boundary conditions for the water column. VEW*Trackgen* (18.19) generates these automatically from standard climatologies of the ocean and atmosphere, or from the user's own data.

The initial conditions describe the state of the ecosystem at the start of the integration, including the profiles of temperature and salinity (and therefore density), nitrate and ammonia concentrations, dissolved inorganic carbon concentration, pH, and the number of plankton in each class and their biological state. It is convenient to initialise an integration before the start of the growing season, in winter when the plankton populations are changing slowly; we normally choose the 1st March. The initial conditions are derived from world ocean climatologies like that of Levitus (1997). The critical variables are density and nutrient concentrations. The other variables can be computed by the model using these two initial conditions and the boundary conditions. As the integration proceeds, errors in the initial plankton populations are forgotten within three years, as the simulation homes in on its demographic attractor (18.12). Mixing layer depth adjusts immediately.

The boundary conditions provide the external information needed to compute the surface fluxes of solar radiation (in 25 spectral bands), net IR radiation, sensible and latent heat, and carbon dioxide. The internal information comes from the simulation (upwelling, irradiance, mixed layer temperature and salinity and partial pressure of carbon dioxide). The external data are derived from an astronomical model (for solar elevation) and atmospheric data. The latter are normally monthly mean values from Bunker's climatology or from ERA (Re-analysis data set of the European Centre for Medium Range Weather Forecasting). It is also possible to use ambient weather from ERA or ship meteorological observations. The external data are computed for each time step before the model integration and stored in a boundary condition file.

18.7 Virtual Plankton Ecosystems

18.7.1 Introduction

The result of integrating the WB model by the LE method is a massive data set called the Virtual Plankton Ecosystem. The VPE describes the state of the plankton ecosystem each time step (typically 30 minutes) for a period of several years. Some of our VPEs describe changes over one hundred years. The data set describes the environment at each time step in terms of values at one metre intervals of depth through the water column, between the sea surface and 500m. This Eulerian time series describes the physical, chemical and biological components of the environment. The biological environment is computed from the plankton sub-populations of the particles in each one metre layer (or passing through it during the time step). Thus the Virtual Plankton Ecosystem derived from LE integration contains all the information in a simulation that has been created by Eulerian integration. The environment contains about one gigabyte of data per simulated year.

In addition, the VPE contains time series of the depth and biological state variables of the plankton sub-population associated with each particle. For a typical

VPE based on one hundred thousand particles, this biological data set contains a few trillion bytes per simulated year. In practice this is too much to diagnose, so the VPE contains only a small sample of these time series; typically one thousand phytoplankton and one thousand zooplankton particles. The user controls the criteria used to select which particles will be held in the VPE. Although only a small fraction are retained for analysis, the information carried by all of the particles is used to simulate the biological environment and biofeedback to the physical and chemical environments, and for the chronological "Parish Register" of demographic events (18.8.1).

Two additional diagnostic products are computed after the VPE data set has been created: demographic time series and audit trails. The demography of each population comprises time series of birth rate, death rate by each cause of death, and life expectancy. These time series are based on the properties of the individuals, including the intra-population variation, which is an emergent property of the VPE. Such information is not available in simulations derived by Eulerian integration.

Audit trails are time series of three properties of each particle: its depth, its ambient environment, and the biological state variables of its sub-population. The ambient environment of a plankton sub-population is defined as the value of each environmental variable interpolated to its exact depth (resolved to one millimetre). Audit trails are useful for discovering what caused puzzling emergent properties of the environment. They are not available in simulations based on Eulerian integration.

18.7.2 The Azores Virtual Plankton Ecosystem

Much of the research with VPE has been designed to simulate the plankton ecosystem at a location (41N, 27W) just north of the Azores. This lies on the zero line of annual surface heat budget, i.e. where solar heating exactly balances the oceanic cooling to the atmosphere, according to Bunker's climatology. (The corresponding zero line for ERA climatology lies within 100km.) At such sites, the ocean environment exhibits a stationary annual cycle when driven by a stationary atmospheric climatology. For example, every year the turbocline at the base of the surface turbulent (mixing) layer has the same depth on any given day of the year and time of day. The annual divergence of the horizontal flux of heat is zero. The Azores also offers other simplifying advantages: the permanent circulation is relatively slow, the geostrophic eddy energy is low, as is annual Ekman pumping, while the upstream gradients of the ocean environment are slack. So we can ignore the impact of advection on the plankton ecosystem to first order, which simplifies analysis and verification of VPEs created by geographically-Eulerian integration at the Azores site. The same is not true at other mid-Atlantic sites, such as Bermuda, where advection and eddying play important roles (McGillicuddy et al 1998).

18.7.3 Environment

The Virtual Plankton Ecosystem describes the changing environment in a water column in terms of time series of the profiles (variation with depth) of each variable in half-hour steps. They are defined by values at one-metre intervals of depth. In addition, there is a time series of the exact depth (to a resolution of one millimetre) of the turbocline, which defines the base of the surface mixing layer, with continuous turbulence above, and laminar flow below. The environment contains physical, chemical and biological variables. The physics includes temperature, turbulent kinetic energy and up- and downwelling solar irradiance in 25 wavebands. The chemistry includes the concentrations of dissolved chemicals (nitrate, ammonium, inorganic carbon), also acidity (pH) and other chemicals required to compute the partial pressure of carbon dioxide. The biology comprises the concentration of plankters (number per cubic metre) of each population; also the layer-mean statistics of the biological state of those plankters, including (for phytoplankton) stored energy and nitrogen and adaptive irradiance, and (for zooplankton) body weight, satiation, egg mass, with derived variables. The variables in this biological environment are computed each time step from the plankton particles in the layer; weighted by the fraction of the time that the particle spends in the layer as it migrates through.

18.7.4 Using audit trails: a case study

Audit trails provide the modeller with a unique tool for clarifying what causes emergent features in a Virtual Plankton Ecosystem. Each audit trail is a time series with two parts: (1) the depth and biological state of the sub-population of identical plankters associated with a particle, and (2) the particle's ambient environment.

A case study shows how analysing audit trails revealed the cause of a mysterious 20-day periodicity in the number of phytoplankton during summer oligotrophy, when the pre-formed nutrients have been consumed above the nutricline (Fig. 18.3). Examination of the environment reveals that the concentration of ammonia above the nutricline also exhibits a 20-day cycle, but in anti-phase to the phytoplankton biomass. The vertical profiles of ammonia and phytoplankton change with this cycle (Fig. 18.3a). The cause is discovered by analysing the audit trails of zooplankton active during the summer. After completing their diurnal descent to avoid being seen by visual predators, they feed on phytoplankton through the night (Figs.18.3c and 18.11). The zooplankton population feeds alternately in the mixed layer and deep chlorophyll maximum, with a 20-day periodicity. While in the mixed layer they take ten days to graze down the phytoplankton. As they feed they excrete ammonia, which provides a source of nitrogen to nourish future regrowth. After reducing the phytoplankton biomass in the mixed layer to below the threshold for grazing, the zooplankton switch to feeding in the thermocline. During the next ten days, the ungrazed phytoplankton in the mixed layer recovers, but the biomass in the thermocline is depleted. The zooplankton then switch back to feeding in the mixed layer. So analysis of the audit trails of zooplankton revealed that they were responsible for fertilizing summer growth in the mixed layer. We shall see later (18.9) that this process contributes to the co-existence of competing populations of phytoplankton.

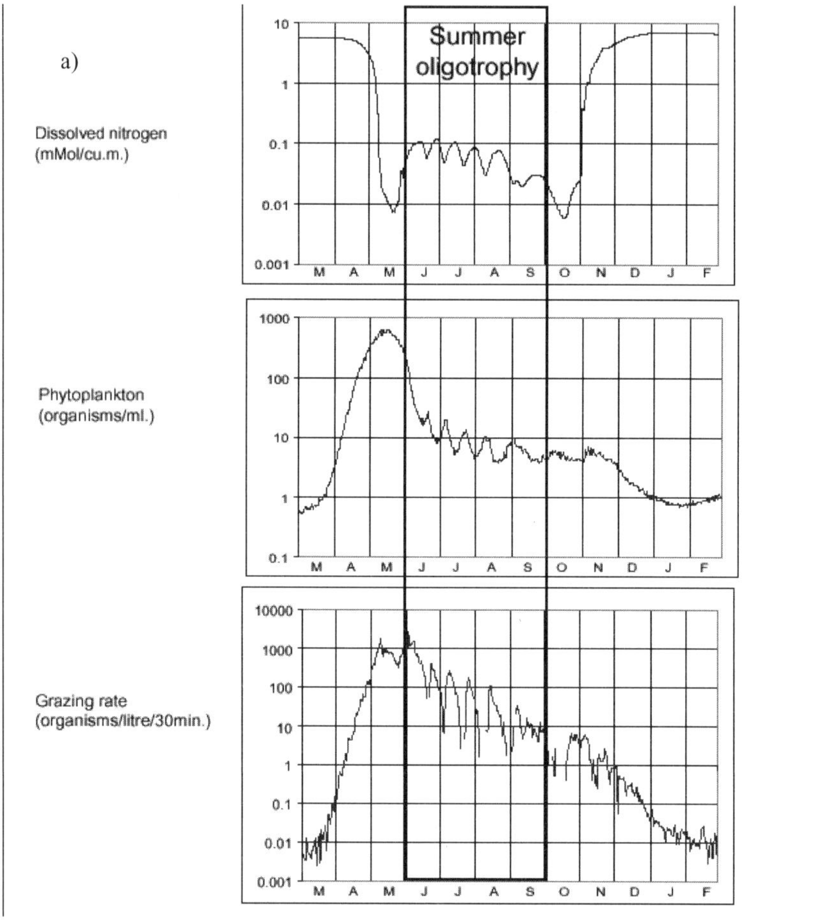

Fig. 18.3 A circa 20-day cycle of primary production in the mixed layer during summer oligotrophy. a) Variation of dissolved nitrogen, phytoplankton concentration and grazing rate in the mixed layer.

18.7.5 Do it yourself

The case study presented above was illustrated using a CD-ROM, which contains the Azores VPE and a tool to analyse it, called VEWdata (see 18.19). This package is designed for users without programming skills. It allows you to create graphs and contour plots of the environment and audit trails of a representative sample of plankton particles. You can obtain a copy of this CD-ROM by registering on http://www.huxley.ic.ac.uk/Oceans/.

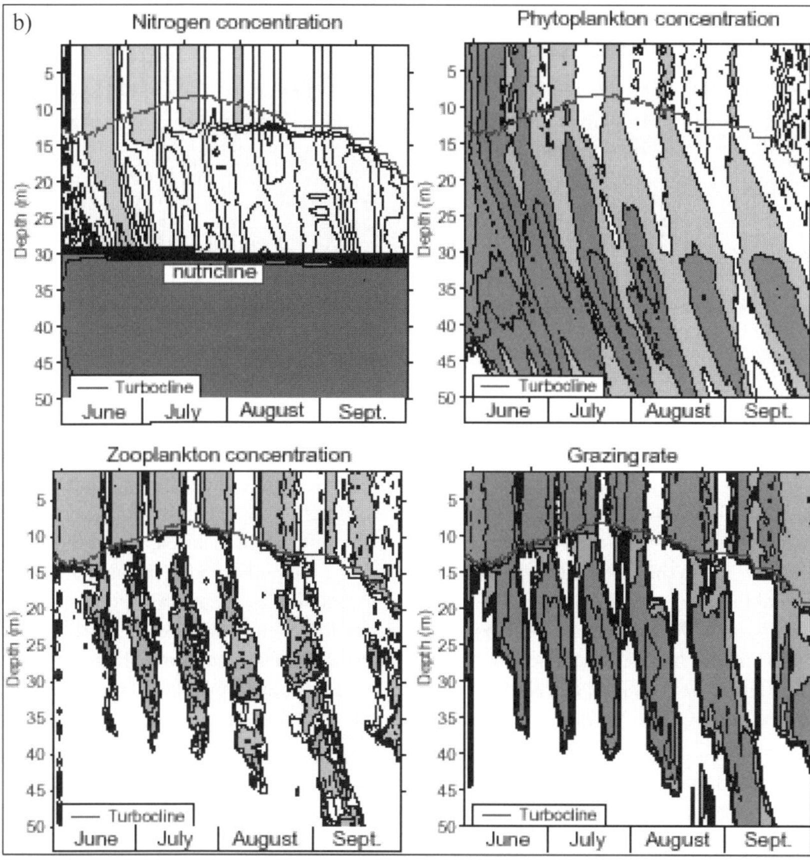

Fig. 18.3 (continued) b) Vertical distribution of dissolved nitrogen, phytoplankton and grazing rate during the period of summer oligotrophy (June to September).

18.8 Population Ecology

18.8.1 Steele's dilemma

The extensive literature on theoretical ecology (e.g. May 1981) is mainly concerned with modelling spatial patterns and temporal changes in terrestrial ecosystems. Steele (1995) has drawn attention to the failure of biological oceanographers and limnologists to adopt the sophisticated concepts and methods that have been developed in terrestrial population ecology. He suggests that the lack of transference between the two communities of researchers arose from the fact that ecosystems are fundamentally different on land and in the sea. If so, then the reason must be that fluid motion plays a dominant role in the population dynamics of life in the

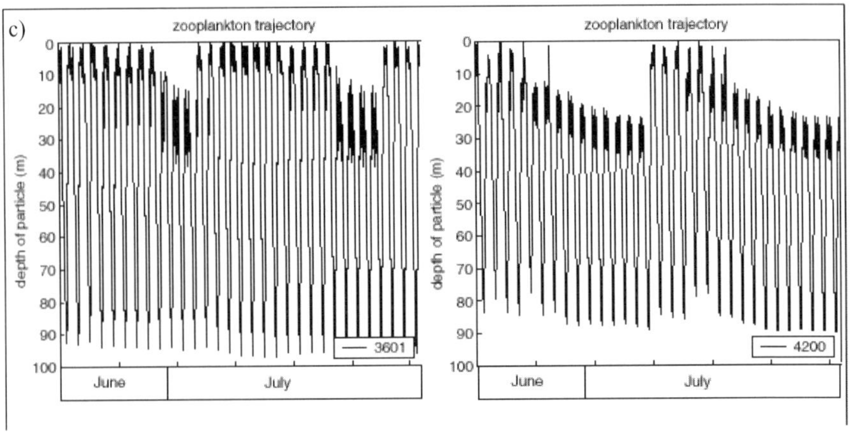

Fig. 18.3 (continued) c) Audit trails of two zooplankton migrating diurnally (to reduce the risk of being seen by visual predators) and foraging nocturnally. The difference between the behaviour of these two plankters arises from their different body weights. Both were initialized with the same body weight five years earlier. Random disturbances of their trajectories led to the different body weights, which are then inherited by primogeniture. Zooplankton 3601 was the first to be born in the water column and put on weight fast. Zooplankton 4200 was the last to be born, and put on less weight because its food supply (phytoplankton) had been depleted by the thousands of zooplankton born earlier. Despite their different weights (and therefore swimming speeds), both zooplankton contribute to periodic summer fertilization of phytoplankton in the mixed layer.

sea. That conjecture has not been tested. In order to do so, it would be necessary to add the effects of fluid motion to population ecology models of the kind used by terrestrial ecologists. In practice, biological oceanographers have not found that a useful approach. Why not?

I do not believe that the need to take account of fluid motion in our models of the plankton ecosystem prevents us from adopting the techniques of terrestrial population ecology. In this lecture I show how it can be done. The WB model (18.5.2) fully describes the impact of vertical motion (upwelling and turbulence) on plankton; while the impact of horizontal motion is taken into account by geographically-Lagrangian integration (18.11) at large scale (>100km), and by three-dimensional simulation (18.14), at small scale (1-100km).

My solution to Steele's dilemma focuses on the difficulty encountered in observing life in the sea. The empirical database available to modellers of marine ecosystems is not sufficient to provide empirical underpinning of the kind that population ecologists depend on for their theories. As I pointed out at the start of my lecture, the database is also inadequate for testing the models used to simulate the plankton ecosystem. And, I shall show later (18.17) that the empirical descriptions of features of the plankton ecosystem contain so much uncertainty that they cannot be used effectively to "tune" models of the kind developed by plankton ecologists (Hassell, Lawton and May 1976). I shall discuss such tuning later (18.15).

However, there is more to say about the differences between modelling practice in marine and terrestrial ecology. I have emphasized the need to divide ecosystems into complex non-adaptive ecosystems and complex adaptive ecosystems (18.3). The former are concerned with micro-organism which can be modelled with biological primitive equations. The latter are concerned with larger organisms, which can change their minds in unpredictable ways; we cannot devise reproducible experiments to establish primitive equations for that process. Plankton ecology falls into the first category; and so does a significant fraction of terrestrial ecology. The techniques of primitive equation modelling that I present in this lecture are equally applicable to both marine and terrestrial modelling of the population ecology of micro-organisms.

Steele's dilemma need not arise if we limited ourselves to comparing concepts applicable to plankton and terrestrial microbes, because they are both examples of complex non-adaptive ecosystems. However, the population ecology of larger organisms, whether on land (e.g. ants or deer) or in the ocean (e.g. fish or seals), cannot be addressed usefully by primitive equation modelling. They both fall into the category of complex adaptive ecosystems. The theoretical ecologist's response to that difficulty has been to adopt either empirical models fitted to data describing the animal populations, or idealized population-based models tuned to the data (Solow 1995). The extensive literature on theoretical population ecology is based on one or other of those two approaches. The literature on the population ecology of fish and sea mammals is different from that on terrestrial animals, because the respective databases are different. The marine databases do not support the kind of sophisticated modelling that can be achieved with terrestrial databases. That is fons et origo of Steele's dilemma.

However, if we restrict ourselves to micro-organisms, we have the much more powerful option of primitive equation modelling. Biological oceanographers are motivated to pursue that approach for two reasons. Firstly, it replaces difficult-to-collect data from observations at sea by data acquired from reproducible experiments in the laboratory. And, secondly, it deals with most of the biomass in the sea, which provides the food for fish and mammals. Lacking such motivation, theoretical ecologists concerned with terrestrial micro-organisms have tended to adapt the methods developed for complex adaptive ecosystems, rather than to develop primitive equation modelling. So Steele's dilemma persists in micro-ecology for historical rather than scientific reasons.

Having introduced theoretical population ecology, I shall now describe how it is practised in primitive equation modelling of the plankton ecosystem.

18.8.2 Demographic time series

Demographic time series are the raw material of population ecology. The demographic variables are birth rate, death rate for each cause of death (starvation, being eaten, fatal disease, old age or after reproduction), and life expectancy. In classical theoretical ecology these are predicted by models comprising equations for the rate of change of the population as a whole. In Eulerian simulations the equations describe contributions to demographic change in each cell of the Eulerian grid (each layer in a one-dimensional simulation). But individual-based models do not have equations for populations: the demographic time series are emergent properties of the simulation.

18.8.3 Computing demography

We noted in 18.5 that the primary task of individual-based modelling is to compute the demographic time series of each population. That is done in two stages. As the model is being integrated, a chronological record is created of every demographic event in every sub-population, with the name of its particle and the time. These events include reproduction (with the number of offspring), the number of plankters in the particle's sub-population that are eaten, or lost to starvation, disease or other cause of death), and the number of plankters emigrating to another particle's sub-population (a technical procedure used in particle management, 18.10.4). This chronological register is called the Parish Register, by analogy with the record used in human demography (the water column replacing the parish). A computer program called Lifespan (Lythgoe, 1990) is used to analyse the Parish Register. It generates the demographic time series for each population.

18.8.4 Emergent demography

As is often the case in primitive equation simulations of complex systems, the emergent properties can be counter-intuitive. That is because our intuition is coloured by simple models in which we can easily discern cause and effect. Changes in the demographic time series of Virtual Plankton Ecosystems often exhibit features that seem unreasonable at first glance. Consider, for example the time series in Fig. 18.4. The only change in the environment is a slow rise in the concentration of dissolved nitrogen in the mixed layer on 1st March each year. Annual primary production is nutrient limited, so it increases with time. But, we see that there is a steady decline in the phytoplankton biomass on 1st March; while there is no significant change in the zooplankton biomass at that time of year.

This counter-intuitive result is explained by a change in the foraging behaviour of zooplankton triggered by the rise in phytoplankton production. This response is an emergent property of the VPE. The model equations include a simple control rule for foraging behaviour that was not designed with any particular scenario in mind. The information available in the demographic time series stimulated conjectures about possible causes for this unexpected response of the demographic attractor to a change in nitrogen stock. But the demographic data were not sufficient to test those conjectures. The actual cause was found by studying the audit trails of individual particles. In this case we found that there was a complex sequence of events involving changes in the pattern of diel migration of zooplankton, summer fertilization of the phytoplankton and the success of carnivorous predation, which together led to the observed multi-year trends in the winter populations. Detailed

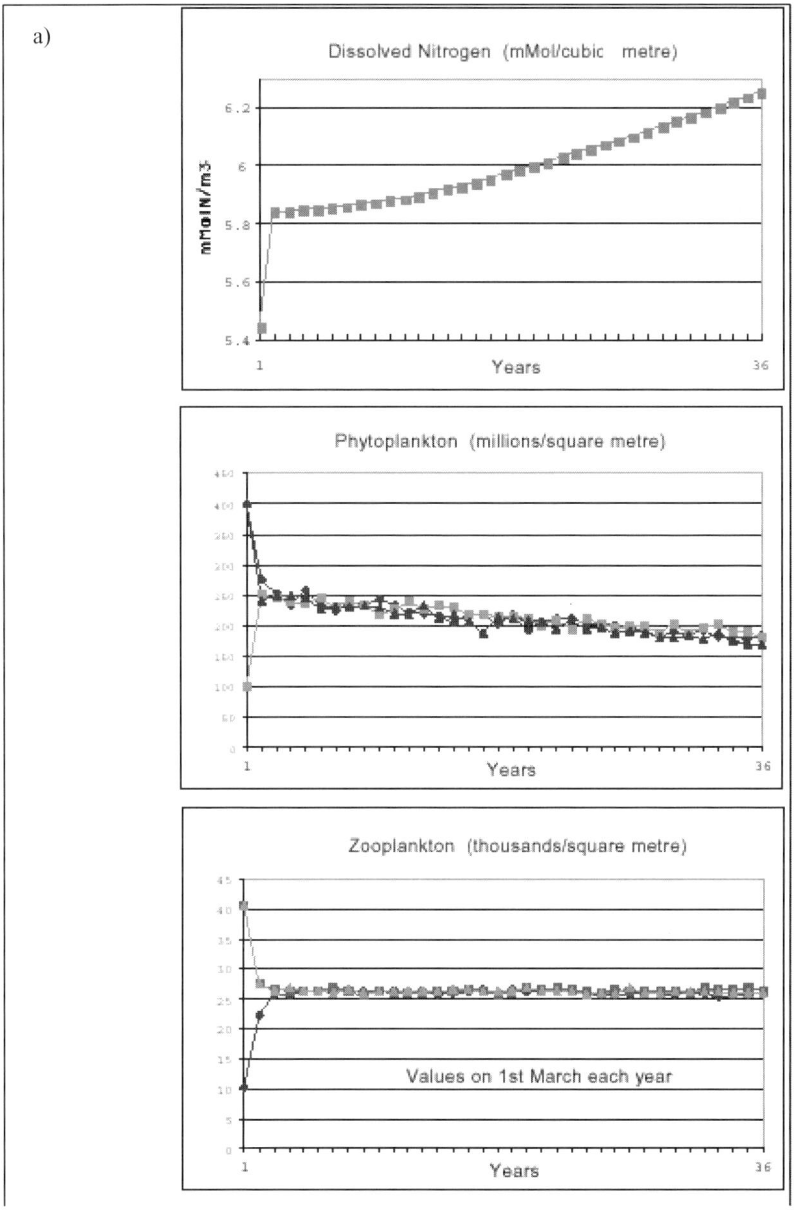

Fig. 18.4 Emergent demography during 36 years in the Azores Virtual Plankton Ecosystem. a) Time series of values on 1st March of dissolved nitrogen, phytoplankton and zooplankton.

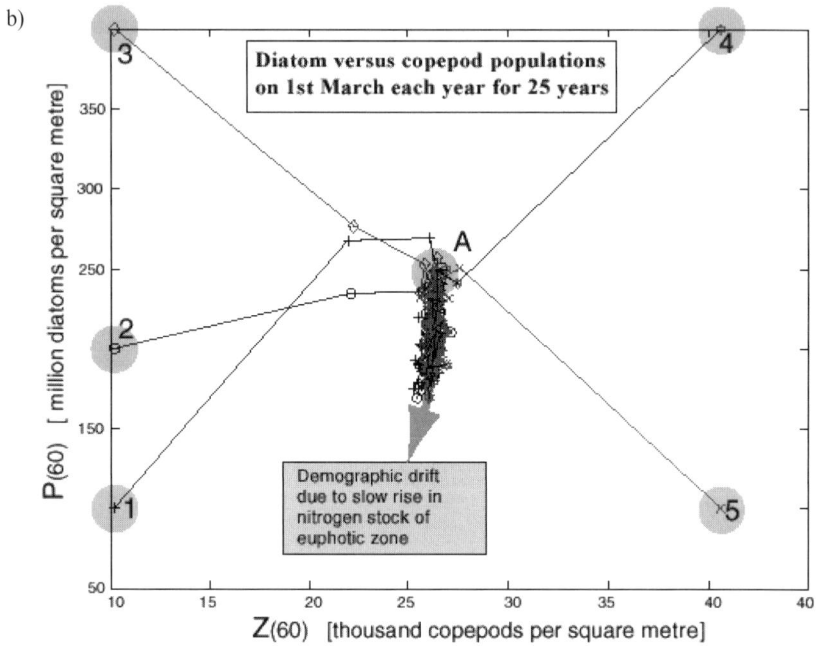

Fig. 18.4 (continued) b) The demographic attractor (A) showing its independence of demographic initial conditions (1), (2), (3), (4), (5), demographic noise and the slow response to changing nitrogen concentration.

analysis also revealed that intra-population variability in zooplankton weight played a significant role in the result. Such intra-population variability is an emergent property of individual-based modelling. It does not occur in classical population ecology or in Eulerian simulation.

18.9 Competition

Long-term monitoring of the plankton ecosystem reveals a seasonal succession of species (Sommer 1989; Lochte et al. 1993). One species often multiplies rapidly to dominate the biomass at a particular time of year. (A phenomenon first explained by Leeuwenhoeck in 1716; see Egerton 1968.) However, these observations also reveal substantial variation from year to year (Colebrook 1982). It is supposed that the ultimate cause of these seasonal and inter-annual variations are the seasonal and inter-annual variation of the upper ocean environment, controlled by surface fluxes and modulated by variation in the circulation of nutrients. One of the fundamental goals of plankton ecology is to predict seasonal succession and inter-annual variation, and to show how it is related to changes in the environment. Only when that is achieved will it be possible to predict changes due to artificial disturbance by man, whether through pollution, fisheries, or climate change.

The particular sequence of plankton abundance encountered in any year reflects the result of competition between many species for the limited resources available. The pre-requisite for simulating observed changes is to demonstrate that our modelling technique can realistically simulate the demographic consequences of competition. In particular we need to show how competitive advantage translates into competitive exclusion and extinction, or co-existence with seasonally changing dominance, depending on the ambient environment. We start this investigation at a generic level, with numerical experiments based on simple scenarios. These experiments use variants of the WB model for which we have acquired a deep understanding of how the virtual ecosystem works. The numerical experiments described below were performed by members of my research group.

In the WB model, all the phytoplankton have identical equations and parameter values, so none has a competitive advantage. The same applies to the zooplankton. The numerical experiments extended the WB model by introducing competing populations of phytoplankton of the same function group (i.e. the same equations), while the zooplankton were kept as one population which grazed on all phytoplankton without discrimination. The competing populations of phytoplankton were defined by differences in the values of one or more parameters in their biological equations. In the first experiment, intra-population variation was defined by only one parameter, which was inherited through lineages. The second experiment was also based on the same parameter, which was allowed to vary by mutation. In the third experiment, there was no mutation, but intra-population variation was defined by a large number of biological parameters.

18.9.1 Competitive advantage

Competitive advantage is measured by the relative demographic response of populations with different parameter values when they are exposed to the same environment, which changed diurnally and seasonally according to the ambient climate. Because phytoplankton reproduce many times during the growing season, there is scope for large demographic differences to established within one year. However, in our simulations it is rare for uncompetitive populations to be driven to extinction within one year. So competitive advantage is best assessed on an annual basis, if the environment performs a stationary annual cycle (as in our experiments at the Azores); or over a longer period, if the ambient climate undergoes a multi-year periodicity (as in the second experiment). The point being made is that the ranking of competitive advantage is an emergent property of competition in a virtual plankton ecosystem. It depends on a combination of the differences in biological properties of the competing plankton, and on the environment in which they are competing. We distinguish between (1) synoptic competitive advantage, the potential for demographic gain at the expense of competitors, and (2) annual competitive advantage, the demographic benefit of year-long competition.

18.9.2 Competitive exclusion

In the first experiment (Al Battran et al. 1998), we changed the values of only one parameter. This is called the "infant mortality" parameter. It determines the amount of energy (e microJoules) stored in the energy pool of a newly born phytoplankter. Reproduction tends to occur in the afternoon, followed by the hours of darkness when energy loss by respiration is not compensated by energy gain from photosynthesis. So having a sufficient energy reserve at birth, e, ensures that the newly born phytoplankter does not die by energy starvation during the first night (hence infant mortality). But there is a demographic cost: more energy must be stored between successive reproductions. So varying e trades birth rate against death rate. In the default version of the WB model, e = 0, so primary production is not optimised. A first guess at the optimal value of e suggests it should equal the nocturnal respiration, which is 12 microJoules. The numerical experiment was designed to discover the optimal value as an emergent property of the Azores VPE.

The experiment was initialised with 16 populations of phytoplankton, differing only in their value of e. That was the basis for the inter-population variation in competitive advantage. Each population started with the same number of plankters, but after five years, the relative abundance had changed, with the greatest number in those populations that had the value of e that was best fitted to the ambient climate. Fig. 18.5a shows the synoptic abundances on the 1 March of year 5 for an ensemble of five VPEs (created with different seed values, see 18.10). The best fitted value of e was 16 μJ, but the effect of demographic noise is to reduce the demographic consequences of competition between the populations with e in the range 14 to 18 μJ. The abundances of phytoplankton populations in this range had increased at the expense of those with either lower or higher values. The less well-fitted plankton were driven to extinction with twenty years. This first numerical experiment illustrated the capability of the LE method to simulate natural selection in a Virtual Plankton Ecosystem.

The Azores VPE was recomputed using a revised version of the WB model in which e has its best-fitted value of 16μJ. With primary production optimised in this way, the spring bloom peaked five days earlier that in the e=0 version. That contributed 10% towards eliminating the error identified by comparison with Sea-WIFS observation (18.17).

18.9.3 Evolution by mutation

The second experiment (Woods & Barkmann 1995) was performed with a single population of phytoplankton all initialised with e = 0. This experiment allowed mutation to occur in the value of e. Mutation events occurred randomly with a mean frequency of once every 100 reproductions. (This is much faster than occurs in nature, the acceleration being necessary for computational reasons.) We see in Fig. 18.5b that the distribution of e values diffused through e-space over twenty years, with a peak at the optimal value of 14 μJ. Note the similarity of the distributions in Fig. 18.5a and the last panel of Fig. 18.5b. Natural selection yields the same distribution in parameter space, whether starting from a prescribed set of

a)

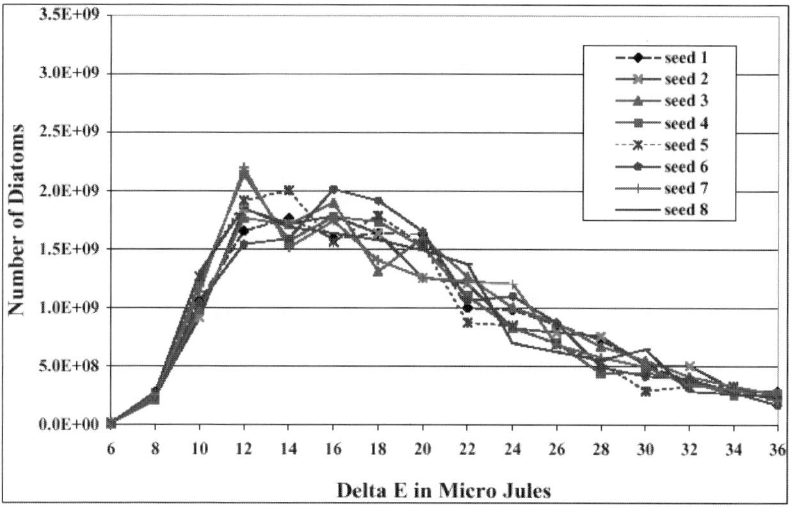

Fig. 18.5 Numerical experiments on competition between phytoplankton in a Virtual Plankton Ecosystems. a) Competitive exclusion (from Al Battran et al 1998). The abscissa refers to parameter "e" in the text.

varieties, or from one variety that mutates. The peak values in the two experiments are slightly different because the histories of ambient climate were different. In the mutation experiment the ambient climate varied with a five-year periodicity as the water column drifted around the Sargasso Sea (see 18.11). This experiment exploits an important property of the LE method, namely the inheritance of biological parameter values from one generation to the next. So a mutation is inherited by succeeding generations. This capability is also useful in modelling the demographic consequences of diseases passed on to offspring.

18.9.4 Co-existence

The third experiment (Nogueira et al 2001) was performed at the fixed location near the Azores, where the ambient climate exhibits a stationary annual cycle (see 18.12). In this experiment, the virtual ecosystem was initialised with three populations of phytoplankton of the same functional group (like diatoms) but differing in size (20, 40 and 60 μm ESD respectively). Unlike the first two experiments, in which only one parameter was allowed to vary, in this experiment all the biological parameters had different values, according to allometric relations derived from laboratory experiments. With several parameters varying, there was scope for the ranking of synoptic competitive advantage to vary with the environment at different seasons. What matters for long term changes in the relative abundance of the three competing populations was the net annual competitive advantage. Fig. 18.5c shows that this was small, despite major changes in the relative abundance at dif-

b)

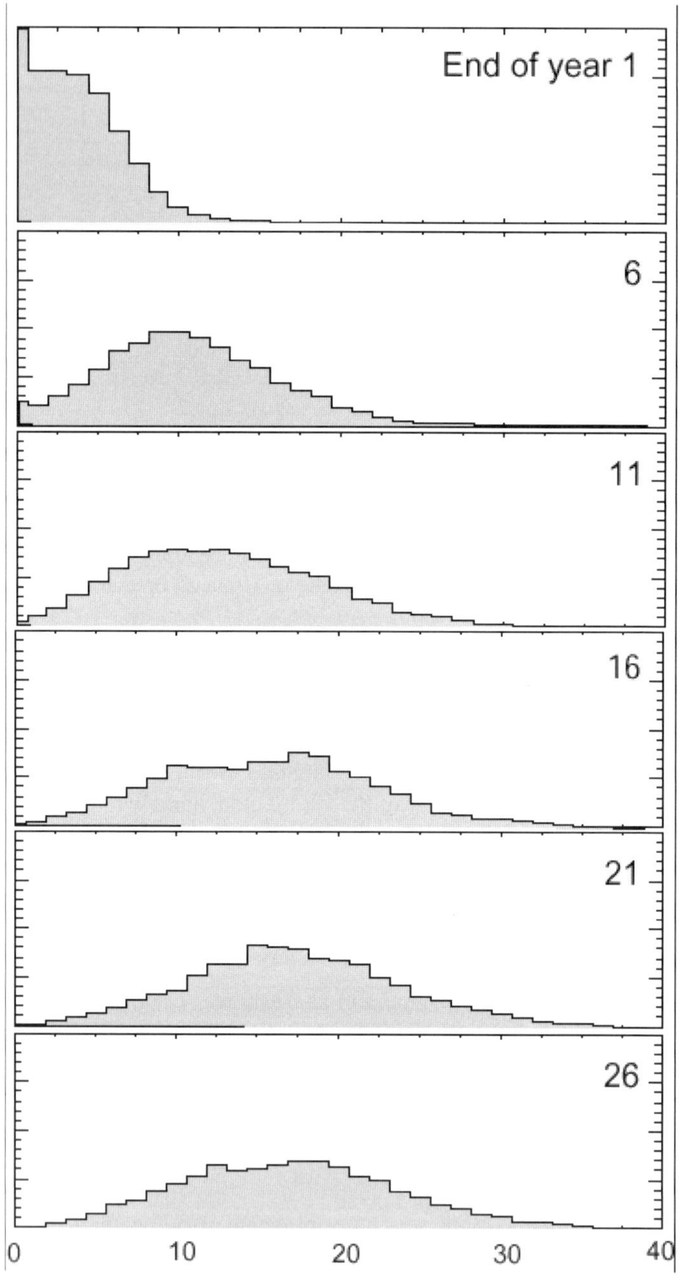

Fig. 18.5 (continued) b) Mutation experiment (from Woods and Barkmann 1985). The abscissa refers to parameter "e" in the text. Distribution of varieties in parameter space at the end of the first year in each lap of the five-year circuit around the Sargasso Sea (Fig. 18.7).

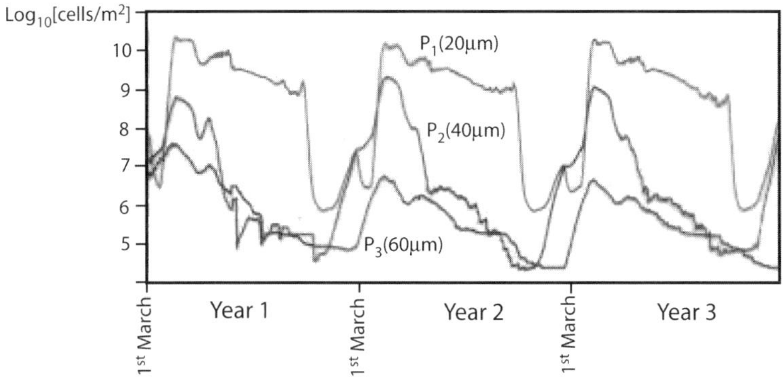

Fig. 18.5 (continued) c) Co-existence (from Nogueira et al 2001). The abscissa indicates time in days and P_1, P_2 and P_3 are three phytoplankton populations.

ferent seasons, with first one then another size-class dominating. In fact the variation in annual competitive advantage was so small that the three populations coexisted for many years. Detailed analysis, supported by audit trails, revealed that co-existence depended critically on processes occurring during the summer, including fertilization by diurnally-migrating zooplankton, and emigration from and later re-entrainement into the mixed layer.

18.9.5 Conclusion

These numerical experiments demonstrate that the LE method is capable of simulating the demographic consequences of competition. That makes it useful for studying both the natural ecosystem, and for predicting the consequences of its disturbance by man.

18.10 Ensemble integrations

18.10.1 Turbulence

The LE method of integration is deterministic: it is *not* an example of Monte Carlo modelling (Mangel & Clark 1988). However, one phenomenon in the upper ocean ecosystem is intrinsically stochastic, namely turbulence. In the WB model we adopt the simplifying assumption that turbulence occurs only in the mixing layer above the turbocline; the flow is assumed to be laminar in the diurnal, seasonal and permanent thermoclines. This assumption is supported by observations (Woods 2001). The depth of the turbocline is diagnosed each time step. Plankton particles below the turbocline lie in laminar flow; their depth is changed by upwelling/downwelling and their own motility (sinking or swimming through the water). Particles above the turbocline are also displaced by turbulence. The half hour time step is usually longer than the integral time scale of the turbulence, so there is no correlation between the depths of a particle in the mixing layer in successive time steps. The turbulent displacement of such particles is computed in the WB model by choosing at random a new depth in the mixing layer. (A random walk procedure is more appropriate when the time step is shorter, see Barkmann and Woods 1996.) A random number generator is used to determine the turbulent displacement of each particle in the mixed layer.

18.10.2 Reproducibility

The (pseudo-) random sequence of numbers is prescribed by the seed value used to initialise the random number generator. If the same seed value is used to initialise two simulations, all other conditions being identical, the set of plankton particles will follow identical trajectories, their plankton sub-populations will experience the same histories of ambient environment, and they will develop identically. So repeating a simulation with the same seed value for the random number generator produces VPEs that are identical in every detail. This property is useful in our research. It allows us to repeat a simulation with a different choice of output variables. This reproducibility is important because data storage, rather than computer time, is often the limiting factor in creating VPEs.

18.10.3 Different instances of the same VPE

Two simulations based on identical conditions apart from the seed value, produce different instances of the same VPE. The audit trails of the plankton particles are all different. Even if we had a computer powerful enough to use one particle per plankter in the water column, this difference in audit trails would lead to a small difference in the demographic time series, and in the biofeedback to the environment. In practice, we do not have such powerful computers. At best we can use a few million particles: a water column of only one square metre cross-section will contain billions of plankters. The LE method resolves that conflict by using dynamic sub-populations of some thousands of plankters per particle (Fig. 18.8). We expect the magnitude of the demographic difference between two instances of the same VPE to depend on the number of particles used in the simulation. Using more particles will reduce the difference.

We compute statistics of the differences between an ensemble of VPE instances. The variance in the demographic variables is called the demographic noise of the VPE. Each VPE is merely one instance of a potentially infinite set. None is more realistic than any other. However, we can describe an ensemble mean VPE based on the set of instances. Fig. 18.6 shows the annual variation of daily mean and standard deviation in an ensemble of eight instances of the Azores VPE. The standard deviation provides a measure of the variability between instances. It is the population's demographic noise due to turbulence in the mixing layer. Some features of the ecosystem are clearer in the ensemble mean than in the noisier instances. For example, compare the autumn bloom starting on 1st November in Fig. 18.6 and Fig. 18.3a.

If the ergodic hypothesis were valid for VPEs we would expect the demographic noise computed from an ensemble to be identical to that computed from a single VPE of long duration. Comparing Figs.18.4 and 18.6 suggests that VPEs are ergodic.

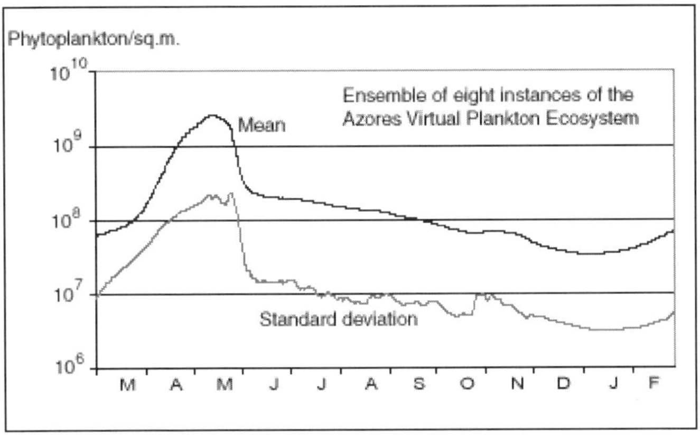

Fig. 18.6 The annual cycles of phytoplankton biomass Virtual Plankton Ecosystem at the Azores (a) ensemble mean and (b) demographic noise.

18.10.4 Demographic noise

The demographic noise is an important property of a VPE. If we want to simulate the relative success of two competing populations of plankton, the difference between their respective demographic time series must be greater than the demographic noise. Otherwise the wrong population might win the competition for spurious reasons that are an artefact of the simulation. Demographic noise was carefully controlled in the competition experiments described in 18.9.

The magnitude of the demographic noise can be reduced by using more particles. But the noise declines more slowly than one over the square root of the number of particles. We reach a practical limit at a few million particles. When that is not sufficient, we exploit particle management rules, which reduce demographic noise by automatically splitting and recombining sub-populations among a changing set of particles. Fig. 18.8 shows the fluctuation in the number of particles in a simulation using a particle management rule, which ensured that the number of particles never fell below twenty in each one-metre thick layer. All other figures were derived from simulations with 200 particles per layer.

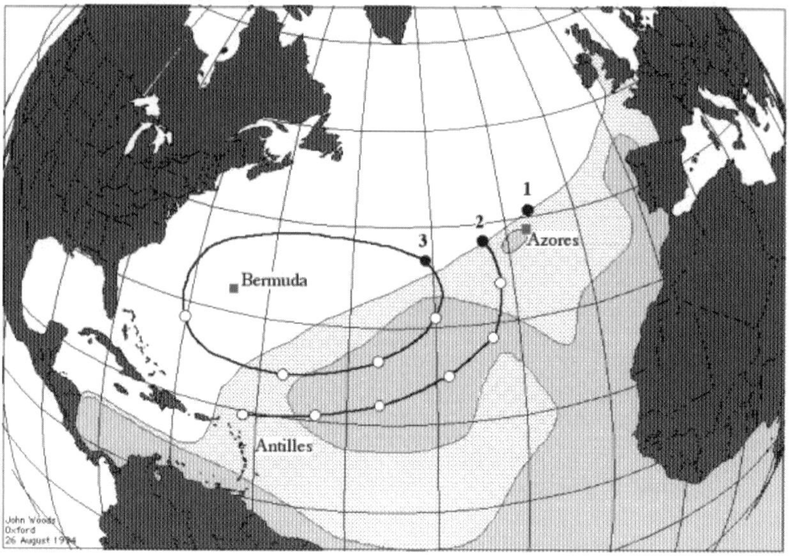

Fig. 18.7 Geographically-Lagrangian integration of a one-dimensional Virtual Plankton Ecosystem. (1) the Azores site on the balance line for annual heat budget, (2) the track from the Azores to the Antilles, in which decline in phytoplankton causes the zooplankton to become extinct, (3) the five-year closed circuit around Bermuda. The water column drifts for one year between successive circles on the tracks. The contours show the annual heat budget at zero and 10 W/m^2

18.11 Geographically-Lagrangian integration

So far I have described VPEs in a vertical water column located near the Azores, where the simulated environment exhibits a stationary annual cycle because the annual surface heat budget is in balance. The next step is to simulate the plankton ecosystem at other locations, where those conditions do not apply. It is not meaningful to perform that simulation at fixed locations where the annual surface heat budget is not in balance. The upper ocean environment does not exhibit a stationary annual cycle at such locations: the mixed layer becomes progressively shallower in regions of annual heating, and progressively deeper in regions of annual cooling. The solution is to use geographically-Lagrangian integration, a technique invented by Woods & Barkmann (1985) to compute the annual production of 18C water in the Sargasso Sea.

The method assumes that the water column containing the VPE is advected barotropically by the permanent circulation, so that it remains vertical, and there is no net mass flow of water through the sidewalls. The baroclinic advective flux divergence of every environmental variable is assumed to be zero at all depths. The boundary conditions are then computed from maps of the circulation U(x,y,t) and the climate C(x,y,t). Temporal change is usually described by monthly mean values interpolated to each time step, and the geographical variation is described by values every degree of latitude and longitude, interpolated to the position of the water column at that time. As the water column drifts with the ocean currents its ambient climate changes at a rate given by dC*/dt = U.DC.

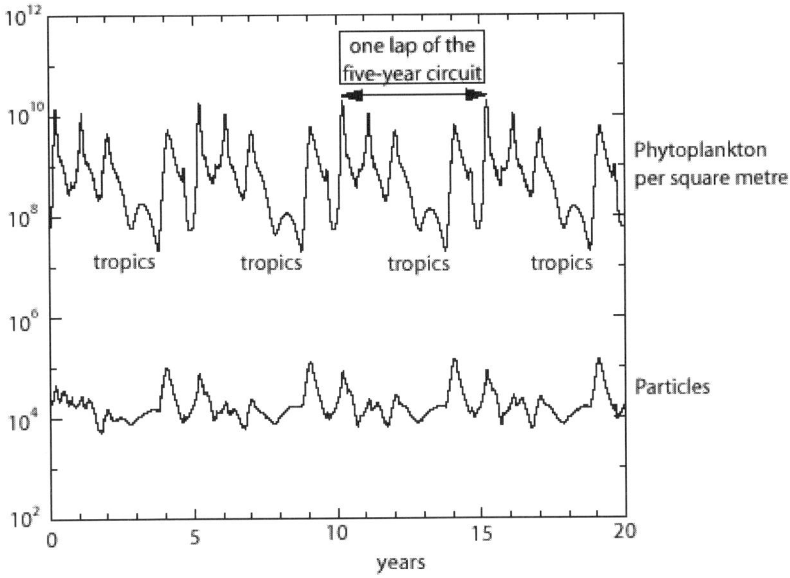

Fig. 18.8 Phytoplankton population (plankters/m^2) and the number particles during four laps of the closed circuit around Bermuda, starting on 1st March. Note the five-year periodicity. In this simulation the particle management rule ensured that the number of particles in the mixed layer never fell below 20 per metre.

This procedure is automated in VEWTrackgen, which creates the initial and boundary condition files to be used in model integration (18.19). When the water column is held at a fixed location, Trackgen creates the initial and boundary condition files by interpolation of climatological tables to that location. For Geographically-Lagrangian integration Trackgen interpolates the initial condition to the starting position, and the boundary conditions to a succession of geographical locations occupied by the water column at each time step as it drifts along its track under the control of the permanent circulation. For this purpose, Trackgen has a data set generated by an ocean general circulation model, with velocity values every month in one degree of latitude/longitude. Fig. 18.7 shows two geostrophic tracks and the Azores site.

Geographically-Lagrangian integration allows us to pass the water column through any geographical location, having started several years upstream, where the ambient climate is quite different. We start the integrations on the lines where the annual heat budget is in balance, and the ecosystem is in balance at its attractor (defined in the next section). By starting at different times on the same track, we can ensure that the water column passes through the target location at different times of year, and so build up the Eulerian-equivalent annual cycle at that location. Before we can trust that simulation of annual cycle of the ecosystem at a target location, we need to consider the stability of a drifting VPE. That will be the subject of the next section.

18.12 Stability

There is a substantial literature on the conditions under which models incorporating non-linear equations may yield simulations that exhibit oscillations or chaos. Systems that are chaotically unstable produce quite different simulations for tiny differences in the initial conditions used to start the integration of the model. Another characteristic of unstable solutions is the existence of an Attractor (sometimes called a Strange Attractor), which comprises all the locations occupied by the simulation in phase space, which has axes based on two orthogonal variables. The existence of an attractor is good news in the sense that the simulation, while unstable, is constrained to particular configurations. But it is bad news, in the sense that it limits predictability. The existence of chaos in simulations from simple models has stimulated the search for similar fluctuations in observations of the natural system being simulated. It has been found in some cases (Gleick 1998; Lorenz 1993).

It has been shown that very simple models of predator-prey interactions exhibit chaotic instability for some ranges of parameter values (May 1976). That is also true for simple NPZD (Nitrogen, Phytoplankton, Zooplankton and Detritus) models of the plankton ecosystem (Asciotti et al 1993). This suggests that the plankton ecosystem may itself be chaotically unstable. The available observations are not sufficient to confirm or reject that hypothesis. So our knowledge of the stability of the plankton system depends critically on the properties of mathematical models. Simple models have unstable solutions, but what about full-blown simulations? This is an important issue for the value of plankton ecosystem modelling in addressing practical problems; for example, in fisheries, toxic algal blooms, epidemics and climate changes. If the simulations are chaotically unstable they will have limited (if any) value in making predictions of phenomena of concern to society. It is a matter of some urgency to establish the stability of complex simulations of the plankton ecosystem, including Virtual Plankton Ecosystems.

Eulerian simulations do seem to exhibit signs of chaos (Popova et al 1997). However that result may be an artefact of the model's population-based equations, which are strongly non-linear, not least in the closure term (Caswell and Neubert 1998). Furthermore Eulerian integration is known to cause instability in engineering models. Fluid dynamicists find that, because the Lagrangian integration follows characteristics, it avoids the instability created by Eulerian integration

(Degond and Mas-Gallic 1989). These findings suggest that the chaotic instability encountered in Eulerian simulation of the plankton ecosystem may be an artefact of the method, and that it may not be present in Virtual Plankton Ecosystems derived from Lagrangian Ensemble integration of primitive equation models. In order to test that hypothesis, my colleagues and I have been exploring the stability of Virtual Plankton Ecosystems derived from the WB model.

18.12.1 Water column at a fixed location

The first numerical experiment (Woods & Perilli 2001) was designed to discover whether the plankton populations in the Azores VPE are stable when the physical and chemical environment exhibits a stationary cycle of diurnal/annual variation. (The environment is exactly the same every year at any chosen hour of the day on any chosen day of the year.) The chosen criterion for stability was the number of phytoplankton and zooplankton per square metre on the 1st March (day 60) in successive years, P(60,y) and Z(60,y) respectively. This date was chosen as being in a biologically quiescent time of year, before the start of the growing season for the zooplankton, and when the phytoplankton had barely begun their population growth which culminates two months later in the spring bloom. The demographic attractor A(60) of this VPE was defined by the locations on 1st March each year in a plot of P(60,y) versus Z(60,y). The result is shown in Fig. 18.4. We see that the demographic attractor is independent of initial conditions, and has a scope that is smaller than the demographic noise in the VPE. Furthermore, we see that the demographic attractor tracks slow changes in the environment; in this case a slow increase in the nitrogen content of the water column, deliberately introduced as part of the numerical experiment. We conclude that predator-prey interactions in the Azores VPE do not exhibit spontaneous chaotic instability. The ecosystem is robustly stable, with the demographic attractor being reached within three years of initialisation by plankton populations that are wildly out of balance with the ambient climate. This VPE responds rationally to changes of boundary conditions. It can therefore be used to make predictions.

18.12.2 Drifting water column

That numerical experiment failed to refute my hypothesis that predator-prey interactions are stable in a VPE with a stationary cycle of environment. The experiment also showed that when the boundary conditions are changed slowly, the fluctuations about the demographic attractor are smaller than the demographic noise. A more realistic test would be to use geographically-Lagrangian integration to change the ambient climate. Woods and Barkmann (1995) used geographically-Lagrangian integration to perform this numerical experiment. They showed that the plankton populations remained stable when the water column drifted around the Sargasso Sea along a closed track (Fig. 18.7), which gave a stationary five-year cycle in ambient climate, and therefore in environment. Although P(60) and Z(60) experienced massive decline in the annual warming sector of this circuit, they recovered in the annual cooling sector. Each location on the five-year circuit was passed on exactly the same day of the year (and time of day). The plankton populations at each location were the same in successive laps five years apart (Fig. 18.8).

Does this mean that the ecosystem was on its demographic attractor at each location around the circuit? The answer must be "No", because the ecosystem is in a permanent state of transient response to the changing ambient climate caused by drifting through the geographical variation in the stationary annual cycle of climate. We know from the numerical experiment at the Azores, that the transient response to demographic imbalance decays within three years. So, as the water column proceeds around the circuit, its plankton populations will always be recovering from the different ambient climate experienced as far back as three years upstream.

The definition of demographic attractor introduced for ecosystems with a stationary annual cycle of environment does not work at geographical locations where the annual heat budget is not in balance. So we cannot think of the plankton populations as merely deviating from a formal balance condition because of their upstream history of ambient climate. Nevertheless, if the patterns of ocean circulation and atmospheric climate both exhibited stationary cycles at every geographical location, then the VPE does experience a unique annual cycle at each location. (Remember that this "Eulerian" annual cycle at a fixed location is synthesized from a set of geographically-lagrangian integrations running down the same track with different phase lags.) This unique cycle is the consequence of two constraints: (1) the track passing through the target site is fixed, and (2) the history of ambient climate experienced upstream along that track is also fixed.

In reality both the track and the ambient climate history vary with the weather and inter-annual variation of climate. So the annual cycle of the VPE at the target location will not be fixed. The magnitude of variation due to inter-annual fluctuations of forcing could be discovered by performing many integrations sampling upstream histories of ambient climate/weather encountered in ERA data along the track. The ensemble mean derived from such a set of simulations will provide a "most probable annual cycle". That is the best description of what to expect when we observe the plankton ecosystem at that location. The variance of the ensemble will be used in Ecological Turing testing (18.17).

18.12.3 Extinction and re-population

In this numerical experiment the multi-year nutrient stress in the oligotrophic (tropical) sector caused the plankton populations to decline, but it was not sufficient to make them go extinct. In the absence of extinction, the VPE is robustly stable to quite large environmental stress extending over years. Other geostrophic tracks keep the water column in the Tropics for so long that zooplankton extinction does occur, for example along the Azores-Antilles track in Fig. 18.7 (see ecosystem plots in my website). Note that the conditions of the numerical experiment (barotropic advection) did not allow horizontal emigration and immigration, which allows the possibility of repopulation after extinction in the water column. Repopulation allows the ecosystem to recover from extinctions during passage through the permanently oligotrophic zone. Remember that vertical emigration from the mixed layer and re-entrainment (immigration) played a key role in the Nogueira at al (2001) simulation of co-existence of competing diatoms (see 18.9.4).

18.12.4 Conclusion

Virtual Plankton Ecosystems created by Lagrangian Ensemble integration of the WB food chain model do not exhibit spontaneous (chaotic) instability in plankton populations. On the contrary, they are robustly stable, relaxing to the demographic attractor within three years of a quite severe demographic or environmental stress. This makes the LE method suitable for use in applications that depend on predicting the response of the plankton ecosystem to external disturbances.

18.13 Shallow water VPEs

The Virtual Plankton Ecosystems described above simulate what happens in a 500 metre water column located in the open ocean, where the depth is much greater. The bottom boundary condition was open, so detritus and faecal pellets dropped out into the unresolved deep ocean, carrying carbon and nitrogen out of the VPE. Many applications are concerned with shallow water, where the sea floor lies at a depth shallower than 500 metres. Detailed modelling of the shallow water plankton ecosystem has been undertaken in recent years (see the review by Hofmann & Lascara 1998). A good example of modern practice is ERSEM (Baretta et al 1995), which is discussed in Chapter 17.

Modifying the LE method to simulate the plankton ecosystem in shallow water remains a challenge for the future. A starting point will be to modify the existing WB model. The physics must include a dynamic turbulent boundary layer powered by tidal or other currents. At some locations the turboclines delimiting the surface and seabed boundary layers will meet, giving a continuous turbulence throughout the water column. The shallow water model will use LE particles to simulate both the plankton (dead and alive) and the inorganic sediments. New equations will describe how both classes of particle can settle on the sea floor, and become re-suspended. Extending the Lagrangian Ensemble method in this way will permit Biological Primitive Equation modelling of the plankton ecosystem in shallow water.

18.14 Three-dimensional simulation

18.14.1 Computing

All the Virtual Plankton Ecosystems described so far have been designed to describe what goes on in a water column, based on LE integration of a one-dimensional model. A single-processor UNIX workstation suffices for simple versions of the WB models. However, when we want to simulate competition, it is necessary to use many more particles per population (to reduce demographic noise), so the model must be integrated on a high performance computer. My colleague Dr. Tony Field and his students in the Department of Computing have developed efficient parallel versions of the WB model code for this purpose. As a rough guide, we assume that each processor can handle about 100,000 particles, at two hours per simulated year. The simulation of competition between 16 varieties of diatom (Fig. 18.5a) used 16 processors on a Fujitsu AP3000 computer.

These examples show that powerful computers are needed to achieve the demographic signal-noise ratio required for simulating competition in one dimension. So extending LE modelling to three dimensions involves substantial computational expense. With machines available today, it is not feasible to simulate competition in three dimensions, because of the large number of particles involved. But it is possible to simulate many million particles moving through a three-dimensional environment, comprising horizontal variation of the (3D vector) velocity profile, plus the usual boundary layer dynamics (light, turbulence). The greatest technical challenge is to simulate predator-prey interaction and biofeedback. These are computed in the LE method by using the biological environment, which is derived from ensemble statistics of the sub-populations. We know from one-dimensional modelling that we need a minimum particle density of 20 per cell in the Eulerian grid. If the 3D grid has, say, 500 layers (as in our 1D modelling), in a 100 x 100 horizontal array, the grid has 5 million cells and we need at least 100 million particles. That is just feasible with a 100 processor parallel computer, taking one day per simulated year.

18.14.2 Patchiness

This approach is being used by Maggiore (2001) to investigate the response of a WB plankton ecosystem to a meandering mesoscale jet. Such jets contain most of the enstrophy (variance of relative vorticity) of mesoscale turbulence, which is believed to be responsible for much of the plankton patchiness observed in satellite images (Martin, Richards & Fasham 2001). This first attempt to extend the LE method to three-dimensions is severely constrained by available computing power. However, that constraint will decline as much more powerful computers become available, for example Petaflops computers like IBM's *Blue Gene*, which will be capable of integrating billions of particles. Large-scale modelling with biological primitive equations in three-dimensions will become more affordable by A.D. 2020.

18.15 Eulerian modelling

For some applications, such as the global carbon cycle, it will always be computationally too expensive to use individual-based modelling. The solution is to use Eulerian integration of the biological equations (Evans & Fasham 1993). This means that the plankton are treated not as individual organisms, but as a set of continuum fields for biomass and biological state variables. These fields are defined by values in each cell of the Eulerian grid. Usually (but not necessarily) these cells are the same as those used to describe the environmental fields. Such Eulerian integration of the biological equations has disadvantages, including the following:

1) The Eulerian simulation cannot describe intra-population variability, and is therefore liable to demographic bias.
2) Plankton movement relative to the water is described by biomass advection equations, which can lead to numerical instability.
3) It is not possible to simulate inheritance of biological attributes through plankton lineages.

But more fundamental than these is the fact that the equations used in Eulerian models describe changes in populations and not individual organisms. They are demographic equations, rather than biological primitive equations. So Eulerian modelling lacks the intrinsic scientific credibility that BPEs provide. Although the equations in Eulerian and Lagrangian models address the same biological processes, such as photosynthesis and respiration in phytoplankton or diel migration and foraging in zooplankton, the values of the parameters in the Eulerian equations cannot be established from laboratory experiments. In practice, the values are normally treated as open to tuning by comparison with observations of emergent properties of the Eulerian simulation (Hassell et al 1976), a very fraught process (Solow 1995 and 18.17).

18.15.1 Using BPE simulations to tune Eulerian models

Nevertheless, Eulerian integration is Hobson's choice for some practical applications. It is therefore worth asking whether the parameter values used in Eulerian models could be tuned by comparison with the emergent properties of an equivalent Lagrangian model, which is securely based on Biological Primitive Equations. In principle this is an attractive option. A Virtual Plankton Ecosystem is much more complete than data sets from observations, and offers many more features for comparison with equivalent Eulerian simulations. However, no attempt has yet been made to tune Eulerian models to VPEs, mainly because the number of tuneable biological parameters is so large: twenty-seven in the WB model. Tuning involves optimisation in 27-dimensional parameter space. Standard procedures exist for multi-dimensional optimisation (based, for example, on the use of genetic algorithms) and can easily be implemented on affordable computer clusters. But, to be effective the user must have a good *a priori* knowledge of the reference data set, its "landscape" in parameter space (Kauffman 1993). We do now have a sufficient knowledge of the Azores VPE. Fig. 18.9 shows how we might tune an Eulerian model based on the Azores VPE. It will reveal how close it is possible to make Eulerian and Lagrangian simulations of a WB plankton ecosystem.

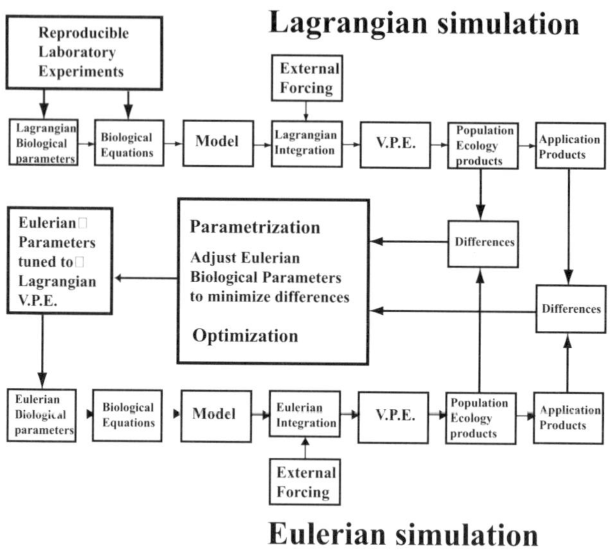

Fig. 18.9 The procedure for tuning parameters in an Eulerian version of the WB model against features in the Azores Virtual Plankton ecosystem created by Lagrangian Ensemble integration of the same WB model

18.16 Biodiversity

Biodiversity, the great number of plankton species in the ocean, poses a major challenge for the modeller. In the last century, available computer power was so limited that all models described only a few plankton types. Their demography was somehow meant to represent the "average" demography of the many species found in the water column simulated by a one-dimensional VPE. This can be rationalized by broad brush arguments. Both simulation and observation are subject to the same global constraints of resource availability, which is controlled by the environment. Hence the success of Sverdrup's (1953) model of light-limited growth. And the ability of quite simple NPZD models to predict the date on which nitrogen is exhausted in the mixed layer, which determines the date of the annual maximum of phytoplankton biomass, the peak of the spring bloom. Food chain models like that of Fasham et al. (1990) and the closely-related WB model used to develop the LE method do quite a good job in simulating such features.

However, there are other features that depend on the changing relative abundance of several competing species. For example, the shape of the rise in phytoplankton biomass cannot be described by a model with just one "average" population, because in nature it involves a succession from smaller to larger species (Ducklow and Harris 1993). And while one group of plankton, diatoms, may dominate the spring bloom overall, they stop reproducing when the supply of silicate is exhausted. Other species can then flourish until the nitrogen runs out. Later in the year, the oligotrophic environment gives a temporary competitive advantage to summer species. Notably dinoflagellates, which have evolved to grow in the seasonal thermocline, where they move up and down, getting nutrients below the nutricline, and energy above it where the light is brighter. Similar specializations exist in the zooplankton.

We showed earlier (18.9) how the changing abundances of competing species are simulated in Virtual Plankton Ecosystems. We were content to perform numerical experiments with idealized diatoms for that technical demonstration. The next step is to simulate competition between real species. Within twenty years we shall have computers that permit us to perform LE integrations with billions of particles. That will allow us to create VPEs with a thousand populations, each represented by one million particles (in a one-dimensional simulation). By the time marine biologists have produced biological primitive equations for each of those species, we shall already have in place the modelling technology to simulate demographic changes in the plankton ecosystem at the level of species.

The value of such biodiverse simulations will depend on the stability, demographic noise and errors in the computation. Acquiring knowledge about the population ecology of biodiverse VPEs will take many years. Meanwhile, we shall move ahead cautiously from the bridgehead secured with the WB food chain model, and its variants described in 18.9. Our biodiversity strategy for the next decade will be progressively to increase the number of populations of phytoplankton and zooplankton. In doing so, we shall have to assume that each population is representative of hundreds or even thousands of species. The question to be considered here is how can we relate the populations in our models to all the species they are meant to represent? The answer lies in the concept of "functional groups" (sometimes called "guilds", see May 1981).

18.16.1 Functional groups

All the species in a functional group share the same biological primitive equations. Within the group, individual species and varieties are distinguished by the values of the parameters found in those equations. Species in different functional groups have different physiology and behaviour, and therefore different BPEs. This is obviously true between trophic levels: only phytoplankton have equations for photosynthesis, and only zooplankton have equations for foraging. But there are also distinct functional groups within the same trophic level. For example, diatoms use silicon, while other phytoplankton do not. And some phytoplankton can control their motion through the water, others cannot.

Different functional groups have different equations, and therefore a different set of parameters. It is helpful to describe the difference between functional groups by their different locations in multi-dimensional parameter space. (If this concept is new to you, read the books by Kaufmann 1993, 1995.) The phytoplankton and zooplankton in the WB model are defined by sets of 15 and 22 parameters respectively. So the phytoplankton are defined by a single location in 15-dimensional parameter space.

We can see the value of this description by considering the effect of competition. In 18.9.2 I described a numerical experiment of Al Battran et al (1998) in which there was competition between 16 varieties of the WB phytoplankton, the difference lying in values of just one parameter. The locations of these varieties in parameter space are fixed, apart from a spread in that one dimension. As time marches on, the less well fitted varieties become extinct, so the spread in parameter space becomes narrower. In the numerical experiment of Nogueira et al (2001), described in 18.9.4, the three competing types of diatoms had different sizes. All 15 parameters had different values, determined allometrically. So they do not share a common position in any dimension of parameter space. But they (and similar types of different sizes) are constrained to lie on the allometric locus. In the mutation experiment (18.9.3), the simulation was initialised with a population of identical phytoplankton, which then diffused one-dimensionally through parameter space. Natural selection limited the spread, as in Al Battran et al. (1998). This dispersion of varieties through parameter space is small compared with the distance between the locations of different species. The scope of intra-population distribution in parameter space will be much less than the spacing between species. (See also the discussion of evolution in 18.3.2.)

18.16.2 Parametrizing biodiversity

Suppose that marine biologists have performed all the experiments needed to establish the biological primitive equations for, say, the thousand most common species of plankton in the North Atlantic. How can we use that information to design function groups for our LE models? The first step is to sort the species into groups that have the same equations. The species in each group have different values of the parameters in the shared equations. We want to reduce them to a single set of parameters for the functional group (FG) to be used in the model. The value of each FG parameter is an average of the values for all the species in the set for which we have experimental values. Its location in parameter space will lie within the cluster formed by the species in the same functional group. But where exactly?

One answer is to assume that the functional group occupies a fixed location in parameter space. That is Fasham's (1990) "robust" parametrization. A more sophisticated approach takes account of the relative abundance of the species contributing to the functional group. This can be done by making each parameter an average of the species values weighted by its biomass (b_i). Thus $\mathbf{P_j}$ (the FG value of the jth parameter) will be computed from the species values $p_{j,i}$ where i denotes the species.

$$\mathbf{P_j} = \Sigma_i[b_i p_{j,i}]/\Sigma_i[b_i]$$

We saw earlier (18.9) that the demographic consequences of competition between species in the same functional group is determined by two factors: (1) the values of their biological parameters, and (2) the ambient environment, which is determined externally by the ambient climate (C^*). The biomasses b_i adjusts to change in the environment, seasonally or from year to year. So the weighted average of each parameter will change with the ambient climate. The location of the functional group in multi-dimensional parameter space depends on the ambient climate.

$$\mathbf{P}_j = \mathbf{P}_j(C^*)$$

We can determine the functional relationship between $\mathbf{P}_j$ and C^* by performing a numerical experiment in which the all the species of that functional group are described explicitly (i.e. by LE particles). The first step is to create a stable virtual plankton ecosystem fixed at the Azores site (27W, 41N), with all the species of the functional group co-existing in balance with the ambient climate. Each species will then lie at its stable demographic attractor $b_i(C^*_{Azores})$. Starting from that balanced ecosystem, the second step is to allow the water column to drift barotropically from (27W, 41N) along the track defined by the mean circulation. The ecosystem in this drifting water column is simulated by geographically-lagrangian integration (18.11). The ambient climate changes as it drifts along the track. The biomasses of the set of species in the functional group adjust in response to their changing competitive advantage (18.9.1). The resulting data set $b_i\{(C^*(t)\}$ can then be used to determine $\mathbf{P}_j\,(C^*)$.

This procedure assumes that each functional group can be parametrized separately. It uses a model in which only one of the model's functional groups is expanded to its constituent species. A more comprehensive parametrization would be based on a model in which all the functional groups are expanded into their constituent species. That demands more computer power than is currently available, but it should become possible within twenty years. Meanwhile, we can parametrize each functional group separately. That will involve a substantial programme of numerical experimentation. It cannot begin until marine biologists have created a data base of the primitive equations and associated parameters for all the common species. The value of establishing a formal procedure for parametrizing biodiversity is that it provides a rational goal for research. It is based on the discovery that LE simulations are intrinsically stable.

18.16.3 Models with several functional groups

Meanwhile, the functional groups in our models are not based on formal parametrization, but on each modeller's personal review of the available data. Sometimes, $\mathbf{P}_{j,k}$ (the value of each parameter j in a functional group k) is adjusted to make the simulation of particular features of the ecosystem reproduce observations better. But such "tuning" is problematic (see below 18.17). Whatever the rationale for choosing $\mathbf{P}_{j,k}$ there is great interest in increasing the number of functional groups from the minimal set represented in a food chain model like the WB model. In this model k has only two values for explicit population of "diatoms" and "copepods" and two implicit populations of carnivorous zooplankton and bacteria attached to detritus and faecal pellets. As usual, the biological sophistication of models designed for LE integration lags several years behind current practice in Eulerian modelling. Here we give an indication of contemporary research aimed at improving both the specification and number of functional groups in LE modelling.

Carlotti and Wolf (1998) replaced the simple parametrization of herbivorous zooplankton used in the WB model (based on body weight) by one based on structured growth. (For a general discussion of this approach see Tuljapurkar & Caswell 1997.) Broekhuizen (1999) developed an LE model with two functional groups of phytoplankton based on diatoms and dinoflagellates respectively. Partridge (2000) developed a more ambitious LE food web model with two groups each of phytoplankton and herbivorous zooplankton. It is too early to report on the multi-year stability of virtual plankton ecosystems based on these models, but experience gained with the WB model shows how to proceed.

18.16.4 Conclusion

Although the long term goal is to simulate individual species explicitly in plankton ecosystem models, that lies well into the future. That is for two reasons: (1) we have biological primitive equations for only a handful of species, and (2) integrating models with thousands of species will require petaflops computers (which can do one billion sums per microsecond). Both these barriers to explicit simulation of biodiversity can be overcome, and probably will be overcome before your career ends. Meanwhile, we must be pragmatic and base our models on functional groups each of which represents many species. In this section I have discussed the scientific basis for those functional groups. The stability of VPEs based on the LE method and their ability to simulate competition provide a sound basis for parametrizing functional groups in terms of biological primitive equations and ambient climate.

18.17 The Ecological Turing Test

Testing the credibility of computer simulation goes back to the earliest days of computing. Turing (1950) famously proposed a test to discover whether one was communicating with a computer or a person. This "Turing test" involved asking a series of questions devised to discriminate between man and machine. The machine was deemed to have a mind if no question could be devised that showed that it was not a person. We can extend that idea to testing an ecosystem model by asking a succession of questions designed to discover whether a particular data set originated in observations or in a computer simulation. I call this the "Ecological Turing Test" (ETT).

18.17.1 Verification

The ETT is related to the classical procedure of verifying primitive equation models by comparing selected features described by the simulation and by observations. The model is said to have been verified if there is no statistically significant difference between simulation and observation. In that case it is impossible to claim that the simulation could not have been observed, given the uncertainties in the data set based on observations (attributable to the instruments, data processing and inadequate sampling). The observations have then failed to reveal that the model is deficient, at least in its ability to simulate the feature chosen for comparison. The model has survived to fight another day, until more challenging observations are available, either to repeat the test for that feature, or to perform a similar test for some other feature.

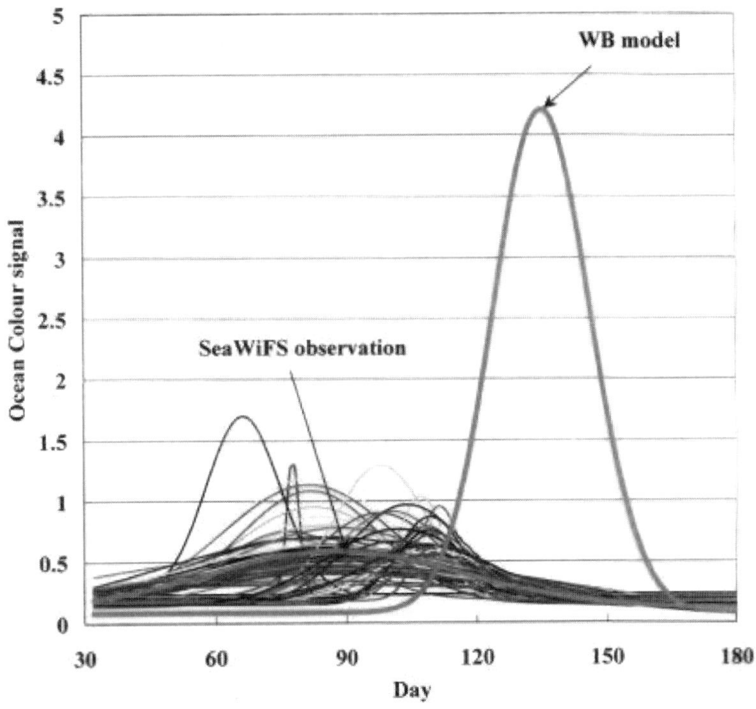

Fig. 18.10 The Ecological Turning Test. Comparison of SeaWIFS ocean colour data in a 100km x 100km box centred on the Azores site (41N 27W) and optical Monte Carlo simulation of ocean colour by from the Azores Virtual Plankton Ecosystem (from Liu et al 2001). Gaussian curves have been fitted to the time series of variation of phytoplankton number/square metre in (a) the VPE derived from the WB model, and (b) the SeaWIFS data in each 10 km square area within the 10km x 100km box. Note the considerable variation within the 100 km box. This is due to patchiness (18.14.1). The ensemble mean of these 100 samples is shown as a heavy curve (labeled SeaWIFS observations). Taking the ensemble mean eliminates patchiness to give a product commensurate with the one-dimensional simulation, and therefore more suitable for the ecological Turing test.

18.17.2 Features of the ecosystem

The choice of feature is limited by the design of the model. For example, a food chain model, with only one functional group of plankton at each trophic level, contains no information about the seasonal succession of species. And a one-dimensional model contains no information about patchiness due to mesoscale turbulence. Such limitations can in principle be overcome by processing the observations to create products commensurate with the simulation. However such processing is in itself a source of uncertainty. In the first example, it is difficult to recreate the time series of biomass for a single function group of plankton (with the specification used in the model, where there is no competition with other functional groups), from observed time series of the abundance of many species, including some from other functional groups. How, for example, can one extract information from observations about a functional group like diatoms, the growth of which is limited in the model by nitrogen, but in the ocean by nitrogen and silicon? In the second example, it is difficult to establish the time series of phytoplankton biomass in the water column of a one-dimensional simulation, from samples made in the ocean where the phytoplankton is patchy. It is not sufficient merely to average the latter, because the patchiness results from complex interaction between the mesoscale turbulence and the plankton, which leads to significant deviation from one-dimensional behaviour. The processes that cause patchiness bias biological production (Martin et al 2001). The demographic attractor of a one-dimensional simulation will not be the same as that computed by horizontal averaging of the plankton populations in a three-dimensional simulation of a patchy ecosystem.

18.17.3 SeaWIFS ocean colour data

In practice, the uncertainties inherent in any feature described by simulation and observation are so great that it is rare for the Ecological Turing Test to reject a simulation as being so unrealistic that it could not have occurred in Nature. Only when the modelling process is severely flawed will there be clear blue water between the uncertain locations of the simulation and observation in feature space. A good example is provided by Liu et al (2001) who compared a time series of mixed layer chlorophyll concentration during the 1999 spring bloom just North of the Azores (Fig. 18.10). In this case, the WB model driven by Bunker climatology produces significant differences in the timing and amplitude of the spring bloom from those extracted from SeaWIFS observations (Bahenfeld et al 2001), after statistical processing to eliminate patchiness. Stimulated by the results of this ETT, the authors subsequently investigated a number of possible deficiencies in the model and in the data used to integrate it. The first hypothesis was that the optical sub-program in the WB model caused the trouble, but replacing it with a more accurate (and more costly) RTE (Radiative Transfer Equation) code only made the matter worse. However, using a more accurate parametrization of cloud-radiation interaction halved the gap between observation and simulation. And further research showed that the remaining gap could be closed by a small number of changes to the model, including optimising the infant mortality parameter (Al Battran et al 1998), improving the accuracy of the routine used for calculating bio-optical feedback, and changing from one to three size classes of phytoplankton (Nogueira et al 2001).

18.17.4 Lessons learnt

What do we learn from this Ecological Turing Test of the WB model focusing on the spring bloom observed by SeaWIFS? Firstly, the source of data was rich enough to permit processing to extract products that were commensurate with the simulation. Secondly, the ETT did reveal deficiencies in the original version of the WB model, which was based on equations and parameters derived from the literature, and had never been "tuned" to improve comparison with observations. Thirdly, the gap between observation and simulation was easily closed by improving the model in a rational way, namely by replacing deficient components (e.g. optical biofeedback, cloud-radiation parametrization) with versions that were known to be superior. Fourthly, there was no need to "tune" the model parameters to close the gap between observation and simulation; such tuning runs counter to the principles of primitive equation modelling, and should only be adopted as a last, desperate measure when all else has failed. (See Fasham and Evans 2000 for the conventional view that tuning Eulerian models to observations is useful.) Finally, the uncertainty in the observations is so large (mainly due to the need to deal with patchiness) that after a few improvements the model was no longer vulnerable to further testing by the SeaWIFS observations (of that feature at that site). That left the design of the model in limbo: it was clearly not perfect, but the remaining deficiencies lay beyond detection by this particular SeaWIFS ETT. After such correction there is no scope for assimilating SeaWIFS data in forecasts.

18.17.5 Imbalance

One final point about Ecological Turing testing of plankton ecosystem models. We have seen that Lagrangian Ensemble modelling is remarkably stable, because it integrates the biological equations along the trajectories of individual organisms. There is a stable demographic attractor for balanced climate and a variety of nutrient loads (Fig. 18.4). To reveal the deficiency of an ecosystem model the Ecological Turing test must focus on that attractor. The test will be less useful if the simulation is off the attractor. This can happen if the simulation has not yet settled after erroneous initialisation, or because it describes a water column that has drifted into the site of observation along an upstream gradient of ambient climate (18.12.2). The WB model takes about three years to settle under stationary ambient climate, so the ETT test described above was performed on the fifth year of the simulation at a fixed location where the annual cycle of the environment is stationary. However, the SeaWIFS observations described a natural ecosystem, which is inevitably in a state of transient response to changing ambient climate and weather. It is very difficult to devise a procedure that will ensure that the natural ecosystem was at its demographic attractor (or most probable demography, see 18.12.2) when the observations were made. The best one can do is to choose a location and time of year for which external stress is low, and has been low over the previous few years.

That will reduce the off-attractor uncertainty in the data set used to represent the natural ecosystem.

18.17.6 The Azores

For that reason, I prefer to perform Ecological Turing Tests with observations made at the Azores, where the mean circulation is weak, which reduces the annual change in ambient climate, and where the Azores Anticyclone reduces weather stress. The Azores also lies on the line of annual balance in the surface heat flux, which makes it possible to ignore the advective heat flux divergence in simulating the ecosystem at the demographic attractor. None of these advantages is found at Bermuda, where data are collected for testing plankton ecosystem models (Michaels and Knap 1996; McGillicudy et al. 1998). Nor are they found at the Ocean Weather Stations in the North Atlantic (Fasham 1993).

18.17.7 Conclusion

In conclusion, Ecological Turing Testing provides a formal way to discover when a model is deficient. The choice of feature for testing may suggest which aspects of the model need to be improved. However, experience with SeaWIFS observations suggests that the uncertainties in the feature as described by analysis of observations is so large that only gross inadequacies of the model will be revealed by existing observations. Urgent attention should be given to designing new experiments that will yield observations that can be used for more stringent testing of ecosystem models. That is the most important task for biological oceanographers.

18.18 Source errors in Primitive Equation modelling

18.18.1 Introduction

Priorities for R&D aimed at improving ecosystem models depend on accurate analysis and ranking of the sources of errors. In the previous section I showed how we quantify the output errors, by comparing selected features in the simulation with observations. That process is limited by uncertainties in the observations. In practice there are not many features that can be tested with currently available observations, because they provide a rather uncertain description of any chosen feature, and many features cannot be observed with present techniques. Furthermore, the difference between simulated and observed versions of a feature seldom tell us what caused the error. Nevertheless, it is essential to identify the causes of errors in simulating the ecosystem. If we cannot learn enough from verification of external features, we must start at the other end, by assessing the source errors in the modelling procedure. That is the subject of this section.

18.18.2 Experience in engineering

It is helpful to follow the procedure used by engineers in analysing the source errors in their systems (Ghanem and Knio 2001). Engineers analyse source errors in three broad classes described by the following equation

$$\varepsilon = \varepsilon^{d} + \varepsilon^{h} + \varepsilon^{s}$$

where:

ε is the total uncertainty in a particular feature of the simulation.

ε^{d} refers to the errors in the simulation arising form the inadequacies of the model equations and in the values of their parameters. Ultimately these are attributable to errors in the data from laboratory experiments used to establish the primitive equations.

ε^{h} refers to the errors arising from limitations of the method of integration.

ε^{s} refers to the errors arising from uncertainty in the initial and boundary conditions.

18.18.3 Source errors in the Azores VPE

We can follow that procedure in assessing the relative importance of each class of source error in primitive equation modelling of the plankton ecosystem. Let us illustrate the procedure with the Azores VPE. Recall that it is based on Lagrangian Ensemble (LE) integration of the WB model (18.5.3). As is normal in simulations based on food chain (as against food web) models, all aspects of the Virtual Ecosystem are tightly coupled; changes in any model parameter lead to changes in all aspects of the ecosystem. Biological feedback (through optics and nutrient and carbon chemistry) means that changes in biological parameters affect the environment as well as the plankton populations. The demography responds to a change in ambient climate. This also applies to ecosystem response to a change in top predators, a phenomenon known as the Trophic Cascade (Carpenter & Kitchell 1993), which is observed in the Azores VPE (Woods & Barkmann 1996).

Any sources error will therefore lead to errors throughout the simulated ecosystem, including features that can be observed. The complexity of the system makes it impossible to predict these ecosystem errors *a priori* from the source errors; they are emergent properties of the simulation. Ranking the magnitudes of sources errors may suggest priorities for improving the model. However, experience has taught us to be cautious. We can illustrate this with a case study. The aim was to find out which source error contributed most to the error in the spring bloom error in the Azores VPE, revealed by ecological Turing testing with SeaWIFS data (see 18.19.3).

18.18.4 A case study: optics in the Azores VPE

It has often been suggested that one of the highest priorities for improving ecosystem models is to concentrate on reducing the errors in the simulation of the profile of solar irradiance (Fasham 1990). So the case study focused on ocean optics. The three sources of optical errors are:

ε^d the optical code used in the WB model to compute irradiance profile from the surface irradiance and the chlorophyll profile.

ε^h this has two components in the LE method: (a) the resolution of the grid, and (b) the number of particles used to simulate the plankton populations, and therefore the chlorophyll profile.

ε^s the optical boundary conditions were computed from data on monthly mean cloud cover, using a well-known cloud-radiation code from Paltridge and Platt (1976).

The question is: How large are these three errors, and which contributed most to the error in the Spring bloom error in the Azores VPE? We follow the analysis of Liu et al (2001).

The optical code in the WB model (ε^d)

Errors in the model equations can be assessed by using independent data sets that isolate one function, in this case bio-optical feedback, which computes the profile of irradiance from the profile of chlorophyll concentration. A suitable data set had been collected during the JGOFS spring bloom experiment (Ducklow and Harris 1993). It comprised simultaneous profiles of chlorophyll concentration and solar irradiance (upwelling and downwelling in the photosynthetically active range 400-700 nm). Liu et al (1999) developed a new optical code based on the radiative transfer equation, which fitted the JGOFS data much better than the simpler Morel (1991) code used in the WB model. Substituting this new code reduced the optical contribution to ε^d from 100% to less than 5%. The SeaWIFS ETT was repeated for a new version of the Azores VPE based on this new optical code. It turned out that the gap between observation and simulation of the spring bloom was increased slightly by this reduction in ε^d (for reasons diagnosed in Liu et al 2001).

Optimising the infant mortality parameter (18.9) increased primary production so that the available nitrogen was consumed faster, with the result that the spring bloom peaked earlier. This reduced the SeaWIFS ETT error by 10% (five days in fifty).

The LE method (ε^h)

The Eulerian grid used in the WB model to define the profiles of chlorophyll concentration and solar irradiance has a layer spacing of one metre. Higher resolution is used in the top metre. That is needed to reduce optically-driven errors in computing the turbocline depth. As the result there is no significant contribution to ε^h from grid resolution.

However, the accuracy of the computed chlorophyll profile depends on the number of particles used to simulate the plankton. The number of particles varies according to the particle management rules (see 18.6.1). Tests showed that the default rule, which ensured that there are never fewer than 20 particles per one-metre-thick layer, gave an error (ε^h) which significantly affected the simulation of chlorophyll concentration. This source error was reduced by increasing the number of particles to 200 per layer. Then it no longer contributed significantly to the SeaWIFS ETT error.

The optical boundary condition (ε^s)

The error in the optical boundary condition ε^s was reduced significantly by introducing more modern treatments of radiation in the clear air (Gregg and Carder (1990; Anderson 1993). Using this improved method to calculate the optical boundary condition led to a substantial change in the Azores VPE. The mismatch between simulation and observation of the timing of the peak in the spring bloom was halved.

Summary

Liu et al (2001) concluded that, as expected, the errors in the optics of the original Azores VPE did make a significant contribution to the mismatch between simulation and observation of the spring bloom. Comparison with independent data showed that there were significant source errors in the model, the boundary conditions and the particle management rules. When these were corrected, by using improved codes and more particles, the new version of the Azores VPE featured a spring bloom that better matched the observations from SeaWIFS. The difference was halved by improving the optics in the simulation. That was encouraging. It was perhaps surprising that reducing ε^d by an order of magnitude made no contribution to that improvement. However, it makes sense that the larger impact is with the surface boundary condition, which determines the total flux of solar energy entering the water, rather than with the optical code, which merely controls the penetration of that flux. The lesson is that ranking the magnitude of source errors may not indicate which are most important in determining the accuracy of features in the virtual plankton ecosystem.

18.19 The Virtual Ecology Workbench (VEW)

The Virtual Ecology Workbench provides user-friendly access to LE modelling. It requires no skill in computer programming. The VEW has four levels of operation:

1) VEWdata (Fig. 18.11) supports analysis of prescribed Virtual Plankton Ecosystems
2) VEWcontroller allows the user to change the values of parameters in a prescribed model. Also to create the initial and boundary conditions used to integrate it; these conditions are generated automatically by VEWTrackgen.
3) VEWmodeller allows the user to create new LE models from prescribed science and technology modules.

4) VEWmoduler allows the user to create new science modules from biological primitive equations.

I plan to make available the Virtual Ecology Workbench for general use. The VEWdata plus the Azores Virtual Plankton Ecosystem in a set of 5 CD-ROMs will be released in 2001. The other levels are currently being used operationally in the Oceanography Laboratory at Imperial College. They will be released after further engineering and testing. Consult my website for information on availability http://www.huxley.ic.ac.uk/Oceans/.

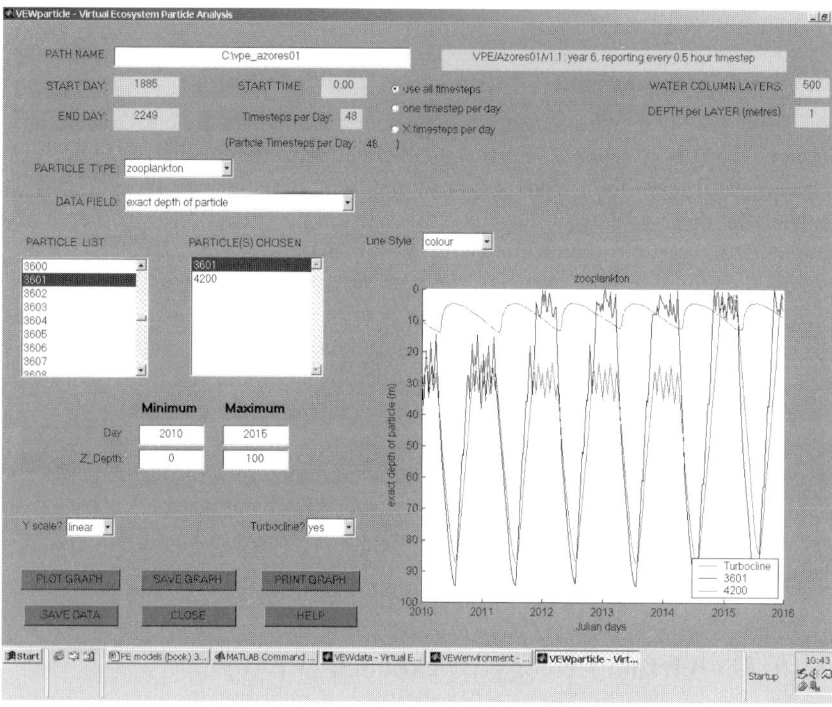

Fig. 18.11 The VEWdata Graphical User Interface showing the trajectories of zooplankton 3601 and 4200 from 3rd to 9th July in the Azores virtual plankton ecosystem. Note that these are the same zooplankton as in Fig. 18.3c. We see here that the heavier zooplankton (3601) switch to foraging in the mixed layer on the 5th July, while the lighter zooplankton (4200) continues to forage each night in the deep chlorophyll maximum (30 metres deep) until the 8th July. Note also the heavier zooplankton swim faster (slope of trajectory during diurnal descent and ascent) and that they get deeper at noon.

18.20 Conclusion

In this article I have shown the potential of primitive equation modelling in plankton ecology, and described the technology needed to implement it. That technology was developed during the last twenty years of the 20th century. It is an example of individual-based modelling, using Biological Lagrangian Integration. The goal for the new modelling technology is to use the life histories of a large number of individual plankton to compute (1) the demographic history of each plankton population, and (2) its influence of the physical and chemical environment (i.e. biofeedback). These are the emergent properties of virtual ecosystems created from biological primitive equation models.

The first technology to achieve that specification is my Lagrangian Ensemble (LE) method, which simulates the life histories of millions of particles, each of which behaves like an individual plankter, while carrying information about a dynamic sub-population of identical plankton. Methods have been developed to monitor and control the errors in the LE method. Tests show that it realistically simulates competition between populations of different plankton, demonstrating competitive exclusion and co-existence. And, unlike Eulerian simulations, Virtual Plankton Ecosystems derived from the LE method are stable, with demographic attractors that respond rationally to changes in external conditions. That stability makes the method useful for practical application.

The LE technology is now fully engineered, and powerful parallel computers are available to run it. The Virtual Ecology Workbench provides user-friendly tools for the non-programmer wishing to use the method. The Lagrangian Ensemble technology is being adapted to address a wide range of practical problems requiring accurate prediction of the plankton ecosystem. These applications include fisheries recruitment, pollution, epidemiology and climate change. Biological primitive equation modelling promises useful predictions for these important problems.

The limiting factor is the availability of biological primitive equations. BPEs are, by definition, derived from reproducible laboratory experiments with plankton. The scarcity of true BPEs made it necessary to develop LE technology with pseudo-BPEs similar those used in Eulerian modelling, but adapted for individual-based modelling. The success of LE technology opens up exciting prospects for prediction. To exploit that opportunity it will be necessary to increase funding to marine biologists, so that they can accelerate their programme of laboratory experiments designed to establish Biological Primitive Equations.

Acknowledgements

My research on the plankton ecosystem modelling has been supported by the Deutsche Forschung Gemeinschaft, the UK Natural Environment Research Council, the Southampton Oceanography Centre, the International Marine Centre (Oristano), the European Commission (MAST), and the Imperial College Parallel

Computing Centre. The work benefited the contributions of my colleagues and students including: Wolfgang Barkmann, Alan Brice, Tony Field, Enrique Nogueira, Lucas Partridge, Angelo Perilli, Sarah Talbot, Roger Wiley (academic and research staff); Cheng-Chien Liu, Uli Wolf and Angelo Maggiore (PhD students); Reza Adams, Samir Al Battran, Tim Barrell, Matt Booth, Cheng-Hua Chang, James Duggin, Pak-Wing Fok, Sam Gratrix, Chris Harris, Wes Hinsley, Chris Hurt, Ben Jefferys, Katrina Lythgoe, Darren Osborne, Simon Smith, Jana Tharmaratnam, Dave Turner, Stephen Warren (MSc and MEng students). I also wish to acknowledge the encouragement and advice received from many distinguished colleagues around the world, including: John Casti, David Cushing, Michael Fasham, Omar Knio, Juergen Lenz, Alan Longhurst, Trevor Platt, Howard Roe, John Shepherd, John Steele and Bernd Zeitzschell.

References

Adams R M, MacCarl B. A., Bryant K., Solow A., Legler D., O'Brien J. and Weiher R., 1995: Value of improved long-range weather information. *Contemporary Economic Policy*, v.13, pp.10-19.

Al Battran, S., A.J.Field, R.L.Wiley & J.D.Woods. ,1998: "Parallel simulation of plankton ecology". In: *Modelling & Simulation* Proceedings of the IASTED International Conference, Philadephia.

Anderson, T.R. ,1993: A spectrally-averaged model of ligh penetration and photosynthesis. *Limnology and Oceanography,* 38 (7), 1403-1419.

Anderson, D. L. T. and J. Willebrand, 1996: Decadal climate variability: Dynamics and Predictability. NATO ASI Series, Series I: Global Environmental Change, Vol.44, Springer, 493 pp.

ARGO, 1998: On The Design and Implementation of Argo - A Global Array of Profiling Floats. The Argo Science Team. 26 pp.

Artegiani, A., D. Bregant, E. Paschini, N. Pinardi, F. Raicich and A. Russo, 1997: The Adriatic Sea General Circulation. Part I: air-sea interactions and water mass structure. *Journal of Physical Oceanography*, Vol. 27, 1492-1514.

Asciotti, F.A., E.Beltroni, T.O.Carroll, C.Wirrick ,1993: Is there chaos in plankton dynamics? *Journal of Plankton Research,* 15, 603-617.

Auclair F., 1999: Modélisation Océanographique Cotière: initialisation et forçage. Ph-D thesis of the University of Toulouse III.

Auclair F., Casitas S. and Marsaleix P., 2000a: Application of an inverse method to coastal modeling. *Journal of Atmospheric and Oceanic Technology.* Vol. 17, 1368-1391.

Auclair F., Marsaleix P. and Estournel C., 2000b: Sigma pressure gradient errors: evaluation and reduction by an inverse method. *Journal of Atmospheric and Oceanic Technology.* Vol. 17, 1348-1367.

Ayoub, N., P.Y. Le Traon and P. De Mey, 1998: A description of the Mediterranean variable circulation from combined ERS-1 and TOPEX/POSEIDON altimetric data, *J. Mar. Sys.*, 18, 3-40.

Bacon,F. ,1605: Advancement of Learning. Ed.by G.W.Kitchen ,(1861) London, Bell & Daldy, 373pp.

Balls, P.W. 1989: The partition of trace metals between dissolved and particulate phases in European coastal waters. a compilation of field data and comparison with laboratory studies. *Neth. J. Sea Res.*, 23, 7-14.

Baretta, J.W., W.Ebenhoh and P.Ruardij 1995: The European Regional Seas Ecosystem Model, a complex marine ecosystem model. *Neth. Jour. Sea Res.*, 33, 233-246.

Barkmann, W. and J. D. Woods ,1996: On using a Lagrangian model to calibrate primary production determined from in vitro incubation measurements. *Journal of Plankton Research,* 18: 767-788.

Barnett T.P. and K. Hasselman 1979: Techniques on linear prediction, with applications to oceanic and atmospheric fields in the tropical Pacific. *Reviews of Geophys. and Space Phys.*, 17, 949-968.

Barnett T.P., M. Latif, N. Graham, M. Flügel, S. Pazan, and W. White 1993: ENSO and ENSO related predictability. Part 1 : Prediction of equatorial Pacific sea surface temperature with a hybrid coupled ocean-atmosphere model. *J. Climate*, 6(8), 1545-1566

Barnett T.P., N.E. Graham, M.A. Cane, S.E. Zebiak, S.C. Dolan, J.J. O'Brien, and D. Legler 1988 : On the prediction of the El Niño 1986/87. *Science*, 241, 192-196

Bates, J.R. 1984: An efficient semi-Lagrangian and alternating-direction implicit method for integrating the shallow-water equations. *Mon. Wea. Rev.,* 112, 2033-2047.

Bates, J.R. and A. McDonald, 1982: Multiply-upstream, semi-Lagrangian advective schemes: Analysis and application to a multilevel primitive equation model. *Mon. Wea. Rev.,* 112,1831-1842.

Bedford, K. and D. J. Schwab, 1998: Recent developments in the Great Lakes Forecasting System (GLFS) - performance evaluation. Proceedings: Second Conference on Coastal Atmospheric and Oceanic Prediction and Processes, American Meteorological Society, Boston MA, 45-50.

Belcher, S.E. and J.C.R. Hunt, 1993: Turbulent shear flow over slowly moving waves. *J. Fluid Mech.* 251, 109-148.

Bell, M. J. 1997: Vortex stretching and bottom torque in the Bryan-Cox ocean circulation model. Applications T. N. 17, The Met. Office, London Road, Bracknell, Berkshire, RG122SY, U. K.

Bengtsson, L., and H-J. Wang 1998: Climate system modelling. In: Proc WCRP Conference.

Benkiran, M., 1997 : Assimilation de données altimétriques dans un modèle de la circulation de la Méditerranée aux équations primitives. Ph.D. Thesis, Univ. Paul Sabatier, Toulouse, France, 168pp.

Benkiran, M., P. De Mey, and N. Ayoub, 1997 : Comparaison des données altimétriques TOPEX/POSEIDON en Méditerranée avec un modèle de circulation générale. *C. R. Acad. Sci.*, 325, 201-205.

Bernstein, S., and D. Hyland, 1985: The optimal projection equations for reduced-order state estimation. *IEEE Trans. Autom. Contr.,* AC-30(6), 583-585.

Beron-Vera, F.J., J.Ochoa, P.Ripa, 1999: A note on the boundary conditions for salt and freshwater balances. *Ocean modelling*, 1, 111-118.

Bjerknes, J., 1964: Atlantic air-sea interaction. Adv. in Geophys., Academic Press, 10, pp 1-82.

Blackmon, M. L., 1976: A climatological spectral study of the 500mb geopotential height of the northern hemisphere, *J. Atmos. Sci.*, 33, 1607-1623.

Blackmon, M. L., Wallace, J.M., Lau, N-C and Mullen, S.L., 1977: An observational study of the Northern Hemisphere, *J. Atmos. Sci.*, 33, 1607-1623.

Blanchet, I., C. Frankignoul and M.A. Cane, 1997: A comparison of adaptive Kalman filters for a Tropical Pacific ocean model. *Mon. Wea. Rev.,* 125, 40-58.

Bleck, R. 1984: An isentropic coordinate model suitable for lee cycrogenesis simulation. *Rev. Meteor. Aeron.*, 43, 189-194.

Bleck, R., and L. T. Smith, 1990: A wind_driven isopycnic coordinate model of the North and Equatorial Atlantic Ocean. 1. Model development and supporting experiments, *J. Geophys. Res.*, 95, 3273--3285.

Bleck, R., C. Rooth, D. Hu, and L. T. Smith, 1992: Salinity_driven thermohaline transients in a wind- and thermohaline_forced isopycnic coordinate model of the North Atlantic, *J. Phys. Oceanogr.*, 22, 1486--1515.

Blumberg, A.F. and G. L. Mellor, 1987: A description of a three-dimensional coastal ocean circulation model. Coastal Estuarine Science, 4, Three-dimensional Coastal Ocean Models, N. Heaps, Ed., Amer. Geophys. Union, Washington, DC, 1 - 16.

Blumberg, A.F., and G.L.Mellor, 1983: Diagnostic and prognostic circulation studies of the south Atlantic bight. *Jour. Geophys. Res.*, 88(C8), 4579-4592.

Boone C., and P.Y. Le Traon, 1997: Near real time processing of TOPEX/POSEIDON and ERS-2 for climate studies. Proceedings of the Symposium "Monitoring the oceans in the 2000s: an integrated approach", Biarritz.

Bourke, W., 1972: An efficient, one level, primitive-equations spectral model., *Mon. Wea. Rev* 100, 683-689.

Bourke, W., 1974: A multi-level spectral model I. Formulation and hemispheric integrations. *Mon. Wea. Rev.*, 102, 687-701.

Bouttier, F., 1994: A dynamical estimation of forecast error covariances in an assimilation system, *Mon. Weather Rev.*, 122, 2376--2390.

Bouws, E., D. Jannink and G.J. Komen, 1996: The increasing wave height in the North Atlantic Ocean. *Bulletin of the American Meteorological Society*, 77, 2275-2277.

Brankovic, C. and T.N. Palmer, 1997: Atmospheric seasonal predictability and estimates of ensemble size. *Mon. Wea. Rev.,* 125,859-874.

Bretherton, C.S., C. Smith and J.M. Wallace, 1992: An intercomparison of methods for finding coupled patterns in climate data. *J. Climate.*, 5, 541-560.

Bricaud, A., 1995: Satellite observation of ocean color: some prospects. Proceedings of the Symposium "Operational oceanography and satellite observation", Biarritz.

Brink, K.H. and A. R. Robinson, 1998: The SEA. The Global Coastal Ocean. Processes and Methods. Wiley, Vol. 10, 604 pp

Broecker W S, 1991: The great ocean conveyor. *Oceanography* 4(2), pp. 79-89.

Broekhuizen, N. ,1999: Simulating motile algae using a mixed Eulerian-Lagrangian approach: does motility promote dinoflagellate persistence or co-existence with diatoms? *J.Plankton Research,* 21, 1191-1216.

Brown M, 2000: Valuing marine activities in Europe: provisional estimates, concepts and data sources. Proceedings of Second EuroGOOS Conference, Rome, March, 1999. In Press.

Bryan, F.O., B.C. Kauffman, W.G. Large, and P.R. Gent, 1997: The NCAR/TN424+STR, NCAR Techn. Note.

Bryan, K. and L. J. Lewis, 1979: A water mass model of the World Ocean. *J. Geophys. Res.* 84, 250-2517.

Bryan, K. and M.D. Cox, 1968: A nonlinear model of an ocean driven by wind and differential heating: Part I. Description of the three-dimensional velocity and density fields. *J. Atmos. Sci.*, 25,945-967.

Bryan, K., 1969: A numerical model for the study of the circulation of the world oceans. *J. Comput. Phys.* 4, 347-376.

Bryan, K., S. Manabe and R. Pacanowski, 1975: A global ocean-atmosphere climate model. Part II. The oceanic circulation. *J. Phys. Oceanogr.*, 5, 30-46.

Buch E. and Dahlin H. (eds), 2000: The BOOS Plan: Baltic Operational Oceanographic System, 1999-2003. EuroGOOS Publication No.14, Southampton Oceanography Centre, Southampton. ISBN 0-904175-41-3. 51 pp.

Burchard, H. and Baumert, H. 1995: On the performance of a mixed-layer model based on the k-e turbulence closure. *J. Geophys. Res.,* 100, 8523-8540.

Burgers, G., P.A.E.M. Janssen and D.L.T. Anderson, 1996: Inpact of sea-state dependent fluxes on the tropical ocean circulation. Proceedings of the International Scientific Conference on the Tropical Ocean Global Atmosphere (TOGA) Programme (2-7 April 1995, Melbourne, Australia) WCRP-91 - WMO/TD No. 717, p 643-646.

Cane M.A., S.E. Zebiak and S.C. Dolan 1986 : Experimental forecasts of El Niño, *Nature*, 321, 827-832.

Carlotti, F., J. Giske and F.Werner ,2000: Modeling zooplankton dynamics. In: *ICES Zooplankton Methodology Manual.* R. Harris, P. Wiebe, J. Lenz, H. R. Skjoldal and M. Huntley. London, Academic Press: 571-667 of 684.

Carlotti, F. and U.Wolf ,1998: A Lagrangian Ensemble model of *Calanus finmarchicus* coupled with a 1-D ecosystem model. *Fisheries Oceanography,* 7 191-204.

Carpenter, S.R. and J.F.Kitchell, 1993: The trophic cascade in lakes. Cambridge University Press, 385pp.

Carter, D.J.T., and L. Draper, 1988: Has the North-east Atlantic become rougher? *Nature*, 332, 494.

Carton, J.A. and E.C. Hackert, 1989: Applications of multi-variate statistical objective analysis to the circulation in the Tropical Atlantic. *Dyn. Atm. Oceans,* 13, 491-515.

Carton, J.A., G. Chepurin, X.Cao, 2000: A simple ocean data assimilation analysis of the global upper ocean 1950-1995. Part II: results. *Jour. Phys.Ocean.*, 30, 311-326.

Castellari, S., N.Pinardi and K.Leaman, 1998: A model study of air-sea interactions in the Mediterranean Sea. *Journal of Marine Sciences*, 18, 89-114.

Castelucci L, 2000: Economic Assessment of the value of Marine Industries and Services and user requirements. Proceedings of Second EuroGOOS Conference, Rome, March 1999. In Press.

Casti, J. ,1997: Would be Worlds: How simulation is changing the frontiers of science. New York, John Wiley 241pp.

Caswell, H. and M.G.Neubert ,1998: Chaos and closure terms in plankton food chain models. *J.Plankton Res.,* 20, 1837-1845.

Chang, P., L. Ji, and H. Li, 1997: A decadal climate variation in the tropical Atlantic Ocean from thermodynamic air-sea interactions. *Nature*, 385,516-518.

Charney, J. G., 1947: The dynamics of long waves in a baroclinic westerly current, *J. Meteor.*, 4, 135-162.

Chassignet, E. P., H. Arango, D. Dietrich, T. Ezer, D.B. Haidvogel, M. Ghil, C.C. Ma, A. Mehra, A.M. Paiva and Z. Sirkes, 2000: DAMEE-NAB: The base experiment. In: Ocean Circulation Model Evaluation Experiments for the North Atlantic Basin, E.P. Chassignet and P. Malanotte-Rizzoli (Eds.), *Dyn. Atmos. Oceans*, 32, 155-184.

Chelton, D.B., 1988: WOCE/NASA altimeter algorithm workshop, U.S. WOCE Tech. Rep., N°2.

Chelton, D.B., and M.G. Schlax, 1996: Global observations of oceanic Rossby waves, *Science*, 272, 234-238.

Chen D., S.E. Zebiak, M.A. Cane, and A.J. Busalacchi 1997 : Initialization and predictability of a coupled ENSO forecast model. *Mon. Wea. Rev. Vol.* 125, No. 5, 773 - 788

Chen, Q.-S. and Y.-H. Kuo, 1992: A harmonic-sine series expansion and its application for partitioning and reconstruction problems in a limited area. *Mon. Wea. Rev.*, 120,91-112.

Cheney, R., L. Miller, C.K. Tai, J. Lillibridge, J. Kuhn, M. Ji and D. Berhinger, 1997: Operational altimeter data processing and assimilation for El Niño forecasts. Proceedings of the Symposium "Monitoring the oceans in the 2000s: an integrated approach", Biarritz.

Cheng, X., G. Nitsche, and J. M. Wallace, 1995: Robustness of low-frequency circulation patterns derived from EOF and rotated EOF analyses. *J. Climate*, 8, 1709-1713.

Colebrook, J.M. ,1982: Continuous Plankton Records: Seasonal variation in the distribution and abundance of plankton in the North Atlantic Ocean and North Sea. *J.Marine Research,* 4, 435-462.

Cook, E. R., R.D. D'Arrigo, and K. R. Briffa, 1997: A reconstruction of the North Atlantic Oscillation using tree-ring chronologies from North America and Europe. *The Holocene*, 8(1), 9-17.

Cooper , M. and K. Haines, 1996 : Altimetric assimilation with water property conservation. *J. Geophys. Res.* 101, C3, 1059 - 1077.

Costanza R, D'Arge R, de Groot R, Farber S, Grasso M, Hannon B, Limburg K, Naeem S, O'Neill R V, Paruelo J, Raskin R G, Sutton P and van den Belt M, 1997: The value of the world's ecosystem services and natural capital. *Nature*, 387, pp. 253-260.

Courtier, P., J.-N. Thépaut and A. Hollingsworth, 1994: A strategy for operational implementation of 4D-Var using an incremental approach. *Q. J. R. Meteorol. Soc.,* 120**,** 1367-1387.

Cox, M. D., 1975: A baroclinic numerical model of the world ocean: preliminary results. Numerical Models of Ocean Circulation, National Academy of Sciences, Washington D.C.

Cox, M. D., 1984: A primitive equation, 3-dimensional model of the ocean. GFDL Ocean Group Tech. Rep. No. 1, GFDL, Princeton NJ, 143 pp .

Cox, M. D., 1987: An eddy_resolving numerical model of the ventilated thermocline: Time dependence, *J. Phys. Oceanogr.*, 17, 1044-1056.

Crise, A., G. Crispi, E. Mauri, 1998: A seasonal three-dimensional study of the nitrogen cycle in the Mediterranean Sea PartI. Model implementation and numerical results. *Journal of Marine Systems*, 18, 287-312.

Csanady, G. T., 1982: Circulation in the Coastal Ocean. D. Reidel, Norwell, Mass., 279 pp.

Cubasch U., R. Voss, G. Hegerl, J. Waszkewitz, T.J. Crowley, 1997: Simulation of the influence of solar radiation variations on the global climate with an ocean-atmosphere general circulation model. *Clim. Dyn.*, 13, 757-767.

Curtin, T.B., 1999: Preface, in Naval Research Reviews (T.B. Curtin, ed.), Office of Naval Reserch, 51(2), i.

Curtin, T.B., J. G. Bellingham, J. Catipovic and D. Webb, 1993: Autonomous ocean sampling networks. *Oceanography*, 6(3), 86-94.

Cushing, D. ,1995: Population, Production and Regulation in the Sea: A Fisheries Perspective. Cambridge, Cambridge University Press, 354pp.

Cushing, D. H. ,1996: Towards a science of recruitment in fish populations. Oldendorf/Luhe (Germany), Ecology Institute, 175pp.

Cushman-Roisin, B., 1994: Introduction to Physical Oceanography, Prentice Hall, Inc., 320pp.

Daley, R., 1991: Atmospheric Data Analysis. Cambridge University Press, Cambridge, England, 457pp.

Davies, H.C. 1976: A lateral boundary formulation for multi-level prediction models. *Quart.J.Roy.Meteor.Soc.*, 102,405-418.

DeAngelis, D. L. and L. J. Gross ,1992: Individual-based models and approaches in ecology: populations, communities and ecosystems. New York, Chapman & Hall, 525pp.

Degond, P. and S.Mas-Gallic ,1989: The weighted particle method for convection-diffusion equations. Part 1: The case of an isotropic viscosity. *Mathematics of Computation,* 53, 485-507.

Delworth, T., S. Manabe, and R. J. Stouffer, 1993: Interdecadal variations of the thermohaline circulation in a coupled ocean-atmosphere model. *J. Climate*, 6, 1993-2011.

DeMadron, 1994: Hydrography abd Nepheloid structures in the Grand Rhone canyon. *Continental Shelf Research*,14,457-477.

De Mey, P. and A. Robinson, 1987: Assimilation of Altimeter Eddy Fields in a Limited-Area Quasi-Geostrophic Model. *J. Phys. Oceanogr.*, 17, 2280-2293.

De Mey, P., 1994 : Optimal interpolation in a model of the Azores Current in 1986-88. In: Data Assimilation: Tools for modelling the ocean in a global change perspective, P. Brasseur and C. J. Nihoul, Eds, NATO ASI Series 1, Vol. 19, Springer-Verlag, Berlin, 253 pp.

De Mey, P., 1997: Data assimilation at the oceanic mesoscale: a review. *J. Met. Soc. Japan*, Special issue on “Data assimilation in meteorology and oceanography: Theory and practice”, 75, 415-425.

De Mey, P., and M.Benkiran, 2001: A miltivariate reduced order optimal interpolation method and its application to the Mediterranean basin-scale circulation. This volume.

Derber, J.C. and A. Rosati, 1989: A global ocean data assimilation system. *J. Phys. Oceanogr.,* 19**,** 1333-1347.

Descartes, R. ,1664: A Treatise on Man. In: The Philosophical Writings of Descartes. Edited by J. Cottingham, R. Stoothoff and D. Murdoch. Cambridge 1985, Cambridge University Press. 1**:** 99-108 of 418.

Deser, C. and M. L. Blackmon, 1993: Surface climate variations over the North Atlantic ocean during winter: 1900-1989. *J. Climate*, 6, 1743-1753.

Dharssi, I., Lorenc, A.C. and Ingleby, N.B., 1992: Treatment of gross errors using maximum probability theory. *Quart.J.Roy.Met.Soc.*, 118, 1017-1036.

Dickey, T., D.Frye, H.Jannash, E.Boyle, D.Manov, D. Sigurdson, J.McNeil, M.Stramska, A.Michaels, N.Nelson, D.Siegel, G. Chang, J. Wu and A. Knap, 1998: Initial results from the Bermuda Testbed Mooring Program, *Deep Sea Res.*, 771-794.

Dickson, R. R., J. Meincke, S.-A. Malmberg, A. J. Lee, 1988: The "Great Salinity Anomaly" in the northern North Atlantic 1968-1982. *Prog. Oceanogr.*, 20, 103-151.

Dickson, R., J. Lazier, J. Meincke, P. Rhines, and J. Swift, 1996: Long-term coordinated changes in the convective activity of the North Atlantic. *Prog. Oceanogr.*, 38, 241-295.

Donelan, M.A., F.W. Dobson, S.D. Smith and R.J. Anderson, 1993: On the dependence of sea surface roughness on wave development. *J.Phys. Oceanogr.*, 23, 2143.

Doyle, J.D., 1994: Air-sea interaction during marine cyclogenesis, p61-67 in: Proc. of the Int Symposium on Life Cycles of Extratropical Cyclones, Bergen, volume 3. Geophys. Inst. Un. Bergen, 388p.

Doyle, J.D., 1995: Coupled ocean wave/atmosphere mesoscale model simulations of cyclogenesis. *Tellus*, 47A, 766-778.

Drago A F, 1998: Report on the impact of marine-related activities on the Maltese economy. 2nd meeting of the EuroGOOS Economy Working Group (EG-EcoWG), 21-22 September 1998, Genova, Italy. 13 pp.

Ducklow,H.W. and R.P.Harris, Eds. ,1993: JGOFS: The North Atlantic Spring Bloom Experiment. *Deep-Sea Research,* II, 40 (1/2), 1-641.

Drakopoulos, P. G., Haines, K. and Wu, P., 1997: Altimetric assimilation in a Mediterranean general circulation model, *J. Geophys. Res.* 102, 10,509--10,523.

Dukowicz, J.K. and R.D. Smith, 1994: Implicit free-surface method for the Bryan-Cox-Semtner ocean model. *J. Geophys. Res.*, 99, No. C4, 7991-8014.

DYNAMO Group, 1997: DYNAMO: Dynamics of North Atlantic models: Simulation and assimilation with high resolution models. Tech. Rep. Nr. 294, Institut FMeereskunde an der Universität Kiel, Kiel, 339 pp.

Eady, E.T., 1949: Long waves and cyclone waves, *Tellus*, 1, No.3, 33-5.

Echevin, V., P. De Mey, and G. Evensen, 2000 : Horizontal and vertical structure of the representer functions for sea surface measurements in a coastal circulation model. *J. Phys. Oceanogr.*, Vol. 30, No. 10, 2627-2635.

Eckert, C. and M. Latif, 1997: Predictability of a stochastically forced hybrid coupled model of El Niño. *J. Climate*, 10(7), 1488-1504.

Egerton, F.N. ,1968: Leeuwenhoek as a founder of animal demography. *J.Hist. Biol.*, 1, 1-22.

Ehrendorfer, M. 1997: Predicting the uncertainty of numerical weather forecasts: a review. Meteorologische Zeitschrift, 6 (4) 147-183.

Ehrendorfer, M., and J.J. Tribbia, 1997: Optimal prediction of forecast error covariances through singular vectors. *J. Atmos. Sci.,* 54**,** 286-313.

Eliasen, E., B. Machenhauer, and E. Rasmussen, 1971: On a numerical method for integration of the hydrodynamical equations with a spectral representation of the horizontal fields. Tech. Rep., U. of Copenhagen.

EuroCLIVAR (Anderson D, et al.), 1998: Climate Variability and Predictability Research in Europe, 1999-2004: EuroCLIVAR Recommendations. KNMI, de Bilt, The Netherlands, xxiv + 120pp.

EuroGOOS, 1998: Report of the second meeting of the EuroGOOS Economics WG, Genova Italy, 21-22 September 1998. EuroGOOS report EG98.44. 13pp.

EuroGOOS, 1999: EuroGOOS Forward Look. EuroGOOS document EG98.22 v5. 10pp.

Evans, G.T. and M.J.R.Fasham ,1993: Towards a model of ocean biogeochemical processes. Springer-Verlag, Berlin 350pp.

Evensen, G., 1992: Using the extended Kalman filter with a multilayer quasi_geostrophic ocean model, *J. Geophys. Res.*, 97(C11), 17,905-17,924.

Evensen, G., 1994a: Inverse methods and data assimilation in nonlinear ocean models, *Physica* D, 77, 108-129.

Evensen, G., 1994b: Sequential data assimilation with a nonlinear quasi_geostrophic model using Monte Carlo methods to forecast error statistics, *J. Geophys. Res.*, 99(C5), 10,143-10,162.

Evensen, G. and N. Fario, 1997: A weak constrain variational inverse for the Lorenz equations using substitutional methods. *J. Meteor. Soc. Japan*, 7(1B), 229-243.

Evensen, G., 1997: Advanced data assimilation for strongly nonlinear dynamics, *Mon. Weather Rev.*, 125, 1342-1354.

Evensen, G., and P. J. van Leeuwen, 1996: Assimilation of Geosat altimeter data for the Agulhas current using the ensemble Kalman filter with a quasi_geostrophic model, *Mon. Weather Rev.*, 124, 85--96.

Ezer, T. and G.L. Mellor, 2000: Sensitivity studies with the North Atlantic sigma coordinate Princeton Ocean Model. In: Ocean Circulation Model Evalutation Experiments for the North Atlantic Basin, E.P. Chassignet and P. Malanotte-Rizzoli (Eds.), *Dyn. Atmos. Ocean*, 32,185-208.

Ezer, T., 1999: Decadal variabilities of the upper layers of the subtropical North Atlantic: An ocean model study. *J. Phys. Oceanogr.*, 29(12), 3111-3124.

Ezer, T., and G. L. Mellor, 1994: Continuous assimilation of Geosat altimeter data into a three dimensional primitive equation Gulf stream model, *J. Phys. Oceanogr.*, 24, 832-847.

Ezer, T., and G. L. Mellor, 1997: Simulations of the Atlantic Ocean with a free surface sigma coordinate ocean model. *J. Geophys. Res.*, 102, 15,647-15,657.

Falkowski, P.G. and J.A.Raven ,1997: Aquatic photosynthesis. Oxford, Blackwell 375pp.

Farrell, B. F, 1985: Transient growth of damped baroclinic waves, *J. Atmos Sci.*, 42, 2718-2727.

Farrell, B. F., 1988: Optimal excitation of neutral Rossby waves, *J. Atmos.Sci.*, 45, 163–172.

Farrell, B. F., 1989: Optimal excitation of baroclinic waves, *J. Atmos.Sci.*, 46, 1193–1206.

Farrell, B. F., 1990: Small error dynamics and the predictability of atmospheric flows, *J. Atmos. Sci.*, 47, 2409–2416.

Farrell, B., and A.M.Moore, 1992: An adjoint method for obtaining the most rapidly growing perturbation to oceanic flows. *J. Phys. Ocean.*, 22(4), 338-349.

Fasham, M.J.R., H.W.Ducklow and S.M.McKelvie, 1990: A nitrogen based model of plankton dynamics in the oceanic mixed layer. *J. Mar. Res.*, 48, 591-639.

Fasham, M.J.R., 1978: The statistical and mathematical analysis of plankton patchiness. *Oceanography and Marine Biology. An Annual Review,* 16, 43-79.

Fasham, M. J. R., 1993: Ecosystem behavior at Bermuda station "S" and ocean weather station "India": a general circulation model and observational analysis. *Global Biogeochemical Cycles,* 7(2): 379-415.

Fasham,M.J.R. and G.T.Evans, 2000: Advances in ecosystem modelling within JGOFS In: R.B.Hanson, H.W.Ducklow and J.G.Field (eds.). The Changing Ocean Carbon Cycle: a midterm synthesis of the Joint Global Ocean Flux Study. Cambridge University Press, 417-446 of 514pp.

Faucher, P., M. Gavart and P. De Mey, 2000 : Isopycnal EOFs and their observability from altimetry in the North and Tropical Atlantic. *J. Phys. Oceanogr.*, in press.

Fischer J. and N.C. Flemming, 1999: "Operational Oceanography: Data Requirements Survey", EuroGOOS Publication No. 12, Southampton Oceanography Centre, Southampton. ISBN 0-904175-36-7. 59 pp.

Fischer M. and M. Latif, 1995 : Assimilation of temperature and sea level observations into a primitive equation model of the tropical Pacific. *J. Mar. Sys.* 6(1995), 31-46.

Fischer M., M. Latif, M. Flügel, and M. Ji 1997 : The impact of data assimilation on ENSO simulations and predictions. *Mon. Wea. Rev.*, Vol. 125, No. 5, 819-829.

Fischer, M., 2001: ENSO predictions with coupled ocean atmosphere models. This volume.

Flather, R.A. 1987: Estimates of extreme conditions of tide and surge using a numerical model of the north-west European continental shelf. *Estuarine, Coastal and Shelf Science*, 24, 69-73.

Flemming N C, 1994: Analytical report (The role and objectives of megascience in oceanography, developing GOOS). pp. 71-126 in: Megascience: the OECD Forum. Oceanography. Paris: Organisation for Economic Cooperation and Development. 167 pp.

Flemming N C, 1999: Dividends from investing in ocean observations: a European Perspective. Proceeding of OceanObs99 (The Ocean Observing System For Climate) Conference, Saint Raphael, October 1999. CNES, Vol. 2.

Flügel M. 1994 : Untersuchungen zu Mechanismus und Vorhersagbarkeit von ENSO mit einem einfachen gekoppelten Ozean-Atmosphären-Modell. Examensarbeit Nr. 23, Max-Planck-Institut für Meteorologie, Hamburg.

Fofonoff, N.P. and R.C.Millero, 1993: Algorthms for computation of fundamental properties of seawater. Unesco technical papers in marine science, 44.

Folland, C. K., T. N. Palmer, and D. E. Parker, 1986: Sahel rainfall and worldwide sea temperatures. *Nature*, 320, 602-607.

Forbes, C., and O. Brown, 1996: Assimilation of sea surface height data into an isopycnic ocean model, *J. Phys. Oceanogr.*, 26, 1189--1213.

Fox, A. D. and Haines, K., 1996: Altimeter data assimilatroin in the OCCAM model, *Sigma*, 21, 3-5.

Fox, A. D., K. Haines, B. De Cuevas and D. J. Webb, 2000a: Altimeter assimilation in the OCCAM global model, Part I: A twin experiment. *J. Marine Sys.*, 26, 303-322.

Fox, A. D., K. Haines, B. De Cuevas and D. J. Webb, 2000b: Altimeter assimilation in the OCCAM global model, Part II: A twin experiment. *J. Marine Sys.*, 26, 323-347.

Frankignoul, C. and K. Hasselmann, 1977: Stochastic climate models. Part II. Application to sea surface temperature variability and thermocline variability. *Tellus*, 29, 284-305.

Frankignoul, C. and R. W. Reynolds, 1983: Testing a dynamical model for mid-latitude sea surface temperature anomalies. *J. Phys. Oceanogr.*, 13, 1131-1145.

Frankignoul, C., P. Müller, and E. Zorita, 1997: A simple model of the decadal response of the ocean to stochastic wind forcing. *J. Phys. Oceanogr.*, 27, 1533-1546.

Frederiksen, J. S., 1983: Disturbances and eddy fluxes in Northern Hemisphere flows: Instability of three-dimensional January and July flows, *J. Atmos. Sci.*, 40, 836–855.

Frey, H., M. Latif and T. Stockdale, 1997: The coupled GCM ECHO-2. Part I: the tropical Pacific. *Mon. Wea. Rev.*, 125, 703-720.

Fu, L.LL and R.D. Smith, 1997: Global ocean circulation from satellite altimetry and high-resolution computer simulations, *Bull. Am. Meteorol. Soc.*, 77, 2625-2636.

Fukumori, I., J. Benveniste, C. Wunsch and D.B. Haidvogel, 1993: Assimilation of sea-surface topography into an ocean circulation model using a steady-state smoother. *J. Phys. Oceanogr.*, 23, 1831-1855.

Funkquist, L. and E. Kleine, 2000: HIROMB, an introduction to an operational baroclinic model for the North Sea and the Baltic Sea. Manuscript available at request to: lfunkquist@smhi.se.

Fryxell, J.M. and P.Lundberg ,1998: Individual behaviour and Community Dynamics. Chapman & Hall, 202pp.

Galilei, Galileo, 1638: Discourses on two new sciences. Translated by H.Crew & A. de Salvio, (New York, 1914).

Gall, R., 1976: A comparison of linear baroclinic instability theory with the eddy statistics of a general circulation model, *J. Atmos. Sci.*, 33, 349-373.

Gallagher F., H. v.Storch, R. Schnur, and G. Hannoschöck 1991 : The POP manual. *Deutsches Klima Rechenzentrum. Technical Report* No.1.

Galperin, B. L., L. H. Kantha, S. Hassid and A. Rosati, 1988: A quasi-equilibrium turbulent energy model for geophysical flows. *J. Atmos. Sci.*, 45, 55-62.

Gandin Lev S., L.L.Morone and W G Collins 1993: Two years of operational comprehensive quality control at the National Meteorological Center, *Weather and Forecasting*, 8, 57-72.

Gaspar, P., and R. M. Ponte, 1997. Relation between sea level and barometric pressure determined from altimeter data and model simulations, *J. Geophys. Res.*, 102, 961-971.

Gauss 1823: Theoria combinationis observationum erroribus minimis obnoxiae.

Gauthier, P., P. Courtier, and P. Moll, 1993: Assimilation of simulated wind lidar data with a Kalman filter, *Mon. Weather Rev.*, 121, 1803--1820.

Gavart, M., and P. De Mey, 1997: Isopycnal EOFs in the Azores Current Region: A Statistical Tool for Dynamical Analysis and Data Assimilation. *J. Phys. Oceanogr.*, 27, 2146–2157.

GCOS, 1994: Report of the GCOS Working Group on Socio-Economic benefits. First Session, Washington D.C., 6 pp. + annexes and extensive bibliography.

Geider, R. J., H. L. MacIntyre & T.M.Kana ,1997: A dynamic model of phytoplankton growth and acclimation: response of the balanced growth rate and the *a*:carbon ratio to light, nutrient-limitation and temperature. *Marine Ecology Progress Series.* 148: 187-200.

Gelb, A., 1974: Applied Optimal Estimation. M.I.T. Press, Cambridge, Ma., 374 pp.

Gelci, R., H. Cazalé and J. Vassal, 1956: Bull. Inform. Comité Central Oceanogr. Etude Côtes 8, 170-187.

Gerdes, R., 1993: A primitive equation ocean circulation ocean model using a general vertical coordinate transformation. Part 1: Description and testing of the model. *J. Geophys. Res.*, 98, 14683-14701.

Gerling, T.W., 1991: A comparative anatomy of the LEWEX wave system, p182-192 in: Directional ocean wave spectra, R.C. Beal (ed); The Johns Hopkins University Press, Baltimore, 218p.

Ghanem, R. and O.Knio ,2001: A probabilistic framework for the validation and certification of computer simulations Unpublished manuscript, Department of Mechanical Engineering, Johns Hopkins University.

Ghil, M., and P. Malanotte-Rizzoli, 1991: Data Assimilation in Meteorology and Oceanography, *Advances in Geophysics*, 33, 141-266

Gill, A. E., 1980: Some simple solutions for heat-induced tropical circulation, *Quart. J. Roy. Meteor. Soc.*, 106,447-462.

Gill, A.E., and P.P. Niiler, 1973. The theory of the seasonal variability in the ocean, *Deep Sea Res.*, 20, 141-177.

Giorgi, F. and G.T. Bates, 1989: The climatological skill of a regional model over complex terrain. *Mon. Wea. Rev.*, 117, 2325-2347.

Glantz M H (ed), 1992: Climate variability, Climate Change and Fisheries. Cambridge University Press, 450 pp.

Gleick, J. ,1998: Chaos. London, Heinemann, 352pp.

Glenn, S.M., T.D.Dickey, B.Parker, W.Boicourt, 2000a: Long term real time coastal ocean observation networks, *Oceanography*, 13,1, 24-34.

Glenn, S.M., W.Boicourt, B.Parker, T.D.Dikey, 2000b: Operational observation networks for ports, a large estuary and an open shelf, *Oceanography*, 13,1, 12-23.

Golnaraghi M., Lozano C. J. and Robinson A. R., 1996: Dynamical adjustment of quasi-synoptic data sets in the eastern Mediterranean: testing a full basin coastal methodology. *Journal of Marine Systems*, 7(1), pp. 9-42.

Golub, G.H., and C. F. van Loan, 1989: Matrix Computations, Johns Hopkins University Press, 642pp.

GOOS, 1999: GOOS News. Newsletter of the GOOS Project Office. 12 pp.

Goswami B.N. and Shukla J., 1991: Predictability of a coupled ocean atmosphere model. *J. Climate*, 4, (1), 3-22.

Grant, W.E., 1986: System analysis and simulation in wildlife and fisheries sciences. ISBN-0-471-89236-X. John Wiley and Sons Inc.

Green, J. S. A., 1960: A problem in baroclinic instability, *Quart. J. Roy. Meteor. Soc.*, 86, 237-251.

Gregg, W.W. and K.L.Carder, 1990: A simple spectral solar irradiance model for cloudless marine atmospheres. *Limnology and Oceanography,* 35 (8), 1657-1675.

Griffies, S. M. and E. Tziperman, 1995: A linear thermohaline oscillator driven by stochastic atmospheric forcing. *J. Climate*, 2440-2453.

Griffies, S. M. and K. Bryan, 1997: Ensemble predictability of simulated North Atlantic interdecadal climate variability. *Science*, 275, 181-184.

Grötzner, A., M. Latif, A. Timmermann, and R. Voss, 1999: Interannual to decadal predictability in a coupled ocean-atmosphere general circulation model. *J. Climate*, 12, 2607-2624.

Grötzner, A., M. Latif, and T. P. Barnett, 1998: A decadal climate cycle in the North Atlantic Ocean as simulated by the ECHO coupled GCM. *J. Climate*, 11, 831-847.

Gu, D. and S.G.H. Philander, 1997: Interdecadal climate fluctuations that depend on exchanges between the tropics and extratropics. *Science*, 275, 805-807.

Guenther, H., W. Rosenthal, M. Stawarz, J.C. Carretero, M. Gomez, I. Lozano, O. Serrano and M. Reistad, 1998: The wave climate of the Northeast Atlantic over the period 1955-1994: the WASA wave hindcast. *The Global Atmosphere and Ocean System*, 6, 121-163.

Günther, H. and W. Rosenthal, 1983: Shallow water surface wave model based on the Texas-Marsen-Arsloe (TMA) wave spectrum, 1983 In: Proceedings of the 20th Congress of the Internat. Association of Hydraulic Research (IAHR), Moscow/SU.

Günther, H., G. J. Komen and W. Rosenthal, 1984: A Semi-Operational Comparison of Two Parameterical Wave prediction Models, *Deutsche Hydrographische Zeitung*., 37, pp 89-106.

Gustafsson, N., L. Nyberg and A. Omstedt, 1997: Coupling of High-Resolution Atmospheric Model and an Ocean Model for the Baltic Sea, *Mon. Wea. Rev.*, 126, 2822-2846.

Gustavsson, N., 1981: A review of methods for objective analysis. In: Dynamic Meteorology – Data assimilation methods, Bengtsson, Ghil, Källen, Eds, Springer-Verlag, New York, 330pp.

Haidvogel, D.B. and A. Beckmann, 1998: Numerical models of the coastal ocean, In: THE SEA: The Global Coastal Ocean, Volume 10: Processes and

Methods (K.H. Brink and A.R. Robinson, eds.), John Wiley and Sons, NY, 457-482.

Haines, K., and P. Wu, 1998: GCM studies of intermediate and deep waters in the Mediterranean. *Journal of Marine Systems*, 18, 197-214.

Hall, A., and S. Manabe, 1997: Can local linear stochastic theory explain sea surface temperature and salinity variability? *Climate Dynamics*, 13, 167-180.

Hammond, N., 1997: Rapid Response 96 - a demonstration of NATO's Rapid Environmental Assessment capability, Proceedings of a conference held in Lerici (SP), Italy, 10-14 March 1997, SACLANTCEN Conference Proceedings Series CP-44, 21-24.

Haney, R. L., 1991: On the pressure gradient force over steep topography in sigma coordinate models. *J. Phys. Oceanogr.*, 21, 610-619.

Harris, R.P., P.H.Wiebe, J.Lenz, H.R.Skjoldal and M.Huntley, 2000: Editors. ICES Zooplankton Methodology Manual. London, Academic Press, 684pp.

Hassell, M.P., J.H.Lawton and R.M.May, 1976: Patterns of dynamical behaviour in single species models. *J.Animal Ecology,* 45, 471-486.

Hasselmann K.H. 1988 : PIPs and POPs : The reduction of complex dynamical systems using Principal Interaction and Oscillation patterns. *Geophys. Res.* 93, 11.015 - 11.021

Hasselmann, K., 1976: Stochastic climate models. Part I: Theory. *Tellus*, 28, 473-485.

Hasselmann, K., T.P. Barnett, E. Bouws, H. Carlson, D.E. Cartwright, K. Enke, J.A. Ewing, H. Gienapp, D.E. Hasselmann, P. Kruseman, A. Meerburg, P. Müller, D.J. Olbers, K. Richter, W. Sell and H. Walden, 1973: Measurements of wind-wave growth and swell decay during the Joint North Sea Wave Project (JONSWAP), Dtsch. Hydrogr. Z. Suppl. A 8(12), 95p.

Hayes S.P., Mangum L.J., Picaut J., Sumi A., Kateuchi K., 1991: TAO: a moored array for real time measurements in the tropical Pacific Ocean. *Bulletin of American Meteorological Society*, 72, 339-347.

Heaps, N.S. 1967: Storm surges. Oceanographhy and Marine Biology, Annual Review, 5, (ed. H. Barnes). Allen and Unwin, London 653pp.

Hernandez, F., P.Y. Le Traon and R. Morrow, 1995: Mapping mesoscale variability of the Azores current using Topex/Poseidon and ERS-1 altimetry, together with hydrographic and Lagrangian measurements, *J. Geophys. Res.*, 100, 24995-25006.

Hersbach, H., 1998: Application of the adjoint of the WAM model to inverse wave modeling. *J. Geophys. Res*, 103, 10469-10487.

Hibler, W. D., 1979: A dynamic thermodynamic sea ice model, *J. Phys. Oceanogr.*, 9(4).

Hoffman,E.E. and C.M.Lascara, 1998: Overview of interdisciplinary modeling for marine ecosystems. In: The Sea. The global coastal ocean. Processes and methods. K.H.Brink and A.R.Robinson (eds), Volume 10, J.Wiley and Sons Inc.

Hoskins, B. J., 1990: Theory of extratropical cyclones, in Extratropical Cyclones, C. W. Newton and E.O. Holopainen, eds., American Meteorological Society,Boston, 262pp.

Houghton J T, Jenkins G J and Ephraums J J (eds), 1990: Climate change, The IPCC Scientific Assessment. Cambridge University Press. Cambridge. 365 pp.

Houghton, J., 1991: The Bahesian Lecture, 1991: The predictability of weather and climate. Phil. Trans. Roy. Soc. London A, 337, 521-572.

Hsu, Y.-J. and A. Arakawa 1990: Numerical modeling of the atmosphere with an isentropic vertical coordinate. *Mon. Wea. Rev.*, 118, 1933-1959.

Hurlburt, H.E., 1986: Dynamic transfer of simulated altimeter data into subsurface information by a numerical ocean model. *J. Geophys. Res.*, 91(C2), 2372-2400.

Hurrell, J. W., 1996: Influence of variations in extratropical wintertime teleconnections on Northern Hemisphere temperature. *Geophys. Res. Lett.*, 23, 665-668.

Hurrell, J. W.,1995: Decadal trends in the North Atlantic Oscillation regional temperatures and precipitation. *Science*, 269, 676-679.

Huthnance, J.M. , 1982: On the formation of sand banks of finite extent. *Estuarine, Coastal and Shelf Science*, 15(3), 277-299.

Huthnance, J.M., 1984: Slope currents and 'JEBAR'. *J. Phys. Ocean.*, 14, 795-810.

Ide, K., P. Courtier, M. Ghil, and A.C. Lorenc, 1997: Unified notation for data assimilation: operational, sequential and variational. *J. Met. Soc. Japan,* Special issue on "Data assimilation in meteorology and oceanography: Theory and practice", 75, 181-189.

IFREMER, 1997: Données économiques maritimes françaises, Service d'Économie maritime. IFREMER, France. ISBN 2 905434 83 X. (ed R Kalymdjian). 95 pp.

IFREMER, 1999: Données économiques maritimes françaises, Service d'Économie maritime. IFREMER, France. ISBN 2 84433 020 7. (ed R Kalaydjian). 95 pp.

Ikeda, M. and F. W. Doodson, 1995: Oceanographic applications of remote sensing, Edited by M. Ikeda and F. Dobson, CRC press, 492 pp.

Ingleby, N.B., and Lorenc, A.C., 1993: Bayesian quality control using multivariate normal distributions. *Quart.J.Roy.Met.Soc.*, 119, 1195-1225

Intergovernmental Oceanographic Commission, 1993: "The case for GOOS", IOC/INF-915 Corr. UNESCO, Paris. 60 pp.

Intergovernmental Oceanographic Commission, 1998: The GOOS 1998 Prospectus. IOC, Paris 168 pp.

Intergovernmental Oceanographic Commission, 1999: Summary Report of the Fourth Session of the IOC-WMO-UNEP Committee of GOOS (I-GOOS IV). IOC-WMO-UNEP/ I-GOOS-IV/3. 47 pp.

IOC, 1995: OceanPC manual. Software for management and international exchange of oceanographic data. IOC/INF-943 rev., 130 pp.

Jäger J and Ferguson H L (eds), 1991: Climate Change: Science, Impacts and Policy. Proceedings of the Second World Climate Conference. Cambridge University Press, Cambridge, 578 p.

James, I. N. and P. M. James, 1989: Ultra-low-frequency variability in a simple atmospheric model. *Nature*, 342,53-55.

JAMP, 1995: Assessment and Monitoring. The Oslo and Paris Commissions, London. 45pp.

Janic, Z.I., 1984: Non-linear advection schemes and energy cascade on semi-staggared grids. *Mon. Wea. Rev.*, 112, 1234-1245.

Janssen, P.A.E.M. and P. Viterbo, 1995:. Ocean waves and the atmospheric climate. *J. of Climate*, 9, 1269 - 1287.

Janssen, P.A.E.M., 1992: Experimental evidence of the effect of surface waves on the airflow. *J. Phys. Oceanogr.* 22, 1600.

Janssen, P.A.E.M., 1994: Wave growth by wind. In: Dynamics and Modelling of Ocean Waves; G.J. Komen, L. Cavaleri, M.Donelan, K. Hasselmann, S. Hasselmann and P.A.E.M. Janssen; Cambridge University Press, 532p.

Janssen, P.A.E.M., B. Hanssen and J. Bidlot, 1996: Verification of the ECMWF Wave Forecasting System against Buoy and Altimeter Data. ECMWF, Technical Memorandum No. 229.

Janssen, P.A.E.M., G.J. Komen and W.J.P. de Voogt, 1984: An operational coupled hybrid wave prediction model. *J. Geophys. Res.* C89, 3635-3654.

Janssen, P.A.E.M., G.J. Komen and W.J.P. de Voogt, 1987: Friction velocity scaling in wind-wave generation. *Boundary Layer Meteorol.* 38, 29-35.

Jazwinski, A. H., 1970: Stochastic Processes and Filtering Theory, Academic, San Diego, Calif.

Jeftic L., Keckes S. and Pernetta J. C. (eds), 1996: Climatic Change and the Mediterranean, Volume 2. Edward Arnold, London, 564 pp.

Jeftic L., Milliman J. D. and Sestini G. (eds), 1993: Climatic Change and the Mediterranean, Edward Arnold, London, 673 pp.

Jerlov, N.G., 1976: Marine Optics, Elsevier, New York, 231 pp.

Ji M. and A. Leetmaa 1997 : Impact of data assimilation on ocean initialization and El Nino prediction. *Mon. Wea. Rev.*, Vol. 125 No. 5, 742 - 753

Ji M., A. Leetma, and J. Derber 1995 : An ocean analysis system for seasonal to interannual climate studies. *Mon. Wea. Rev.*, Vol 123, 460-481

Ji M., A. Leetmaa, and V.E. Kousky 1996 : Coupled model predictions of ENSO during the 1980s and the 1990s at the National Centers for Environmental Prediction. *J. Climate* 9, 3105-3120

Ji, V. and A.D.Vernekar, 1997: Simulation of the Asian summer monsoon of 1987 and 1988 with a regional model nested in a global GCM. *J. Climate*, 10,1965-1979.

Jiang, S., F.-F. Jin, and M. Ghil. 1996: Multiple equilibria, periodic, and aperiodic solutions in a wind-driven, double gyre, shallow-water model. *J. Phys. Oceanogr.*, 25, 764-786.

Johannessen, J.A., 1995: The potential in using synthetic aperture radar for coastal environmental monitoring: the ERS-1 experience. Proceedings of the Symposium "Operational oceanography and satellite observation", Biarritz.

Johannessen, J.A., H. Rebhan, C. Le Provost, H. Drange, M. Srokosz, P. Woddworth, P. Schlussel, P. Le grand, Y. Kerr, and D. Wingham, 1999: Emerging new earth observation capabilities in the context of the ocean observing system for climate. OceanObs 99 Conference, Saint Raphael, France.

Johannessen, O.M., M.W. Miles, S. Sandven, E. Bjorgo, V.V. Melentyev, L.P. Bobylev and K.Y. Kondratyev, 1995: Sea ice monitoring using satellite-borne microwave sensors. Proceedings of the Symposium "Operational oceanography and satellite observation", Biarritz.

Juang, H.-M. H. and M. Kanamitsu, 1994: The NMC nested regional spectral model. *Mon. Wea. Rev.*, 122, 3-26.

Juang, H.-M. H., S-Y. Hong, and M. Kanamitsu, 1997: The NCEP regional spectral model: An update. *Bull. Amer. Meteor. Soc.*, 78, 2125-2143.

Kanamitsu, M. and coauthors., 1991: Recent change implemented into the global forecast system at NMC. *Wea. Forecasting*, 6, 425-435.

Katz R W and Murphy A H (eds), 1997: Economic value of weather and climate forecasts. Cambridge University Press. 222 pp.

Kauffman, S., 1993: The Origins of Order: Self-organisation and selection in Evolution. Oxford University Press, 709pp.

Kauffman, S., 1995: At home in the Universe. Penguin (Viking), 321pp.

Kawai, S., 1979: Generation of initial wavelets by instability of a coupled shear flow. *J. Fluid Mech.* 9, 661 - 703.

Kelly, C. and P.T. Strub, 1992: Comparison of velocity estimates from advanced very-high-resolution radiometer, *J. Geophys. Res.*, 97, 9653.

Kirtman B.P., J. Shukla, B. Huang, Z. Zhu, and E.K. Schneider 1997 : Multiseasonal predictions with a coupled tropical ocean / global atmosphere system. *Mon. Wea. Rev.*, Vol. 125 No. 5, 789 - 808

Kleeman R., A.M. Moore, and N.R. Smith 1995 : Assimilation of subsurface thermal data into a simple ocean model for the initialization of an intermediate tropical coupled ocean-atmosphere forecast model. *Mon. Wea. Rev.* Vol. 123, 3103-3113

Kleine, E., 1994: Das operationelle Modell des BSH für Nordsee und Ostsee, Konzeption und Übersicht. Bundesamt für Seeschiffahrt und Hydrographie, Hamburg, Germany.

Komen, G.J., L. Cavaleri, M.Donelan, K. Hasselmann, S. Hasselmann and P.A.E.M. Janssen, 1994: Dynamics and Modelling of Ocean Waves. Cambridge University Press, 532p.

Komen, G.J., P.A.E.M. Janssen, V. Makin, and W. Oost, 1997: On the sea state dependence of the Charnock parameter. The Global Atmosphere and Ocean System, 5, 367-388.

Komen, G.J., S. Hasselmann and K. Hasselmann, 1984: On the existence of a fully developed windsea spectrum. *J. Phys. Oceanogr.* 14, 1271-1285.

Kourafalou, V. H., 1999: Process studies on the Po river plume, North Adriatic Sea. *Jour. Geophys. Res.*, 104(C12), 29963-29985.

Kurihara, Y., C. L. Kerr and M. A. Bender, 1989: An improved numerical scheme to treat the open lateral boundary of a regional model. Mon.Wea. Rev., 117, 2714-2722.

Kurihara, Y., G. J. Tripoli, and M. A. Bender, 1979: Design of a movable nested-mesh primitive equation model. *Mon.Wea. Rev.*, 107, 239-249.

Kurihara, Y., R. E. Tuleya and M. A. Bender, 1997: The GFDL hurricane prediction system and its performance in the 1995 hurricane season. To be publ.

Kushnir, Y., 1994: Interdecadal variations in the North Atlantic sea surface temperature and associated atmospheric conditions. *J. Climate*, 7, 141-157.

Labitzke, K., 1987: Sunspots, the QBO, and the stratospheric temperature in the north polar region. *Geophys. Res. Lett.*, 14, 535-537.

Lacarra, J., and O. Talagrand, 1988: Short range evolution of small perturbations in a barotropic model, *Tellus*, 40A, 81–95.

Laevestu, T., 1993: Marine Climate, Weather and Fisheries: The effects of weather and climate changes on fisheries and ocean resources. Oxford, Blackwell, 204pp.

Lalli, C.M. and T.R.Parsons, 1997: Biological Oceanography. London, Butterworth-Heinemann 314pp.

Lane, A., Prandle, D., Harrison, A.J., Jones, P.D. and Jarvis, J. 1997: Measuring fluxes in tidal estuaries: sensitivity to instrumentation and associated data analyses. Estuarine, Coastal and Shelf Science, 45, 433-451.

Large, W.G., J.C.McWilliams and S.C.Doney, 1994: Oceanic vertical mixing: a review and a model with nonlocal boundary layer parametrization. Reviews of geophysics, 32(4), 363-403.

Larnicol, G., P.-Y. Le Traon, N.Ayoub and P. De Mey, 1995: Mean sea level and surface circulation variability of the Mediterranean Sea from 2 years of Topex/Poseidon altimetry. Jour. Geophys. Res., 100(C2), 25163-25177.

Latif M. and N.E. Graham 1992 : How much predictive skill is contained in the thermal structure of an OGCM ? *J. Phys. Oceanogr.*, 22, 951-962.

Latif M., 1987: Tropical ocean circulation experiments. *J. Phys. Oceanogr.*, 17, 246-263

Latif M., A. Sterl, E. Maier-Reimer, and M.M. Junge 1992 : Structure an predictability of the El Niño/Southern Oscillation phenomenon in a coupled ocean-atmosphere circulation model. *J. Climate* 6, 700-708.

Latif M., D. Anderson, T. Barnett, M. Cane, R. Kleeman, A. Leetmaa, J.J. O' Brien, and E. Schneider, 1997 : TOGA review paper : Predictability and Prediction. Submitted to *J. Geophys. Res.*, Special TOGA issue 1997.

Latif M. and H. Frey, and T. Stockdale, 1997 : The coupled ocean atmosphere model ECHO-2. Part I: The tropics. *Mon. Wea. Rev.*,Vol. 125, No. 5 703 – 720.

Latif M., M. Flügel, 1991 : An investigation of short-range predictability in the tropical Pacific. *J. Geophys. Res.* , Vol. 96, 2662-2673.

Latif M., T.P. Barnett, M.A. Cane, N.E. Graham, H. von Storch, J-S Xu, S.E. Zebiak, 1994 : A review of ENSO prediction studies. *Climate Dynamics*, 9, 167-179.

Latif, M. and T.P. Barnett, 1994: Causes of decadal climate variability over the North Pacific and North America. *Science*, 266, 634-637.

Latif, M., 1998: Dynamics of interdecadal variability in coupled ocean-atmosphere models. *J. Climate*, 11, 602-624.

Latif, M., D. Anderson, T. Barnett, M. Cane, R. Kleeman, A. Leetmaa, J. O'Brien, A. Rosati, and E. Schneider, 1998: A review of the predictability and prediction of ENSO. *J. Geophys. Res. (Oceans)*, 103, 14,375-393.

Latif, M., A. Timmerman, A. Grötzner, C. Eckert and R. Voss, 2001: On North Atlantic Intedecadal Variability: A Stochastic View. This volume.

Lau N-C, and E. O. Holopainen, 1984: Transient eddy forcing of the time-mean flow as identified by geopotential tendencies, *J. Atmos. Sci.*, 41, 313-328.

Lean, J., J. Beer, and R. Bradley, 1995: Reconstruction of solar irradiance since 1600: Implications for climate change. *Geophys. Res. Lett.*, 22, 3195-3198.

Leetmaa A., M. Ji 1989 : Operational hindcasting of the tropical Pacific. *Dyn. Atmos. Oceans*, 13, 465-490.

Leeuwenhoeck, A, van , 1677: *Phil.Trans.Roy.Soc.London*, 12, 821-831.

Leeuwenhoeck, A, van , 1716: Letter with calculation of geometric rise in population. See Egerton (1968).

Legendre, L., and F. Rassoulzadegan, 1995: Plankton and nutrient dynamics in marine waters. *Ophelia*, 41, 153-172

Lemke, P., E. W. Trinkl, and K. Hasselmann, 1980: Stochastic dynamic analysis of polar sea ice variability. *J. Phys. Oceanogr.*, 10, 2100-2120.

Le Provost, C. and N.C. Flemming (eds.), 1998: The EuroGOOS Atlantic Workshop Report. EuroGOOS Publication 9. Southampton Oceanography Centre. ISBN 0-904175-33-2. 76 pp.

Lermusiaux, P.F.J. and A.R. Robinson, 1999: Data Assimilation via Error Subspace Statistical Estimation, Part I: Theory and Schemes. *Monthly Weather Review*, Vol. 127(8), 1385-1407.

Lermusiaux, P.F.J., 1999: Estimation and Study of Mesoscale Variability in the Strait of Sicily. *Dynamics of Atmospheres and Oceans*, Special Issue in honor of Prof. A.R. Robinson, 29, 255- 303.

Lermusiaux, P.F.J. 2001. Evolving the subspace of the three-dimensional multiscale ocean variability: Massachusetts Bay. *J. Marine Systems*, Special issue on ``Three-dimensional ocean circulation: Lagrangian measurements and diagnostic analyses", May 2001, in press.

Lermusiaux P.F.J., D.G.M. Anderson and C.J. Lozano, 2000: On the mapping of multivariate geophysical fields: error and variability subspace estimates. *Q.J.R. Meteorol. Soc.*, April B, 1387-1430.

Le Traon, P. Y., P. Gaspar, F. Bouyssel and H. Makhmara, 1995: Using Topex/Poseidon data to enhance ERS-1 data, *J. Atm. Ocean. Tech.*, 12, 161-170.

Le Traon, P.Y. and F. Ogor, 1998: ERS-1/2 orbit improvement using Topex/Poseidon: the 2 cm challenge, *J. Geophys. Res.*, 103, 8045-8057.

Le Traon, P.Y., 1995: Basin-scale oceanic circulation from satellite altimetry, pages 79 to 96, in Oceanographic applications of remote sensing, Edited by M. Ikeda and F. Dobson, CRC press, 492 pp.

Le Traon, P.Y., F. Nadal and N. Ducet, 1998: An improved mapping method of multi-satellite altimeter data, *J. Atm. Ocean. Tech.*, 15, 522-533.

Le Traon, P.Y., M. Rienecker, N. Smith, P. Bahurel, M. Bell, H. Hurlburt, P. Dandin and M. Clancy, 1999: Operational oceanography and prediction - a GODAE perspective. OceanObs 99 Conference, Saint Raphael, France.

Levitus, S. ,1997: World Ocean Atlas, Washington DC, NOAA.

Levitus, S. and J. Antonov, 1995: Observational evidence of interannual to decadal-scale variability of the subsurface temperature-salinity structure of the world ocean. *Climate Change*, 31, 495-514, Kluwer Academic Publishers, The Netherlands.

Levitus, S., 1994: World ocean atlas 1994 cd-rom sets., National Oceanographic Data Center Informal Report.

Levitus, S., T. P. Boyer, and J. Antonov, 1994: World Ocean Atlas 1994. Volume 5: Interannual variability of upper ocean thermal structure structure. U.S. Department of Commerce, NOAA, Washington, D. C.

Levitus, S.: 1989: Interpentadal variability of temperature and salinity at intermediate depths in the North Atlantic Ocean, 1970-1974 versus 1955-1959. *J. Geophys. Res.*, 94, 6091--6131.

Levy, M., L. Mémery, G. Madec, 2000: Combined effects of mesoscale processes and atmospheric high-frequency variability on the spring bloom in the MEDOC area. *Deep-Sea Research*, I 47, 27-53.

Lindzen, R. S.(ed.), 1990: The Atmosphere - A Challenge: The Science of Jule Gregory Charney, American Meteorological Society Monograph, Boston, 321pp.

Lionello, P., H. Günther and P.A.E.M. Janssen, 1992: Assimilation of altimeter data in a global third generation wave model. *J. Geophys. Res. C97*, 14453-14474.

Liu, C.-C. and J.D.Woods, 2001: Retrieving information from patchy observations of ocean color for validating a plankton ecosystem model. *J.Geophysical Research,* (submitted).

Liu, C.C., J.D.Woods and C.D.Mobley, 1999: Optical model for use in oceanic ecosystem models. *J.Applied Optics,* 38, 4475-84.

Lochte, K., H.W.Ducklow, M.J.R.Fasham, C.Stienen, 1993: Plankton succession and carbon cycling at 47N, 20W during the JGOFS North Atlantic Spring Bloom Experiment. *Deep-Sea Research,* II 40, 91-114.

Lomnicki, A., 1988: Population Ecology Of Individuals. Princeton, N.J., Princeton University Press, 223pp.

Lomnicki, A., 1999: Individual-based models and the individual-based approach to population ecology. *Ecological Modelling*, 115 (2-3): 191-198.

Longhurst, A., 1998: Ecological Geography of the Sea. Academic Press, 398pp.

Longuet-Higgins, M.S. , 1969: On the transport of mass by time-varying ocean currents. *Deep-Sea Research*, 16, 431-447.

Lorenc, A.C. , 1981: A global three-dimensional multivariate statistical analysis scheme, *Mon. Wea. Rev.*, 109, 701-721.

Lorenc, A.C. , 1986: Analysis methods for numerical weather prediction, *Quart. J. Roy. Met. Soc.*, 112, 1177-1194.

Lorenc, A.C. and Hammon, O., 1988: Objective quality control of observations using Bayesian methods - Theory, and a practical implementation, *Quart. J. Roy. Met. Soc.,* 114, 515-543

Lorenc, A.C., 1988: Optimal nonlinear Objective Analysis, *Quart. J. Roy. Met. Soc.,* 114, 205-240.

Lorenc, Andrew 1992: Iterative analysis using covariance functions and filters, *Quart. J. Roy. Met. Soc.*, 118, 569-591.

Lorenz, E.N., 1993: The Essence of Chaos , London, UCL Press, 227pp.

Lorenz E.N., 1956 : Empirical orthogonal functions and statistical weather prediction. *Sci. Rep. No. 1, Stat. Forecast Project, Dept. of Meteo., MIT, 49 pp.*

Lorenz, E. N., 1963: Deterministic non periodic flow. *J. Atmos. Sci*., 20, 130-141.

Lozano, C.J., A.R.Robinson, H.G.Arango, A.Gangopadhyay, N.Q.Sloan, P.J.Haley and W.G.Leslie, 1996: An interdisciplinary ocean prediction system: assimilation strategies and structured data models. Modern Appraoches to data assimilation in Ocean Modeling, P.Malanotte-Rizzoli, ed., Elsevier Oceanography Series, Elsevier Science, The Netherlands.

Lozier, M. S., Owens, W. B. and Curry, R. G., 1995: The climatology of the North Atlantic. *Progress in Oceanography* ,36, 1-44.

Lythgoe, K., 1990: Lifespan: a computer program for computing demography in the Lagrangian Ensemble simulations Unpublished report, Oceanography Laboratory Imperial College.

Maggiore, A., 2001: Plankton Patchiness: A three-dimensional Lagrangian Ensemble Ecosystem Model Unpublished Report, Oceanography Laboratory, Imperial College. 82pp.

Makin, V.K., V.N. Kudryatsev and C. Mastenbroek, 1995: Drag of the sea surface. *Boundary-Layer Meteorol*. 73, 159 - 182.

Malanotte-Rizzoli, P. (editor), 1996: Modern Approaches to Data Assimilation, Elsevier Oceanography Series, Elsevier Science, The Netherlands.

Malanotte-Rizzoli, P., B.B. Manca, M.R. D'Alcala, A. Theocharis, A. Bergamasco, D. Bregant, G. Budilon, G. Civaterese, D. Georgopoulos, A. Michelato, E. Sansone, P. Scarazzato and E. Souvermezoglou, 1997: A synthesis of the Ionian Sea hydrography, circulation and water mass pathways during POEM-Phase I, *Prog. Oceanogr.*, 39, 153-204.

Malanotte-Rizzoli, P., Manca, B.B., d'Alcala, M.R., Theocharis, A., Brenner, S., Budillon, G., and E. Ozsoy, 1999: The Eastern Mediterranean in the 80s and in the 90s: the big transition in the intermediate and deep circulations, *Dynamics of Atmospheres and Oceans*, 29: (2-4), 365-395.

Malone T., N. Andersen, P. Brewer, E. Buckley, H. Frey, F. Grassie , G. Gross, K. Tenore, L. Walstad, C. Woody, J. Yoder, 1999: An ocean observing system for the U.S. coastal waters first step. C-GOOS report, UMCES contr. 3217, 47pp.

Manabe, S. and R. J. Stouffer, 1996: Low-frequency variability of surface air-temperature in a 1000-year integration of a coupled atmosphere-ocean-land surface model. *J. Climate*, 9, 376-393.

Mangel, M. and C.W.Clark, 1988: Dynamic Modeling in Behavioural Ecology, Princeton University Press, 308pp.

Mann, M. E. and J. Park, 1994: Global-scale modes of surface temperature variability on interannual to century timescales. *J. Geophys. Res.*, 9, 25819-25833.

Mantua, N.J., S. R. Hare, Y. Zhang, J. M. Wallace, and R C. Francis, 1997: A Pacific interdecadal climate oscillation with impacts on salmon production. *Bull. Amer. Meteorol. Soc.*, 78, 1067-1079.

Manzella G.M.R. and MFS-VOS Group, 2001: A marine information system for ocean prediction. This volume.

Manzella G.M.R. and R. Nair, 1998: Availability and access to data on Europes Marine Environment. European Environment Agency Technical report n.4, 32 pp.

Martin, A.P., K.J.Richards and M.J.R.Fasham, 2001: Phytoplankton production and community structure in an unstable frontal region. *J.Marine Systems,* 28,65-89.

Masina, S., N.Pinardi and A. Navarra, 2001: A global ocean temperature and altimeter data assimilation system for studies of climate variability. In press on *Climate Dynamics*.

Mastenbroek, C., G.J.H. Burgers and P.A.E.M. Janssen, 1993: The dynamical coupling of a wave model and a storm surge model through the atmospheric boundary layer. *J. Phys. Oceanogr.* 23, 1856-1866.

Mauchline, J., 1998: The Biology of Calanoid Copepods. London, Academic Press.

May, R.M., 1976: Simple mathematical models with very complicated dynamics. *Nature,* 261, 459-467.

May, R.M., Ed., 1981: Theoretical Ecology: Principles and Applications. Oxford, Blackwell 489pp.

McGillicuddy , D. J. J., A. R. Robinson, D. A. Siegel. H. W. Jannash, R. Johnson, T. D. Dickey, J. McNeil, A. F. Michael, and A. H. Knap, 1998: Influence of mesoscale eddies on new production in the Sargasso Sea, *Nature*, 392, 263-265.

McManus, J.P. and D. Prandle, 1994: Comparisons between observations and model simulations of dissolved trace metal concentrations in the southern North Sea. *Marine Pollution Bulletin*. 28(7), 451-455.

McManus, J.P. and D. Prandle, 1996: Development of a model to reproduce observed suspended sediment in the southern North Sea using principle component analyses with multiple linear regression. *Continental Shelf Research* 17(7), 761-778.

McPhaden M.J. 1993 : TOGA-TAO and the 1991-1993 El Niño/Southern Oscillation event. *Oceanography*, Vol. 6, No. 2: 36-44

McPhaden, M.J., 1995: The tropical atmosphere ocean array is completed. Bulletin American Meteorological Society, 75, 5, 739-741.

MedGOOS, 1999: Workshop: 'Benefits of the implementation of the Global Ocean Observing System in the Mediterranean region'. 1-3 November 1999, Rabat, Morocco. Abstracts. 54 pp.

Mellor, G.L., S.Hakkinen, T.Ezer and R.Patchen, 2001: A generalization of a sigma coordinate ocean model and an intercomparison of model vertical grids. This volume.

Mellor, G.L. and S. X. H. Wang, 1996: Pressure compensation and the bottom boundary layer. *Journal of Physical Oceanography*, 26, 2214-2222.

Mellor, G. L. and A. F. Blumberg, 1985: Modeling Vertical and horizontal diffusivities with the sigma coordinate system. *Mon Wea. Rev.* 113, 1379-1383.

Mellor, G. L. and T. Yamada, 1982: Development of a turbulent closure model for geophysical fluid problems. *Rev. Geophys.*, 20, 851-875.

Mellor, G. L., 1991: An equation of state for numerical models of ocean and estuaries. *J. Atmos. Ocean Technol.*, 8 , 609-611.

Mellor, G. L., 1996: User's guide for a three-dimensional, primitive equation, numerical ocean model. Report, Prog. in Atmos. and Ocean. Sci, Princeton University, Princeton NJ 08540, 35 pp.

Mellor, G. L., L.-Y. Oey and T. Ezer, 1998: Sigma coordinate pressure gradient errors and the seamount problem. *J. Atmos. Oceanic. Technol.*, 15, 1122-1131.

Mellor, G. L., T. Ezer and L. Y. Oey, 1994: The pressure gradient conundrum of sigma coordinate ocean models. *J. Atmos. Oceanic. Technol.*, 11, 1126-1134.

Mellor, G.L. and T. Yamada, 1974: A hierarchy of turbulence closure models for planetary boundary layers. *J. Atmos. Sci.*, 1791-1806.

Mellor, G.L., and T. Ezer, 1991: A Gulf Stream model and an altimetry assimilation scheme. *J. Geophys. Res.*, 96(C5), 8779-8795.

Mesinger, F., 1984: A blocking technique for representation of muontains in atmospheric models. *Rev. Meteor. Aeronautica*, 44, 195-202.

Mesinger, F., Z. I. Janjic, S. Nickovic, D. Gavrilov, and D. G. Deaven, 1988: The step-mountain coordinate: model description and performance for cases of Alpine lee cyclogenesis and for a case of an Applacian redevelopment. *Mon. Wea. Rev.*, 116, 177-181.

Michaels, A. F. and A. W. Knap, 1996: Overview of the U.S. JGOFS Bermuda Atlantic Times-Series Study and the Hydrostation S program. *Deep-Sea Research,* II 43: 157-198.

Mikolajewicz, U. and E. Maier-Reimer, 1990: Internal secular variability in an ocean general circulation model. *Climate Dynamics*, 4, 145-156.

Miller, A.J., P.-M. Poulain, A.R. Robinson, H.G. Arango, W.G. Leslie and A. Warn-Varnas, 1995: Quantitative skill of quasigeostrophic forecasts of a baroclinically unstable Icealand-Faeroe Front, *J. Geophys. Res.*, 100 (C6), 10833-10849.

Miller, R. N., E. F. Carter, and S. T. Blue, Data assimilation into nonlinear stochastic models, *Tellus*, To appear, 1999.

Miller, R. N., M. Ghil, and F. Gauthicz, 1994: Advanced data assimilation in strongly nonlinear dynamical systems, *J. Atmos. Sci.*, 51, 1037--1056.

Milliff R. , M. Freilich, T. Liu, B. Atlas and B. Large, 1999: Global ocean surface vector wind observations from space. Proceeding of OceanObs99 (The Ocean Observing System For Climate) Conference, Saint Raphael, October 1999. CNES, Vol. 2.

Molcard A., N. Pinardi, M. Iskandarani and D.B. Haidvogel, 2001: Wind driven general circulation of the Mediterranean Sea simulated with a Spectral Element Ocean Model. Submitted.

Moler, C. B., and C. F. van Loan, 1978: Nineteen dubious ways to compute the exponential of a matrix, SIAM Review, 20, 801-836.

Molinari R.L. , 1999: Lessons learned from operating global ocean observing networks. *Bulletin American Meteorological Society*, 80,7, 1413-1419.

Moll, A., 1986: Regional distribution of primary production in the North Sea simulated by a three-dimensional model. *Journal of Marine Systems*, Special Volume Liege Colloquium 1996.

Molteni, F.,R. Buizza, T. N. Palmer, and T. Petroliagis, 1996: The ECMWF ensemble prediction system: Methodology and validation, *Quart. J. Roy. Meteor. Soc.*, 122, 73–119.

Monbaliu, J., Zhang, M.Y., Backer, K. De, Hargreaves, J., Luo, W., Flather, R., Carretero, J.C., Gomez Lahoz, M., Lozano, I., Stawarz, M., GÜNTHER, H., Rosenthal, H. and Ozer, J. 1997: WAM Model Intercomparisons - North Sea. POL (Proudman Oceanographic Laboratory, UK) Report No. 47.

Mooers, C.N.K. (editor), 1999: Coastal Ocean Prediction, AGU *Coastal and Estuarine Studies* Series #56.

Morel, A., 1991: Light and marine photosynthesis: a spectral model with geochemical and climatological implications. *Progress in Oceanography,* 26, 263-306.

Mysak, L. A., D. K. Manak, and R. F. Mardsen, 1990: Sea-ice anomalies in the Greenland and Labrador Seas 1900-84 and their relation to an interdecadal Artctic climate cycle. *Climate Dynamics*, 5,111-133.

National Research Council, 1989. Committee on Opportunities to Improve Marine Observations and Forecasting (1989). Opportunities to Improve Marine Forecasting. National Academy Press, Washington, DC., 125 pp.

Navarra A., 2001 : Ensembles, Forecasts and Predictability. This volume.

Navarra, A., 1987: An application of the Arnoldi method to a geophysical fluid dynamics problem, *J. Comput. Phys.*, 69, 143-162.

NEAR-GOOS, 1999: Physical, chemical and biological studies on monitoring of marginal seas for ocean forecasting - A fundamental research project for NEAR-GOOS. Scientific research of priority areas, 1999-2002. Ministry of Education, Science, Sports and Culture, Japan. 5 pp.

Neelin, J. D., M. Latif, and F.-F. Jin, 1994: Dynamics of coupled ocean-atmosphere models: The tropical problem. *Ann. Rev. Fluid Mech.*, 26, 617-659.

Nichols, J.H. and Brander, K.M. 1989: Herring larval studies in the west central North Sea. Rapports et Process-Verbaux des Reunions ICES, 191, 160-168.

Nihoul, J.C.J. and Ronday, F.C. 1975: The influence of the 'tidal stress' on the residual circulation. Application to the Southern Bight of the North Sea. *Tellus*, 27, 484-489.

Nittis, K., and A. Lascaratos, 1998: Diagnostic and prognostic numerical studies of LIW formation. *Journal of Marine Systems*, 18, 179-195.

Nogueira, E., J.D.Woods, C.Harris, A.J.Field and S.Talbot, 2001: Phytoplankton co-existence in an individual-based plankton ecosystem model. *Marine Ecology Progress Series*, (submitted).

OECD, 1994: Megascience: The OECD Forum. Oceanography. Organisation for Economic Co-operation and Development, Paris, France. ISBN 92-64 14205-3. 167 pp.

OECD, 1998: Report of the OECD Megascience Forum Workshop on Global Scale Issues, Grand Hôtel Saltsjöbaden, March 4-6, 1998. OECD. Paris. 43 pp.

Omstedt A. and L. Nyberg, 1995: A coupled ice-ocean model supporting winter navigation in the Baltic Sea. Part 2: Thermodynamics and meteorological coupling. Reports Oceanography 21, SMHI, S-601 76 Norrköping, Sweden, 38 pp.

Omstedt A., L. Nyberg and M. Leppäranta, 1994: A coupled ice-ocean model supporting winter navigation in the Baltic Sea. Part 1: Ice dynamics and water levels. Reports Oceanography 17, SMHI, S-601 76 Norrköping, Sweden, 17 pp.

OOSDP, 1995: Scientific design for the common Module of the Global Ocean Observing System and the Global Climate Observing System: An ocean observing system for climate. Department of Oceanography, Texas A and M University, College Station, Texas. 265 pp.

Ormond, R.F.G., J.D.Gage and M.V.Angel, 1997: Marine Biodiversity: patterns and processes. Cambridge University Press, 449pp.

Orlanski, I. 1975: A rational subdivision of scales for atmospheric processes. *Bull. Amer. Meteor. Soc.*, 56, 527-530.

Orlanski, I. 1976: A simple boundary condition for unbounded hyperbolic flows. *J. Comp. Phys.*, 21, 251-269.

Orzag, S. A. 1970: Transform method for the calculation of vector-coupled sum application to the spectral form of the vorticity equation. *J. Atmos. Sci.*, 27, 890-895.

Oschlies, A., and J. Willebrand, 1996: Assimilation of Geosat altimeter data into an eddy_resolving primitive equation model of the North Atlantic Ocean, *J. Geophys. Res.*,101,14,175-14,190.

Pacanowski, R.C., K. Dixon, and A. Rosati, 1993: The GFDL Modular Ocean Model User's Guide, version 1.0, GFDL Ocean group, Tech. Rep. no. 2 [available from GFDL/NOAA, Princeton University, Princeton, NJ 08542, USA]

Palmer, T. N. and Z. Sun, 1985: A modelling and observational study of the relationship between sea surface temperature in the northwest Atlantic and the atmospheric general circulation. *Quart. J.R. Meteorol. Soc.*, 111,947-975.

Palmer, T. N., 1993: A nonlinear dynamical perspective on climate change. *Weather*, 48, 313-348.

Paltridge,G.W. and C.M.R.Platt, 1976: Radiative processes in meteorology and climatology. Amsterdam, Elsevier, 318pp.

Partridge, L., 2000: A Lagrangian Ensemble food web model Unpublished Report, Oceanography Laboratory, Imperial College.

Parrilla G., A. Lavin, H. Bryden, M. Garcia and R. Millard , 1994: Rising temperatures in the subtropical North Atlantic Ocean over the past 35 years. *Nature*, 396 (6475) p.48-51.

Pedlosky, 1979: Geophysical fluid dynamics. Springer Verlag, New York, 624 pp.

Perkey, D.J. and C.W. Kreitzberg, 1976: A time-dependent lateral boundary scheme for limited-area primitive equation models. *Mon.Wea. Rev.*, 104, 744-755.

Pham, D.T., J. Verron and M.-C. Roubaud, 1998: Singular Evolutive Extended Kalman filter with EOF initialization for data assimilation in oceanography. *J. Marine Sys.,* 16**,** 323-340.

Philander, S.G. H., 1990: El Niño, La Niña, and the Southern Oscillation. Academic Press, Inc., San Diego, 293 pp.

Phillips, N. A. 1957: A coordinate system having some special advantages for numerical forecasting. *J. Meteor.*, 14, 184-185.

Phillips, N. A. 1979: The nested grid model. NOAA Tech. Rept. NWS 22, 89pp.

Phillips, O.M., 1977: The dynamics of the upper ocean, Cambridge University Press, Cambridge, 336p.

Picaut, J., Busalacchi, A.,J., Mc Phaden, M.J., and N. Camusat, 1990. Validation of the geostrophic method for estimating zonal currents at the equator from GEOSAT altimeter data. *J. Geophys. Res.*, 95, 3015.

Pierson, W.J., G. Neumann and R.W. James, 1955: Practical methods for observing and forecasting ocean waves by means of wave spectra and statistics. H.O. Pub 603, US Navy Hydrographic Office.

Pinardi, N., E.Demirov and M.Tonani, 2001: The Mediterranean Forecasting System Pilot Project: the first phase. EuroGOOS Publication 14, Southampton, UK.

Pinardi N and Flemming N C, 1998: The Mediterranean Forecasting System Science Plan. EuroGOOS Publication no. 11. Southampton Oceanography Centre, 48 pp. ISBN 0-904175-35-9.

Pinardi, N., A. Rosati and R.C. Pacanowski, 1995: The sea-surface pressure formulation of rigid-lid models. Implications for altimetric data assimilation studies. *J. Marine Sys.,* 16**,** 109-119.

Pinardi, N., and E. Masetti, 2000: Variability of the large scale general circulation of the Mediterranean Sea from observations and modelling: a review. *Palaeogeography, Palaeoclimatology, Palaeoecology,* 158**,** 153-173.

Pingree, R.D. and Maddock, L. 1980: Tidally induced residual flows around an island due to both frictional and rotational effects. *Geophys. J. Roy. Astronomical Society*, 63, 533-546.

Planton, S., 1995: Interface fluxes retrieval from satellite data. Proceedings of the Symposium "Operational oceanography and satellite observation", Biarritz.

Platt,T., J.D.Woods, S.Sathyendranth, W.Barkmann, 1994: Net primary productiuon and stratification in the ocean. In: A.M.Johannessen, R.D.Muench and J.E.Overland (Eds.) The Polar Oceans and their role in shaping global climate. American Geophysical Union, Geophysical Monograph 85, 247-254.

Pollard R T, 1994. Megachallenges in physical oceanography. p.27-48. in: Oceanography, Megascience the OECD Forum. OECD, Paris, 167 pp.

Popova, E. E., M. J. R. Fasham, et al., 1997: Chaotic behaviour of an ocean ecosystem model under seasonal external forcing. *Journal of Plankton Research,* 19 (10): 1495-1515.

Pouliquen, E., A.D. Kirwan and R.T. Pearson (eds.), 1997: Rapid Environmental Assessment, Proceedings of a conference held in Lerici (SP), Italy, 10-14 March 1997, SACLANTCEN Conference Proceedings Series CP-44.

Prandle, D., 2001: Predictions in the North Sea. This volume.

Prandle, D. 1984: A modelling study of the mixing of 137Cs in the seas of the European continental shelf. Philosophical Transactions of the Royal Society of London, A, 310, 407-436.

Prandle, D. 1987: Response of the N.W. European shelf seas at subtidal frequencies. *Continental Shelf Research*, 7, 267-284.

Prandle, D. 1991b: A new view of near-shore dynamics based on observations from HF radar. *Progress in Oceanography*, 27, 403-438.

Prandle, D. 1997a: The influence of bed friction and vertical eddy viscosity on tidal propagation. *Continental Shelf Research*, 17(11), 1367-1374.

Prandle, D. 1991a: Tides in estuaries and embayments (review). pp.125-152 in, Tidal hydrodynamics, (ed.B.B.Parker). New York: John Wiley and Sons. 883pp.

Prandle, D. 1997b: Tidal characteristics of suspended sediment concentrations. *Amer. Soc. Civ. Eng*., J. Hyd. Div. 123(4), 341-350.

Prandle, D. 1998: Global expressions for seasonal temperatures of the sea surface and ambient air - the influence of tidal currents and water depth. *Oceanologica Acta,* 21(3):419-428.

Prandle, D. and A. Lane, 1995: The annual temperature cycle in shelf seas. *Continental Shelf Research*, 15(6), 681-704.

Prandle, D. and J. Beechey, 1991: Marine dispersion of caesium 137 released from Sellafield and Chernobyl. *Geophysical Research Letters*, 18, 1723-1726.

Prandle, D., C.F. Jago, S.E. Jones, D.A. Purdie and A. Tappin, 1993: The influence of horizontal circulation on the supply and distribution of tracers. *Phil. Trans. Roy. Soc. London*, A. 343, 405-421.

Prandle, D., D.J. Hydes, J. Jarvis and J. McManus, 1997: The seasonal cycles of temperature, salinity, nutrients and suspended sediments in the southern North Sea in 1988 and 1989. *Estuarine, Continental Shelf Research,* 45 (5) 669-680.

Proudman, J. and A.T. Doodson, 1924: The principal constituent of the tides in the North Sea. *Phil. Trans. Roy. Soc. London*, A. 224, 185-219.

Pugh, D. P. and L. Skinner, 1996: An analysis of marine-related activities in the UK economy and supporting science and technology. IACMST Information Document No. 5. IACMST, Southampton, UK. 52 pp.

RASCL, 1998: Canada's Ocean Industries: Contribution to the Economy 1988 - 1996. Prepared for Economic and Policy Analysis Division, Department of Fisheries and Oceans. Roger A Stacey Consultants Ltd. 77 pp.

Ratier, A., 1999: Space-based observations in the global ocean observing system: the operational transition issue. OceanObs 99 Conference, Saint Raphael, France.

Rhines, P.B., 1979: Geostrophic turbulence. *Annual Reviews of Fluid Mechanics,* 11, 401-441.

Richardson, L.F., 1910: The approximate arithmetical solution by finite differences of physical problems involving differential equations. *Phil Trans Roy Soc London*, A 210, 307-357.

Richardson, L. F., 1922: Weather Prediction by Numerical Process. Cambridge University Press. 236pp.

Rider, K., G.J. Komen and J.J. Beersma, 1996: Simulations of the response of the ocean waves in the North Atlantic and North Sea to CO2 doubling in the atmosphere. KNMI Scientific Report, 96-05, 55p.

Rienecker, M.M., and D. Adamec, 1995: Assimilation of altimeter data into a quasigeostrophic ocean model using optimal interpolation and EOFs. *J. Marine Sys.,* 6**,** 125-143.

Robert, A. 1982: A semi-Lagrangian and semi-implicit numerical integration scheme for the primitive meteorological equations. *J. Japan. Meteor. Soc.*, 60, 315-325.

Robert, A., T. L. Yee, and H. Richie, 1985: A semi-Lagrangian and semi-implicit numerical integration scheme for multilevel atmospheric models. *Mon. Wea. Rev.*, 113, 388-394.

Robinson, A.R. and P.Malanotte-Rizzoli (eds.), 1993: Physical Oceanography of the Eastern Mediterranean, *Deep Sea Research*, 40(6), 1332 pp.

Robinson, A.R. and S.M. Glenn, 1999: Adaptive sampling for ocean forecasting, *Naval Research Reviews*, 51(2), 28-38.

Robinson, A.R. and the LOOPS Group, 1999: Realtime forecasting of the multidisciplinary coastal ocean with the Littoral Ocean Observing and Predicting System (LOOPS), to appear in the Preprint Volume of the Third Conference on Coastal Atmospheric and Oceanic Prediction and Processes, 3-5 November 1999, New Orleans, LA, American Meteorological Society, Boston, MA.

Robinson, A.R., 1997. On the theory of advective effects on biological dynamics in the sea. Proc. R. Soc.Lond., A 453, 2295-2324.

Robinson, A.R., 1996: Physical processes, field estimation and interdisciplinary ocean modeling, *Earth-Science Reviews*, 40, 3-54.

Robinson, A.R., 1999: Forecasting and Simulating Coastal Ocean Processes and Variabilities with the Harvard Ocean Prediction System. In: Coastal Ocean Prediction, (C.N.K. Mooers, editor), AGU Coastal and Estuarine Studies Series, 77-100, American Geophysical Union.

Robinson, A.R., H.G. Arango, A.J. Miller, A. Warn-Varnas, P.-M. Poulain and W.G. Leslie, 1996: Realtime operational forecasting on shipboard of the Iceland-Faeroe Frontal Variability, *Bull. of the Amer. Met. Soc.*, 72(2), 243-259.

Robinson, A.R., J. Sellschopp, A. Warn-Varnas, W.G. Leslie, C.J. Lozano, and P.J. Haley, Jr., 1999b: The Atlantic Ionian Stream, Special Issue in Memory of Dr. Antonio Michelato. *Journal of Marine Systems*, 20, 129-156.

Robinson, A.R., J.J. McCarthy, and B.J. Rothschild, 1999a: Interdisciplinary Ocean Science is Evolving and a Systems Approach is Essential. *Journal of Marine Systems*, 22, 231-239.

Robinson, A.R. and W.G. Leslie, 1985: Estimation and prediction of oceanic fields, Progress in Oceanography, Vol. 14, J. Crease, W.J. Gould and P.M. Saunders, eds. Essay on Oceanography. Pergamon Press, Oxford, pp. 485-510.

Robinson, A.R., P.F.J. Lermusiaux, and N.Q. Sloan, 1998: Data Assimilation. In THE SEA: The Global Coastal Ocean, Volume 10: Processes and Methods (K.H. Brink and A.R. Robinson, eds.), John Wiley and Sons, NY, 541-594.

Robinson, A.R. and L.J.Walstad, 1987: The Harvard ocean model: calibration and applications to dynamical processes, forecasting and data assimilation studies. *J.Appl. Num. Mathem.*, 3, 89-131.

Robinson, I.S. 1985: Satellite oceanography : an introduction for oceanographers and remote-sensing scientists. Ellis Horwood Series in Marine Science, Chichester, England.

Robock, A. D. and J. Mao, 1995: The volcanic signal in surface temperature observations. *J.Climate*, 8, 1086-1103.

Rodwell, M. J. and B. J. Hoskins, 1996: Monsoon and the dynamics of desert. *Quart. Roy. Meteor. Soc.*, 122, 1385-1404.

Rodwell, M. J., D. P. Rowell, and C. K. Folland, 1999: Oceanic forcing of the wintertime North Atlantic Oscillation and European climate. *Nature*, 398, 320-323.

Roemmich, D., O. Boebel, Y. Desaubies, H. Freeland, B. King, P.-Y. Le Traon, R. Molinari, W.B. Owens, S. Riser, U. Send, K. Takeuchi, W. Wijffels, 1999: ARGO: The Global Array of Profiling Floats. Proceeding of OceanObs99 (The Ocean Observing System For Climate) Conference, Saint Raphael, October 1999. CNES, Vol. 2.

Ropelewski C.F. and M.S. Halpert, 1987 : Global and regional scale precipitation patterns associated with the El Niño/Southern Oscillation. *Mon. Wea. Rev.*, Vol. 115, 1606-1626

Rosati A., K. Miyakoda, and R. Gudgel 1996 : The impact of ocean initial conditions on ENSO forecasting with a coupled model. *Mon. Wea. Rev.*, Vol. 125 , 754 - 772

Rosati, A. and K. Miyakoda, 1988: A general circulation model for upper ocean simulation. J. P*hys. Oceanogr.*, 18, 1601-1626.

Rossby, T. , G. Siedler, W. Zenk, 1995: The volunteer observing ship and future ocean monitoring. Bulletin American Meteorological Society, 76(1), 5.

Roughgarden, J., 1996: Theory of Population Genetics and Evolutionary Dynamics - An Introduction. Upper Saddle River, NJ, Prentice Hall, 612pp.

Roussenov, V. E.Stanev, V.Artale and N. Pinardi, 1995: A seasonal model of the Mediterranean Sea. *J. Geophys. Res.*, 100, C7, 13515-13538.

Rowell, D.P., 1997: Using an ensemble of Multi-decadal GCM simulations to assess potential seasonal predictability. *J Climate*, 11, 109-120.

Rummel R. and F. Sanso, 1993: Satellite altimetry in geodesy and oceanography, Lecture Notes in Earth Sciences, Editors R. Rummel and F. Sanso, Springer Verlag.

Saravanan, R. and J. C. McWilliams, 1997: Stochasticity and spatial resonance in interdecadal climate fluctuations. *J. Climate*, 10, 2299-2320.

Sassone, P. G. and R. F. Weiher, 1997: Cost benefit analysis of TOGA and the ENSO Observing System. Proceedings of First International Conference on EuroGOOS, The Hague, The Netherlands, October 1996. Elsevier, The Netherlands. ISBN 0 444 82892 3. pp. 36-50.

Schlitzer, R. (2000). Ocean Data View, http://www.awi-bremerhaven.de/GEO/ODV.

Schopf, P.S. and M.J. Suarez, 1988 : Vacillations in a coupled ocean-atmosphere model. *J. Atmos. Sci*, 45, 549-566

Schott F., Boning C., Bryden H., Molinari R., Schlosser P., Wunsch C., Stramma L. (2000). Report of the WOCE North Atlantic Workshop, WOCE report n. 169/2000, 114 pp.

Schott F., M. Visbeck and U. Send , 1994: Open ocean deep convection, Mediterranean and Greenland Seas, in: Ocean Processes in Climate Dynamics: Global and Mediterranean Examples. Malanotte P and Robinson A R, (eds), volume 419 of NATO ASI series C: Mathematical and Physical Sciences. pp. 203-226. Kluwer Academic Publishers.

Sellner, K.G. and S.Fonda-Umani, 1999: Dinoflagellate blooms and mucilage production. In: Ecosystems at the Land-Sea margin: Drainage basin to Coastal Sea. Malone T.C., Malej A., Harding L.W.Jr., Smodlaka N., Turner R.E. (eds). *Coastal and Estuarine Studies*, 55: 173-206.

Sellschopp, J. and A.R. Robinson, 1997: Describing and Forecasting Ocean Conditions during Operation Rapid Response Rapid Environmental Assessment, Proceedings of a conference held in Lerici (SP), Italy, 10-14 March 1997, SACLANTCEN Conference Proceedings Series CP-44, 35- 42.

Sellschopp, J., 1998: Experience gained and lessons learnt in three years of Rapid Response, UDT 1998, 23-25 June 1998 in London, Conference Proceedings, 69-73.

Sellschopp, J., 1999: Rapid Environmental Assessment, Naval Forces, International Forum for Maritime Power, June 1999.

Semtner, A. J. 1974: An oceanic general circulation model with bottom topography. Numerical simulation of weather and climate. Tech. Rep., No.9, Univ. California, Los Angels, 99 pp.

Skiena, S.S., 1997: The algorithm design manual. TELOS, The Electronic Library Of Science, Springer-Verlag, New York, 486pp.

Shindell D. T., Miller R. L., Schmidt G. A. and Pandolfo L., 1999: Simulation of recent northern winter climate trends by greenhouse-gas forcing. *Nature* 399. pp. 452-454.

Shukla J., 1999: Predictability in the midst of chaos: a scientific basis for climate forecasting, *Science*, 282, 728-731.

Siedler, G. and R. Onken, 1996: Eastern recirculation, in The Warmwatersphere of the North Atlantic Ocean (W. Krauss, ed.), Gebruder Borntraeger, Berlin, Stuttgart, 339-364.

Simmons, A. J., J. M. Wallace and G. W. Branstator, 1983: Barotropic wave propagation and instability and atmospheric teleconnection patterns, *J. Atmos. Sci.,* 40, 1363-1392.

Simmons, A., J. and B. J. Hoskins, 1976: Baroclinic instability on the sphere - normal modes of the primitive and quasigeostrophic equations, *J. Atmos. Sci.,* 33, 1454-1477.

Simpson, J.H. and J.R. Hunter, 1974: Fronts in the Irish Sea. *Nature*, Lond. 250, 404-406.

Simpson, J.H., W.R. Crawford, T.P. Rippeth, A.R. Campbell and J.V.S. Cheok, 1996: The vertical structure of turbulent dissipation in Shelf Seas. *J. Physical Oceanography*, 26(8), 1579-1590.

Smagorinsky,J., 1993: Some historical remarks on the use of nonlinear viscosities. In: Large eddy simulation of complex engineering and geophysical flows. Proceedings of an international workshop in Large eddy simulation. St.Petersburg, FL, B.Galperin and S.A.Orszag (eds), University Press, 1-36.

Smith N., and M. Lefebvre, 1997: The Global Data Assimilation Experiment (GODAE), Biarritz International Symposium, Proceedings, October 1997.

Smith N.R., D.E. Harrison, R. Bailey, O. Alves, T. Delcroix, K. Hanawa, B. Keeley, G. Meyers, B. Molinari, D. Roemmich (1999) The role of XBT sampling in the ocean thermal network. OCEANOBS

Smith, S.D., R.J. Anderson, W.A. Oost, C.Kraan, N. Maat, J. DeCosmo, K.B. Katsaros, K.L. Davidson, K. Bumke, L. Hasse and H.M. Chadwick, 1992: Sea surface wind stress and drag coefficients: the HEXOS results. *Boundary-Layer Meteorol.* 60, 109.

Solow, A.R., 1995: Fitting population ecology models to time series data In: T.M.Powell and J.H.Steele (eds.) Ecological Time Series Chapman & Hall, 20-27.

Sommer, U. Ed., 1989: Plankton Ecology, Berlin, Springer Verlag 369pp.

Song. Y. and D. Haidvogel, 1994: A semi-implicit ocean circulation model using a generalized topography-following coordinate system. *J. Comput. Physics*. 115, 228-244.

Souza, A.J. and J.H. Simpson, 1996: The modification of tidal ellipses by stratification in the Rhine ROFI. *Continental Shelf Research* (16), 997-1008.

Spall, M. A. and A. R. Robinson, 1990: Regional primitive equation studies of the Gulf Stream meander and ring formation region. *J. Phys. Oceanogr.*, 20, 985-1016.

Spall, M. A., 1996: Dynamics of the Gulf Stream/Deep western boundary current crossover. Part II: Low-frequency internal oscillations. *J. Phys. Oceanogr.*, 26, 2169-2182.

Stammer, D., 1997: Steric and wind-induced changes in TOPEX/POSEIDON large-scale sea surface topography observations, *J. Geophys. Res.*, 102, 20987-21009.

Stammer, D., R. Tokmakian, A. Semtner, and C. Wunsch, 1996: How well does a ¼° global circulation model simulate large-scale oceanic observations?, *J. Geophys. Res.*, 101, 25779 - 25812.

Steele, J.H., Ed., 1977: Fisheries Mathematics. London, Academic Press, 198pp.

Steele, J.H. ,Ed., 1978: Spatial patterns in plankton communities. Plenum Press.

Steele, J.H., 1995: Can ecological concepts span the land and ocean domains? In: T.W.Powell and J.H.Steele (Eds.) Ecological Time Series Chapman and Hall, 5-19 of 491pp.

Stel J. H. (ed), 1996: Proceedings of the First EuroGOOS Conference, 7-11 October 1996, The Hague, The Netherlands. Elsevier Oceanography Series, 62. ISBN 0 444 82892 3. 757 pp.

Stern, W., and K. Miyakoda, 1995: Feasibility of Seasonal Forecasts Inferred from Multiple GCM Simulations. *J. Climate*, 8, 1071-1085.

Stewart, R.H., 1985: Methods of satellite oceanography, Univ. of California Press, Berkeley, CA. 360pp.

Storch H.v, G. Bürger, R. Schnur, and J.S. von Storch 1995 : Principal oscillation patterns: A review. *J. Climate*, Vol. 8, No.3, 377-400.

Sutherland,W.J., 1997: From Individual Behaviour to Population Ecology. Oxford University Press 213pp.

Sutton, R. T. and M. R. Allen, 1997: Decadal predictability in Gulf Stream sea surface temperatures. Nature, 388, 563-567.

Sverdrup, H.U. and W.H. Munk, 1947: Wind sea and swell: Theory of relations for forecasting. H.O. Pub. 601, US Navy Hydrographic Office, Washington, DC, 44p.

Sverdrup.H., 1953: On the conditions for vernal blooming of phytoplankton. *J.Cons.Perm.Int.Explor.Mer.*, 18, 287-295.

SWAMP group: J.H. Allender, T.P. Barnett, L. Bertotti, J. Bruinsma, V.J. Cardone, L. Cavaleri, J. Ephraums, B. Golding, A. Greenwood, J. Guddal, H. Günther, K. Hasselmann, S. Hasselmann, P. Joseph, S. Kawai, G.J. Komen, L. Lawson, H. Linné, R.B. Long, M. Lybanon, E. Maeland, W. Rosenthal, Y. Toba, T. Uji and W.J.P. de Voogt, 1985: Sea wave modeling project (SWAMP). An intercomparison study of wind wave predictions models, part 1: Principal results and conclusions, in: Ocean wave modeling; Plenum, New York, 256p.

Tarantola,A. 1987: Inverse Problem Theory - Methods for data fitting and model parameter estimation, publ. Elsevier. 613pp. ISBN 0-444-42765-1

Tatsumi, Y., 1986: A spectral limited area model with time-dependent lateral boundary conditions and its application to a multi-level primitive equation model. *J. Meteor. Soc. Japan*, 64, 637-664.

Taylor, G.I. 1921: Tidal oscillations in Gulfs and rectangular basins. *Proc. Lond. Math. Soc. Ser.*, 2.20. 148-81.

Timmermann, A., 1996: Modes of variability as simulated by a global climate model. PhD thesis. Available from Max-Planck-Institut für Meteorologie, Bundesstr.55, D-20146 Hamburg, Germany.

Timmermann, A., M. Latif, R. Voss, and A. Grötzner, 1998: North Atlantic interdecadal variability: A coupled air-sea mode. *J. Climate*, 11, 1906-1931.

Toth, Z., and E. Kalnay, 1993: Ensemble forecasting at NMC: The generation of perturbations, *Bull. Amer. Meteor. Soc.*, 74, 2317–2330.

Trenberth, K. E. and J. W. Hurrell, 1994: Decadal atmosphere-ocean variations in the Pacific. *Climate Dynamics*, 9, 303-319.

Tuljapurkar, S. and H.Caswell , 1997: Structured-population models in marine, terrestrial & freshwater systems. New York, Chapman and Hall, 643pp.

Turing, A. , 1950: Computing machinery and Intelligence. *Mind,* 59 (236).

Tyce, R., F. de Strobel, V. Grandi and L. Gualdesi, 1998: Shallow water expendable environmental profiler (sweep) design with control and data recovery via cellular phone and low earth orbit satellite networks, In: Oceanology International, Brighton, September, 1998. New Malden, Surrey, Spearhead, 199-207.

UNESCO (1994a). Report on diagnostic procedures and a definition of minimum requirements for providing information services on a national and/or regional level. IOC Manual and Guides n. 30, Vol. 1, 4 pp.

UNESCO (1994b). Information network: the development of national or regional scientific information exchange. IOC Manuals and Guides n. 3°, Vol. 2, 20 pp.

UNESCO (1994c). Standard directory record structure for organisations, individuals and their research interest. IOC Manuals and guide n. 30, Vol. 3, 22 pp.

UNESCO, 1984: Ocean Science for the year 2000. IOC/UNESCO Paris, 95 pp.

UNESCO, 1997: Guide for establishing a national oceanographic data centre (revised version). Manual and guides 5, 37 pp.

UNESCO, 1999: Guide to operational procedures for the collection and exchange of JCOMM oceanographic data - third revised version. Manuals and guides 3, 35 pp.

United Nations, 1992: Agenda 21: United Nations Conference on Environment and Development. Rio de Janeiro, 3-14 June. A/CONF. 151/4.

van Loon, H. and J. C. Rogers, 1978: The seasaw in winter temperatures between Greenland and Northern Europe. Part I: General description. *Mon. Wea. Rev.*, 106, 296-310.

Vichi, M., M.Zavatarelli, N.Pinardi, 1998: Seasonal Modulation of Microbial Mediated Carbon Fluxes in the Northern Adriatic Sea. A Model Study. *Fisheries Oceanography,* 7, 182-190.

Vichnevetsky, R., 1981: Propagation through numerical mesh refinement for hyperbolic equations. Mathematics and Computers in Simulation XXIII, 344-353, North-Holland Publ. Co.

Vintzileos, A. and R. Sadourny, 1997: A general interface between an atmospheric general circulation model and underlying oceanic and land surface models: Delocalised physics scheme. *Mon. Wea. Rev.,*125, 926-941.

Vollenweider, R.A., F.Giovanardi, G. Montanari, A. Rinaldi, 1998: Characterization of the trophic conditions of marine coastal waters with special reference to the North Western Adriatic Sea: proposal for a trophic scale, turbidity and generalized water quality index. *Environmetrics*, 9, 329-357.

Voorrips, A.C., V.K. Makin and S. Hasselmann, 1997: Assimilation of wave spectra from pitch-and-roll buoys in a North Sea wave model. *J. Geophys. Res.* 102 (C3), 5829-5849.

Voss, R., R. Sausen, and U. Cubasch,1998: Periodically synchronously coupled integrations with the atmosphere-ocean general circulation model ECHAM3/ LSG. *Climate Dynamics*, 14, 249-266.

Walker, G. T. and E. W. Bliss, 1932: World Weather V. Mem. Roy. Meteor. Soc., 4, 53-84.

Walker, G. T., 1924: Correlations in seasonal variations of weather, IX. Mem. Indian. Meteor. Dept., 24, 275-332.

Wallace, J.M. and D.S. Gutzler, 1981: Teleconnections in the geopotential height field during the northern hemisphere winter. *Mon. Wea. Rev.*, 109, 784-812.

WAMDI group: S. Hasselmann, K. Hasselmann, E. Bauer, P.A.E.M. Janssen, G.J. Komen, L. Bertotti, P. Lionello, A. Guillaume, V.C. Cardone, J.A. Greenwood, M. Reistad, L. Zambresky and J.A. Ewing, 1988: The WAM model - a third generation ocean wave prediction model. *J. Phys. Oceanogr.*, 18, 1775-1810.

Ward, M.N. and A. Navarra 1997: Pattern analysis of SST-forced variability in ensemble GCM simulations: Examples over Europe and the tropical Pacific. *J. Climate*, 10, 2210-2220.

Ward, M.N., 1996: Local and remote climate variability associated with East Mediterranean sea-surface temperature anomalies. Proc. of European Research Conf. for Mediterranean forecasting. La Londe, France, 21-26 Oct. 1995, 16-25.

Warner, T.T., R.A. Petersen, and R.E. Treadon, 1997: A tutorial on lateral boundary conditions as a basic and potentially serious limitation to regional numerical weather prediction. *Bull. Amer. Meteor. Soc.*, 78, 2599-2617.

Warn-Varnas, A., J. Sellschopp, P.J. Haley, Jr., W.G. Leslie and C.J. Lozano, 1999: Strait of Sicily water masses, *Dyn. Atmos. and Oceans*, 29(2-4), 437-469.

WASA group: J.C. Carretero, M. Gomez, I. Lozano, A. Ruiz de Elvira, O. Serrano, K. Iden, M. Reistad, H. Reichardt, V. Kharin, M. Stolley, H. von Storch, H. Günther, A. Pfizenmayer, W. Rosenthal, M. Stawarz, T. Schmith, E. Kaas, T. Li, H. Alexandersson, J. Beersma, E. Bouws, G.J. Komen, K. Rider, F. Flather, J. Smith, W. Bijl, J. de Ronde, M. Mietus, E. Bauer, H. Schmidt, H. Langeberg, 1998: Changing waves and storms in the Northeast Atlantic? *Bulletin of the American Meteorological Society*, 79, 741-760.

Weiher R F (ed.), 1999: Improving El Niño Forecasting: The potential economic benefits. U.S. Department of Commerce. NOAA Office of Policy and Strategic Planning. 57pp.

Weisse, R., U. Mikolajewicz, A. Sterl, and S.S. Drijfhout, 1997: Stochastically forced variability in the Antarctic Circumpolar Current. *J. Geophys. Res. (Oceans)*, 104, 11,049-064.

Weisse, R., U. Mikolajewicz, and E. Maier-Reimer, 1994: Decadal variability of the North Atlantic in an ocean general circulation model. *J. Geophys. Res. (Oceans)*, 99, 12411-12421.

Weller, R. A. and P. K. Taylor, 1993: Surface conditions and air-sea fluxes. CCCO-JSC Ocean Observing System Development Panel, 131 pp., Texas A&M University, College Station, TX 77843-3146.

White, W. B and R. G. Peterson, 1996: An Antarctic Circumpolar Wave in surface pressure, wind, temperature, and sea-ice extent. *Nature*, 380, 699-702.

Wilhelmsson T. and J Schüle, 1998: Parallelizing the Operational Ocean Model HIROMB. Technical Report TRITA-NA-9816, Department of Numerical Analysis and Computing Science, Royal Institute of Technology, Stockholm, Sweden.

Willems, R. C., S. M. Glenn, M. F. Crowley, P. Malanotte-Rizzoli, R. E. Young, T. Ezer, G. L. Mellor, H. G. Arango, A. R. Robinson, and C.-C. A. Lai, 1994: Experiment evaluates ocean models and data assimilation in the Gulf Stream. EOS, Transactions, American Geophysical Union, 75, 385-394.

Winton, M., R. Hallberg and A. Gnanadesikan, 1998: Simulation of density-driven frictional downslope flow in z - coordinate ocean model, *J. Phys. Oceanogr.*, 28, 2163-2174.

WMO, 1988: Guide to wave analysis and forecasting. WMO-No. 702.

Walstad, L.J. and A.R. Robinson, 1990: Hindcasting and forecasting of the POLYMODE data set with the Harvard Open Ocean Model. *J. Phys. Ocean.*, 20(11), 1682-1702.

Wajsowicz, R.C., 1986: Free planetary waves in finite difference numerical models. *J.Phys.Ocean.*, 16(4), 773-789.

Wood, R A, A.P. Keen, J.F.B. Mitchell and J.M. Gregory, 1999: Changing spatial structure of the thermohaline circulation in response to atmospheric CO2 forcing in a climate model. *Nature* 399, p.572-575.

Woods Hole Oceanographic Institution, 1993: Preliminary analysis of benefits and costs of a proposed Coastal Forecast System. Report on workshop held at Woods Hole Oceanographic Institution, 7-8 June 1993. Marine Policy Center and Coastal Research Centre, Woods Hole, Massachusetts. 35 pp.

Woods, J. D., H. Dahlin, L. Droppert, M. Glass, S. Vallerga and N.C. Flemming, 1996: The Strategy for EuroGOOS. EuroGOOS Publication No. 1. Southampton Oceanography Centre. 132 pp. + Annexes. ISBN 0-904175-22-7.

Woods, J.D., 1988: Mesoscale upwelling and primary production In: B.J.Rothschild (Ed.) Towards a theory of Biological-Physical interactions in the World Ocean. Dordrecht, Kluwer, 7-38.

Woods, J.D., 2001: Laminar flow in the Ekman Layer. In: R.Pearce (Ed.) Meteorology at the Millenium. Academic Press.

Woods, J. and W. Barkmann, 1985: A Lgrangian mixed layer model of Atlantic 18C water formation. *Nature,* 319, 574-576.

Woods, J. and W. Barkmann, 1994. Simulating plankton ecosystems by the Lagrangian Ensemble method. *Philosophical Transactions of the Royal Society,* B 343: 27-31.

Woods, J.D. and W.Barkmann, 1996: Diatom demography in winter. *Fisheries Oceanography,* 2, 202-222.

Woods, J.D. and W.Barkmann, 1995: Decrease in the biological pump in IPCC Business as Usual scenario for the greenhouse gases, 1995-2045 (unpublished manuscript).

Woods, J.D. and R.Onken, 1982: Diurnal variation and primary production in the ocean. *J.Plankton Research,* 4, 735-756.

Woods, J.D. and A.Perilli, 2001: Stability of a Virtual Plankton Ecosystem. *Nature,* (submitted).

Wyatt, L. 1996: OSCR wave measurements - some preliminary results. IEE *Journal of Oceanic Engineering*, 21, 64-76.

Wyrtki, K. 1985 : Water displacements in the Pacific and the genesis of El Niño cycles. *J. Geophys. Res.*, 90(C4), 7129-7132.

Xu, J.-S. and H. von Storch, 1990: Predicting the state of the Southern Oscillation using the POP analysis. *J. Climate*, 3(12), 1316-1329.

Yelland, M. and P.K. Taylor, 1996: Wind stress measurements from the open ocean. *J. Phys. Oceanogr.,* 26, 541 - 558.

Zavatarelli, M., Baretta, J.W, Baretta-Bekker, J.G, N. Pinardi, 2000.: The Dynamics of the Adriatic Sea ecosystem. Part I: an idealised model study. *Deep Sea Research*, Vol.47, 437-470

Zavatarelli, M., N. Pinardi, V.H. Kourafalou, A. Maggiore, 2001: Diagnostic and prognostic model studies of the Adriatic Sea general circulation. Part 1: The seasonal variability. In press on *Journal of Geophysical Research.*

Zavatarelli, M. and G.L. Mellor, 1995: A numerical study of the Mediterranean Sea circulation. *J. Phys. Oceanogr.*, 25, 1384-1414.

Zebiak, S.E. and M.A. Cane, 1987: A model ENSO. *Mon. Wea. Rev.* , 115, 2262-2278.

Zhang ,Y., J.M. Wallace, and D. S. Battisti, 1997: ENSO-like interdecadal variability: 1900-93. *J. Climate*, 10, 1004-1020.

Zhu, Z-G, J. Thuburn, B.J. Hoskins, and P.H. Haynes, 1992: A vertical finite difference scheme based on a hybrid σ-θ-p coordinate. *Mon. Wea. Rev.*, 120, 851-862.

Subject Index

S

T

U

V

W

X

Z

Printing: Mercedes-Druck, Berlin
Binding: Stürtz AG, Würzburg